Lecture Notes in Mathematics 1729

Subseries:
Institut de Mathématiques, Université de Strasbourg
Adviser: J.-L. Loday

Springer
Berlin
Heidelberg
New York
Barcelona
Hong Kong
London
Milan
Paris
Singapore
Tokyo

J. Azéma M. Émery M. Ledoux M. Yor (Eds.)

Séminaire de Probabilités XXXIV

Springer

Editors

Jacques Azéma
Laboratoire de Probabilités
Université Pierre et Marie Curie
4, Place Jussieu
75252 Paris cedex 05, France
E-mail: jaze@ccr.jussieu.fr

Michel Émery
Institut de Recherche Mathématique
Avancée
Université Louis Pasteur
7, rue René Descartes
67084 Strasbourg, France
E-mail: emery@math.u-strasbg.fr

Michel Ledoux
Laboratoire de Statistiques
et Probabilités
Université Paul Sabatier
118, route de Narbonne
31601 Toulouse cedex, France
E-mail: ledoux@cict.fr

Marc Yor
Laboratoire de Probabilités
Université Pierre et Marie Curie
4, Place Jussieu
75252 Paris cedex 05, France

Cataloging-in-Publication Data applied for

Die Deutsche Bibliothek - CIP-Einheitsaufnahme
Séminaire de probabilités .... - 1. 1966/67-. - Berlin ; Heidelberg ;
New York ; Barcelona ; Hong Kong ; London ; Milan ; Paris ; Singapore
; Tokyo : Springer, 1967
(Lecture notes in mathematics ; ...)
ISSN 0720-8766
34 . - 2000
(Lecture notes in mathematics ; Vol. 1729)
ISBN 3-540-67314-8

Mathematics Subject Classification (2000): 60GXX, 60HXX, 60JXX

ISSN 0075-8434
ISBN 3-540-67314-8 Springer-Verlag Berlin Heidelberg New York

Typesetting: Camera-ready TEX output by the author
Printed on acid-free paper SPIN: 10724949 41/3143/du

SÉMINAIRE DE PROBABILITÉS XXXIV

Table des Matières

Cours spécialisé

Exposés

Branching and Interacting Particle Systems Approximations of Feynman-Kac Formulae with Applications to Non-Linear Filtering

P. Del Moral L. Miclo

LSP-CNRS and Université Toulouse III

abstract

This paper focuses on interacting particle systems methods for solving numerically a class of Feynman-Kac formulae arising in the study of certain parabolic differential equations, physics, biology, evolutionary computing, nonlinear filtering and elsewhere. We have tried to give an "exposé" of the mathematical theory that is useful for analyzing the convergence of such genetic-type and particle approximating models including law of large numbers, large deviations principles, fluctuations and empirical process theory as well as semigroup techniques and limit theorems for processes.
In addition, we investigate the delicate and probably the most important problem of the long time behavior of such interacting measure valued processes.
We will show how to link this problem with the asymptotic stability of the corresponding limiting process in order to derive useful uniform convergence results with respect to the time parameter.
Several variations including branching particle models with random population size will also be presented. In the last part of this work we apply these results to continuous time and discrete time filtering problems.

Keywords:

Interacting and branching particle systems, genetic algorithms, weighted sampling Moran processes, measure valued dynamical systems defined by Feynman-Kac formulae, asymptotic stability, chaos weak propagation, large deviations principles, central limit theorems, nonlinear filtering.

A.M.S. codes:

60G35, 60F10, 60H10, 60G57, 60K35, 60F05, 62L20, 92D25, 92D15, 93E10, 93E11.

Contents

1 Introduction

1.1 Background and Motivations

The aim of this set of notes is the design of a branching and interacting particle system (abbreviate **BIPS**) approach for the numerical solving of a class of Feynman-Kac formulae which arise in the study of certain parabolic differential equations, physics, biology, evolutionary computing, economic modelling and nonlinear filtering problems.

Our major motivation is from advanced signal processing and particularly from optimal nonlinear filtering problems. Recall that this consists in computing the conditional distribution of a partially observed Markov process.

In discrete time and in a rather general setting the classical nonlinear filtering problem can be summarized as to find distributions of the form

$$\forall\, f \in \mathcal{B}_b(E),\, \forall\, n \geq 0, \qquad \eta_n(f) \;=\; \frac{\gamma_n(f)}{\gamma_n(1)} \tag{1}$$

where $\mathcal{B}_b(E)$ is the space of real bounded measurable functions over a Polish state space E and $\gamma_n(f)$ is a Feynman-Kac formula given by

$$\gamma_n(f) = \mathbb{E}\left(f(X_n) \prod_{m=1}^{n} g_m(X_{m-1}) \right) \tag{2}$$

where $\{X_n \,;\, n \geq 0\}$ is a given time inhomogeneous Markov chain taking values in E and $\{g_m \,;\, m \geq 1\}$ is a given sequence of bounded positive functions.

In continuous time, the computation of the optimal pathwise filter can be summarized as to find the flow of distributions

$$\forall\, f \in \mathcal{B}_b(E)\,, \forall\, t \in \mathbb{R}_+, \qquad \eta_t(f) \;=\; \frac{\gamma_t(f)}{\gamma_t(1)} \tag{3}$$

where $\gamma_t(f)$ is again defined through a Feynman-Kac formula of the following form

$$\gamma_t(f) = \mathbb{E}\left(f(X_t) \exp\left(\int_0^t U_s(X_s)\, ds \right)\right) \tag{4}$$

This time $\{X_t \,;\, t \in \mathbb{R}_+\}$ denotes an E-valued càdlàg inhomogeneous Markov process and $\{U_t \,;\, t \in \mathbb{R}_+\}$ is a measurable collection of locally bounded (in time) and measurable nonnegative functions.

Even if equations (1) and (3) look innocent they can rarely be solved analytically and their solving require extensive calculations. More precisely, with the notable exception of the so-called "linear-Gaussian" situation (Kalman-Bucy's filter [15]) or wider classes of models (Benes' filters [12]) optimal filters have no finitely recursive solution [20]. To obtain a computationally feasible solution some kind of approximation is needed.

Of course, there are many filtering algorithms that have been developed in mathematics and signal processing community. Until recently most works in this direction were based on fixed grid approximations, conventional linearization (Extended Kalman Filters) or determining the best linear filter (in least squares sense). These various numerical methods have never really cope with large scale systems or unstable processes. Comparisons and examples when the extended Kalman-Bucy filter fails can be found for instance in [15]. In addition all these deterministic schemes have to be handled very carefully mainly because they are usually based on specific criteria and rates of convergence are not always available.

The particle algorithms discussed in these lectures belong to the class of Monte Carlo methods and they do not use regularity informations on the coefficients of the models. Thus, large scale systems and nonlinear models with non sufficiently smooth coefficients represent classes of nonlinear filtering problems to which particle methods might be applied. These methods are in general robust and very efficient and many convergence results are available including uniform convergence results with respect to the time parameter. But, from a strict practical point of view, if there exists already a good specialized method for a specific filtering problem then the BIPS approach may not be the best tool for that application.

Let us briefly survey some distinct approaches and motivate our work.

In view of the functional representations (1) and (3) the temptation is also to apply classical Monte-Carlo simulations based on a sequence of independent copies of the process X. Unfortunately it is well known that the resulting particle scheme is not efficient mainly because the deviation of the particles may be too large and the growth of the exponential weights with respect to the time parameter is difficult to control (see for instance [28, 34, 57]).
In [34] we propose a way to regularize these weights and we give a natural ergodic assumption on the signal semigroup under which the resulting Monte-Carlo particle scheme converges in law to the optimal filter uniformly with respect to the time parameter.
In more general situations, complications occur mainly because this particle scheme is simply based on a sequence of independent copies of the signal. This is not surprising: roughly speaking the law of signal and the desired conditional distribution may differ considerably and they may be too few particles in the space regions with high probability mass.

Among the most exciting developments in nonlinear filtering theory are those centering around the recently established connections with branching and interacting particle systems. The evolution of this rapidly developing area of research may be seen quite directly through the following chains of papers [23, 21] [25, 24], [30, 33, 31] [35, 37, 36], [42], [41, 40], [45], [32, 47] as well as [11, 18, 46, 73, 64] and finally [59, 85, 84].

Instead of *hand crafting* algorithms, often based on the basis of had-hoc criteria, particle systems approaches provide powerful tools for solving a large class of nonlinear filtering problems. In contrast to the first Monte-Carlo scheme the branching and interacting particle approximating models involve the use of a system of particles which evolve in correlation with each other and give birth to a number of offsprings depending on the observation process. This guarantees an occupation of the probability space regions proportional to their probability mass thus providing a well behaved adaptative and stochastic grid approximating model. Furthermore these particle algorithms also belong to the class of resampling methods and they have been made valuable in practice by advances in computer technology [52]. Different adaptative locally refined but deterministic multi-grid methods can be found in [17]. In contrast to BIPS approaches the latter are limited to low dimensional state space examples.

It is hard to know where to start in describing contributions to BIPS approximations of Feynman-Kac formulae.

In discrete time and nonlinear filtering settings the more embryonic form of interacting particle scheme appeared in the independent works [32, 47], [64] and [73]. The first proof of convergence of these heuristics seems to be [30, 31]. The analysis of the convergence has been further developed in [33, 35, 37, 36, 42].

In continuous time settings the origins of interacting particle schemes is a more delicate problem. The first studies in continuous time settings seem to be [23] and [21]. These works were developed independently of the first set of referenced papers. The authors present a branching particle approximating model without any rates of convergence and the main difference with previous interacting particle schemes comes from the fact that the number of particle is not fixed but random. Moreover the authors made the crucial assumptions that we can exactly simulate random transitions according to the semigroup of the continuous time signal and stochastic integrals arising in Girsanov exponentials are exactly known. Therefore these particle algorithms do not applied directly to the continuous time case. On the other hand these branching particle models are based on a time discretization procedure. As a result the corresponding nonlinear filtering problem can be reduced to a suitably defined discrete time filtering problem. The corresponding discrete time version of such branching and interacting particle schemes as well as the first convergence rates are described in [25] and [24].

The studies [41] and [40, 39] discuss several new interacting particle schemes for solving nonlinear filtering problems where the signal is continuous but the number of observations is finite. To get some feasible solution which can be used in practice several additional levels of approximations including time discretizations are also analyzed. In contrast to previous referenced papers these schemes can also be used for solving numerically filtering problems with correlated signal and observation noise sources.

As we shall see in the further development of section 1.2 the interacting or branching particle schemes based on an additional time discretization procedure are not really efficient for solving continuous time filtering problems. The authors presented in [45] a genuine continuous time genetic type particle scheme for solving the robust optimal filter. This scheme will be discussed in section 1.3 and section 3.

The connections between this IPS model, the classical Moran IPS and the Nanbu IPS (arising respectively in the literature of genetic algorithms and Boltzmann equations) are discussed in section 1.3.

The modelling and the analysis of such particle approximating models has matured over the past ten years in ways which make it much more complete and rather beautiful to learn and to use. One objective of these notes is to introduce the reader to branching and interacting particle interpretations of Feynman-Kac formulae of type (1)-(4).
We have also tried to give an "exposé" of the mathematical theory that it is useful in analyzing the convergence of such approximating models including law of large numbers, large deviations, fluctuations and empirical process theory, as well as semigroup techniques and functional limit theorems for stochastic processes.

Although only a selection of existing results is presented, many results appear here for the first time and several points have been improved. The proofs of existing results are only sketched but the methodologies are described carefully. Deeper informations are available in the list of referenced papers.

The material for this paper has also been chosen in order to give some feel of the variety of the theory but the development is guided by the classical interplay between theory and detailed consideration of application to specific nonlinear filtering models.

This set of notes is very far from being exhaustive and only surveys results that are closely related to BIPS-approximations of Feynman-Kac formulae and non linear filtering problems. Among the topics omitted are those centering around evolutionary computing and numerical function optimization problems. Among the huge literature on evolutionary computing and genetic algorithms we refer to [6, 7, 14], [19], [44], [60, 61, 62], [63] and [111].
We emphasize that the so-called simple genetic algorithm is a special case of the BIPS models presented in this work.
In this connection, the measure valued distribution flows (1)-(4) and the corresponding interacting particle approximating models can be regarded as the so-called infinite and finite population models. Therefore the methodologies presented in this work can be used for establishing the most diverse limit theorems on the long time behavior of these models as well as the asymptotics of the finite population model as the number of individuals tends to infinity.

An overview of the material presented in these notes was presented in a three one hour lectures for the Symposium/Workshop on Numerical Stochastics (April 1999) at the Fields Institute for Research in Mathematical Sciences (Toronto). At the same period they were presented at the University of Alberta Edmonton with the support of the Canadian Mathematics of Information Technology and Complex Systems project (MITACS).
We would like to thank Professor M. Kouritzin from the University of Alberta for stimulating discussions.

A part of this material presented in this set of notes results from collaborations of one of the authors with D. Crisan and T. Lyons [25, 24], with A. Guionnet [35, 37, 36], with J. Jacod [40] and Ph. Protter [41], and with M. Ledoux [42].

We also heartily thank Gérard Ben Arous, Carl Graham and Sylvie Méléard for encouraging and fruitful discussions and Michel Ledoux for inviting us to write these notes for *Le Séminaire de Probabilités*.

We gratefully acknowledge CNRS research fellowship No 97N23/0019 "Modélisation et simulation numérique", European community for the CEC Contracts No ERB-FMRX-CT96-0075 and INTAS-RFBR No 95-0091 and the Canadian Network MITACS-Prediction In Interacting Systems.

1.2 Motivating Examples

The interacting particle interpretation of the Feynman-Kac models (1)-(4) has had numerous applications in many nonlinear filtering problems; to name a few, radar signal processing ([46, 47]), global positioning system ([18]) and tracking problems ([73, 85, 84, 64]). Other numerical experiments are also given in [21] and [41].

The purpose of these notes is to expose not only the theory of interacting particle approximations of Feynman-Kac formulae but also to provide a firm basis for the understanding and solving nonlinear filtering problems. To guide the reader and motivate this study we present here two generic models and the discrete and continuous time Feynman-Kac formulations of the corresponding optimal filter.

The distinction between continuous and discrete time will lead to different kind of interacting particle approximating models. Intuitively speaking continuous time models correspond to processes of classical physics while discrete time models arise in a rather natural way as soon as computers are part of the process. More general and detailed models will be discussed in the further development of section 4.

In discrete time settings the state signal $X = \{X_n ; \; n \geq 0\}$ is an $\mathbb{R}^p$-valued Markov chain usually defined through a recursion of the form

$$X_n = F_n(X_{n-1}, W_n)$$

where $W = \{W_n \; ; \; n \geq 1\}$ is a noise sequence of independent and $\mathbb{R}^q$-valued random variables. For each $n \geq 1$ the function $F_n : \mathbb{R}^p \times \mathbb{R}^q \to \mathbb{R}^p$ is measurable and the initial value X_0 is independent of W. The above recursion contains the laws of evolution for system states such as the laws of evolution of a target in tracking problems, an aircraft in radar processing or inertial navigation errors in GPS signal processing. The noise component W models the statistics of unknown control laws of an aircraft or a cruise control in an automobile or a non cooperative target as well as uncertainties in the choice of the stochastic mathematical model.

For future reference it is convenient to generalize the definition of the state signal X. More precisely an alternative way to define X consists in embedding the latter random dynamical system through its transition probabilities. This approach gives some insights into ways of thinking the evolution of the marginal laws of X and it also allows to consider signals taking values on infinite dimensional spaces. For these reasons we will systematically assume that the sequence $X = \{X_n \; ; \; n \geq 0\}$ is a Markov process taking values in a Polish state space E with transition probabilities $\{K_n \; ; \; n \geq 1\}$ and initial distribution η_0. As we shall see in the further development this formulation also allows a uniform treatment of filtering problems with continuous time signals and discrete time observations. The signal X is not known but partially observed. Usually we assume that the observation process $Y = \{Y_n \; ; \; n \geq 1\}$ is a sequence of $\mathbb{R}^r$-valued random variables given by

$$Y_n = h_n(X_{n-1}) + V_n$$

where $V = \{V_n \; ; \; n \geq 1\}$ are independent and $\mathbb{R}^r$ valued random variables whose marginal distributions possess a density $\varphi_n(v)$ with respect to Lebesgue measure on $\mathbb{R}^r$ and for each $n \geq 1$, $h_n : E \to \mathbb{R}^r$ is a measurable function.
Here again the design of the disturbance sequence V depends on the class of sensors at hand. For instance noise sources acting on sensors model thermic noise resulting from electronic devices or atmospheric propagation delays and/or received clock bias in GPS signal processing. For more details we refer the reader to the set of referenced articles.

Given the stochastic nature of the pair signal/observation process and given the observation values $Y_n = y_n$, for each $n \geq 1$, the nonlinear filtering problem consists in computing recursively in time the one step predictor conditional probabilities η_n and the filter conditional distributions $\widehat{\eta}_n$ given for any bounded Borel test function f by

$$\begin{aligned}
\eta_n(f) &= \mathbb{E}(f(X_n)|Y_1 = y_1, \ldots, Y_n = y_n) \\
\widehat{\eta}_n(f) &= \mathbb{E}(f(X_n)|Y_1 = y_1, \ldots, Y_n = y_n, Y_{n+1} = y_{n+1})
\end{aligned}$$

As usually the n-step filter $\widehat{\eta}_n$ is written in terms of η_n as

$$\widehat{\eta}_n(f) = \frac{\displaystyle\int_E f(x)\, \varphi_{n+1}(y_{n+1} - h_{n+1}(x))\, \eta_n(dx)}{\displaystyle\int_E \varphi_{n+1}(y_{n+1} - h_{n+1}(x))\, \eta_n(dx)}$$

and the n-step predictor may be defined in terms of the Feynman-Kac type formula

$$\eta_n(f) = \frac{\gamma_n(f)}{\gamma_n(1)}$$

with

$$\gamma_n(f) = \mathbb{E}\left(f(X_n) \prod_{m=1}^{n} \varphi_m(y_m - h_m(X_{m-1})) \right)$$

We will return to this model with more detailed examples in section 5.

The second filtering model presented hereafter is a continuous time (signal/observation) Markov process $\{(S_t, Y_t) \; ; \; t \in \mathbb{R}_+\}$ taking values in $\mathbb{R}^p \times \mathbb{R}^q$. It is solution of the Itô's stochastic differential equation

$$\begin{cases} dS_t &= A(t, S_t)\, dt + B(t, S_t)\, dW_t + \displaystyle\int_{\mathbb{R}^m} C(t, S_{t-}, u)\, (\mu(dt, du) - \nu(dt, du)) \\ dY_t &= h(S_t)\, dt + \sigma\, dV_t \end{cases}$$

(V, W) is an $(q+r)$ dimensional standard Wiener process and μ is a Poisson random measure on $\mathbb{R}_+ \times \mathbb{R}^m$ with intensity measure $\nu(dt, du) = dt \otimes F(du)$ and F is a positive σ-finite measure on $\mathbb{R}^m$. The mappings $A : \mathbb{R}_+ \times \mathbb{R}^p \to \mathbb{R}^p$, $B : \mathbb{R}_+ \times \mathbb{R}^p \to \mathbb{R}^p \otimes \mathbb{R}^r$, $C : \mathbb{R}_+ \times \mathbb{R}^p \times \mathbb{R}^m \to \mathbb{R}^p$, and $h : \mathbb{R}^p \to \mathbb{R}^q$ are Borel functions, S_0 is a random variables independent of (V, W, μ) and $Y_0 = 0$.
Here again the first equation represents the evolution laws of the physical signal process at hand. For instance the Poisson random measure μ may represent jumps variations of a moving and non cooperative target (see for instance [47]).

Next we examine three situations. In the first one we assume that observations are given only at regularly spaced times $t_0, t_1, \ldots, t_n, \ldots \in \mathbb{R}_+$ ($t_0 = 0, Y_0 = 0$) and we are interested in the conditional distributions given for any bounded Borel function $f : \mathbb{R}^p \to \mathbb{R}$ by

$$\tilde{\pi}_n(f) \stackrel{\text{def.}}{=} \mathbb{E}(f(S_{t_n}) \,|Y_{t_1} = y_1, \ldots, Y_{t_n} = y_n)$$

where $y_1, \ldots, y_n \in \mathbb{R}^q$ is a given sequence of observations.
If we denote $E = D([0, t_1[, \mathbb{R}^p)$ the Polish space of càdlàg paths from $[0, t_1[$ into $\mathbb{R}^p$ then the discrete time sequence

$$\forall\, n \geq 0, \qquad X_n \stackrel{\text{def.}}{=} S_{[t_n, t_{n+1}[} \circ \theta_{t_n}$$

(where $(\theta_t)_{t \geq 0}$ stands for the usual family of time shifts), is an E-valued Markov chain. On the other hand if $H : E \to \mathbb{R}^q$ is the mapping defined by

$$\forall x \in E\,, \qquad H(x) = \int_0^{t_1} h(x_s)\, ds$$

then using the above notations for any $n \geq 1$:

$$Y_{t_n} - Y_{t_{n-1}} = H(X_{n-1}) + \sigma(V_{t_n} - V_{t_{n-1}})$$

This description is interesting in that it shows that the latter filtering problem is equivalent to the previous discrete time model. In this connection it is also worth noting that for any bounded Borel test function $f : E \to \mathbb{R}$

$$\eta_n(f) \stackrel{\text{def.}}{=} \mathbb{E}\left(f(S_{[t_n, t_{n+1}[}) \,|Y_{t_1} = y_1, \ldots, Y_{t_n} = y_n\right) = \frac{\gamma_n(f)}{\gamma_n(1)}$$

with

$$\gamma_n(f) = \mathbb{E}\left(f(X_n) \prod_{m=1}^{n} \varphi_m(y_m - y_{m-1} - H(X_{m-1}))\right)$$

where, for any $m \geq 1$, φ_m is the density of the Gaussian variable $\sigma(V_{t_m} - V_{t_{m-1}})$. This observation also explain why it is necessary to undertake a study of Feynman-Kac formula of type (1) with Polish valued Markov processes X.

A more traditional question in continuous time nonlinear filtering settings is to compute the conditional distributions $\{\pi_t \; ; \; t \in \mathbb{R}_+\}$ given for any bounded Borel functions $f : \mathbb{R}^p \to \mathbb{R}$ by setting

$$\pi_t(f) = \mathbb{E}\left(f(S_t) \,\middle|\, \mathcal{Y}_{[0,t]}\right) \tag{5}$$

where $\mathcal{Y}_{[0,t]}$ is the filtration generated by the observations up to time t. Roughly speaking this continuous time problem is close to the previous discrete time model when the time step $\Delta t = t_n - t_{n-1}$ is sufficiently small and when observations are delivered continuously in time. For instance in radar processing the measurements derived from the return signal of a moving target can be regarded as a pulse train of rectangular or Gaussian pulses with period 10^{-4} seconds.
In real applications one usually consider the discrete time signal model

$$\{S_{t_n} \; ; \; n \geq 0\}$$

as $\mathbb{R}^p$-valued Markov chain and one replaces the continuous time observation process by the discrete time sequence

$$\Delta Y_{t_n} \stackrel{\text{def.}}{=} h\left(S_{t_{n-1}}\right) \Delta t + \sigma(V_{t_n} - V_{t_{n-1}}) \tag{6}$$

This first level of approximation is commonly used in practice and the error caused by the discretization of the time interval is well understood (see for instance [76] and references therein as well as section 4.3 in these notes).
One consequence of the previous discretizarion is that the filtering problem is now reduced to find a way of computing the conditional distributions given for any bounded Borel function $f : \mathbb{R}^p \to \mathbb{R}$ by

$$\pi_n^{\Delta}(f) \stackrel{\text{def.}}{=} \mathbb{E}(f(S_{t_n}) \,|\Delta Y_{t_1} = y_1 - y_0, \ldots, \Delta Y_{t_n} = y_n - y_{n-1}) = \frac{\gamma_n^{\Delta}(f)}{\gamma_n^{\Delta}(1)}$$

with $y_0 = 0$ and

$$\gamma_n^{\Delta}(f) = \mathbb{E}\left(f(S_{t_n}) \prod_{m-1}^{n} \varphi_m(y_m - y_{m-1} - h(S_{t_{m-1}})\Delta t)\right)$$

One drawback of this formulation is that the corresponding IPS approximating model is not really efficient. As we shall see in the further development of section 1.3 and section 4.3 the evolution of this scheme is decomposed into two genetic type selection/mutation transitions. During each selection stage the system of particles takes advantage of the current observation data in order to produce an adaptative grid.
The underlying n^{th} step selection transition is related to the fitness functions g_n defined by

$$g_n(x) = \varphi_n(y_n - y_{n-1} - h(S_{t_{n-1}})\Delta t)$$

but the physical noise in sensors as well as the choice of the short time step Δt critically corrupt the information delivered between two dates (recall that the current observation at time n has the form (6)). One consequence is that the resulting particle

scheme combined with the previous time discretization converges slowly to the desired conditional distributions (5) (see for instance section 4.3 and [24] as well as [23, 21] for a branching particle alternative scheme).
One way to improve these rates is to use a genuine continuous time and interacting particle approximating model. In this alternative approach the information used at each selection date is not the "increments" of the observation process but the current observation value at that time.

The key idea presented in [45] and further developed in this work is to study the robust and pathwise filter defined for any $y \in \mathcal{C}(\mathbb{R}_+, \mathbb{R}^q)$ (and not only on a set of probability measure 1) and for any bounded Borel function $f : \mathbb{R}^p \to \mathbb{R}$ by a formula of the form

$$\pi_{y,t}(f) = \frac{\displaystyle\int_{\mathbb{R}^p} f(x)\, e^{h^\star(x)y_t}\, \eta_{y,t}(dx)}{\displaystyle\int_E e^{h^\star(x)y_t}\, \eta_{y,t}(dx)} \qquad \text{with} \quad \eta_{y,t}(f) \stackrel{\text{def.}}{=} \frac{\gamma_{y,t}(f)}{\gamma_{y,t}(1)}$$

and $\gamma_{y,t}$ is again defined through a Feynman-Kac type formula

$$\gamma_{y,t}(f) = \mathbb{E}\left(f(X_t^y)\, \exp \int_0^t V_s(X_s^y, y_s)\, ds \right)$$

For any $s \in \mathbb{R}_+$, $V_s : \mathbb{R}^p \times \mathbb{R}^q \to \mathbb{R}_+$ is a Borel measurable function which depends on the coefficients of the filtering problem at hand and $\{X_t^y \; ; \; t \in \mathbb{R}_+\}$ is a Markov process which depends on the observations. To describe precisely these mathematical models we need to introduce several additional notations. We will return to this model with detailed examples of signal processes which can be handled in our framework in section 4.2.

The convergence results for the resulting interacting particle approximating model will improve the one presented in section 4.3 and in the articles [23, 21, 24]. From a practitioner's view point the main difference between these two approaches lies in the fact that in the former the selection and interaction mechanism only depends on the current observation y_t and on the fitness function $U_t(x) \stackrel{\text{def.}}{=} V_t(x, y_t)$. Moreover under natural stability assumptions the resulting IPS scheme converges uniformly in time to the desired optimal filter.

1.3 Description of the Models

To provide a red line in this work and to point out the connections with classical mean-field interacting particle system theory, the models are focused around two approximating interacting particle systems (abbreviate **IPS**). The research literature abounds with variation of these two IPS models. The interested reader will also find a detailed description of several variants including branching schemes with random population size, periodic selection schemes and conditional mutations.

Most of the terminology we have used is drawn from mean field IPS and measure valued processes theory. We shall see that the flows of distributions (1) and (3) are solutions of deterministic measure valued and nonlinear evolution equations. These equations will be used to define the transition probability kernel of the corresponding IPS approximating models. In this sense the deterministic evolution equations will be regarded as the limiting measure valued process associated with a sequence of IPS schemes.

The resulting IPS models can also be regarded as genetic type algorithms as those arising in numerical function optimization, biology and population analysis (cf. [62, 67, 87, 63]).
The range of research areas in which genetic algorithms arise is also quite board, to name a few: machine learning [61], control systems [60], electromagnetics [69, 110], economics and finance [8, 80, 97], aircraft landing [1, 2], topological optimum design [72] and identification of mechanical inclusions [70, 71].

In continuous time settings the corresponding IPS can be viewed as a weighted sampling Moran particle system model. Moran particle models arise in population genetics. They usually model the evolution in time of the genetic structure of a large but finite population (see for instance [29] and references therein). In the classical theory of measure valued processes the limiting process is random and it is called the Fleming-Viot process. In this setting the limiting process is commonly used to predict the collective behavior of the system with a finite but large number of particles.
In contrast to the above situation the desired distributions (1) and (3) are not random (except, in filtering settings, through the observation process). It is therefore necessary to find a new strategy to define an IPS scheme that will approximate (1) and (3).

In time homogeneous settings and in the case where $E = \mathbb{R}^d$, $d \geq 1$, the evolution equation (11) and the corresponding IPS can also be regarded, in some sense, as a simple generalized and spatially homogeneous Boltzmann equation and as the corresponding Nanbu type IPS (see for instance [65, 83] and references therein). At this juncture many results presented in this work can be applied, but this is outside the scope of these notes, to study the fluctuations or to prove uniform convergence results for a class of Nanbu IPS.

Let us finally mention that the Feynman-Kac models considered in this study can be regarded as the distributions of a random Markov particle X killed at a given rate and conditioned by non-extinction (see for instance [96]). In this connection the asymptotic stability properties of the limiting processes associated with the Feynman-Kac models can be used to give conditions under which a killed particle conditioned by non-extinction forgets exponentially fast its initial condition (see for instance [43]).

In view of the above discussion the objects on which the limiting processes (8), (11) and the corresponding IPS schemes are sought may vary considerably. It is therefore necessary to undertake a study of the convergence in an abstract and time inhomogeneous setting.
For instance, in nonlinear filtering applications the time-inhomogeneous assumption is essential since the objects depend on the observation process.

The genetic type algorithms arising in numerical function analysis or nonlinear control problems usually also depends on a specific cooling schedule parameter (see for instance [19, 44])

1.3.1 The Limiting Measure Valued Models

To describe precisely the limiting measure valued models, let us introduce some notations. Let (E, r) be a Polish space, ie E is a separable topological space whose topology is given by a metric r which is supposed to be complete.
Let $\mathbf{B}(E)$ be the σ-field of Borel subsets of E. We denote by $\mathbf{M}(E)$ the space of all finite and signed Borel measures on E. Let $\mathbf{M}_1(E) \subset \mathbf{M}(E)$ be the subset of all probability measures. As usual, both spaces will be furnished with the weak topology. We recall that weak topology is generated by the Banach space $\mathcal{C}_b(E)$ of all bounded continuous functions, endowed with the supremum norm, defined by

$$\forall\, f \in \mathcal{C}_b(E), \qquad \|f\| \;=\; \sup_{x\in E} |f(x)|$$

(since $\mathbf{M}(E)$ can be regarded naturally as a part of the dual of $\mathcal{C}_b(E)$). More generally, the norm $\|\cdot\|$ is defined in the same way on $\mathcal{B}_b(E)$, the set of all bounded measurable functions, which is also a Banach space.

We denote by μK the measure given by $\mu K(A) = \int_E \mu(dx)\, K(x, A)$ where K is any integral operator on $\mathcal{B}_b(E)$, $\mu \in \mathbf{M}(E)$ and $A \in \mathbf{B}(E)$. We also write

$$\forall\, f \in \mathcal{B}_b(E), \qquad \mu K(f) = \int \mu(dx)\, K(x, dz)\, f(z). \tag{7}$$

If K_1 and K_1 are two integral operators on $\mathcal{B}_b(E)$ we denote by $K_1 K_2$ the composite operator on $\mathcal{B}_b(E)$ defined for any $f \in \mathcal{B}_b(E)$ by

$$K_1 K_2 f(x) = \int_E K_1(x, dy)\, K_2(y, dz)\, f(z)$$

We also denote by $\{K_n \;;\; n \geq 1\}$ the transition probability kernels (respectively $\{L_t \;;\; t \geq 0\}$ the family of pregenerators) of the discrete time (resp. the continuous time) Markov process X.

To get formally the nature of such schemes we first note that the distribution flow $\{\eta_n \;;\; n \geq 0\}$ defined by (1) page 3 is a solution of the following measure valued dynamical system

$$\forall\, n \geq 1, \qquad \eta_n \;=\; \Phi_n(\eta_{n-1}) \tag{8}$$

For all $n \geq 1$, $\Phi_n : \mathbf{M}_1(E) \to \mathbf{M}_1(E)$ is the mapping defined by

$$\begin{aligned} \Phi_n(\eta) &= \Psi_n(\eta) K_n \\ \forall\, f \in \mathcal{B}_b(E), \qquad \Psi_n(\eta)(f) &= \frac{\eta(g_n f)}{\eta(g_n)} \end{aligned} \tag{9}$$

We note that the recursion (8) involves two separate transitions:

$$\eta_{n-1} \xrightarrow{\text{Updating}} \widehat{\eta}_{n-1} \stackrel{\text{def}}{=} \psi_n(\eta_{n-1}) \xrightarrow{\text{Prediction}} \eta_n = \widehat{\eta}_{n-1} K_n \tag{10}$$

The first one is nonlinear and it will be called the updating step and the second one is linear and it will be called the prediction transition with reference to filtering theory.

In the continuous time situation, the distributions flow $\{\eta_t \; ; \; t \geq 0\}$ defined by (3) page 3 satisfies for any regular test function f the following nonlinear evolution equation

$$\frac{d}{dt}\eta_t(f) = \eta_t(\mathcal{L}_{t,\eta_t}(f)) \tag{11}$$

where $\mathcal{L}_{t,\eta}$, for $t \geq 0$ and $\eta \in \mathbf{M}_1(E)$ fixed, is a pregenerator on E, defined on a suitable domain by

$$\mathcal{L}_{t,\eta}(f)(x) = L_t f(x) + \int (f(z) - f(x))\, U_t(z)\, \eta(dz) \tag{12}$$

1.3.2 Interacting Particle Systems Models

In this section we introduce the IPS approximating models of the previous evolution equations (8) and (11). To give an initial introduction and to illustrate the idea in a simple form we emphasize that the particle approximation approach described here is not restricted to the particular form of mappings $\{\Phi_n \; ; n \geq 1\}$ or to the nature of the pregenerators $\{\mathcal{L}_{t,\eta} \; ; \; t \geq 0, \;\; \eta \in \mathbf{M}_1(E)\}$.

In discrete time, starting from a collection of mappings

$$\Phi_n : \mathbf{M}_1(E) \to \mathbf{M}_1(E), \qquad n \geq 1,$$

we consider an N-IPS, $\xi_n = (\xi_n^1, \ldots, \xi_n^N)$, $n \geq 0$, which is a Markov chain on the product state space E^N with transition probability kernels satisfying

$$P\left(\xi_n \in dx \,|\, \xi_{n-1} = z\right) = \prod_{p=1}^{N} \Phi_n(m(z))(dx^p) \tag{13}$$

where $dx \stackrel{\text{def}}{=} dx^1 \times \cdots \times dx^N$ is an infinitesimal neighborhood of the point $x = (x^1, \ldots, x^N) \in E^N$, $z = (z^1, \ldots, z^N) \in E^N$, δ_a stands for the Dirac measure at $a \in E$ and

$$\forall\, z = (z^1, \ldots, z^N) \in E^N, \qquad m(z) \;=\; \frac{1}{N}\sum_{i=1}^{N} \delta_{z^i} \;\in\; \mathbf{M}_1(E) \tag{14}$$

The initial system $\xi_0 = (\xi_0^1, \ldots, \xi_0^N)$ consists in N independent particles with common law η_0.

Intuitively speaking it is quite transparent from the above definition that if Φ_n is sufficiently regular and if $m(\xi_{n-1})$ is close to the desired distribution η_{n-1} then one expects that $\Phi_n(m(\xi_{n-1}))$ is a nice approximating measure for η_n. Therefore at the next step the particle system $\xi_n = (\xi_n^1, \ldots, \xi_n^N)$ looks like a sequence of independent random variables with common law η_n. This general and abstract IPS model first appeared in [33, 31] and its analysis was further developed in [35].

In much the same way starting from a family of pregenerators

$$\{\mathcal{L}_{t,\eta}\ ;\ t \geq 0,\ \eta \in \mathbf{M}_1(E)\}$$

we consider an interacting N-particle system $(\xi_t)_{t\geq 0} = ((\xi_t^1, \ldots, \xi_t^N))_{t\geq 0}$, which is a time-inhomogeneous Markov process on the product space E^N, $N \geq 1$, whose pregenerator acts on functions ϕ belonging to a good domain by

$$\forall\,(x_1, \ldots, x_N) \in E^N, \quad \mathcal{L}_t^{(N)}(\phi)(x_1, \ldots, x_N) \;=\; \sum_{i=1}^{N} \mathcal{L}_{t,m(x)}^{(i)}(\phi)(x_1, \ldots, x_N) \tag{15}$$

The notation $\mathcal{L}_{t,\eta}^{(i)}$ has been used instead of $\mathcal{L}_{t,\eta}$ when it acts on the i-th variable of $\phi(x_1, \ldots, x_N)$. This abstract and general formulation is well known in mean field IPS literature (the interested reader is for instance referred to [83] and [98] and references therein).
In section 3.2 we will give a more detailed description of (15) when the pregenerators $\mathcal{L}_{t,\eta}$ are defined by (12).

In discrete time settings, a more explicit description of (13) in terms of a two stage genetic type process can already been done. More precisely, using the fact that the mappings Φ_n under study are given by (9) and

$$\Psi_n\Big(\frac{1}{N}\sum_{i=1}^{N}\delta_{x^i}\Big) = \sum_{i=1}^{N}\frac{g_n(x^i)}{\sum_{j=1}^{N} g_n(x^j)}\,\delta_{x^i}\,,$$

we see that the resulting motion of the particles is decomposed into two separate mechanisms

$$\xi_{n-1} \xrightarrow{\text{Selection/Updating}} \widehat{\xi}_{n-1} \xrightarrow{\text{Mutation/Prediction}} \xi_n$$

These mechanisms can be modelled as follows:

Selection/Updating:

$$P(\widehat{\xi}_{n-1} \in dx \,|\, \xi_{n-1} = z) = \prod_{p=1}^{N}\sum_{i=1}^{N}\frac{g_n(z^i)}{\sum_{j=1}^{N} g_n(z^j)}\,\delta_{z^i}(dx^p).$$

Mutation/Prediction:

$$P(\xi_n \in dz \,|\, \widehat{\xi}_{n-1} = x) = \prod_{p=1}^{N} K_n(x^p, dz^p). \tag{16}$$

Thus, we see that the particles move according to the following rules. In the selection transition, one updates the positions in accordance with the fitness functions $\{g_n;\, n \geq 1\}$ and the current configuration. More precisely, at each time $n \geq 1$, each particle examines the system $\xi_{n-1} = (\xi_{n-1}^1, \ldots, \xi_{n-1}^N)$ and chooses randomly a site

ξ^i_{n-1}, $1 \leq i \leq N$, with a probability which depends on the entire configuration ξ_{n-1} and given by

$$\frac{g_n\left(\xi^i_{n-1}\right)}{\sum_{j=1}^N g_n\left(\xi^j_{n-1}\right)}.$$

This mechanism is called the selection/updating transition as the particles are selected for reproduction, the most fit individuals being more likely to be selected. In other words, this transition allows particles to give birth to some particles at the expense of light particles which die.
The second mechanism is called mutation/prediction since at this step each particle evolves randomly according to a given transition probability kernel.

The preceding scheme is clearly a system of interacting particles undergoing adaptation in a time non-homogeneous environment represented by the fitness functions $\{g_n;\, n \geq 1\}$. Roughly speaking the natural idea is to approximate the two step transitions (10) of the system (8) by a two step Markov chain taking values in the set of finitely discrete probability measures with atoms of size some integer multiple of $1/N$. Namely, we have that

$$\eta^N_{n-1} \stackrel{\text{def.}}{=} \frac{1}{N}\sum_{i=1}^N \delta_{\xi^i_{n-1}} \xrightarrow{\text{Selection}} \widehat{\eta}^N_{n-1} \stackrel{\text{def.}}{=} \frac{1}{N}\sum_{i=1}^N \delta_{\widehat{\xi}^i_{n-1}} \xrightarrow{\text{Mutation}} \eta^N_n = \frac{1}{N}\sum_{i=1}^N \delta_{\xi^i_n}.$$

As it will be obvious in subsequent sections the continuous time IPS model is again a genetic type model involving selection and mutation transitions.

Another interest feature of the IPS defined by (13) and (22) is that they can be used to approximate the Feynman-Kac formulae (2) and (4). One of the best ways for introducing the corresponding particle approximating models is through the following observation. In discrete time and on the basis of the definition of the distributions $\{\eta_n, \gamma_n \;;\; n \geq 0\}$ we have that

$$\forall n \geq 1, \qquad \eta_{n-1}(g_n) = \frac{\gamma_{n-1}(g_n)}{\gamma_{n-1}(1)} = \frac{\gamma_n(1)}{\gamma_{n-1}(1)}$$

Therefore for any $f \in \mathcal{B}_b(E)$ and $n \geq 0$

$$\gamma_n(f) = \gamma_n(1)\, \eta_n(f) \qquad \text{with} \qquad \gamma_n(1) = \prod_{m=1}^n \eta_{m-1}(g_m)$$

(with the convention $\prod_\emptyset = 1$). Taking in consideration the above formula the natural particle approximating scheme for the "unnormalized" distributions $\{\gamma_n \;;\; n \geq 0\}$ is simply given by

$$\gamma^N_n(f) \stackrel{\text{def.}}{=} \gamma^N_n(1)\, \eta^N_n(f) \qquad \text{with} \qquad \gamma^N_n(1) \stackrel{\text{def.}}{=} \prod_{m=1}^n \eta^N_{m-1}(g_m) \tag{17}$$

Similarly, in continuous time settings we have that

$$\eta_t(U_t) = \frac{d}{dt} \log E\left[\exp \int_0^t U_s(X_s)\, ds\right]$$

in the Radon-Nykodim sense and a.s. for $t \geq 0$. Thus, one gets

$$\gamma_t(f) = \gamma_t(1)\,\eta_t(f) \qquad \text{with} \qquad \gamma_t(1) = \exp \int_0^t \eta_s(U_s)\,ds$$

Therefore the natural particle approximating model for the flow of "un-normalized" distributions is given by

$$\gamma_t^N(f) = \gamma_t^N(1)\,\eta_t^N(f) \qquad \text{with} \qquad \gamma_t^N(1) = \exp \int_0^t \eta_s^N(U_s)\,ds \tag{18}$$

where η_t^N is again the empirical measure of the system ξ_t.

1.4 Outline and Contents

These notes are essentially divided into three main parts devoted respectively to discrete time and continuous time models and applications to nonlinear filtering problems.

The first part (section 2) concerns discrete time models. Most of the material presented in this section is taken from the papers [25], [35, 37, 36] and [42].

Section 2.1 focuses on the standard properties of the limiting process needed in the further development of section 2.2. In the preliminary subsection 2.1.1 we give a brief exposition of basic terminologies and properties of the limiting process. In subsection 2.1.2 we discuss the delicate and important problem of the asymptotic stability of the nonlinear semigroup associated with the limiting process. These properties are treated under several type of hypothesis on the functions $\{g_n \; ; \; n \geq 1\}$ and on the transition semigroup associated with the Markov process X. We will also connect in section 2.2.3 these stability properties with the study of the long time behavior of the IPS approximating models.
In nonlinear filtering settings these properties are also essential in order to check whether or not the so-called nonlinear filtering equation forgets erroneous initial conditions. Applications to discrete time filtering problems are discussed in section 5.4.

Section 2.2 is concerned with the asymptotic behavior of the IPS models as the number of particles tends to infinity. This section is essentially divided into four parts. Each of them is devoted to a specific technique or notion to analyze the convergence of the IPS models. This section covers $\mathbb{L}^p$-mean error bounds and uniform convergence results with respect to the time parameter as well as functional central limit theorems, a Glivenko-Cantelli Theorem and large deviations principles.
The quick sketch of the contents of this section will be developed more fully in the preliminary and introductive section 2.2.1.

In section 3 we propose to extend some of the above results to continuous time models. This analysis involves different techniques and it is far from being complete. Among these techniques two are of great importance: semigroup techniques and limit theorems for processes.

In the preliminary section 3.1 we discuss the main hypothesis on the Feynman-Kac formula needed in the further development of section 3.3. In the first subsections 3.1.1 and 3.1.2 we present several regularity assumptions on the fitness functions $\{U_t \; ; \; t \geq 0\}$ and on the Markov process $\{X_t \; ; \; t \geq 0\}$. Section 3.1.3 is concerned with extending the techniques presented in section 2.1.2 to derive stability properties of the Feynman-Kac limiting process. These results will be important later to prove uniform convergence results with respect to the time parameter.

In section 3.2 we give a rigorous mathematical model for the continuous time IPS approximating scheme.

The asymptotic behavior of these IPS numerical schemes as the number of particles tends to infinity is discussed in section 3.3. In the first subsection 3.3.1 we propose $\mathbb{L}^p$-mean errors as well as uniform results. In the last subsection 3.3.2 we prove a functional central limit theorem. Many important problems such as that of the fluctuations and large deviations for the empirical measures on path space remain to be answered.

In view of their kinship to discrete time models the results presented in section 3 can be seen as a good departure to develop a fairly general methodology to study the same asymptotic properties as we did for the discrete time IPS models.

To guide the reader we give in all the development of section 2 and section 3 several comments on the assumptions needed in each specific situation. In section 5 we give detailed and precise examples for each particular assumption.
The applications of the previous results to nonlinear filtering problems are given in section 4. We will discuss continuous time as well as time discretizations and discrete time filtering problems.

The study of the asymptotic stability properties of the limiting system and the investigations of $\mathbb{L}^p$-mean errors, central limit theorems and large deviation principles will of course require quite specific tools. To each of these approaches and techniques corresponds an appropriate set of conditions on the transition probability kernels $\{K_n \; ; \; n \geq 1\}$ and on the so-called fitness functions $\{g_n \; ; \; n \geq 1\}$.

We have tried to present these conditions and to give results at a relevant level of generality so that each section can be read independently of each other. The reader only interested in IPS and BIPS approximating models is recommended to consult section 1.3.2 and section 2.3 as well as section 3 for continuous time models.

The $\mathbb{L}^p$-mean errors presented in section 2.2.2 as well as the central limit theorems for processes exposed in section 3.3.2 are only related to the dynamics structure of the limiting system studied in section 2.1.1. Section 2.2.4 and section 2.2.5 complement previous results of section 2.2.2 and section 2.2.3 by providing asymptotic but precise estimates of $\mathbb{L}^p$-mean errors and exponential rates.

The specific conditions needed in each section are easy to interpret. Furthermore they can be in surprisingly many circumstances be connected one each other. For all these reasons we have no examples in these sections. The interplay and connections between the employed conditions will be described in full details in section 5. We will also connect them to classical examples arising in nonlinear filtering literature. We hope that this choice of presentation will serve our reader well.

2 The Discrete Time Case

2.1 Structural Properties of the Limiting Process.

The investigations of law of large numbers, fluctuations and large deviation principles require quite specific mathematical tools. We shall see in the forthcoming development that these properties are also strongly related to the dynamics structure of the limiting measure valued process (8).
In this section we introduce some basic terminology and a few results on the dynamics structure and the stability properties of (8). In our study a dynamical system is said to be asymptotically stable when its long time behavior does not depends on its initial condition.

In section 2.2 we will prove that the asymptotic stability of the limiting system (8) is a sufficient condition for the uniform convergence of the density profiles of the IPS.

Asymptotic stability properties are also particularly important in filtering settings mainly because the initial law of the signal is usually unknown. In this situation it is therefore essential to check whether the optimal filter is asymptotic stable otherwise all approximating numerical schemes will almost necessarily fails to describe the real optimal filter with the exact initial data.

The genetic type scheme (13) is also used as a random search procedure in numerical function optimization. As for most of stochastic search algorithms, a related question which is of primary interest is to check that their long time behavior does not depend on its initial value.

The study of the long time behavior of the nonlinear filtering equation has been started in [77, 78, 102, 103]. These papers are mainly concerned with the existence of invariant probability measures for the nonlinear filtering equation and do not discuss asymptotic stability properties. A first attempt in this direction was done in [88]. The authors use the above results to prove that the optimal filter "forgets" any erroneous initial condition if the unknown initial law of the signal is absolutely continuous with respect to this new starting point. In the so-called linear-Gaussian situation the optimal filter is also known to be exponentially asymptotically stable under some suitable detectability and observability assumptions (see for instance [88]).
In [5, 48] the authors employ Lyapunov exponent and Hilbert projective techniques to prove exponential asymptotic stability for finite state nonlinear filtering problems. More recently an exponentially asymptotic stability property has been obtained in [16]

for real transient signals, linear observations and bounded observation noise. Here again the proof entirely relies on using Hilbert's projective metrics and showing that the updating transition is a strict contraction with respect to this metric. As pointed out by the authors in [16] the drawback of this metric is its reliance on the assumption of bounded observation noise and does not apply when the optimal filter solution has not bounded support.
Another approach based on the use of Hilbert projective metric to control the logarithmic derivatives of Zakai's kernel is described in [26].
In [4] the authors extend their earlier Lyapunov exponent techniques to Polish-valued signals. Here again the technique consists in evaluating the rate of contraction in the projective Hilbert metric under a mixing type condition on the Markov process X. In discrete time settings this assumption is that the transition probability kernel K_n is homogeneous with respect to time (that is $K_n = K$) and it satisfies the following condition.

$(\mathcal{K})_\varepsilon$ ***There exist a reference probability measure*** $\lambda \in \mathrm{M}_1(E)$ ***and a positive number*** $\epsilon \in (0,1]$ ***so that*** $K(x, \bullet) \sim \lambda$ ***for any*** $x \in E$ ***and***

$$\epsilon \leq \frac{dK(x, \bullet)}{d\lambda} \leq \frac{1}{\epsilon}$$

Here we have chosen to present the novel approach introduced in [36, 38] and further developed in [43]. It is based on the powerful tools developed by R.L. Dobrushin to study central limit theorems for nonstationnary Markov chains [53]. We believe that this approach is more transparent than the previous ones and it also allows to relax considerably the assumption $(\mathcal{K})_\varepsilon$. The continuous time version of this approach will be given in section 3.1.3.

2.1.1 Dynamics Structure

Let us introduce some additional notations. Let $\{Q_{p,n} \ ; \ 0 \leq p \leq n\}$ be the time-inhomogeneous semigroup defined by the relations

$$Q_{p,n} = Q_{p+1} \cdots Q_{n-1} Q_n \quad \text{where} \quad Q_n(f)(x) \stackrel{\text{def.}}{=} g_n(x)\ (K_n f)(x)$$

for any $f \in \mathcal{B}_b(E)$, $0 \leq p \leq n$ and with the convention $Q_{n,n} = \mathrm{Id}$. It is transparent from the definition of "un-normalized" distributions $\{\gamma_n \ ; \ n \geq 0\}$ that

$$\forall\, 0 \leq p \leq n, \qquad \gamma_n \ = \ \gamma_p Q_{p,n}$$

Similarly, we denote by $\{\Phi_{p,n} \ ; \ 0 \leq p \leq n\}$ the composite mappings

$$\Phi_{p,n} = \Phi_n \circ \Phi_{n-1} \circ \ldots \circ \Phi_{p+1}$$

with the convention $\Phi_{n,n} = \mathrm{Id}$. A clear backward induction on the parameter p shows that the composite mappings $\{\Phi_{p,n} \ ; \ 0 \leq p \leq n\}$ have the same form as the one step mappings $\{\Phi_n \ ; \ n \geq 0\}$. This is the content of next lemma.

Lemma 2.1 ([30]) *For any $0 \leq p \leq n$ and $f \in \mathcal{B}_b(E)$ we have that*

$$\Phi_{p,n}(\mu)(f) = \frac{\mu(g_{p,n}\, K_{p,n}(f))}{\mu(g_{p,n})} \tag{19}$$

where

$$K_{p,n}(f) = \frac{Q_{p,n}(f)}{Q_{p,n}(1)} \quad \text{and} \quad g_{p,n} = Q_{p,n}(1)$$

The fitness functions $\{g_{p,n}\ ;\ 0 \leq p \leq n\}$ and the transition probability kernels $\{K_{p,n}\ ;\ 0 \leq p \leq n\}$ satisfy the backward formulae

$$K_{p-1,n}(f) = \frac{K_p(g_{p,n}\, K_{p,n}(f))}{K_p(g_{p,n})}, \quad g_{p-1,n} = g_p\ .\ K_p(g_{p,n}) \tag{20}$$

for any $f \in \mathcal{B}_b(E)$, $1 \leq p \leq n$ and conventions $g_{n,n} = 1$, $K_{n,n} = \mathrm{Id}$.

Lemma 2.2 *For any $0 \leq p \leq n$, $\mu, \nu \in \mathbf{M}_1(E)$ and $f \in \mathcal{B}_b(E)$*

$$\begin{aligned}
&\Phi_{p,n}(\mu)(f) - \Phi_{p,n}(\nu)(f) \\
&= \frac{\nu(g_{p,n})}{\mu(g_{p,n})}\ \mu\Big[\frac{g_{p,n}}{\nu(g_{p,n})} K_{p,n}\left(f - \Phi_{p,n}(\nu)(f)\right)\Big] \\
&= \frac{1}{\nu(g_{p,n})}\Big[(\mu(f_{p,n}) - \nu(f_{p,n})) + \Phi_{p,n}(\mu)(f)\ (\nu(g_{p,n}) - \mu(g_{p,n}))\Big]
\end{aligned} \tag{21}$$

where

$$f_{p,n} \stackrel{\text{def}}{=} g_{p,n}\ .\ K_{p,n}(f).$$

Unless otherwise stated we will always assume that the fitness functions $\{g_n\ ;\ n \geq 1\}$ are continuous and satisfy the following condition.

($\mathcal{G}$) ***For any** $n \geq 1$, **there exists** $a_n \in [1, \infty)$ **with***

$$\forall\, x \in E, \forall\, n \geq 1, \qquad \frac{1}{a_n}\ \leq\ g_n(x)\ \leq\ a_n \tag{22}$$

Under this assumption and using the above notations we notice that

- For any $x \in E$ and $0 \leq p \leq n$

$$\frac{1}{a_{p,n}} \leq g_{p,n}(x) \leq a_{p,n} \quad \text{where} \quad a_{p,n} = \prod_{q=p+1}^{n} a_q$$

- For any $f \in \mathcal{B}_b(E)$ and $0 \leq p \leq n$

$$\|Q_{p,n}(f)\| \leq a_{p,n}\ \|f\| \qquad a_{0,n}^{-1} \leq \gamma_n(1) \leq a_{0,n}$$

In order that our paper is both broad and has some coherence we have chosen to concentrate on the analysis of the particle density profiles $\{\eta_n^N \; ; \; n \geq 0\}$ and the corresponding limiting system $\{\eta_n \; ; \; n \geq 0\}$.

Most of the results presented in section 2.1.2 and section 2.2 can be used to obtain analog results for $\{\widehat{\eta}_n^N \; ; \; n \geq 0\}$ and the flow $\{\widehat{\eta}_n \; ; \; n \geq 0\}$. The interested reader is recommended to use the decompositions

$$\Phi_n(\mu)f - \Phi_n(\nu)f = \frac{1}{\nu(g_n)}\left([\mu(g_n K_n f) - \nu(g_n K_n f)] + (\Phi_n(\mu)f)\,[\nu(g_n) - \mu(g_n)]\right)$$

and

$$\widehat{\eta}_{n-1}^N f - \widehat{\eta}_{n-1} f = \left[\widehat{\eta}_{n-1}^N f - \Psi_n(\eta_{n-1}^N)f\right] + \left[\Psi_n(\eta_{n-1}^N)f - \Psi_n(\eta_{n-1})f\right]$$

and to recall that $\widehat{\eta}_{n-1}^N$ is the empirical measure associated with N conditionally independent random variables with common law $\Psi_n(\eta_{n-1}^N)$.
There also exists a different strategy to approximate the flow of distributions $\{\widehat{\eta}_n \; ; \; n \geq 0\}$. This alternative approach is simply based on the observation that the dynamics structure of the latter is again defined by a recursion of the form (8). More precisely, in view of (10) we have that

$$\forall\, n \geq 1, \qquad \widehat{\eta}_n \;=\; \widehat{\Phi}_n(\widehat{\eta}_{n-1}) \tag{23}$$

where $\widehat{\Phi}_n(\eta) = \widehat{\Psi}_n(\eta)\widehat{K}_n$ and for any $f \in \mathcal{B}_b(E)$,

$$\widehat{\Psi}_n(\eta)(f) \stackrel{\text{def.}}{=} \frac{\eta(\widehat{g}_n f)}{\eta(\widehat{g}_n)}, \qquad \widehat{K}_n f \stackrel{\text{def.}}{=} \frac{K_n(g_{n+1} f)}{K_n(g_{n+1})}, \qquad \widehat{g}_n \stackrel{\text{def.}}{=} K_n(g_{n+1}).$$

Noticing that the new fitness functions $\{\widehat{g}_n \; ; \; n \geq 0\}$ again satisfy $(\mathcal{G})$ we see that the study of (23) and its corresponding IPS approximating model is reduced to the former by replacing the fitness functions $\{g_n \; ; \; n \geq 1\}$ and the transitions $\{K_n \; ; \; n \geq 0\}$ by $\{\widehat{g}_n \; ; \; n \geq 0\}$ and $\{\widehat{K}_n \; ; \; n \geq 1\}$ and the initial data η_0 by the distribution $\widehat{\eta}_0$.

Nevertheless the above description shows that the formulation of the new fitness functions involves integrations over the whole state space E. In practice these integrals are not known exactly and another level of approximation is therefore needed. We also remark that we use the well known "rejection/acceptation" sampling method to produce transitions of the IPS approximating scheme. But, when the constants $\{a_n \; ; \; n \geq 1\}$ are too large it is well known that the former sampling technique is "time-consuming" and not really efficient.

We have chosen to restrict our attention to the IPS approximating model $\{\eta_n^N \; ; \; n \geq 0\}$ for several reasons:

- First of all it is defined as a two stage transition which can serve as a Markov model for classical genetic algorithms arising in biology and nonlinear estimation problems.
- Secondly, some results such as the fluctuations on path space for the IPS $\{\widehat{\xi}_n \; ; \; n \geq 0\}$ are not yet available.

- Another important reason is that the limiting system $\{\eta_n \; ; \; n \geq 0\}$ is exactly the discrete time approximating model of the Kushner-Stratonovitch equation (see for instance [24] and references therein as well as section 4.3 page 122).

- Finally the evolution equation of the distributions $\{\widehat{\eta}_n \; ; \; n \geq 0\}$ is a special case of (8) so that the results on (8) will also include those on (23).

2.1.2 Asymptotic Stability

In this section we discuss the asymptotic stability properties of (8). The main difficulty lies in the fact that (8) is a two stage process and it may have completely different kinds of long time behavior.

For instance, if the fitness functions are constant functions then (8) is simply based on prediction transitions. In this situation the recursion (8) describes the time evolution of the distributions of the Markov process $\{X_n \; ; \; n \geq 0\}$. In this very special case the theory of Markov processes and stochastic stability can be applied.

On the other hand, if the transition probability kernels $\{K_n \; ; \; n \geq 1\}$ are trivial, that is $K_n = \text{Id}$, $n \geq 1$, then (8) is only based on updating transitions. In this case its long time behavior is strongly related on its initial value. For instance, if $g_n = \exp(-U)$, for some $U : E \to \mathbb{R}_+$, then for any bounded continuous function $f : E \to \mathbb{R}_+$ with compact support

$$\eta_n(f) = \frac{\eta_0(f \, e^{-nU})}{\eta_0(e^{-nU})} \xrightarrow[n \to \infty]{} \frac{\eta_0(f \, 1_{U^\star})}{\eta_0(U^\star)}$$

where $U^\star \stackrel{\text{def.}}{=} \{x \in E \; ; \; U(x) = \text{essinf}_{\eta_0} U\}$ (at least if $\eta_0(U^*) > 0$).

If we write $\mathbf{M}_0(E) \subset \mathbf{M}(E)$ the subset of measures μ such that $\mu(E) = 0$ then any transition function $T(x, dz)$ on E maps $\mathbf{M}_0(E)$ into $\mathbf{M}_0(E)$. We recall that its norm is given by

$$\beta(T) \stackrel{\text{def.}}{=} \sup_{\mu \in \mathbf{M}_0(E)} \frac{\|\mu T\|_{tv}}{\|\mu\|_{tv}} = \sup_{\mu, \nu \in \mathbf{M}_1(E)} \frac{\|\mu T - \nu T\|_{tv}}{\|\mu - \nu\|_{tv}}$$

The quantity $\beta(T)$ is a measure of contraction of the total variation distance of probability measures induced by T. It can also be defined as

$$\beta(T) = \sup_{x, y \in E} \|\delta_x T - \delta_y T\|_{tv} = 1 - \alpha(T) \tag{24}$$

The quantity $\alpha(T)$ is called the Dobrushin ergodic coefficient of T defined by the formula

$$\alpha(T) = \inf \sum_{i=1}^{m} \min\left(T(x, A_i), T(z, A_i)\right) \tag{25}$$

where the infimum is taken over all $x, z \in E$ and all resolutions of E into pairs of non-intersecting subsets $\{A_i \; ; \; 1 \leq i \leq m\}$ and $m \geq 1$ (see for instance [53]). Our analysis is based on the following observation: In view of (20) the Markov transition probability kernels $\{K_{n,p}\ ;\ 0 \leq p \leq n\}$ are composite operators of time-inhomogeneous but linear Markov operators. More precisely it can be checked directly from (20) that

$$K_{p-1,n} = S_p^{(n)} K_{p,n} = S_p^{(n)} S_{p+1}^{(n)} \dots S_{n-1}^{(n)} S_n^{(n)}, \qquad S_p^{(n)} f = \frac{K_p(g_{p,n}\ f)}{K_p(g_{p,n})} \tag{26}$$

The usefulness of Dobrushin ergodic coefficient in the study of the asymptotic stability of the nonlinear dynamical system (8) can already been seen from the next theorem.

Theorem 2.3 *Assume that the fitness functions* $\{g_n\ ;\ n \geq 1\}$ *is a sequence of bounded and positive functions on* E. *For any* $n \geq p$ *we have that*

$$\|\Phi_{p,n}(\mu) - \Phi_{p,n}(\nu)\|_{\text{tv}} \leq \beta(K_{p,n})\ \|\Psi_{p,n}(\mu) - \Psi_{p,n}(\nu)\|_{\text{tv}} \tag{27}$$

with

$$\forall \mu \in \mathbf{M}_1(E)\,, \forall f \in \mathcal{B}_b(E)\,, \qquad \Psi_{p,n}(\mu)(f) = \frac{\mu(g_{p,n}\ f)}{\mu(g_{p,n})}$$

and

$$\sup_{\mu,\nu} \|\Phi_{p,n}(\mu) - \Phi_{p,n}(\nu)\|_{\text{tv}} = \beta(K_{p,n}) \leq \prod_{q=1}^{n-p} \left[1 - \alpha(S_{p+q}^{(n)})\right] \tag{28}$$

Assume that the transition probability kernels $\{K_n\ ;\ n \geq 1\}$ *satisfy the following condition.*

$(\mathcal{K})_1$ ***For any time*** $n \geq 1$, ***there exist a reference probability measure*** $\lambda_n \in \mathbf{M}_1(E)$ ***and a positive number*** $\epsilon_n \in (0,1]$ ***so that*** $K_n(x, \bullet) \sim \lambda_n$ ***for any*** $x \in E$ ***and***

$$\epsilon_n \leq \frac{dK_n(x,\bullet)}{d\lambda_n} \leq \frac{1}{\epsilon_n}$$

Then, we have for any $\mu, \nu \in \mathbf{M}_1(E)$

$$\begin{aligned} \sum_{n\geq 1} \epsilon_n^2 = \infty \quad &\Longrightarrow \quad \lim_{n\to\infty} \|\Phi_{0,n}(\mu) - \Phi_{0,n}(\nu)\|_{\text{tv}} = 0 \\ \lim_{n\to\infty} \frac{1}{n} \sum_{p=1}^{n} \epsilon_p^2 \overset{\text{def.}}{=} \epsilon^2 > 0 \quad &\Longrightarrow \quad \limsup_{n\to\infty} \frac{1}{n} \log \|\Phi_{0,n}(\mu) - \Phi_{0,n}(\nu)\|_{\text{tv}} \leq -\epsilon^2 < 0 \\ \inf_{n\geq 1} \varepsilon_n \overset{\text{def.}}{=} \epsilon > 0 \quad &\Longrightarrow \quad \frac{1}{T} \log \sup_{p\geq 0} \|\Phi_{p,p+T}(\mu) - \Phi_{p,p+T}(\nu)\|_{\text{tv}} < -\epsilon^2 \end{aligned} \tag{29}$$

Proof:(*Sketch*) Since, for any $\mu \in \mathbf{M}_1(E)$ we have $\Phi_{p,n}(\mu) = \Psi_{p,n}(\mu) K_{p,n}$ then, using (24) and (26) one proves (28). Under $(\mathcal{K})_1$, the second part of the theorem is a consequence of the fact that $\alpha(S_p^{(n)}) \geq \epsilon_p^2$, for any $0 \leq p \leq n$. To prove these lower bounds we use (25) and the definition of the transitions $S_p^{(n)}$. ∎

Remarks 2.4:

- Theorem 2.3 holds for any sequence of fitness functions $\{g_n ; n \geq 1\}$. One also observe that the last assertion in the Theorem 2.3 holds true when the transition probability kernels $\{K_n ; n \geq 1\}$ satisfy $(\mathcal{K})_\varepsilon$.
  In contrast to [4] (see for instance Corollary 1 p. 706 in [4]) the exponential asymptotic property (29) is valid for any time parameter and it does not depend on the form neither on the regularity of the fitness functions.

- In time homogeneous settings (that is $K_n = K$, $g_n = g$) the mapping $\Phi_n = \Phi$ is again time homogeneous. If there exists some fixed point $\mu = \Phi(\mu) \in \mathbf{M}_1(E)$. Theorem 2.3 gives several conditions underwhich μ is unique and any solution $\{\eta_n ; n \geq 0\}$ of the time homogeneous dynamical system associated to Φ converges exponentially fast to μ as the time parameter $n \to \infty$. For instance if $(\mathcal{K})_1$ holds with $\epsilon_n = \epsilon > 0$ then

$$\forall\, \eta \in \mathbf{M}_1(E), \qquad \sup_{p \geq 0} \|\Phi_{p,p+T}(\eta) - \mu\|_{\text{tv}} \;\leq\; e^{-\epsilon^2 . T}$$

- Let us consider the special case where $\epsilon_n = \epsilon_1 . n^{-\beta/2}$ with $\epsilon_1 > 0$ and $\beta \leq 1$. Theorem 2.3 tells us that (8) is asymptotically stable but we don't know if exponential rates are available in this scale.

- If $(\mathcal{K})_1$ is satisfied for some sequence of numbers $\{\epsilon_n ; n \geq 1\}$ and a collection of distributions $\{\lambda_n ; n \geq 1\}$ then we also have that

$$\forall\, x, z \in E, \forall\, n \geq 1, \qquad \widehat{\epsilon}_n \;\leq\; \frac{d\widehat{K}_n(x, .)}{d\widehat{\lambda}_n}(z) \;\leq\; \frac{1}{\widehat{\epsilon}_n}$$

  where $\widehat{\epsilon}_n \stackrel{\text{def.}}{=} \epsilon_n^2$ and $\widehat{\lambda}_n \in \mathbf{M}_1(E)$ is the distribution on E given for any bounded test function f by

$$\widehat{\lambda}_n(f) \stackrel{\text{def.}}{=} \frac{\lambda_n(\, g_{n+1}\, f\,)}{\lambda_n(\, g_{n+1}\,)}$$

  This observation shows that Theorem 2.3 remains true if we replace the one step mappings $\{\Phi_n ; n \geq 1\}$ and the numbers $\{\epsilon_n^2 ; n \geq 1\}$ by the mappings $\{\widehat{\Phi}_n ; n \geq 1\}$ and the numbers $\{\epsilon_n^4 ; n \geq 1\}$. It is also worth to notice that this results is again independent of the fitness functions.

- By definition of the one step mappings $\{\widehat{\Phi}_n ; n \geq 1\}$ we have the formula

$$\forall\, \mu \in \mathbf{M}_1(E), \forall\, n \geq 1, \qquad \widehat{\Phi}_n(\mu) \;=\; \Psi_{n+1}\left(\Phi_{1,n}(\mu K_1)\right)$$

  Thus, by definition of the mappings $\{\Psi_n ; n \geq 1\}$, if $(\mathcal{G})$ is satisfied then one can check that for any $\mu, \nu \in \mathbf{M}_1(E)$ and $n \geq 1$

$$\left\|\widehat{\Phi}_{0,n}(\mu) - \widehat{\Phi}_{0,n}(\nu)\right\|_{\text{tv}} \leq 2\, a_{n+1}^2 \;\; \|\Phi_{1,n}(\mu K_1) - \Phi_{1,n}(\nu K_1)\|_{\text{tv}}$$

  where $\widehat{\Phi}_{0,n}$ are the composite mappings

$$\widehat{\Phi}_{0,n} = \widehat{\Phi}_n \circ \widehat{\Phi}_{n-1} \circ \ldots \circ \widehat{\Phi}_1$$

Therefore Theorem 2.3 can also be used to derive several conditions for the asymptotic stability of (23).
In contrast to the previous remark the resulting error bound depends on the sets of numbers $\{a_n \ ; \ n \geq 1\}$ and $\{\epsilon_n^2 \ ; \ n \geq 1\}$ instead of $\{\epsilon_n^4 \ ; \ n \geq 1\}$. This fact is important for instance if $\epsilon_n = \epsilon_1 . n^{-1/2}$ and $\sup_n a_n < \infty$. In this specific situation the estimate resulting from this approach will improve the one we would obtain using the previous remark and it allows to conclude that (23) is asymptotically stable even if $\sum_n \epsilon_n^4 < \infty$.

Next condition relax $(\mathcal{K})_1$.

$(\mathcal{K})_2$ ***For any time*** $p \geq 0$ ***there exist some*** $m \geq 1$, $\lambda_p \in \mathbf{M}_1(E)$ ***and*** $\epsilon_p > 0$ ***such that for any*** $x \in E$

$$\varepsilon_p \leq \frac{dK_p^{(m)}(x, \bullet)}{d\lambda_p} \leq \frac{1}{\varepsilon_p} \qquad \text{where} \quad K_p^{(m)} \stackrel{\text{def.}}{=} K_{p+1} \ldots K_{p+m} \tag{30}$$

Under $(\mathcal{K})_2$ and $(\mathcal{G})$ one can check that for any $0 \leq p+m \leq n$

$$\left(\frac{\varepsilon_p}{a_{p+1,p+m}}\right)^2 \alpha\left(K_p\right) \leq \alpha\left(S_p^{(n)}\right) \leq \left(\frac{a_{p+1,p+m}}{\varepsilon_p}\right)^2 \alpha\left(K_p\right) \tag{31}$$

We now turn to still another way to produce useful estimates.

By definition of the integral operators $\{S_p^{(n)} \ ; \ 0 \leq p \leq n\}$ we also have for any positive Borel test function $f : E \to \mathbb{R}_+$

$$\frac{1}{a_{p+2,p+m}^2} \frac{K_p^{(m)}(g_{p+m,n} f)}{K_p^{(m)}(g_{p+m,n})} \leq S_{p+1}^{(n)} \ldots S_{p+m}^{(n)} f \leq a_{p+2,p+m}^2 \frac{K_p^{(m)}(g_{p+m,n} f)}{K_p^{(m)}(g_{p+m,n})}$$

From (30) it is clear that for any $0 < p+m \leq n$

$$\left(\frac{\varepsilon_p}{a_{p+2,p+m}}\right)^2 \leq \alpha\left(S_{p+1}^{(n)} \ldots S_{p+m}^{(n)}\right) \leq \left(\frac{a_{p+2,p+m}}{\varepsilon_p}\right)^2 \tag{32}$$

In contrast to (31) the estimate (32) does not depend anymore on the ergodic coefficients $\{\alpha(K_p) \ ; \ p \geq 1\}$.

Nevertheless the estimate induced by (31) may improve the one we would obtain using (32). For instance, in time homogeneous settings (that is $\varepsilon_p = \varepsilon$, $a_p = a$, $K_p = K$, for all $p \geq 0$) the bounds (32) lead to

$$\left(1 - \alpha\left(S_{p+1}^{(n)} \ldots S_{p+m}^{(n)}\right)\right) \leq 1 - \frac{\varepsilon^2}{a^{2(m-1)}} \leq \exp -\frac{\varepsilon^2}{a^{2(m-1)}} \tag{33}$$

and (31) implies that

$$\left(1 - \alpha\left(S_{p+1}^{(n)} \ldots S_{p+m}^{(n)}\right)\right) \leq \prod_{q=1}^{m} \left(1 - \alpha\left(S_{p+q}^{(n)}\right)\right) \leq \left(1 - \frac{\varepsilon^2}{a^{2m}} \alpha(K)\right)^m$$

and therefore

$$\left(1-\alpha\left(S_{p+1}^{(n)}\dots S_{p+m}^{(n)}\right)\right)\leq \exp-\left(m\frac{\varepsilon^2}{a^{2m}}\alpha(K)\right) \tag{34}$$

One concludes that (34) improves (33) as soon as $m.\alpha(K) \geq a^2$ and therefore

$$\log\left(1-\alpha\left(S_{p+1}^{(n)}\dots S_{p+m}^{(n)}\right)\right)\leq -\frac{\epsilon^2}{a^{2m}}\max\left(m.\alpha(K),a^2\right)$$

Theorem 2.5 *Assume that $(\mathcal{G})$ and $(\mathcal{K})_2$ hold for some $m \geq 1$ and some sequence of numbers $\{\epsilon_p\ ;\ p \geq 0\}$. For any $\mu, \nu \in \mathbf{M}_1(E)$ and $n \geq m$ we have*

$$\|\Phi_{0,n}(\mu)-\Phi_{0,n}(\nu)\|_{\text{tv}} \leq \prod_{p=1}^{n-m}\left(1-\frac{\varepsilon_p^2}{a_{p+1,p+m}^2}\alpha(K_p)\right)$$

and

$$\|\Phi_{0,n}(\mu)-\Phi_{0,n}(\nu)\|_{\text{tv}} \leq \prod_{p=0}^{[n/m]-1}\left(1-\frac{\varepsilon_{pm}^2}{a_{pm+2,pm+m}^2}\right)$$

In addition, if $\inf_n \varepsilon_n \stackrel{\text{def.}}{=} \varepsilon > 0$, $\inf_n \alpha(K_n) \stackrel{\text{def.}}{=} \alpha > 0$ and $\sup_n a_n \stackrel{\text{def.}}{=} a < \infty$ then for any $u \geq 1$ and $T \geq u.m$

$$\frac{1}{T}\log\sup_{p\geq 0}\|\Phi_{p,p+T}(\mu)-\Phi_{p,p+T}(\nu)\|_{\text{tv}} < -\frac{1}{v.m}\left(\frac{\varepsilon}{a^m}\right)^2\max\left(m.\alpha,a^2\right)$$

where $1/u + 1/v = 1$.

Next we present an easily verifiable sufficient condition for $(\mathcal{K})_2$. It also gives some connections between $(\mathcal{K})_1$ and $(\mathcal{K})_2$. It will be used in section 5 to check that some classes of Gaussian transitions satisfy $(\mathcal{K})_2$ for $m = 2$. To clarify the presentation the transition probability kernels K_n here are assumed to be time homogeneous (that is $K_n = K$ for all $n \geq 0$).

$(\mathcal{K})_3$ ***There exist a subset $A \in \mathbf{B}(E)$, a reference probability measure $\lambda \in \mathbf{M}_1(E)$ and a positive number $\epsilon \in (0,1)$ such that $\lambda(A) \geq \varepsilon$ and***

$$\forall\, x \in E, \forall\, z \in A, \qquad \epsilon\ \leq\ \frac{dK(x,\bullet)}{d\lambda}(z)\ \leq\ \frac{1}{\epsilon}$$

In addition there exist a decomposition $A^c = B_1 \cup \ldots \cup B_m$, $m \geq 1$ and $2m$ reference probability measures $\lambda_1,\ldots,\lambda_m,\gamma_1,\ldots,\gamma_m \in \mathbf{M}_1(E)$ such that for any $1 \leq k \leq m$

$$\forall\, x \in B_k, \forall\, z \in E, \qquad \epsilon\ \leq\ \frac{dK(x,\bullet)}{d\lambda_k}(z)\ \leq\ \frac{1}{\epsilon}$$

and

$$\forall\, x \in E, \forall\, z \in B_k, \qquad \gamma_k(B_k) \geq \varepsilon \qquad \text{and} \qquad \frac{dK(x,\bullet)}{d\gamma_k}(z)\ \geq\ \epsilon$$

Under $(\mathcal{G})$ and $(\mathcal{K})_3$ one concludes that $(\mathcal{K})_2$ holds for $m = 2$ (the last condition above is even useless for that).

Furthermore under $(\mathcal{K})_3$ one can also prove the following bounds

$$\alpha(K) \geq m.\varepsilon^2 \quad \text{and} \quad \alpha\left(S_p^{(n)}\right) \geq \frac{\varepsilon^4}{a_{p+1}^2}$$

Corollary 2.6 *Assume that $(\mathcal{G})$ and $(\mathcal{K})_3$ hold. Then, for any $\mu, \nu \in \mathbf{M}_1(E)$ we have that*

$$\sum_{n\geq 1} a_n^{-2} = \infty \implies \lim_{n\to\infty} \|\Phi_{0,n}(\mu) - \Phi_{0,n}(\nu)\|_{\text{tv}} = 0$$

$$\lim_{n\to\infty} \frac{1}{n}\sum_{p=1}^{n} a_p^{-2} \stackrel{\text{def.}}{=} a^{-2} \implies \limsup_{n\to\infty} \frac{1}{n} \log \|\Phi_{0,n}(\mu) - \Phi_{0,n}(\nu)\|_{\text{tv}} \leq -\frac{\varepsilon^4}{a^2}$$

$$\sup_n a_n = a < \infty \implies \frac{1}{T} \log \sup_{p\geq 0} \|\Phi_{p,p+T}(\mu) - \Phi_{p,p+T}(\nu)\|_{\text{tv}} \leq -\frac{\varepsilon^4}{a^2}$$

A more interesting and stronger result for transition probability kernels $\{K_n \,;\, n \geq 1\}$ which do not satisfy $(\mathcal{K})_2$ is given next. We will use the following condition.

$(\mathcal{KG})$ ***There exists some*** $p_0 \geq 0$ ***such that***

$$\forall\, p \geq p_0, \qquad \delta_p \stackrel{\text{def.}}{=} \inf_{n\geq p}\, \inf_{x,y\in E} \frac{g_{p,n}(x)}{g_{p,n}(y)} > 0 \tag{35}$$

Let us observe that under $(\mathcal{KG})$ one has a uniform control of the fitness functions $\{g_{p,n} \,;\, 0 \leq p \leq n\}$ in the sense that for any $n \geq p \geq p_0$ and $x, y \in E$

$$\delta_p \leq \frac{g_{p,n}(x)}{g_{p,n}(y)} \leq \frac{1}{\delta_p}$$

In contrast to previous conditions $(\mathcal{KG})$ depends on both the fitness functions $\{g_n \,;\, n \geq 1\}$ and on the transition probability kernels $\{K_n \,;\, n \geq 1\}$. Note that $(\mathcal{KG})$ is clearly satisfied when the fitness functions are constant or more generally if the fitness functions become constant after a given step p_0 (that is $g_p = 1$ for any $p \geq p_0$). In the same vein $(\mathcal{KG})$ is satisfied if we are allowed to choose the constants $\{a_n \,;\, n \geq 1\}$ such that $\sum_{n\geq 1} \log a_n < \infty$. In this situation we would have

$$\delta_p \geq \exp -\left(2 \sum_{q\geq 1} \log a_{p+q}\right) > 0.$$

Another way of viewing condition $\sum_{n\geq 1} \log a_n < \infty$ is to say that the sequence of functions $\{g_n \,;\, n \geq 1\}$ satisfy

$$\sum_{n\geq 1} \sup_{x\in E} |\log g_n(x)| < \infty$$

which clearly implies that g_n tends to the unit function 1 as n tends to infinity.

It is also noteworthy that if $(\mathcal{KG})$ holds for $p_0 = 0$ then condition $(\mathcal{G})$ is directly satisfied. More precisely in this situation we would get the bounds

$$\forall\, x,y \in E, \forall\, p \geq 0, \qquad \delta_p \;\leq\; \frac{g_{p+1}(x)}{g_{p+1}(y)} \;\leq\; \frac{1}{\delta_p}$$

To see that $(\mathcal{KG})$ also relax $(\mathcal{K})_2$ it suffices to note that $(\mathcal{K})_2$ implies that for any $0 \leq p+m \leq n$

$$\frac{g_{p,n}(x)}{g_{p,n}(y)} \geq a^{-2}_{p+1,p+m}\, \frac{K_{p+1}\ldots K_{p+m}\,(g_{p+m,n})\,(x)}{K_{p+1}\ldots K_{p+m}\,(g_{p+m,n})\,(y)} \geq \epsilon_p^2\; a^{-2}_{p+1,p+m} > 0$$

Since for any $p \leq n \leq p+m$ and $x,y \in E$

$$\frac{g_{p,n}(x)}{g_{p,n}(y)} \geq a^{-2}_{p+1,p+m}$$

we finally have the lower bounds

$$\forall\, p \geq 0, \qquad \delta_p \;\geq\; \epsilon_p^2\; a^{-2}_{p+1,p+m} \;>\; 0$$

Under $(\mathcal{KG})$ and using (27) and the decomposition

$$\Psi_{p,n}(\mu)(f) - \Psi_{p,n}(\nu)(f) = \frac{\nu(g_{p,n})}{\mu(g_{p,n})}\;\mu\left[\frac{g_{p,n}}{\nu(g_{p,n})}\;(f - \Psi_{p,n}(\nu)(f))\right]$$

we also have that

$$\forall n \geq p \geq p_0\,, \qquad \|\Psi_{p,n}(\mu) - \Psi_{p,n}(\nu)\|_{\mathrm{tv}} \leq \frac{2}{\delta_p^2}\;\|\mu - \nu\|_{\mathrm{tv}}$$

Using the same lines of reasoning as before one can prove that for any Borel test function $f : E \to \mathbb{R}_+$ and $m \geq 1$ and $n \geq p+m$

$$\begin{aligned} S^{(n)}_{p+1}\ldots S^{(n)}_{p+m} f \;&\geq\; \delta_{p+1}\ldots\delta_{p+(m-1)}\;\frac{K^{(m)}_p(g_{p+m,n} f)}{K^{(m)}_p(g_{p+m,n})} \\ &\geq\; \delta_{p+1}\ldots\delta_{p+m}\;\; K^{(m)}_p f \end{aligned}$$

from which one deduces the following deeper result.

Theorem 2.7 *If $(\mathcal{KG})$ is satisfied for some parameter $p_0 \geq 0$ then we have for any $n \geq p \geq p_0$*

$$\|\Phi_{p,n}(\mu) - \Phi_{p,n}(\nu)\|_{\mathrm{tv}} \leq \frac{2}{\delta_p^2}\;\beta\left(K_{p,n}\right)\;\;\|\mu - \nu\|_{\mathrm{tv}}$$

and for any $m \geq 1$

$$\sup_{\mu,\nu}\|\Phi_{0,n}(\mu) - \Phi_{0,n}(\nu)\|_{\mathrm{tv}} \leq \prod_{q=0}^{[(n-p_0)/m]-1}\left(1 - \delta^{(m)}_{p_0+q.m}\;\alpha\left(K^{(m)}_{p_0+q.m}\right)\right) \qquad (36)$$

where $[a]$ *denotes the integer part of* $a \in \mathbb{R}$, $\delta_p^{(m)} \stackrel{\text{def.}}{=} \delta_{p+1} \ldots \delta_{p+m}$ *and*

$$K_p^{(m)} \stackrel{\text{def.}}{=} K_{p+1} \ldots K_{p+m}.$$

In addition if $p_0 = 0$ *and* $\inf_p \delta_p \stackrel{\text{def.}}{=} \delta > 0$ *then for* $m = 1$ *(36) leads to*

$$\frac{1}{n} \log \sup_{\mu,\nu} \|\Phi_{0,n}(\mu) - \Phi_{0,n}(\nu)\|_{\text{tv}} \leq -\frac{\delta}{n} \sum_{p=1}^{n} \alpha(K_p) \tag{37}$$

Remarks 2.8:

- The bound (36) is sharp in the sense that if the fitness functions g_n are constant then one may choose $m = n, p_0 = 0$. In this situation (36) reads

$$\begin{aligned} \sup_{\mu,\nu} \|\mu K_1 \ldots K_n - \nu K_1 \ldots K_n\|_{\text{tv}} &\leq 1 - \alpha(K_1 \ldots K_n) = \beta(K_1 \ldots K_n) \\ &= \sup_{x,y} \|\delta_x K_1 \ldots K_n - \delta_y K_1 \ldots K_n\|_{\text{tv}} \end{aligned}$$

- In view of (37) and under $(\mathcal{KG})$, condition $\sum_{n \geq 1} \alpha(K_n) = \infty$ is a sufficient condition for the asymptotic stability of the nonlinear semigroup

$$\{\Phi_{p,n} \,; 0 \leq p \leq n\}$$

  This condition is a familiar necessary and sufficient condition for the temporally inhomogeneous semigroup associated with the transitions $\{K_n \,;\, n \geq 1\}$ to be strongly ergodic (see for instance [53], part I, p. 76).

- In nonlinear filtering settings condition $\sum_n \log a_n < \infty$ is related to the form of the observation noise source. An example of observation process satisfying this condition will be given in the end of section 5. Roughly speaking the assumptions $(\mathcal{K})_i$, $i = 1, 2, 3$ say that the signal process is sufficiently mixing and condition $\sum_n \log a_n < \infty$ says that the observation process is sufficiently noisy.

2.2 Asymptotic Behavior of the Particle Systems

2.2.1 Introduction

In this section we investigate the asymptotic behavior of the IPS as the number of particles tends to infinity. In the first subsection 2.2.2 we discuss $\mathbb{L}^p$-mean error bounds and a uniform convergence theorem with respect to the time parameter. A Glivenko-Cantelli Theorem is described in section 2.2.3. Subsection 2.2.4 is concerned with fluctuations of IPS. In the last subsection 2.2.5 we present large deviation principles.

All of the above properties will of course be stated under appropriate regularity conditions on the transition probability kernels $\{K_n \,;\, n \geq 1\}$ and on the fitness functions $\{g_n \,;\, n \geq 1\}$.

Assumption $(\mathcal{G})$ is the only assumption needed on the fitness functions. Unless otherwise stated we will always assume that $(\mathcal{G})$ holds.

This condition has a clear interpretation in nonlinear filtering settings (see for instance [33, 35, 36, 37] and section 5). It can be regarded as a technical assumption and several results presented here can be proved without this condition. To illustrate this remark we will give in the beginning of section 2.2.2 a very basic convergence result that does not depend on $(\mathcal{G})$. To guide the reader we now give some comments on the assumptions needed on the transition probability kernels $\{K_n\ ;\ n \geq 1\}$.

The uniform convergence result presented in section 2.2.2 is based on the asymptotic stability properties of the limiting measure valued process (8) we have studied in section 2.1.2. This part will then be related to assumptions $(\mathcal{K})_1$,$(\mathcal{K})_2$ and $(\mathcal{K})_3$.

The Glivenko-Cantelli and Donsker Theorems presented in section 2.2.3 and section 2.2.4 extend the corresponding statements in the classical theory of empirical processes. They are simply based on $(\mathcal{G})$. The idea here is to consider a given random measure μ^N as a stochastic process indexed by a collection $\mathcal{F}$ of measurable functions $f : E \to \mathbb{R}$. If μ^N is an empirical measure then, the resulting $\mathcal{F}$-indexed collection $\{\mu^N(f);\ f \in \mathcal{F}\}$ is usually called the $\mathcal{F}$-empirical process associated with the empirical random measures μ^N. The semi-metric commonly used in such a context is the Zolotarev semi-norm defined by

$$\forall\ \mu, \nu \in \mathbf{M}_1(E), \qquad \|\mu - \nu\|_{\mathcal{F}} \ = \ \sup\{|\mu(f) - \nu(f)|;\ f \in \mathcal{F}\}$$

(see for instance [95]). In order to control the behavior of the supremum $\|\eta_n^N - \eta_n\|_{\mathcal{F}}$ as $N \to \infty$, we will impose conditions on the class $\mathcal{F}$ that are classically used in the statistical theory of empirical processes for independent samples. To avoid technical measurability conditions, and in order not to obscure the main ideas, we will always assume the class $\mathcal{F}$ to be countable and uniformly bounded. Our conclusions also hold under appropriate separability assumptions on the empirical process (see [50]).

The Glivenko-Cantelli and Donsker Theorems are uniform versions of the law of large numbers and the central limit theorem for empirical measures. In the classical theory of independent random variables, these properties are usually shown to hold under entropy conditions on the class $\mathcal{F}$. Namely, to measure the size of a given class $\mathcal{F}$, one considers the covering numbers $N(\varepsilon, \mathcal{F}, L_p(\mu))$ defined as the minimal number of $L_p(\mu)$-balls of radius $\varepsilon > 0$ needed to cover $\mathcal{F}$. With respect to classical theory, we will need assumptions on these covering numbers uniformly over all probability measures μ. Classically also, this supremum can be taken over all discrete probability measures. Since we are dealing with interacting particle schemes, we however need to strengthen the assumption and take the corresponding supremum over all probability measures. Several examples of classes of functions satisfying the foregoing uniform entropy conditions are discussed in the book [50].

Denote thus by $\mathcal{N}(\varepsilon, \mathcal{F})$, $\varepsilon > 0$, and by $I(\mathcal{F})$ the uniform covering numbers and entropy integral given by

$$\mathcal{N}(\varepsilon, \mathcal{F}) \ = \ \sup\{N(\varepsilon, \mathcal{F}, L_2(\mu));\ \mu \in \mathbf{M}_1(E)\},$$

$$I(\mathcal{F}) = \int_0^1 \sqrt{\log \mathcal{N}(\varepsilon, \mathcal{F})}\, d\varepsilon.$$

which will be assumed to be finite.

The fluctuations and the large deviations principles for the particle density profiles $\{\eta_n^N \; ; \; n \geq 0\}$ will be simply based on $(\mathcal{G})$. In fact this assumption on the fitness functions will be used to check the asymptotic tightness property in Donsker's Theorem and an exponential tightness property in large deviation settings.

The study of fluctuations and large deviations on path space require more attention. It becomes more transparent if we introduce a more general abstract formulation. Namely, we will assume that $\{\xi_n \; ; \; n \geq 0\}$ is the IPS approximating model (13) associated with a given sequence of continuous functions $\{\Phi_n \; ; \; n \geq 1\}$. We will use the following assumption.

$(\mathcal{P})_0$ ***For any time*** $n \geq 1$ ***there exists a reference probability measure*** $\lambda_n \in \mathbf{M}_1(E)$ ***such that***

$$\forall\, \mu \in \mathbf{M}_1(E), \qquad \Phi_n(\mu) \;\sim\; \lambda_n$$

This condition might seems difficult to check in general. In fact if the functions $\{\Phi_n \; ; \; n \geq 1\}$ are given by (8) then $(\mathcal{P})_0$ holds if and only if the transition probability kernels $\{K_n \; ; \; n \geq 1\}$ satisfy the following condition.

$(\mathcal{K})_0$ ***For any time*** $n \geq 1$ ***there exists a reference probability measure*** $\lambda_n \in \mathbf{M}_1(E)$ ***so that***

$$\forall\, x \in E, \qquad K_n(x, \bullet) \;\sim\; \lambda_n$$

We shall see in section 5 that this condition covers many typical examples of nonlinear filtering problems (see also [35, 37, 36]).
The main reason for our needing to make the assumption $(\mathcal{P})_0$ for the analysis of the fluctuations on path space is that we want to use a reference product probability measure. We also notice that there is no loss of generality in choosing $\lambda_n = \eta_n$. This particular choice of reference probability measure will be used for technical reasons in the study of the fluctuations on path space. Roughly speaking this is the appropriate and natural choice for studying the weak convergence of the resulting Hamiltonian function under the underlying product measure (see for instance Lemma 2.24 and its proof given on page 52).
The main simplification due to assumption $(\mathcal{P})_0$ is the following continuity property. For any $T \in \mathbb{N}$, we denote $P_T^{(N)}$ the law of

$$\xi_{[0,T]} \stackrel{\text{def.}}{=} (\xi_0^i, \xi_1^i \ldots, \xi_T^i)_{1 \leq i \leq N}$$

on the path space $(\Sigma_T^N, \mathbf{B}(\Sigma_T)^{\otimes N})$ where

$$\Sigma_T^N \stackrel{\text{def.}}{=} \underbrace{\Sigma_T \times \ldots \times \Sigma_T}_{N-\text{times}} \qquad \mathbf{B}(\Sigma_T)^{\otimes N} \stackrel{\text{def.}}{=} \underbrace{\mathbf{B}(\Sigma_T) \otimes \ldots \otimes \mathbf{B}(\Sigma_T)}_{N-\text{times}}$$

and $\Sigma_T \stackrel{\text{def.}}{=} E^{T+1}$. Then $P_T^{(N)}$ is absolutely continuous with respect to the product measure $\eta_{[0,T]}^{\otimes N}$ where

$$\eta_{[0,T]} \stackrel{\text{def.}}{=} \eta_0 \otimes \ldots \otimes \eta_T$$

Next we denote by μ_n the marginal at the time $n \in \{0, \ldots, T\}$ of a measure $\mu \in \mathbf{M}_1(\Sigma_T)$ and with some obvious abusive notations by $m(x)$ the empirical measure on path space associated with a configuration $x \in \Sigma_T^N$, that is

$$m(x) \stackrel{\text{def.}}{=} \frac{1}{N} \sum_{i=1}^{N} \delta_{(x_0^i, \ldots, x_T^i)} \in \mathbf{M}_1(\Sigma_T),$$

Under $(\mathcal{P})_0$, it is easily seen that

$$\frac{dP_T^{(N)}}{d\eta_{[0,T]}^{\otimes N}}(x) = \exp H_T^{(N)}(x) \qquad \eta_{[0,T]}^{\otimes N} - \text{a.e.}$$

where $H^{(N)} : \Sigma_T^N \to \mathbb{R}$ is the function defined by

$$H_T^{(N)}(x) = N \sum_{n=1}^{T} \int \log \frac{d\Phi_n(m_{n-1}(x))}{d\eta_n} \, dm_n(x)$$

In addition, if we consider the function $\Phi_{[0,T]} : \mathbf{M}_1(\Sigma_T) \to \mathbf{M}_1(\Sigma_T)$ so that

$$\Phi_{[0,T]}(\mu) = \eta_0 \otimes \Phi_1(\mu_0) \otimes \ldots \otimes \Phi_T(\mu_{T-1})$$

then we see that $H_T^{(N)}$ can also be rewritten in the following form

$$H_T^{(N)}(x) = N \int_{\Sigma_T} \log \frac{d\Phi_{[0,T]}(m(x))}{d\eta_{[0,T]}} \, dm(x) \tag{38}$$

Therefore, the density of $P_T^{(N)}$ only depends on the empirical measure m and we find ourselves exactly in the setting of mean field interacting particles with regular Laplace density. The study of the fluctuations for mean field interacting particle systems via precise Laplace method is now extensively developed (see for instance [3],[13], [66], [79] [109] and references therein).

Various methods are based on the fact that the law of mean field interacting processes can be viewed as a mean field Gibbs measure on path space (see (38)). In such a setting, precise Laplace's method can be developed (see for instance [3], [66], [79]). In [66], the study of the fluctuations for mean field Gibbs measures was extended to analytic potentials which probably includes our setting.
However the analysis of the fluctuations on the path space presented in section 2.2.4 is more related to Shiga/Tanaka's paper [98]. In this article, the authors restrict themselves to dynamics with independent initial data so that the partition function of the corresponding Gibbs measure is equal to one. This simplifies considerably the analysis. In fact, the proof then mainly relies on a simple formula on multiple Wiener integrals and Dynkin-Mandelbaum Theorem [54] on symmetric statistics. Also the pure jump McKean-Vlasov process studied in [98] is rather close to our model.

Using the above notations and, under $(\mathcal{P})_0$, if we denote by $Q_T^{(N)}$ the law of the empirical measures

$$\eta_{[0,T]}^N \stackrel{\text{def.}}{=} m\left(\xi_{[0,T]}\right) = \frac{1}{N}\sum_{i=1}^N \delta_{(\xi_0^i,\ldots,\xi_T^i)} \in \mathbf{M}_1(\Sigma_T),$$

then $Q_T^{(N)}$ is absolutely continuous with respect to the distribution

$$R_T^{(N)} \in \mathbf{M}_1(\mathbf{M}_1(\Sigma_T))$$

given by

$$R_T^{(N)} F = \int_{\Sigma_T^N} F(m(x))\, \lambda_0^{\otimes N}(dx_0)\ldots\lambda_T^{\otimes N}(dx_T)$$

for any $F \in \mathcal{C}_b(\mathbf{M}_1(\Sigma_T))$, with the convention $\lambda_0 = \eta_0$. We notice that the latter formula can be written in the form

$$R_T^{(N)} F = \int_{\Sigma_T^N} F(m(x))\, R_T^{\otimes N}(dx) \quad \text{with} \quad R_T = \lambda_0 \otimes \ldots \otimes \lambda_T$$

Arguing as before, it is also easily seen that

$$\frac{dQ_T^{(N)}}{dR_T^{(N)}} = \exp\left(NF_T\right) \qquad R_T^{(N)} - \text{a.s.} \tag{39}$$

where $F_T : \mathbf{M}_1(\Sigma_T) \to \mathbb{R}$ is the function defined by

$$F_T(\mu) = \sum_{n=1}^T \int_E \log \frac{d\Phi_n(\mu_{n-1})}{d\lambda_n}\, d\mu_n = \int_{\Sigma_T} \log \frac{d\Phi_{[0,T]}(\mu)}{dR_T}\, d\mu \tag{40}$$

The above formulation will be used in section 2.2.5 to obtain large deviation principles for the law of the empirical measures on path space.

It is worth observing immediately that $R_T^{(N)}$ is the law of an empirical measure associated with N independent path-valued random variables. This observation, together with the above formulation clearly shows that Varadhan's Lemma combined with Sanov's Theorem and cut-off techniques are natural tools for deriving the desired large deviation principle (see [35], [49, 51] or [106] and references therein).

We have no examples in this section 2.2. This choice is deliberate. We will give in section 5 a glossary of assumptions and detailed examples for each specific condition.

2.2.2 $\mathbb{L}^p$-mean errors

As announced in the introduction we start with a very basic but reassuring convergence result which holds without assumption $(\mathcal{G})$ on the fitness functions.

Proposition 2.9 *For any time $n \geq 0$ and $p \geq 1$, there exists some finite constant $C_n^{(p)} < \infty$ such that*

$$\forall\, f \in \mathcal{B}_b(E), \qquad \mathbb{E}\left(|\eta_n^N f - \eta_n f|^p\right)^{\frac{1}{p}} \;\leq\; \frac{1}{\sqrt{N}}\, \|f\|\, C_n^{(p)} \tag{41}$$

In particular, for any $f \in \mathcal{B}_b(E)$ and $n \geq 0$, $\{\eta_n^N f \,;\, N \geq 1\}$ converges almost surely to $\eta_n f$ as N tends to ∞.

Proof: The proof of (41) is simply based on Marcinkiewicz-Zygmund's inequality (cf. [100] p. 498). More precisely, by definition of the N-particle system (13), for any $n \geq 0$ and $f \in \mathcal{B}_b(E)$ we have that

$$\mathbb{E}\left(|\eta_n^N f - \Phi_n(\eta_{n-1}^N) f|^p\right)^{\frac{1}{p}} \leq \frac{1}{\sqrt{N}}\, \|f\|\, B_p \tag{42}$$

where B_p is a finite constant which only depends on the parameter $p \geq 1$ and where we have used the convention $\Phi_0(\eta_{-1}^N) = \eta_0$. We end up with (41) by induction on the time parameter. Since the result clearly holds for $n = 0$, we next assume that it holds at rank $(n-1)$ for some constant $C_{n-1}^{(p)}$.
Using Lemma 2.2 and the induction hypothesis at rank $(n-1)$ on can check that

$$\mathbb{E}\left(|\eta_n^N f - \eta_n f|^p\right)^{\frac{1}{p}} \leq \mathbb{E}\left(|\eta_n^N f - \Phi_n(\eta_{n-1}^N) f|^p\right)^{\frac{1}{p}} + \frac{1}{\sqrt{N}} \frac{2\|g_n\|\|f\|}{\eta_{n-1}(g_n)}\, C_{n-1}^{(p)}$$

Consequently, in view of (42) the result holds at rank n with

$$C_n^{(p)} = B_p + \frac{2\|g_n\|}{\eta_{n-1}(g_n)}\, C_{n-1}^{(p)}$$

The last assertion is a clear consequence of Borel-Cantelli Lemma. ∎

One important drawback of the above inductive proof is that it does not present any information about the "exact values" of the $\mathbb{L}^p$-mean errors. For instance, $C_n^{(p)}$ is roughly of the form $B_p(n+1)\prod_{l=1}^{n}(\frac{2\|g_l\|}{\eta_{l-1}(g_l)})$ and therefore tends to ∞ as $n \to \infty$. In order to get an idea of the exact values of these bounds some numerical simulations have been done in [41]. Precise asymptotic estimates will also be given in section 3.3.2.

When $(\mathcal{G})$ holds then (41) can be proved without using an induction. This approach is based on a natural decomposition of the errors. Since this decomposition will also be used to prove a uniform convergence result with respect to time, Glivenko-Cantelli and Donsker's Theorem and elsewhere we have chosen to present this alternative proof. The decomposition we will use now is the following:

$$\forall f \in \mathcal{B}_b(E)\,,\ \forall n \geq 0\,, \qquad \eta_n^N f - \eta_n f = \sum_{p=0}^{n} \left[\Phi_{p,n}(\eta_p^N) f - \Phi_{p,n}\left(\Phi_p(\eta_{p-1}^N)\right) f\right] \tag{43}$$

with the convention $\Phi_0(\eta_{-1}^N) = \eta_0$. Using Lemma 2.2 we see that each term

$$\left|\Phi_{p,n}(\eta_p^N) f - \Phi_{p,n}\left(\Phi_p(\eta_{p-1}^N)\right) f\right|, \qquad 0 \leq p \leq n, \tag{44}$$

is bounded by

$$a^2_{p,n} \quad \left[\left|\eta^N_p(f_{p,n}) - \Phi_p\left(\eta^N_{p-1}\right)(f_{p,n})\right| + \|f\| \ \left|\eta^N_p(\overline{g}_{p,n}) - \Phi_p\left(\eta^N_{p-1}\right)(\overline{g}_{p,n})\right| \right] \tag{45}$$

with

$$f_{p,n} = \overline{g}_{p,n} \ \ K_{p,n}(f) \qquad \text{and} \qquad \overline{g}_{p,n} = \frac{g_{p,n}}{a_{p,n}}$$

so that $\|f_{p,n}\| \leq \|f\|$ and $\|(\overline{g}_{p,n})\| \leq 1$. Using Marcinkiewicz-Zygmund's inequality it is then easy to conclude that, for any $p \geq 1$, $n \geq 0$ and $f \in \mathcal{B}_b(E)$

$$E\left(|\eta^N_n f - \eta_n f|^p\right)^{\frac{1}{p}} \leq \frac{B_p}{\sqrt{N}} \ \|f\| \ \sum_{q=0}^{n} a^2_{q,n}$$

where B_p is a finite universal constant. Here again, a clear deficiency in the preceding result is the degeneracy of the constants when the time parameter is growing, that is $\sum_{q=0}^n a^2_{q,n}$ tends to ∞ as $n \to \infty$. To connect this problem with the stability properties discussed in section 2.1.2 we note that each term (44) can be rewritten as follows

$$\left|\eta^N_p\left(\ \tilde{g}^N_{p,n}\ \tilde{K}^N_{p,n}(f)\ \right) / \eta^N_p\left(\ \tilde{g}^N_{p,n}\ \right)\right| \quad \text{with} \quad \tilde{g}^N_{p,n} = \frac{g_{p,n}}{\Phi_p(\eta^N_{p-1})(g_{p,n})}, \qquad 0 \leq p \leq n$$

and

$$\begin{aligned}\tilde{K}^N_{p,n}(f)(x) &= K_{p,n}\left(f - \Phi_{p,n}\left(\Phi_p(\eta^N_{p-1})\right)(f)\right)(x)\\ &= \int \left(K_{p,n}f(x) - K_{p,n}f(y)\right)\tilde{g}^N_{p,n}(y)\Phi_p(\eta^N_{p-1})(dy)\end{aligned}$$

Since $\Phi_p(\eta^N_{p-1})\left(\ \tilde{g}^N_{p,n}\ \tilde{K}^N_{p,n}(f)\ \right) = 0$ and $\|\tilde{K}^N_{p,n}(f)\| \leq \beta(K_{p,n})\ \|f\|$ we use the same line of arguments as before to prove that

$$E\left(|\eta^N_n f - \eta_n f|^p\right)^{\frac{1}{p}} \leq \frac{B_p}{\sqrt{N}} \ \|f\| \ \sum_{q=0}^{n} \sup_{x,y} \left|\frac{g_{q,n}(x)}{g_{q,n}(y)}\right|^2 \beta(K_{q,n})$$

Next we would like to extend these results in two different ways. First we would like to be able to turn this result into a statement uniform in f varying in a suitable class of functions $\mathcal{F} \subset \mathcal{B}_b(E)$. Second we would like to obtain a uniform $\mathbb{L}^p$-error bound with respect to the time parameter without any assumptions on the mutation transition kernels K_n but only on the stability properties of the limiting system (8). We will also use the following extension of Marcinkiewicz-Zygmund's inequality to empirical processes [50].

Lemma 2.10 *Let $(X^1, \ldots, X^N)$ be independent E − valued random variables with common law $P^{(N)}$ and let $\mathcal{F}$ be a countable sequence of functions $f : E \to \mathbb{R}$ such that $\|f\| \leq 1$. Then, for any $p \geq 1$ there exists a universal constant B_p such that*

$$\mathbb{E}\left(\|\frac{1}{N}\sum_{i=1}^{N}\delta_{X^i} - P^{(N)}\|^p_{\mathcal{F}}\right)^{\frac{1}{p}} \leq \frac{B_p}{\sqrt{N}}\ I(\mathcal{F}) \tag{46}$$

Theorem 2.11 *Let $\mathcal{F}$ be a countable collection of functions f such that $\|f\| \leq 1$ and satisfying the entropy condition $I(\mathcal{F}) < \infty$. Assume moreover that the limiting dynamical system (8) is asymptotically stable in the sense that*

$$\lim_{T\to\infty} \sup_{\mu,\nu\in\mathbf{M}_1(E)} \sup_{p\geq 0} \|\Phi_{p,p+T}(\mu) - \Phi_{p,p+T}(\nu)\|_{\mathcal{F}} = 0 \tag{47}$$

When the fitness functions $\{g_n \,;\, n \geq 1\}$ satisfy $(\mathcal{G})$ with $\sup_{n\geq 1} a_n \stackrel{\text{def.}}{=} a < \infty$ then we have the following uniform convergence result with respect to time

$$\lim_{N\to\infty} \sup_{n\geq 0} E\left(\|\eta_n^N - \eta_n\|_{\mathcal{F}}\right) = 0 \tag{48}$$

In addition, let us assume that the limiting dynamical system (8) is exponentially asymptotically stable in the sense that there exist some positive constant $\gamma > 0$ and $T_0 \geq 0$ such that,

$$\forall\, \mu,\nu \in \mathbf{M}_1(E),\, \forall\, T \geq T_0, \qquad \sup_{p\geq 0} \|\Phi_{p,p+T}(\mu) - \Phi_{p,p+T}(\nu)\|_{\mathcal{F}} \leq e^{-\gamma T} \tag{49}$$

Then we have for any $p \geq 1$, the uniform $\mathbb{L}^p$-error bound given by

$$\sup_{n\geq 0} \mathbb{E}\left(\|\eta_n^N - \eta_n\|_{\mathcal{F}}^p\right)^{\frac{1}{p}} \leq \frac{C_p\, e^{\gamma'}}{N^{\frac{\alpha}{2}}}\; I(\mathcal{F}) \tag{50}$$

for any $N \geq 1$ so that

$$T(N) \stackrel{\text{def}}{=\!=} \left[\frac{1}{2}\frac{\log N}{\gamma + \gamma'}\right] + 1 \geq T_0$$

where C_p is a universal constant which only depends on $p \geq 1$ and α and γ' are given by

$$\alpha = \frac{\gamma}{\gamma + \gamma'} \quad \text{and} \quad \gamma' = 1 + 2\log a.$$

Proof: We use again the decomposition (43). There is no loss of generality to assume that $1 \in \mathcal{F}$. Then, by the same line of arguments as before and using the extended version of Marcinkiewicz-Zygmund's inequality one can prove that for any $0 \leq q \leq n$ and $p \geq 1$

$$\mathbb{E}\left(\left\|\Phi_{q,n}(\eta_q^N) - \Phi_{q,n}\left(\Phi_q(\eta_{q-1}^N)\right)\right\|_{\mathcal{F}}^p \Big|\eta_{q-1}^N\right)^{\frac{1}{p}} \leq \frac{B_p}{\sqrt{N}}\; a_{q,n}^2\; I(\mathcal{F}_{q,n})$$

where B_p is a universal constant and

$$\mathcal{F}_{q,n} \stackrel{\text{def.}}{=} \overline{g}_{q,n}\, K_{q,n}(\mathcal{F}) \stackrel{\text{def.}}{=} \{\overline{g}_{q,n}\, K_{q,n}(f) \,;\, f \in \mathcal{F}\} \qquad \overline{g}_{q,n} = \frac{g_{q,n}}{a_{q,n}} \tag{51}$$

Now, using the fact that $I(\mathcal{F}_{q,n}) \leq I(\mathcal{F})$ (cf. Lemma 2.3, p. 9, [42]) and $a_{q,n} \leq a_{0,n} \leq a^n$ one concludes that

$$\mathbb{E}\left(\left\|\Phi_{q,n}(\eta_q^N) - \Phi_{q,n}\left(\Phi_q(\eta_{q-1}^N)\right)\right\|_{\mathcal{F}}^p\right)^{\frac{1}{p}} \leq \frac{B_p}{\sqrt{N}}\; a^{2n}\; I(\mathcal{F})$$

and therefore for any $T \geq T_0$

$$\sup_{0 \leq n \leq T} \mathbb{E}\left(\|\eta_n^N - \eta_n\|_{\mathcal{F}}^p\right)^{\frac{1}{p}} \leq \frac{B_p}{\sqrt{N}} \; (T+1)\; a^{2T} \;\; I(\mathcal{F}) \tag{52}$$

Similarly, for any $q \geq 0$ and $T \geq T_0$ we have

$$\|\eta_{q+T}^N - \eta_{q+T}\|_{\mathcal{F}}$$

$$\leq \sum_{r=q+1}^{q+T} \left\|\Phi_{r,q+T}(\eta_r^N) - \Phi_{r,q+T}\left(\Phi_r(\eta_{r-1}^N)\right)\right\|_{\mathcal{F}} + \left\|\Phi_{q,q+T}(\eta_q^N) - \Phi_{q,q+T}(\eta_q)\right\|_{\mathcal{F}}$$

Under our assumptions, this implies that

$$\|\eta_{q+T}^N - \eta_{q+T}\|_{\mathcal{F}} \leq \sum_{r=q+1}^{q+T} \left\|\Phi_{r,q+T}(\eta_r^N) - \Phi_{r,q+T}\left(\Phi_r(\eta_{r-1}^N)\right)\right\|_{\mathcal{F}} + e^{-\gamma T}$$

Thus, using the same line of arguments as before, one gets for any $T \geq T_0$

$$\sup_{q \geq 0} \mathbb{E}\left(\|\eta_{q+T}^N - \eta_{q+T}\|_{\mathcal{F}}^p\right)^{\frac{1}{p}} \leq e^{-\gamma T} + \frac{B_p}{\sqrt{N}} \; (T+1)\; a^{2T} \;\; I(\mathcal{F}) \tag{53}$$

Combining (52) and (53) leads to a uniform $\mathbb{L}^p$-error bounds with respect to time in the form of the inequality

$$\forall\, T \geq T_0, \qquad \sup_{n \geq 0} \mathbb{E}\left(\|\eta_n^N - \eta_n\|_{\mathcal{F}}^p\right)^{\frac{1}{p}} \;\leq\; e^{-\gamma T} + \frac{B'_p}{\sqrt{N}} \; e^{\gamma' T} \;\; I(\mathcal{F})$$

where $\gamma' = 1 + 2\log a$ and $B'_p > 0$ is a universal constant. Obviously, if we choose $N \geq 1$ and

$$T = T(N) \stackrel{\text{def}}{=} \left[\frac{1}{2}\frac{\log N}{\gamma + \gamma'}\right] + 1 \geq T_0$$

where $[r]$ denotes the integer part of $r \in \mathbb{R}$, we get that

$$\sup_{n \geq 0} \mathbb{E}\left(\|\eta_n^N - \eta_n\|_{\mathcal{F}}^p\right)^{\frac{1}{p}} \leq \frac{1}{N^{\alpha/2}} \left(1 + e^{\gamma'} B'_p \, I(\mathcal{F})\right)$$

where $\alpha = \gamma/(\gamma + \gamma')$. This ends the proof of the theorem. ∎

Remarks 2.12:

- The critical exponent resulting from the proof of Theorem 2.11 is sharp in the following sense: if the transition probability kernels $\{K_n \,;\, n \geq 1\}$ are given by

$$K_n(x, dz) = \lambda_n(dz), \qquad \lambda_n \in \mathbf{M}_1(E)$$

  then we see that $\xi_n = (\xi_n^1, \ldots, \xi_n^N)$ is a sequence of independent random variables with common law λ_n. In this situation the uniform upper bound (50) holds for any choice of $\gamma \in \mathbb{R}_+$. Letting $\gamma \to \infty$ the critical exponent tends to 1/2 which is the characteristic exponent of the weak law of large numbers.

- Several conditions for exponential asymptotic stability of the limiting measure system (8) are given in section 2.1.2. The exponential stability condition (49) can be easily related to the conditions $(\mathcal{K})_i$, $i = 1, 2, 3$ discussed in section 2.1.2.

- The proof of Theorem 2.11 can be used to treat polynomial asymptotically stable limiting dynamical systems (8). The resulting uniform $\mathbb{L}^p$-error bound has roughly the form $(\log N)^{-\beta}$ for some $\beta > 0$.

- The above result can also be used to obtain reliability intervals which are valid uniformly with respect to the time parameter (cf. [36]).

- In nonlinear filtering settings, the fitness functions and therefore the sequence $\{a_{q,n}\ 0 \leq q \leq n\}$ depend on the observation process. The above result can be used to give quenched and/or averaged uniform $\mathbb{L}^p$-mean error bounds with respect to time (cf. section 4.4 and [36]).

2.2.3 Glivenko-Cantelli Theorem

Let $\mathcal{F}$ be a countable collection of functions f such that $\|f\| \leq 1$. Upon carefully examining the proof of Theorem 2.11, we have already proved that for any $p \geq 1$ and for any time $n \geq 0$

$$\mathbb{E}\left(\|\eta_n^N - \eta_n\|_{\mathcal{F}}^p\right)^{\frac{1}{p}} \leq \frac{B_p}{\sqrt{N}}\ (n+1)\ a_{0,n}^2\ \ I(\mathcal{F})$$

Then, as an immediate consequence of Borel-Cantelli Lemma we have that

$$I(\mathcal{F}) < \infty \Longrightarrow \lim_{N\to\infty} \|\eta_n^N - \eta_n\|_{\mathcal{F}} = 0 \qquad \text{P-a.e.}$$

Our aim is now to show that this almost sure convergence remains true if we replace the entropy condition $I(\mathcal{F}) < \infty$ by a boundness condition of the covering numbers, namely $\mathcal{N}(\varepsilon, \mathcal{F}) < \infty$, for any $\varepsilon > 0$.

As usually we make use of the decomposition (43). There is no loss of generality in assuming that $1 \in \mathcal{F}$. Then, by Lemma 2.2, we get for any $0 \leq q \leq n$

$$\|\Phi_{q,n}(\mu) - \Phi_{q,n}(\nu)\|_{\mathcal{F}} \leq 2\, a_{q,n}^2\ \|\mu - \nu\|_{\mathcal{F}_{q,n}} \tag{54}$$

where the class $\mathcal{F}_{q,n}$ is the class of functions defined in (51). It easily follows that, for every $\varepsilon > 0$,

$$\begin{aligned}
&P\left(\|\eta_n^N - \eta_n\|_{\mathcal{F}} > \varepsilon\right) \\
&\quad \leq (n+1) \sup_{0\leq q\leq n} P\left(\|\Phi_{q,n}(\eta_q^N) - \Phi_{q,n}(\Phi_p(\eta_{q-1}^N))\|_{\mathcal{F}} > \frac{\varepsilon}{n+1}\right).
\end{aligned}$$

Using (54), this implies that

$$P\left(\|\eta_n^N - \eta_n\|_{\mathcal{F}} > \varepsilon\right) \leq (n+1) \sup_{0\leq q\leq n} P\left(\|\eta_q^N - \Phi_q(\eta_{q-1}^N)\|_{\mathcal{F}_{q,n}} > \frac{\varepsilon}{\sigma_n}\right) \tag{55}$$

where $\sigma_n = 2(n+1)a_{0,n}^2$.

The Glivenko-Cantelli Theorem may then be stated as follows.

Theorem 2.13 *Assume that $\mathcal{F}$ is a countable collection of functions f such that $\|f\| \leq 1$ and $\mathcal{N}(\varepsilon, \mathcal{F}) < \infty$ for any $\varepsilon > 0$. Then, for any time $n \geq 0$, $\|\eta_n^N - \eta_n\|_{\mathcal{F}}$ converges almost surely to 0 as $N \to \infty$.*

Theorem 2.13 is based on the following standard lemma in the theory of empirical processes (see for instance [50] or [42]).

Lemma 2.14 *Let $\{X^i; 1 \leq i \leq N\}$ be independent random variables with common law $P^{(N)}$ and let $\mathcal{F}$ be a countable collection of functions f such that $\|f\| \leq 1$. Then, for any $\varepsilon > 0$ and $\sqrt{N} \geq 4\,\varepsilon^{-1}$ we have that*

$$P\left(\|\frac{1}{N}\sum_{i=1}^{N} \delta_{X^i} - P^{(N)})\|_{\mathcal{F}} > 8\,\varepsilon\right) \leq 8\,\mathcal{N}(\varepsilon, \mathcal{F})\, e^{-N\varepsilon^2/2}$$

Proof of Theorem 2.13: Let us fixed n throughout the argument. Using (55) and Lemma 2.14 one easily gets that, for $\sqrt{N} \geq 4\,\varepsilon_n^{-1}$ where $\varepsilon_n = \varepsilon/8\sigma_n$,

$$P\left(\left\|\eta_n^N - \eta_n\right\|_{\mathcal{F}} > \varepsilon\right) \leq 8(n+1)\, e^{-N\varepsilon_n^2/2} \sup_{0 \leq q \leq n} \mathcal{N}(\varepsilon_n, \mathcal{F}_{q,n}).$$

Since $\mathcal{N}(\varepsilon, \mathcal{F}_{q,n}) \leq \mathcal{N}(\varepsilon, \mathcal{F})$ for each $\varepsilon > 0$, and $0 \leq q \leq n$ (cf. Lemma 2.3, p. 9, [42]) one concludes that

$$P\left(\left\|\eta_n^N - \eta_n\right\|_{\mathcal{F}} > \varepsilon\right) \leq 8(n+1)\mathcal{N}(\varepsilon_n, \mathcal{F})\, e^{-N\varepsilon_n^2/2}$$

as soon as $\sqrt{N} \geq 4\,\varepsilon_n^{-1}$. The end of proof of Theorem 2.13 is an immediate consequence of the Borel-Cantelli Lemma. ■

The proof of Theorem 2.13 gives an exponential rate of convergence but this result is only valid for a number of particles larger than some value depending on the time parameter. Our aim is now to refine this exponential bound in the case of uniformly bounded classes $\mathcal{F}$ with polynomial covering numbers. More precisely we will use the following assumption

($\mathcal{P}oly.$) ***There exist some constants C and V such that***

$$\forall 0 < \varepsilon < C, \qquad \mathcal{N}(\varepsilon, \mathcal{F}) \leq \left(\frac{C}{\varepsilon}\right)^V$$

Several examples of classes of functions satisfying this condition are discussed in [50]. For instance Vapnik-Cervonenkis classes $\mathcal{F}$ of index $V(\mathcal{F})$ and envelope function $F = 1$ satisfy ($\mathcal{P}oly.$) with $V = 2(V(\mathcal{F}) - 1)$ and a constant C that only depends on V.

Theorem 2.15 *Let $\mathcal{F}$ be a countable class of measurable functions $f : E \to [0,1]$ satisfying ($\mathcal{P}oly.$) for some constants C and V. Then, for any $n \geq 0$, $\delta > 0$ and $N \geq 1$,*

$$P\left(\sqrt{N}\left\|\eta_n^N - \eta_n\right\|_{\mathcal{F}} > \delta\sigma_n\right) \leq (n+1)\left(\frac{D\,\delta}{\sqrt{V}}\right)^V e^{-2\,\delta^2}$$

where D is a constant that only depends on C and $\sigma_n = 2(n+1)\prod_{q=1}^{n} a_q^2$.

Proof: We use again the decomposition (43). Using the same notations as in the proof of Theorem 2.13 and Theorem 2.11, we have

$$P\left(\left\|\Phi_{q,n}(\eta_q^N) - \Phi_{q-1,n}(\eta_{q-1}^N)\right\|_{\mathcal{F}} > \frac{\varepsilon}{n+1}\right) \leq P\left(\left\|\eta_q^N - \Phi_q(\eta_{q-1}^N)\right\|_{\mathcal{F}_{q,n}} > \varepsilon_n\right) \quad (56)$$

where $\varepsilon_n = \varepsilon/\sigma_n$ and $\sigma_n = 2(n+1)a_{0,n}^2$. It is also convenient to note that each class $\mathcal{F}_{q,n}$, $0 \leq q \leq n$, satisfies $(\mathcal{P}oly.)$. Indeed, the class $\mathcal{F}_{q,n}$ is again a countable class of functions $f : E \to [0,1]$ and using Lemma 2.3 in [42] we also have, for every $\varepsilon > 0$ and $0 \leq q \leq n$

$$\mathcal{N}(\varepsilon, \mathcal{F}_{q,n}) \leq \mathcal{N}(\varepsilon, \mathcal{F}) \leq \left(\frac{C}{\varepsilon}\right)^V.$$

We are now in position to apply the exponential bounds of [107] (see also [50]). More precisely, by recalling that η_p^N is the empirical measure associated with N conditionally independent random variables with common law $\Phi_q(\eta_{q-1}^N)$, we get

$$P\left(\left\|\eta_q^N - \Phi_q(\eta_{q-1}^N)\right\|_{\mathcal{F}_{q,n}} > \varepsilon_n \mid \eta_{q-1}^N\right) \leq \left(\frac{D\sqrt{N}\varepsilon_n}{\sqrt{V}}\right)^V e^{-2(\sqrt{N}\varepsilon_n)^2}$$

where D is a constant that only depends on C. The remainder of the proof is exactly as in the proof of Theorem 2.13. Using (56), one gets finally

$$P\left(\left\|\eta_n^N - \eta_n\right\|_{\mathcal{F}} > \varepsilon\right) \leq (n+1)\left(\frac{D\sqrt{N}\varepsilon_n}{\sqrt{V}}\right)^V e^{-2(\sqrt{N}\varepsilon_n)^2}.$$

If we denote $\delta = \sqrt{N}\varepsilon/\sigma_n$ we obtain the desired inequality and the theorem is thus established. ∎

2.2.4 Central Limit Theorems

The study of the fluctuations for the IPS scheme (13) is decomposed into three parts.

- In the first one we present central limit theorems (**CLT**) for a class of processes arising naturally in the study of the convergence of the particle scheme.
- The second part concerns a Donsker's Theorem for the particle density profiles. The identification of the covariance function is based on the convergence results presented in the first part.
- The last part of this section presents a technique for obtaining fluctuations for the empirical distributions on path space.

CLT for Processes

One of the best approaches for obtaining fluctuations of the particle density profiles is through a study of the convergence of some suitably chosen processes. To describe these processes, it is convenient to introduce some additional notations. For any $\mathbb{R}^d$-valued function $f = (f^1, \ldots, f^d)$, $f^i \in \mathcal{B}_b(E)$, $1 \leq i \leq d$, and for any integral operator K on E and $\mu \in \mathbf{M}_1(E)$ we will slightly abuse notations and we write

$$\mu K(f) \stackrel{\text{def.}}{=} (\mu K(f^1), \ldots, \mu K(f^d))$$

Let $F^N = \{F_n^N \; ; \; n \geq 0\}$ be the natural filtration associated with the N-particle system $\{\xi_n^{(N)} \; ; \; n \geq 0\}$. The first class of processes which arises naturally in our context are the $\mathbb{R}^d$-valued and F^N-martingale $\{M_n^{(N)}(f) \; ; \; n \geq 0\}$ defined by

$$\forall\, n \geq 0, \qquad M_n^{(N)}(f) \;=\; \sum_{p=0}^{n} \left[\eta_p^N(f_p) - \Phi_p(\eta_{p-1}^N)(f_p)\right] \tag{57}$$

with the usual convention $\Phi_0(\eta_{-1}^N) = \eta_0$ and where $f : (p,x) \in \mathbb{N} \times E \mapsto f_p(x) \in \mathbb{R}^d$ is a bounded measurable function. Using the above notations, the j-th, $1 \leq j \leq d$, component of the martingale $\{M_n^{(N)}(f) \; ; \; n \geq 0\}$ is the F^N-martingale defined by

$$\forall\, n \geq 0, \qquad M_n^{(N)}(f^j) \;=\; \sum_{p=0}^{n} \left[\eta_p^N(f_p^j) - \Phi_p(\eta_{p-1}^N)(f_p^j)\right]$$

Most of the results presented here are based on the following CLT for the martingale (57).

Lemma 2.16
For any bounded measurable function $f : (p,x) \in \mathbb{N} \times E \mapsto f_p(x) \in \mathbb{R}^d$ and $d \geq 1$, the $\mathbb{R}^d$-valued and F^N-martingale $\{\sqrt{N}\; M_n^{(N)}(f) \; ; \; n \geq 0\}$ converges in law to an $\mathbb{R}^d$-valued and Gaussian martingale $\{M_n(f) \; ; \; n \geq 0\}$ such that for any $1 \leq i,j \leq d$

$$\forall\, n \geq 0, \qquad \langle M(f^i), M(f^j)\rangle_n \;=\; \sum_{p=0}^{n} \eta_p\left(\left(f_p^i - \eta_p(f_p^i)\right)\left(f_p^j - \eta_p(f_p^j)\right)\right)$$

Proof: The proof is based on ideas of J. Jacod (see [40] for another presentation in more general settings). To use the CLT for triangular arrays of $\mathbb{R}^d$-valued random variables (Theorem 3.33, p. 437 in [68]) we first rewrite the martingale $\sqrt{N}\; M_n^{(N)}(f)$ in the following form

$$\sqrt{N}\; M_n^{(N)}(f) = \sum_{i=1}^{N} \sum_{p=0}^{n} \frac{1}{\sqrt{N}} \left(f_p(\xi_p^i) - \Phi_p\left(\eta_{p-1}^N\right)(f_p) \right)$$

If we denote by $[a]$ the integer part of $a \in \mathbb{R}$ and $\{a\} = a - [a]$ this yields

$$\sqrt{N}\; M_n^{(N)}(f) = \sum_{k=1}^{(n+1)N} U_k^N(f)$$

where for any $1 \leq k \leq (n+1)N$,

$$U_k^N(f) = \frac{1}{\sqrt{N}} \left(f_p(\xi_p^i) - \Phi_p\left(\eta_{p-1}^N\right)(f_p) \right) \quad \text{with} \quad i = N\{\frac{k}{N}\} \quad \text{and} \quad p = \left[\frac{k}{N}\right]$$

so that $k = pN+i$. Our aim is now to describe the limiting behavior of the martingale $\sqrt{N}\, M^{(N)}(f)$ in terms of the process

$$X_t^N(f) \stackrel{\text{def.}}{=} \sum_{k=1}^{[Nt]+N} U_k^N(f)$$

To this end, we denote $\mathcal{F}_k^N$ the σ-algebra generated by the random variables ξ_p^j for any pair-index (j,p) such that $pN+j \leq k$. By definition of the IPS transitions (13) and using the fact that $\left[\frac{[Nt]}{N}\right] = [t]$ one gets that for any $i,j \leq d$

$$\sum_{k=1}^{[Nt]+N} E\left(U_k^N(f^i)U_k^N(f^j)\,|\mathcal{F}_{k-1}^N\right)$$

$$= C_{[t]}^N(f^i,f^j) + \tfrac{[Nt]-N[t]}{N}\left(C_{[t]+1}^N(f^i,f^j) - C_{[t]}^N(f^i,f^j)\right)$$

where, for any $n \geq 0$ and $1 \leq i,j \leq d$

$$C_n^N(f^i,f^j) = \sum_{p=0}^{n} \Phi_p(\eta_{p-1}^N)\left(\left(f_p^i - \Phi_p(\eta_{p-1}^N)f_p^i\right)\left(f_p^j - \Phi_p(\eta_{p-1}^N)f_p^j\right)\right)$$

This implies that for any $1 \leq i,j \leq d$

$$\sum_{k=1}^{[Nt]+N} E\left(U_k^N(f^i)U_k^N(f^j)\,|\mathcal{F}_{k-1}^N\right) \underset{N\to\infty}{\longrightarrow^P} C_t(f^i,f^j)$$

with

$$\forall\, n \geq 0, \qquad C_n(f^i,f^j) = \sum_{p=0}^{n} \eta_p\left(\left(f_p^i - \eta_p f_p^i\right)\left(f_p^j - \eta_p f_p^j\right)\right)$$

and

$$\forall\, t \in \mathbb{R}_+, \qquad C_t(f^i,f^j) = C_{[t]}(f^i,f^j) + \{t\}\left(C_{[t]+1}(f^i,f^j) - C_{[t]}(f^i,f^j)\right)$$

Since $\|U_k^N(f)\| \leq \frac{2}{\sqrt{N}}\|f_{[\frac{k}{N}]}\|$ for any $1 \leq k \leq [Nt]+N$, the conditional Linderberg condition is clearly satisfied and therefore one concludes that the $\mathbb{R}^d$-valued martingale $\{X_t^N(f)\,;\, t \in \mathbb{R}_+\}$ converges in law to a continuous Gaussian martingale

$$\{X_t(f)\,;\, t \in \mathbb{R}_+\}$$

such that, for any $1 \leq i,j \leq d$

$$\forall\, t \in \mathbb{R}_+, \qquad \langle X(f^i), X(f^j)\rangle_t = C_t(f^i,f^j)$$

Recalling that $X_{[t]}^N(f) = \sqrt{N}\, M_{[t]}^{(N)}(f)$ the proof of the lemma is completed. ∎

A first consequence of Lemma 2.16 is another CLT for a martingale process related to the "un-normalized" approximating measures $\{\gamma_n^N\,;\, n \geq 0\}$ defined in (17).

Proposition 2.17 *For any* $T \geq 0$ *and* $f = (f^1, \ldots, f^d) \in \mathcal{B}_b(E)^d$, *the* $\mathbb{R}^d$*-valued process*

$$\Gamma_n^N(f) = \gamma_n^N(Q_{n,T}f) - \gamma_n(Q_{n,T}f) \qquad 0 \leq n \leq T \tag{58}$$

is an F^N*-martingale such that, for any* $1 \leq i, j \leq d$ *and* $0 \leq n \leq T$

$$\begin{aligned} &\langle \Gamma^N(f^i), \Gamma^N(f^j) \rangle_n \\ &= \frac{1}{N} \sum_{p=0}^{n} \left(\gamma_p^N(1)\right)^2 \, \Phi_p(\eta_{p-1}^N)\left(\left[Q_{p,T}f^i - \Phi_p(\eta_{p-1}^N)Q_{p,T}f^i\right] \right. \\ &\qquad\qquad \left. \times \left[Q_{p,T}f^j - \Phi_p(\eta_{p-1}^N)Q_{p,T}f^j\right] \right) \end{aligned} \tag{59}$$

Moreover, the F^N*-martingale* $\{\sqrt{N}\Gamma_n^N(f) \; ; \; 0 \leq n \leq T\}$ *converges in law to an* $\mathbb{R}^d$*-valued and Gaussian martingale* $\{\Gamma_n(f) \; ; \; 0 \leq n \leq T\}$ *such that for any* $1 \leq i, j \leq d$ *and* $0 \leq n \leq T$

$$\langle \Gamma(f^i), \Gamma(f^j) \rangle_n = \sum_{p=0}^{n} (\gamma_p(1))^2 \; \eta_p\left(\left[Q_{p,T}f^i - \eta_p Q_{p,T}f^i\right] \left[Q_{p,T}f^j - \eta_p Q_{p,T}f^j\right] \right)$$

Proof: For any $\varphi = (\varphi^1, \ldots, \varphi^d) \in \mathcal{B}_b(E)^d$ we have the decomposition

$$\gamma_n^N(\varphi) - \gamma_n(\varphi) = \sum_{p=0}^{n} \left[\gamma_p^N(Q_{p,n}\varphi) - \gamma_{p-1}^N(Q_p Q_{p,n}\varphi)\right]$$

with the usual convention $\gamma_{-1}^N(Q_0) = \gamma_0 (= \eta_0)$. By definition of $\{\gamma_n^N \; ; \; n \geq 0\}$ this can also be written in the following form

$$\gamma_n^N(\varphi) - \gamma_n(\varphi) = \sum_{p=0}^{n} \gamma_p^N(1) \left[\eta_p^N(Q_{p,n}\varphi) - \Phi_p(\eta_{p-1}^N)(Q_{p,n}\varphi)\right]$$

Therefore one gets (58) by choosing $\varphi = Q_{n,T}f$ and (59) is a clear consequence of the above decomposition. We turn now to the proof of the convergence of the F^N-martingale $\{\sqrt{N}\Gamma_n^N(f) \; ; \; 0 \leq n \leq T\}$. If we put for any $0 \leq n \leq T$

$$\overline{\Gamma}_n^N(f) \stackrel{\text{def}}{=} \sum_{p=0}^{n} \gamma_p(1) \left[\eta_p^N(Q_{p,T}(f)) - \Phi_p(\eta_{p-1}^N)(Q_{p,T}(f))\right]$$

and if we denote $\|a\| = \sum_{i=1}^{d} |a^i|$, for any $a \in \mathbb{R}^d$, then, under our assumptions,

$$\mathbb{E}\left(\sup_{0 \leq n \leq T} \left\|\overline{\Gamma}_n^N(f) - \Gamma_n^N(f)\right\|\right) \leq \frac{C_T(f)}{N} \tag{60}$$

for some finite constant $C_T(f) < \infty$ which does not depend on the parameter N. Lemma 2.16 clearly implies that the F^N-martingale $\{\sqrt{N}\,\overline{\Gamma}_n^N(f) \; ; \; 0 \leq n \leq T\}$ converges in law to the desired Gaussian martingale $\{\Gamma_n(f) \; ; \; 0 \leq n \leq T\}$ and (60) clearly ends the proof of the proposition. ∎

Corollary 2.18
For any time $T \geq 0$, the sequence of random fields $\{W_T^{\gamma,N}(f)\ ;\ f \in \mathcal{B}_b(E)\}$ where

$$W_T^{\gamma,N}(f) \ \stackrel{\text{def.}}{=}\ \sqrt{N}\left(\gamma_T^N(f) - \gamma_T(f)\right)$$

converges in law as $N \to \infty$, in the sense of convergence of finite dimensional distributions, to a centered Gaussian field $\{W_T^{\gamma}(f)\ ;\ f \in \mathcal{B}_b(E)\}$ satisfying for any $f, h \in \mathcal{B}_b(E)$

$$\mathbb{E}(W_T^{\gamma}(f)W_T^{\gamma}(h)) = \sum_{p=0}^{T} (\gamma_p(1))^2\ \eta_p\left(\ [Q_{p,T}f - \eta_p Q_{p,T}f]\,[Q_{p,T}h - \eta_p Q_{p,T}h]\ \right)$$

The analysis of the fluctuations for the particle density profiles $\{\eta_n^N\ ;\ n \geq 0\}$ is more delicate mainly because the limiting measure valued process (8) is not linear. In fact many ideas work equally well when we replace the semigroup $\{Q_{p,n}\ ;\ 0 \leq p \leq n\}$ by the "normalized" one $\{\overline{Q}_{p,n}\ ;\ 0 \leq p \leq n\}$ given by

$$\forall\, 0 \leq p \leq n,\ \forall\, f \in \mathcal{B}_b(E), \qquad \overline{Q}_{p,n}f \ \stackrel{\text{def.}}{=}\ \frac{Q_{p,n}f}{\eta_p(Q_{p,n}1)}$$

To see that $\{\overline{Q}_{p,n}\ ;\ 0 \leq p \leq n\}$ is indeed a semigroup we first note that for any $0 \leq p \leq m \leq n$

$$\overline{Q}_{p,n}f = \frac{Q_{p,m}Q_{m,n}f}{\eta_p(Q_{p,n}1)} = \frac{\eta_m(Q_{m,n}1)}{\eta_p(Q_{p,n}1)}\ Q_{p,m}\overline{Q}_{m,n}f$$

Since

$$\eta_m(Q_{m,n}1) = \frac{\eta_p(Q_{p,m}Q_{m,n}1)}{\eta_p(Q_{p,m}1)} = \frac{\eta_p(Q_{p,n}1)}{\eta_p(Q_{p,m}1)}$$

one concludes that for any $0 \leq p \leq m \leq n$

$$\overline{Q}_{p,n}f = \frac{1}{\eta_p(Q_{p,m}1)}Q_{p,m}\overline{Q}_{m,n}f = \overline{Q}_{p,m}\overline{Q}_{m,n}f$$

For any $0 \leq n \leq T$ and $f = (f^1, \ldots, f^d) \in \mathcal{B}_b(E)^d$ we write

$$f_{n,T} \ =\ \overline{Q}_{n,T}(f - \eta_T f) \tag{61}$$

Using this notation, the analog of Proposition 2.17 for the particle density profiles $\{\eta_n^N\ ;\ n \geq 0\}$ is the following result.

Proposition 2.19 *For any $T \geq 0$ and $f = (f^1, \ldots, f^d) \in \mathcal{B}_b(E)^d$ the $\mathbb{R}^d$-valued process $\{W_n^N(f_{n,T})\ ;\ 0 \leq n \leq T\}$ given by*

$$W_n^N(f_{n,T}) \stackrel{\text{def.}}{=} \sqrt{N}\ \eta_n^N(f_{n,T}) \tag{62}$$

converges in law to an $\mathbb{R}^d$-valued Gaussian martingale $\{W_n(f_{n,T})\ ;\ 0 \leq n \leq T\}$ such that for any $1 \leq i, j \leq d$ and $0 \leq n \leq T$

$$\langle\, W(f^i_{\cdot,T}),\, W(f^j_{\cdot,T})\,\rangle_n = \sum_{p=0}^{n} \eta_p\left(\ f^i_{p,T}\ f^j_{p,T}\ \right)$$

Proof: For any $\varphi = (\varphi^1, \ldots, \varphi^d) \in \mathcal{B}_b(E)^d$ we have the decomposition

$$\eta_n^N(\overline{Q}_{n,T}\varphi) = \eta_0^N(\overline{Q}_{0,T}\varphi) + \sum_{p=1}^{n} \left[\eta_p^N(\overline{Q}_{p,T}\varphi) - \eta_{p-1}^N(\overline{Q}_{p-1,T}\varphi)\right]$$

If we choose $\varphi = (f - \eta_T f)$ with $f = (f^1, \ldots, f^d) \in \mathcal{B}_b(E)^d$ this yields

$$\eta_n^N(f_{n,T}) = B_n^{(N)}(f_{\cdot,T}) + M_n^{(N)}(f_{\cdot,T})$$

where

$$B_n^{(N)}(f_{\cdot,T}) = \sum_{p=1}^{n} [1 - \eta_{p-1}^N(\overline{Q}_{p-1,p}1)] \quad \Phi_p(\eta_{p-1}^N)(f_{p,T}) \tag{63}$$

$$M_n^{(N)}(f_{\cdot,T}) = \sum_{p=0}^{n} \left[\eta_p^N(f_{p,T}) - \Phi_p(\eta_{p-1}^N) f_{p,T}\right] \tag{64}$$

with the usual convention $\Phi_0(\eta_{-1}^N) = \eta_0$. Since for any $0 \leq p \leq T$

$$\eta_p\left(f_{p,T}\right) = \eta_T(f - \eta_T f) = 0 \quad \text{and} \quad \eta_{p-1}\left(\overline{Q}_{p-1,p}1\right) = \eta_p(1) = 1$$

then (63) can also be written in the following form

$$B_n^{(N)}(f_{\cdot,T}) = \sum_{p=1}^{n} \left[\eta_{p-1}(\overline{Q}_{p-1,p}1) - \eta_{p-1}^N(\overline{Q}_{p-1,p}1)\right] \left[\Phi_p(\eta_{p-1}^N) f_{p,T} - \Phi_p(\eta_{p-1}) f_{p,T}\right]$$

Using Proposition 2.9 and Lemma 2.2 one gets after some tedious but easy calculations

$$\mathbb{E}\left(\sup_{0 \leq n \leq T} \|B_n^{(N)}(f_{\cdot,T})\|\right) \leq \frac{C_T}{N} \|f\| \tag{65}$$

for some finite constant $C_T < \infty$ which only depends on the parameter T (we recall that $\|f\| = \sum_{i=1}^d \|f^i\|$, for any $f = (f^1, \ldots, f^d)$). Using the same arguments as in the proof of Proposition 2.17 we ends the proof of Proposition 2.19. More precisely, Lemma 2.16 implies that the F^N-martingale

$$\{\sqrt{N}\, M_n^{(N)}(f_{\cdot,T}) \; ; \; 0 \leq n \leq T\}$$

converges in law to the desired Gaussian martingale $\{W_n(f_{\cdot,T}) \; ; \; 0 \leq n \leq T\}$ and (65) completes the proof of the proposition. ∎

Corollary 2.20
For any time $T \geq 0$, the sequence of random fields $\{W_T^N(f) \; ; \; f \in \mathcal{B}_b(E)\}$ where

$$W_T^N(f) \overset{\text{def.}}{=} \sqrt{N}\left(\eta_T^N(f) - \eta_T(f)\right)$$

converges in law as $N \to \infty$, in the sense of convergence of finite dimensional distributions, to a centered Gaussian field $\{W_T(f) \; ; \; f \in \mathcal{B}_b(E)\}$ satisfying for any $f, h \in \mathcal{B}_b(E)$

$$\mathbb{E}(W_T(f) W_T(h)) = \sum_{p=0}^{T} \eta_p\left[\overline{Q}_{p,T}(f - \eta_T f)\, \overline{Q}_{p,T}(h - \eta_T h)\right]$$

Donsker Theorem

Before getting into the details it is useful to make a couple of remarks. In the first place we note that the covariance functions can be formulated using the fitness functions $\{g_{p,T}\ ;\ 0 \leq p \leq T\}$ and the transitions $\{K_{p,T}\ ;\ 0 \leq p \leq T\}$ defined in Lemma 2.2. More precisely, since

$$\overline{Q}_{p,T}(f - \eta_T f) = \overline{Q}_{p,T}(1) \left(\frac{\overline{Q}_{p,T}(f)}{\overline{Q}_{p,T}(1)} - \eta_T f \right) = \frac{g_{p,T}}{\eta_p(g_{p,T})} \left(K_{p,T} f - \eta_T f \right)$$

for any $f \in \mathcal{B}_b(E)$, we have that

$$\mathbb{E}(W_T(f) W_T(h)) = \sum_{p=0}^{T} \int \left(\frac{g_{p,T}}{\eta_p(g_{p,T})} \right)^2 (K_{p,T} f - \eta_T f)(K_{p,T} h - \eta_T h)\ d\eta_p$$

If the transition probability kernels $\{K_n;\ n \geq 1\}$ are trivial, in the sense that,

$$\forall\, 0 \leq p \leq T, \qquad K_p(x, dz) \;=\; \mu_p(dz) \;\in\; \mathbf{M}_1(E)$$

then one can readily check that $\eta_p = \mu_p$ for any $0 \leq p \leq T$ and

$$\forall 0 \leq p < T\,, \qquad K_{p,T}(x, dz) = \mu_T(dz)$$

In this particular situation $\{W_T(f);\ f \in \mathcal{B}_b(E)\}$ is the classical μ_T-Brownian bridge. Namely, W_T is the centered Gaussian process with covariance

$$\mathbb{E}(W_T(f) W_T(h)) = \mu_T\left((f - \mu_T f)(h - \mu_T h) \right).$$

The second remark is that the random fields $\{W_T^N(f);\ f \in \mathcal{B}_b(E)\}$ can also be regarded as an empirical process indexed by the collection of bounded measurable functions. In this interpretation, Corollary 2.20 simply says that the marginals of the $\mathcal{B}_b(E)$-indexed empirical process weakly converge to the marginals of a centered Gaussian process $\{W_T(f);\ f \in \mathcal{B}_b(E)\}$. One natural question we may ask is whether there exists a functional convergence result for an $\mathcal{F}$-indexed empirical process $\{W_T^N(f);\ f \in \mathcal{F}\}$ where $\mathcal{F} \subset \mathcal{B}_b(E)$.

We recall that weak convergence in $l^\infty(\mathcal{F})$ can be characterized as the convergence of the marginals together with the asymptotic tightness of the process

$$\{W_T^N(f)\ ;\ f \in \mathcal{F}\}$$

As announced in section 2.2.1 the asymptotic tightness is related to the entropy condition $I(\mathcal{F}) < \infty$. The following technical lemma is proved in [42].

Lemma 2.21 *If $\mathcal{F}$ is a countable collection of functions f such that $\|f\| \leq 1$ and $I(\mathcal{F}) < \infty$ then, for any $T \geq 0$, the $\mathcal{F}$-indexed process $\{W_T^N(f);\ f \in \mathcal{F}\}$ is asymptotically tight.*

Theorem 2.22 (Donsker Theorem) *Assume that $\mathcal{F}$ is a countable class of functions such that $\|f\| \leq 1$ for any $f \in \mathcal{F}$ and $I(\mathcal{F}) < \infty$. Then, for any $T \geq 0$, $\{W_T^N(f);\ f \in \mathcal{F}\}$ converges weakly in $l^\infty(\mathcal{F})$ as $N \to \infty$ to a centered Gaussian process $\{W_T(f);\ f \in \mathcal{F}\}$ with covariance function*

$$\mathbb{E}(W_T(f)W_T(h)) = \sum_{p=0}^{T} \int \left(\frac{g_{p,T}}{\eta_p(g_{p,T})}\right)^2 (K_{p,T}(f) - \eta_T(f))(K_{p,T}(h) - \eta_T(h))d\eta_p.$$

Fluctuations on Path Space

In this section we will use notations of section 2.2.1, p. 33 and the following strengthening of $(\mathcal{K})_0$

$(\mathcal{TCL})$ ***For any time*** $n \geq 1$ ***there exist a reference probability measure*** $\lambda_n \in \mathbf{M}_1(E)$ ***and a*** $\mathbf{B}(E)$***-measurable function*** φ_n ***so that*** $K_n(x, \bullet) \sim \lambda_n$ ***and***

$$\forall\, p \geq 1, \qquad \left|\log \frac{dK_n(x,\bullet)}{d\lambda_n}(z)\right| \leq \varphi_n(z) \qquad \textbf{and} \qquad \int \exp\left(p\,\varphi_n\right) d\lambda_n < \infty$$

As we already noticed the distribution $P_T^{(N)}$ induced by $\xi_{[0,T]}$ on path space

$$(\Sigma_T^N, \mathbf{B}(\Sigma_T)^N)$$

is absolutely continuous with respect to the product measure $\eta_{[0,T]}^{\otimes N}$ and

$$\frac{dP_T^{(N)}}{d\eta_{[0,T]}^{\otimes N}}(x) = \exp H_T^{(N)}(x), \qquad \eta_{[0,T]}^{\otimes N} - \text{a.e.},$$

where $H^{(N)} : \Sigma_T^N \to \mathbb{R}$ is the symmetric function given by

$$H_T^{(N)}(x) = N \sum_{n=1}^{T} \int \log \frac{d\Phi_n(m_{n-1}(x))}{d\eta_n}\, dm_n(x)$$

To clarify the presentation, we simplify the notations suppressing the time parameter T in our notations so that we write η, $P^{(N)}$, W^N, $H^{(N)}$, Σ and Σ^N instead of $\eta_{[0,T]}$, $P_{[0,T]}^{(N)}$, $W_{[0,T]}^N$, $H_T^{(N)}$, Σ_T and Σ_T^N.

In what follows we use $\mathbb{E}_{\eta^{\otimes N}}(\,.\,)$ (resp. $\mathbb{E}_{P^{(N)}}(\,.\,)$) to denote expectations with respect to the measure $\eta^{\otimes N}$ (resp. $P^{(N)}$) on Σ^N and, unless otherwise stated, the sequence $\{x^i\ ;\ i \geq 1\}$ is regarded as a sequence of Σ-valued and independent random variables with common law η.

To get the fluctuations of the empirical measures on path space it is enough to study the limit of

$$\{\mathbb{E}_{P^{(N)}}(\exp(iW^N(\varphi)))\ ;\ N \geq 1\} \qquad \text{where} \quad W^N = \sqrt{N}\ (\eta^N - \eta)$$

for functions $\varphi \in L^2(\eta)$. Writing

$$\mathbb{E}_{P^{(N)}}(\exp(iW^N(\varphi))) = \mathbb{E}_{\eta^{\otimes N}}(\exp(iW^N(\varphi) + H^{(N)}(x))),$$

one finds that the convergence of

$$\{\mathbb{E}_{P^{(N)}}(\exp(iW^N(\varphi))) \; ; \; N \geq 1\}$$

follows from the convergence in law and the uniform integrability of

$$\exp(iW^N(\varphi) + H^{(N)}(x))$$

under the product law $\mathbb{E}_{\eta^{\otimes N}}$. The last point is clearly equivalent to the uniform integrability of $\exp H^{(N)}(x)$ under $\mathbb{E}_{\eta^{\otimes N}}$. The proof of the uniform integrability of $\exp H^{(N)}(x)$ then relies on a classical result (see for instance Theorem 5 p. 189 in [100] or Scheffé's Lemma 5.10 p.55 in [112]) which says that, if a sequence of non-negative random variables $\{X_N \; ; \; N \geq 1\}$ converges almost surely towards some random variable X as $N \to \infty$ then we have

$$\lim_{N\to\infty} \mathbb{E}(X_N) = \mathbb{E}(X) < \infty \Longleftrightarrow \{X_N \; ; \; N \geq 1\} \text{ is uniformly integrable}$$

The equivalence still holds if X_N only converges in distribution by Skorohod's Theorem (see for instance Theorem 1 p. 355 in [100]). Since $\mathbb{E}_{\eta^{\otimes N}}(\exp H^{(N)}(x)) = 1$ it is clear that the uniform integrability of

$$\{\exp H^{(N)}(x) \; ; \; N \geq 1\}$$

follows from the convergence in distribution of $H^{(N)}(x)$ towards a random variable H such that $\mathbb{E}(\exp H) = 1$. Thus, it suffices to study the convergence in distribution of

$$\{iW^N(\varphi) + H^{(N)}(x) \; ; \; N \geq 1\}$$

for $L^2(\eta)$ functions φ to conclude.

To state such a result we first need to introduce some notations. Under the assumption $(\mathcal{K})_0$, for any $n \geq 1$ there exists a reference probability measure $\lambda_n \in \mathbf{M}_1(E)$ such that $K_n(x, .) \sim \lambda_n$. In this case we shall use the notation

$$\forall\, (x,z) \in E^2, \qquad k_n(x,z) \stackrel{\text{def.}}{=} \frac{dK_n(x,.)}{d\lambda_n}(z)$$

For any $x = (x_0, \ldots, x_T)$ and $z = (z_0, \ldots, z_T) \in \Sigma$ set

$$q(x,z) = \sum_{n=1}^{T} q_n(x,z) \quad \text{with} \quad q_n(x,z) = \frac{g_n(z_{n-1})\, k_n(z_{n-1}, x_n)}{\int_E g_n(u)\, k_n(u, x_n) \eta_{n-1}(du)}$$

$$a(x,z) = q(x,z) - \int_\Sigma q(x', z)\, \eta(dx')$$

One consequence of $(\mathcal{TCL})$ is that the integral operator A given by

$$\forall\, \varphi \in L^2(\Sigma, \eta), \qquad A_T\varphi(x) = \int a(z,x)\, \varphi(z)\, \eta(dz)$$

is an Hilbert-Schmidt operator on $L^2(\Sigma, \eta)$.

Theorem 2.23 *Assume that condition* $(\mathcal{TCL})$ *is satisfied. For any* $T \geq 0$ *the integral operator* $I - A_T$ *is invertible and the random field*

$$\{W^N_{[0,T]}(\varphi) \; ; \; \varphi \in L^2(\eta_{[0,T]})\}$$

converges as $N \to \infty$ *to a centered Gaussian field*

$$\{W_{[0,T]}(\varphi) \; ; \; \varphi \in L^2(\eta_{[0,T]})\}$$

satisfying

$$\begin{aligned} &\mathbb{E}\left(W_{[0,T]}(\varphi_1)W_{[0,T]}(\varphi_2)\right) \\ &= \left((I - A_T)^{-1}(\varphi_1 - \eta(\varphi_1)), (I - A_T)^{-1}(\varphi_2 - \eta(\varphi_2))\right)_{L^2(\eta_{[0,T]})} \end{aligned}$$

for any $\varphi_1, \varphi_2 \in L^2(\eta_{[0,T]})$*, in the sense of convergence of finite dimensional distributions.*

The basic tools for studying the convergence in law of $\{H^{(N)}(x) \; ; \; N \geq 1\}$ are the Dynkin-Mandelbaum Theorem on symmetric statistics and Shiga-Tanaka's formula of Lemma 1.3 in [98]. The detailed proof of Theorem 2.23 is given in [37]. Here we merely content ourselves in describing the main line of this approach. Here again we simplify notations suppressing the time parameter T and we write A instead of A_T. Let us first recall how one can see that $I - A$ is invertible. This is in fact classical now (see [3] and [98] for instance). First one notices that, under our assumptions, A^n, $n \geq 2$ and $A\,A^*$ are trace class operators with

$$\begin{aligned} \text{Trace} A^n &= \int \cdots \int a(x^1, x^2) \ldots a(x^n, x^1)\, \eta(dx^1) \ldots \eta(dx^n) \\ \text{Trace} AA^* &= \int_{\Sigma_T^2} a(x,z)^2\, \eta(dx)\, \eta(dz) = \|a\|^2_{L^2(\eta \otimes \eta)} \end{aligned}$$

Furthermore by definition of a and the fact that η is a product measure it is easily checked that

$$\forall\, n \geq 2, \qquad \text{Trace} A^n = 0$$

Standard spectral theory (see [101] for instance) then shows that $\det_2(I - A)$ is equal to one and therefore that $I - A$ is invertible.

The identification of the weak limit of $\{H^{(N)}(x) \; ; \; N \geq 1\}$ relies on L^2-techniques and more precisely Dynkin-Mandelbaum construction of multiple Wiener integrals as a limit of symmetric statistics. To state such a result, we first introduce Wiener integrals.

Let $\{I_1(\varphi) \; ; \; \varphi \in L^2(\eta)\}$ be a centered Gaussian field satisfying

$$\mathbb{E}(I_1(\varphi_1)I_1(\varphi_2)) = (\varphi_1, \varphi_2)_{L^2(\eta)}$$

If we set, for each $\varphi \in L^2(\eta)$ and $m \geq 1$

$$h_0^\varphi = 1 \qquad h_m^\varphi(z_1, \ldots, z_m) = \varphi(z_1) \ldots \varphi(z_m),$$

the multiple Wiener integrals $\{I_m(h_m^\varphi)\ ;\ \varphi \in L^2(\eta)\}$ with $m \geq 1$, are defined by the relation

$$\sum_{m\geq 0} \frac{t^m}{m!} I_m(h_m^\varphi) = \exp\left(tI_1(\varphi) - \frac{t^2}{2}\|\varphi\|^2_{L^2(\eta)}\right).$$

The multiple Wiener integral $I_m(\phi)$ for $\phi \in L^2_{\text{sym}}(\eta^{\otimes m})$ is then defined by a completion argument. Theorem 2.23 is therefore a consequence of the following lemma.

Lemma 2.24 ([37])

$$\lim_{N\to\infty} H^{(N)}(x) \overset{\text{law}}{=} \frac{1}{2} I_2(f) - \frac{1}{2}\text{Trace}AA^* \tag{66}$$

where f is given by

$$f(y,z) = a(y,z) + a(z,y) - \int_{\Sigma_T} a(u,y)\, a(u,z)\, \eta(du). \tag{67}$$

In addition, for any $\varphi \in L^2(\eta)$,

$$\lim_{N\to\infty} (H^{(N)}(x) + iW^N(\varphi)) \overset{\text{law}}{=} \frac{1}{2} I_2(f) + iI_1(\varphi) - \frac{1}{2}\text{Trace}AA^*$$

Following the above observations, we get for any $\varphi \in L^2(\eta)$,

$$\begin{aligned}
\lim_{N\to\infty} \mathbb{E}_{P^{(N)}}\left(\exp iW^N(\varphi)\right) &= \lim_{N\to\infty} \mathbb{E}_{\eta^{\otimes N}}\left(\exp\left(iW^N(\varphi) + H^{(N)}(x)\right)\right) \\
&= \mathbb{E}\left(\exp\left(iI_1(\varphi) + \frac{1}{2}I_2(f) - \frac{1}{2}\text{Trace}AA^*\right)\right)
\end{aligned}$$

Moreover, Shiga-Tanaka's formula of Lemma 1.3 in [98] shows that for any $\varphi \in L^2_{\text{sym}}(\eta)$,

$$\mathbb{E}\left(\exp\left(iI_1(\varphi) + \frac{1}{2}I_2(f) - \frac{1}{2}\text{Trace}AA^*\right)\right) = \exp\left(-\frac{1}{2}\|(I-A)^{-1}\varphi\|^2_{L^2(\eta)}\right) \tag{68}$$

The proof of Theorem 2.23 is thus complete. The proof of Lemma 2.24 entirely relies on a construction of multiple Wiener integrals as a limit of symmetric statistics. For completeness and to guide the reader we present this result.

Let $\{\zeta^i\ ;\ i \geq 1\}$ be a sequence of independent and identically distributed random variables with values in an arbitrary measurable space $(\mathcal{X}, \mathcal{B})$. To every symmetric function $h(z_1, \ldots, z_m)$ there corresponds a statistic

$$\sigma_m^N(h) = \sum_{1\leq i_1<\ldots<i_m\leq N} h(\zeta^{i_1}, \ldots, \zeta^{i_m})$$

with the convention $\sigma_m^N = 0$ for $m > N$. Every integrable symmetric statistic $S(\zeta^1, \ldots, \zeta^N)$ has a unique representation of the form

$$S(\zeta^1, \ldots, \zeta^N) = \sum_{m\geq 0} \sigma_m^N(h_m) \tag{69}$$

where $h_m(z_1,\ldots,z_m)$ are symmetric functions subject to the condition

$$\int h_m(z_1,\ldots,z_{m-1},u)\,\mu(du)=0 \tag{70}$$

where μ is the probability distribution of ζ^1.
We call such functions $\{h_m\,;\,m\geq 0\}$ canonical. Finally we denote by $\mathcal{H}$ the set of all sequences $h=(h_0,h_1(z_1),\ldots,h_m(z_1,\ldots,z_m),\ldots)$ where h_m are canonical and

$$\sum_{m\geq 0}\frac{1}{m!}\mathbb{E}(h_m^2(\zeta^1,\ldots,\zeta^m))<\infty$$

As in [98] we will use repeatedly the following

Theorem 2.25 (Dynkin-Mandelbaum [54])
For $h\in\mathcal{H}$ the sequence of random variables $Z_N(h)=\sum_{m\geq 0}\frac{1}{N^{m/2}}\sigma_m^N(h_m)$ converges in law, as $N\to\infty$, to

$$W(h)=\sum_{m\geq 0}\frac{I_m(h_m)}{m!}$$

Proof of Lemma 2.24: (*Sketched*)
It is first useful to observe that for any $\mu\in\mathbf{M}_1(E)$ and $n\geq 1$ we have that

$$\frac{d\Phi_n(\mu)}{d\eta_n}(x)=\frac{d\Phi_n(\mu)}{d\Phi_n(\eta_{n-1})}(x)=\frac{\mu\left(g_n(.)k_n(.,x)\right)}{\eta_{n-1}\left(g_n(.)k_n(.,x)\right)}\Big/\frac{\mu\left(g_n\right)}{\eta_{n-1}\left(g_n\right)}$$

Therefore the symmetric statistics $H^{(N)}(x)$ can be written in the form

$$H^{(N)}(x)=\sum_{n=1}^{T}\sum_{i=1}^{N}\left[\log\left(\frac{1}{N}\sum_{j=1}^{N}q_n(x^i,x^j)\right)-\log\left(\frac{1}{N}\sum_{j=1}^{N}\overline{q}_n(x^j)\right)\right]$$

where

$$\overline{q}_n(x^j)=g_n(x^j_{n-1})/\eta_{n-1}(g_n)=\int_{\Sigma}q_n(y,x^j)\,\eta(dy)$$

By the representation

$$\log z=(z-1)-\frac{(z-1)^2}{2}+\frac{(z-1)^3}{3(\varepsilon z+(1-\varepsilon))^3}$$

which is valid for all $z>0$ with $\varepsilon=\varepsilon(z)$ such that $\varepsilon(z)\in[0,1]$ we obtain the decomposition

$$\begin{aligned}H^{(N)}(x)&=\frac{1}{N}\sum_{i=1}^{N}\sum_{j=1}^{N}a(x^i,x^j)-\frac{1}{2}\sum_{n=1}^{T}\sum_{i=1}^{N}\left(\frac{1}{N}\sum_{j=1}^{N}q_n(x^i,x^j)-1\right)^2\\&\quad+\frac{N}{2}\sum_{n=1}^{T}\left(\frac{1}{N}\sum_{j=1}^{N}\overline{q}_n(x^j)-1\right)^2+R^{(N)}\end{aligned} \tag{71}$$

where the remainder term $R^{(N)}$ cancels as N tends to ∞. The technical trick is to decompose each term as in (69) in order to identify the limit by applying Theorem 2.25. For instance, the first term can readily be written as follows

$$\frac{1}{N}\sum_{i=1}^{N} a(x^i, x^i) + \frac{1}{N}\sum_{i<j}(a + a^*)(x^i, x^j)$$

with $\int_\Sigma a(z,z)\ \eta(dz) = \int_\Sigma a(x,z)\ \eta(dz) = \int_\Sigma a(z,x)\ \eta(dz) = 0$ for any $x \in \Sigma$ and therefore a clear application of Theorem 2.25 yields that it converges in law as $N \to \infty$ to $\frac{1}{2}I_2(a + a^*)$. ∎

2.2.5 Large Deviations Principles

The LDP presented in this section are not restricted to the situation where the functions $\{\Phi_n\ ;\ n \geq 1\}$ have the form (8). In what follows we undertake a more general and abstract formulation and we assume that $\{\xi_n\ ;\ n \geq 0\}$ is the IPS approximating model (13) associated with a given sequence of continuous functions $\{\Phi_n\ ;\ n \geq 1\}$. The LDP for the IPS approximating model (13) for (8) will then be deduced directly from these results.

Large Deviations on Path Space

To prove large deviations principles (**LDP**) for the laws $\{Q_T^{(N)}\ ;\ N \geq 1\}$ of the empirical measures $\eta_{[0,T]}^N$ we will always assume that the continuous functions $\{\Phi_n\ ;\ n \geq 1\}$ satisfy $(\mathcal{P})_0$. As it has already seen in section 2.2.1, p. 34 the main simplification due to this assumption is that $Q_T^{(N)}$ is absolutely continuous with respect to the distribution $R_T^{(N)} \in \mathbf{M}_1(\mathbf{M}_1(\Sigma_T))$ of the empirical measure associated with N independent random variables with common law $\lambda_{[0,T]}$. In addition, we have that

$$\frac{dQ_T^{(N)}}{dR_T^{(N)}} = \exp\left(NF_T\right) \qquad R_T^{(N)} - \text{a.s.}$$

where $F_T : \mathbf{M}_1(\Sigma_T) \to \mathbb{R}$ is the function defined by

$$F_T(\mu) = \sum_{n=1}^{T} \int_E \log \frac{d\Phi_n(\mu_{n-1})}{d\lambda_n}\ d\mu_n = \int_{\Sigma_T} \log \frac{d\Phi_{[0,T]}(\mu)}{dR_T}\ d\mu \tag{72}$$

In a first stage for analysis it is reasonable to suppose that

$(\mathcal{P})_1$ ***For any*** $n \geq 1$ ***there exists a reference probability measure*** $\lambda_n \in \mathbf{M}_1(E)$ ***such that for all*** $\mu \in \mathbf{M}_1(E)$***,*** $\Phi_n(\mu) \sim \lambda_n$ ***and the function*** $\mathbf{M}_1(E)^2 \ni (\mu,\nu) \to \int \log \frac{d\Phi_n(\nu)}{d\lambda_n}\ d\mu$ ***is bounded continuous.***

If $I(\mu|\nu)$ denotes the relative entropy of μ with respect to ν, that is the function

$$I(\mu|\nu) = \int \log \frac{d\mu}{d\nu}\ d\mu$$

if $\mu << \nu$ and $+\infty$ otherwise, Sanov's Theorem and Varadhan's Lemma yields

Theorem 2.26 *Assume that* $\{\Phi_n \ ; \ n \geq 1\}$ *is a sequence of continuous functions such that* $(\mathcal{P})_1$ *holds. Then, for any* $T \geq 0$, $\{Q_T^{(N)}, \ N \geq 1\}$ *satisfies a LDP with good rate function*

$$J_T(\mu) = I(\mu|\Phi_{[0,T]}(\mu))$$

and $\eta_{[0,T]}$ *is the unique minimizer of* J_T.

Proof:(*Sketch*) Under the assumptions of the theorem, F_T is bounded continuous so that $\{Q_T^{(N)}, \ N \geq 1\}$ satisfies a LDP with good rate function

$$J_T(\mu) = I(\mu|R_T) - F_T(\mu)$$

according to Sanov's Theorem and Varadhan's Lemma (see [51] for instance). ∎

Corollary 2.27 *Assume that the functions* $\{\Phi_n \ ; \ n \geq 1\}$ *are given by (8) and the transitions probability kernels* $\{K_n \ ; \ n \geq 1\}$ *are Feller and satisfy the following assumption*

$(\mathcal{K})_1'$ ***For any time*** $n \geq 1$, ***there exists a reference probability measure*** $\lambda_n \in \mathbf{M}_1(E)$ ***such that*** $K_n(x, \bullet) \sim \lambda_n$ ***and***

- ***the function*** $z \mapsto \log \frac{dK_n(x,\bullet)}{d\lambda_n}(z)$ ***is Lipschitz, uniformly on the parameter*** $x \in E$, ***and for any*** $z \in E$ ***the map*** $x \mapsto \frac{dK_n(x,\bullet)}{d\lambda_n}(z)$ ***is continuous***
- ***there exists a positive number*** $\epsilon_n \in (0,1]$ ***such that***

$$\epsilon_n \leq \frac{dK_n(x,\bullet)}{d\lambda_n} \leq \frac{1}{\epsilon_n}$$

Then, for any $T \geq 0$, $\{Q_T^{(N)}, \ N \geq 1\}$ *satisfies a LDP with good rate function* J_T.

Condition $(\mathcal{K})_1'$ is stronger than condition $(\mathcal{K})_1$ which has been used in section 2.1.2, p. 24, as a mixing condition to derive exponential stability properties for the limiting measure valued system (8). In LDP settings this hypothesis is more related to a compactness assumption.

Here we present a way to relax $(\mathcal{P})_1$ based on cut-off arguments.

Let $F_T^M : \mathbf{M}_1(\Sigma_T) \to \mathbb{R}$ be the cut-off transformation of F_T given by

$$F_T^M(\mu) = \sum_{n=1}^{T} \int_E \psi^M \left(\log \frac{d\Phi_n(\mu_{n-1})}{d\lambda_n} \right) \, d\mu_n$$

where

$$\psi^M(x) = x \, 1_{|x| \leq M} + \text{sign}(x) \, M \, 1_{|x| > M}$$

Next assumptions relax $(\mathcal{P})_1$

$(\mathcal{L})_0$ ***For any time*** $n \geq 1$***, there exists a reference probability measure*** $\lambda_n \in \mathbf{M}_1(E)$ ***such that for all*** $\mu \in \mathbf{M}_1(E)$***,*** $\Phi_n(\mu) \sim \lambda_n$ ***and the function***

$$(x,\nu) \mapsto \log \frac{d\Phi_n(\nu)}{d\lambda_n}(x)$$

is uniformly continuous w.r.t. x ***(and uniformly w.r.t.*** ν***) and continuous w.r.t.*** ν***.***

$(\mathcal{L})_1$ ***There exist constants*** $c_T < \infty$***,*** $\alpha_T > 1$ ***such that***

$$R_T^{(N)}\left(e^{\alpha_T N F_T}\right) \leq e^{c_T N}$$

and, for every $\epsilon > 0$ ***there exists a function*** $L_{T,\epsilon}$***, such that*** $L_{T,\epsilon}(M)$ ***goes to infinity when*** M ***goes to infinity, so that***

$$R_T^{(N)}\left(|F_T - F_T^M| > \epsilon\right) \leq e^{-NL_{T,\epsilon}(M)}. \tag{73}$$

$(\mathcal{L})_2$ ***There exist constants*** $\delta_T > 0$ $C_T < \infty$***,*** $D_T < \infty$ ***and a function*** ϵ_T***,*** $\epsilon_T(M)$ ***is going to zero when*** M ***is going to infinity, such that for any*** $\mu \in \mathbf{M}_1(\Sigma_T)$ ***and*** $M \in \mathbb{R} \cup \{\infty\}$

$$\begin{aligned} I(\mu|R_T) - F_T^M(\mu) &\geq \delta_T\, I(\mu|R_T) - C_T \\ |F_T(\mu) - F_T^M(\mu)| &\leq \epsilon_T(M)\,(I(\mu|R_T) + D_T) \end{aligned}$$

Theorem 2.28 *Assume conditions* $(\mathcal{L})_0$*,* $(\mathcal{L})_1$ *and* $(\mathcal{L})_2$ *are satisfied.*
Then, $\{Q_T^{(N)} : N \geq 1\}$ *satisfies a LDP with good rate function* J_T*.*

Proof:(*Sketch*) Under $(\mathcal{L})_0$ one first check that F_T^M is bounded continuous. The proof is now based on the ideas of Azencott and Varadhan and amounts to replace the functions F_T (which are a priori nor bounded nor continuous) by the functions F_T^M to get the LDP up to a small error ϵ in the rate function by $(\mathcal{L})_1$ and then pass to the limit $M \to \infty$ by $(\mathcal{L})_2$ to let finally $\epsilon \downarrow 0$. ■

Conditions $(\mathcal{L})_1$ and $(\mathcal{L})_2$ are hard to work with. It is quite remarkable that an exponential moment condition suffices to check $(\mathcal{L})_1$ and $(\mathcal{L})_2$.

Corollary 2.29 *Suppose the functions* $\{\Phi_n \; ; \; n \geq 1\}$ *satisfy next condition*

$(\mathcal{P})_1'$ ***For any time*** $1 \leq n \leq T$ ***there exists a reference probability measure*** $\lambda_n \in \mathbf{M}_1(E)$ ***such that for all*** $\mu \in \mathbf{M}_1(E)$***,*** $\Phi_n(\mu) \sim \lambda_n$ ***and***

- ***For any*** $1 \leq n \leq T$ ***the function***

$$(x,\nu) \mapsto \log \frac{d\Phi_n(\nu)}{d\lambda_n}(x)$$

is uniformly continuous w.r.t. x ***and continuous w.r.t.*** ν***.***

- ***There exist*** $\mathbf{B}(E)$***-measurable functions*** φ ***and*** ψ ***and constants*** $\alpha, \beta \in]1, \infty]$ ***and*** $\epsilon > 0$ ***such that*** $\frac{1}{\alpha} + \frac{1}{\beta} < 1$ ***and for any*** $1 \leq n \leq T$

$$\left| \log \frac{d\Phi_n(\mu)}{d\lambda_n}(x) \right| \leq \varphi(x) + \mu(\psi)$$

and

$$\int \exp\left(\alpha \varphi^{1+\epsilon}\right) d\lambda_n \vee \int \exp\left(\beta \psi^{1+\epsilon}\right) d\lambda_n < \infty \tag{74}$$

Then, $\{Q_T^{(N)} : N \geq 1\}$ *satisfies the LDP with good rate function* J_T.

Corollary 2.30 *Assume that the functions* $\{\Phi_n \; ; \; n \geq 1\}$ *are given by (8) and the transitions probability kernels* $\{K_n \; ; \; n \geq 1\}$ *are Feller and satisfy the following assumption*

$(\mathcal{K})_1''$ ***For any time*** $1 \leq n \leq T$ ***there exists a reference probability measure*** $\lambda_n \in \mathbf{M}_1(E)$ ***such that*** $K_n(x, \bullet) \sim \lambda_n$ ***and***

- ***For any time*** $1 \leq n \leq T$ ***the function***

$$z \mapsto \log \frac{dK_n(x, \bullet)}{d\lambda_n}(z)$$

is Lipschitz, uniformly on the parameter $x \in E$***, and for any*** $z \in E$ ***the map***

$$x \mapsto \frac{dK_n(x, \bullet)}{d\lambda_n}(z)$$

is continuous.

- ***There exist a*** $\mathbf{B}(E)$***-measurable function*** φ ***and constants*** $\alpha > 1$ ***and*** $\epsilon > 0$ ***such that for any time*** $1 \leq n \leq T$

$$\left| \log \frac{dK_n(x, \bullet)}{d\lambda_n}(z) \right| \leq \varphi(z) \qquad \text{and} \qquad \int \exp\left(\alpha \varphi^{1+\epsilon}\right) d\lambda_n < \infty$$

Then, $\{Q_T^{(N)} : N \geq 1\}$ *satisfies a LDP with good rate function* J_T.

Large Deviations for the Particle Density Profiles

The large deviations results on path space rely largely on the existence of a family of reference distributions $\{\lambda_n : n \geq 1\}$ satisfying condition $(\mathcal{P})_0$ and therefore does not apply to some filtering problems (see section 5). To remove this assumption we shall be dealing with the law $\{P_n^N \; ; n \geq 1\}$, of the particle density profiles $\{\eta_n^N \; ; n \geq 1\}$.

Theorem 2.31 *Assume that the continuous functions* $\{\Phi_n \; ; \; n \geq 1\}$ *satisfy the following condition*

($\mathcal{ET}$) ***For any*** $n \geq 1$, $\epsilon > 0$ ***and for any Markov transition*** M ***on*** $\mathbf{E}$***, there exist a Markov kernel*** $\tilde{M}$ ***and*** $0 < \delta \leq \epsilon$ ***such that***

$$\mu\tilde{M}(A^c) < \delta \Longrightarrow \Phi_n(\mu)M(A^c) < \epsilon$$

for any $\mu \in \mathbf{M}_1(E)$ ***and for any compact set*** $A \subset E$***.***

Then, for any $n \geq 0$, $\{P_n^{(N)} \; : \; N \geq 1\}$ *obeys a LDP with convex good rate function* H_n *given by*

$$\begin{cases} H_n(\mu) &= \sup_{V \in \mathcal{C}_b(E)} \left(\mu(V) + \inf_{\nu \in \mathbf{M}_1(E)} \left(H_{n-1}(\nu) - \log\left(\Phi_n(\nu)(e^V)\right)\right)\right), \quad n \geq 1 \\ H_0(\mu) &= I(\mu|\eta_0) \end{cases}$$

In addition $H_n(\mu) = 0$ *iff* $\mu = \eta_n$*, for any* $n \geq 1$.

Proof:(*Sketch*) First we check that ($\mathcal{ET}$) insures that for any time $n \geq 0$ the sequence

$$\{P_n^N : N \geq 1\}$$

is exponentially tight (cf. Proposition 2.5 in [35]). To get the desired LDP we proceed by induction on the time parameter. For $n = 0$ the result is a trivial by Sanov's Theorem so we assume that it holds for $(n-1)$. Observe that the moment generating function at rank n is given for any $V \in \mathcal{C}_b(E)$ by

$$\mathbb{E}\left(\exp\left(N\eta_n^N(V)\right)\right) = \mathbb{E}\left(\left(\Phi_n(\eta_{n-1}^N)(e^V)\right)^N\right) = \mathbb{E}\left(\exp\left(NG_n(\eta_{n-1}^N)\right)\right)$$

with

$$G_n(\eta) \stackrel{\text{def}}{=} \log[\Phi_n(\eta)(e^V)]$$

Since G_n is bounded continuous then Varadhan's Lemma (see for instance [106], Theorem 2.6 p 24) and the induction hypothesis at rank $(n-1)$ imply that

$$\lim_{N\to\infty} \frac{1}{N} \log \int \exp\left(N\mu(V)\right) P_n^N(d\mu) = \Lambda_n(V)$$

with

$$\Lambda_n(V) = -\inf_{\mu \in \mathbf{M}_1(E)} \left(H_{n-1}(\mu) - \log\left(\Phi_n(\mu)(e^V)\right)\right)$$

Using the exponential tightness property we are now in position to apply Baldi's Theorem (cf. Theorem 4.5.20 p. 139 and Corollary 4.6.14 in [49]). More precisely, it remains to check that Λ_n is finite valued and Gateaux differentiable. The first point is obvious. To check that Λ_n is differentiable we introduce for $\nu \in \mathbf{M}_1(E)$,

$$I_n^V(\nu) := H_{n-1}(\nu) - \log(\Phi_n(\nu)(e^V)).$$

After some calculations one finds that for any $v \in \mathcal{C}_b(E)$, $\|v\| \leq 1$

$$D\Lambda_n(V)[v] = \sup_{\{\nu : I_n^V(\nu) \leq -\Lambda_n(V)\}} \frac{\int v e^V d\Phi_n(\nu)}{\int e^V d\Phi_n(\nu)}.$$

■

To see that $(\mathcal{ET})$ holds if the functions $\{\Phi_n \; ; \; n \geq 1\}$ are given by (8) we simply notice that, under the assumption $(\mathcal{G})$, for any $n \geq 1$ and for any compact set $A \subset E$,

$$\forall \, \eta \in \mathbf{M}_1(E), \qquad \Phi_n(\eta)(A^c) \;\leq\; a_n^2 \, \eta K_n(A^c)$$

Corollary 2.32 *If the functions $\{\Phi_n \; ; \; n \geq 1\}$ are given by (8) and the transitions probability kernels $\{K_n \; ; \; n \geq 1\}$ are Feller then for any $n \geq 0$, $\{P_n^N \; : \; N \geq 1\}$ obeys a LDP with convex good rate function H_n.*

2.3 Branching Particle Systems Variants

The research literature abounds with variations on the IPS model (13). Each of these variants is intended to make the selection and/or the mutation transition more efficient in some sense. As a guide to their usage, this section presents a survey of what is currently the state of the art.

The analysis of the convergence of all these alternative schemes is actually not complete. We therefore gather together here several discussions and results which we hope will assist in giving an illustration of how the analysis developed in section 2.2 may be applied.

The first scheme discussed in section 2.3.1 concerns a genetic type algorithm with periodic selection times. The key idea here is to use the stability properties of the limiting system to produce a more efficient IPS. We will show that the resulting algorithm can be reduced to the latter through a suitable state space basis. In this specific situation all the results developed in section 2.2 may be applied. It will also be shown that the convergence exponent in the uniform convergence result 2.11 improves the one obtained for the generic IPS.

Section 2.3.2 presents a way to produce an IPS whose mutation transitions depends on the fitness functions. In nonlinear filtering settings this scheme is often referred as an IPS with conditional mutations. In this situation the fitness functions and therefore the mutation transitions depend on the observations record so that the particles are more likely to track the signal process.

In section 2.3.3 we present "less time-consuming" selection transitions such as the *Remainder Stochastic Sampling* and other BIPS alternatives including *Bernoulli, Poisson and Binomial* branching distributions. These selection transitions can be related to the classical weighted bootstrap theory as well as genetic algorithms theory (the book [10] includes a useful survey on these two subjects).

In filtering settings the choice of the mutation transitions $\{K_n \; ; \; n \geq 1\}$ is dictated by the form of the signal. It may happen that these transitions are degenerated (i.e. deterministic) and therefore the IPS approximating models will not work in practice since after a finite number of steps we would get a single deterministic path. The last section 2.3.4 presents a way to regularize such degenerated mutation transitions. This regularization step has been introduced in [59]. We shall indicate how the results of section 2.2 are applied.

In practice the most efficient IPS approximating model is the one obtained by combining conditional mutations and periodic selections. The choice of the best selection scheme is more an art form. There is a balance between *time-consuming* and *efficiency.* The less time consuming selections seems to be *Remainder Stochastic Sampling* and *Bernoulli branching selections.* In the last case the size of the system is not fixed but random and explosion is possible (cf. [36, 38] and 4) section 2.3.3).

The most important and unsolved problem is to understand and compare the influence of the various selections schemes on the long time behavior of the BIPS approximating models. The interested reader will find that although we have restricted ourselves to the relatively less complicated generic IPS model (13) most of our general techniques apply across these more complex BIPS variants.

There are many open problems such as that of finding the appropriate analog of the results of section 2.2 for the BIPS approximating models presented in section 2.3.3. This study has been started in [22] and in [25] but many open problems such as the fluctuations remain to be answered. A related question will be to find a criterion for determining the "optimal branching" transition. This problem will probably lead to difficult optimization problems since this criterion should be related to the long time behavior of the BIPS approximating schemes.

2.3.1 Periodic Interactions/Selections

The IPS with periodic selections discussed here has been introduced in [36] as a way to combine the stability properties of the limiting system with the long time behavior of the particle scheme.

The prediction/mutation of the former include exploration paths of a given length $T \geq 1$ and the selection transition is used every T steps and it involves T fitness functions. Our immediate goal is to show that the former genetic algorithm can be reduced to the latter through a suitable state space basis. To this end we need to introduce some additional notations.

To each $p \in \{1, \ldots, T\}$ we associate a sequence of meshes $\{t_n^{(T,p)} \; ; \; n \geq 0\}$ by setting

$$t_0^{(T,p)} = 0 \qquad \text{and} \qquad \forall\, n \geq 1, \qquad t_n^{(T,p)} = (n-1)T + p$$

If we write $\Delta_n = t_n - t_{n-1}$ for any $n \geq 1$ we clearly have that

$$\Delta_1 = p \qquad \text{and} \qquad \forall\, n > 1, \qquad \Delta_n = T$$

The parameter T is called the selection period, n will denote the time step and the parameter p will only be used to cover all the time space basis $\mathbb{N}$. The construction below will depend on the pair parameter (T,p) but to clarify the presentation we simplify the notations suppressing the pair parameter (T,p) so that we simply note t_n instead of $t_n^{(T,p)}$.

We also notice that the distributions given by

$$\forall n \geq 0, \qquad \mu_n = \eta_{t_n} \times K_{t_n+1} \times \ldots \times K_{t_{n+1}-1} \in \mathbf{M}_1(E^{\Delta_{n+1}})$$

are solutions of the measure valued process

$$\mu_n = \Phi_n^{(p)}(\mu_{n-1}) \qquad n \geq 1 \tag{75}$$

where $\Phi_n^{(p)} : \mathbf{M}_1(E^{\Delta_n}) \to \mathbf{M}_1(E^{\Delta_{n+1}})$ is the continuous function given by

$$\forall\, \mu \in \mathbf{M}_1(E^{\Delta_n}), \qquad \Phi_n^{(p)}(\mu) = \Psi_n^{(p)}(\mu)\mathcal{K}_n^{(p)}$$

and $\mathcal{K}_n^{(p)}$ and $\Psi_n^{(p)}$ are defined as follows.

- $\Psi_n^{(p)} : \mathbf{M}_1(E^{\Delta_n}) \to \mathbf{M}_1(E^{\Delta_n})$ is the continuous function defined for any test function $f \in \mathcal{B}_b(E^{\Delta_n})$ by setting

$$\Psi_n^{(p)}(\mu)(f) = \frac{\mu(g_n^{(p)}\, f)}{\mu(g_n^{(p)})} \quad \text{with} \quad g_n^{(p)}(x) = \prod_{q=1}^{\Delta_n} g_{t_{n-1}+q}(x_q).$$

- $\mathcal{K}_n^{(p)}$ is the transition probability kernel from E^{Δ_n} to $E^{\Delta_{n+1}}$ given by

$$\begin{aligned}\mathcal{K}_n^{(p)}((x_1,\ldots,x_{\Delta_n}),d(z_1,\ldots,z_{\Delta_{n+1}})) = K_{t_n}(x_{\Delta_n},dz_1) \times \ldots \\ \ldots \times K_{t_{n+1}-1}(z_{\Delta_{n+1}-1},dz_{\Delta_{n+1}})\end{aligned}$$

The IPS associated with (75) is now defined as a Markov chain $\{\zeta_n\,;\, n\geq 0\}$ with product state spaces $\{(E^{\Delta_{n+1}})^N\,;\, n\geq 0\}$ where N is the number of particles and $\{\Delta_{n+1}\,;\, n\geq 0\}$ the selection periods.
The initial particle system $\zeta_0 = (\zeta_0^1,\ldots,\zeta_0^N)$ takes values in $(E^{\Delta_1})^N = (E^p)^N$ and it is given by

$$P(\zeta_0 \in dx) \;=\; \prod_{i=1}^{N} \mu_0(dx^i)$$

and the transition of the chain is now given by

$$\begin{aligned}P(\zeta_n \in dx|\zeta_{n-1}=z) &= \prod_{i=1}^{N} \Phi_n^{(p)}\left(\frac{1}{N}\sum_{j=1}^{N}\delta_{z^j}\right)(dx^i)\\ &= \prod_{i=1}^{N}\sum_{j=1}^{N}\frac{g_n^{(p)}(z^j)}{\sum_{k=1}^{N} g_n^{(p)}(z^k)}K_n^{(p)}(z^i,dx^i)\end{aligned}$$

where $dx = dx^1 \times \ldots \times dx^N$ is an infinitesimal neighborhood of the point $x = (x^1, \ldots, x^N) \in (E^{\Delta_{n+1}})^N$ and for any $1 \leq i \leq N$, $z^i = (z^i_1, \ldots, z^i_{\Delta_n}) \in E^{\Delta_n}$.

If we denote

$$\forall\, n \geq 0, \qquad \zeta_n \;=\; (\xi_{t_n}, \ldots, \xi_{t_{n+1}-1})$$

we see that the former algorithm is indeed a genetic type algorithm with T-periodic selection/updating transitions. Between the dates t_n and t_{n+1} the particles evolves randomly according to the transition probability kernel of the signal and the selection mechanism takes place at each time t_n , $n \geq 1$.

As announced in the beginning of the section, this IPS scheme with periodic selection times is reduced to the one presented in section 1.3 through a suitable state space basis so that the analysis given in section 2.2 applies to this situation.

The uniform results with respect to time given in Theorem 2.11 can be improved by choosing a suitable period which depends on the stability properties of the limiting system and on the number of particles. More precisely, if we denote by $\eta^N_{t_n}$ the particle density profiles given by

$$\eta^N_{t_n} = \frac{1}{N} \sum_{i=1}^{N} \delta_{\xi^i_{t_n}}$$

then we have the the following theorem.

Theorem 2.33 *Assume that the limiting dynamical system (8) is exponentially asymptotically stable in the sense that there exist some positive constant $\gamma > 0$ such that for any function $f \in \mathcal{B}_b(E)$ with $\|f\| \leq 1$*

$$\forall\, \mu, \nu \in \mathbf{M}_1(E), \forall\, T \geq 0, \qquad \sup_{p \geq 0} |\Phi_{p,p+T}(\mu)(f) - \Phi_{p,p+T}(\nu)(f)| \;\leq\; e^{-\gamma T}$$

If the selection period is chosen so that $T = T(N) \stackrel{\text{def}}{=} \left[\frac{1}{2} \frac{\log N}{\gamma + \gamma'}\right] + 1$ then for any $f \in \mathcal{B}_b(E)$, $\|f\| \leq 1$, we have the uniform bound given by

$$\sup_{n \geq 0} \mathbb{E}\left(|\eta^N_{t_n} f - \eta_{t_n} f|\right) \leq \frac{4e^{2\gamma'}}{N^{\beta/2}} \qquad \text{with} \qquad \beta = \frac{\gamma}{\gamma + \gamma'}, \qquad \gamma' = 2 \log a \tag{76}$$

Remarks 2.34:

- Although we have not yet checked more general $\mathbb{L}^p$-bounds or uniform convergence results over a suitable class of functions, the proof of Theorem 2.33 essentially follows the same arguments as in the proof of Theorem 2.11.
- It is also worth observing that the choice of the selection period depends on the stability properties of the limiting system as well as on the number of particles.
- Another remark is that the critical exponent β resulting from the proof of Theorem 2.33 is now sharp in the following sense: if the fitness functions are constant then, without loss of generality, we may chose $a = 1$. In this specific situation the critical exponent $\beta = 1$ which is again the characteristic exponent of the weak law of large numbers.

- Our last remark is that periodic selections are very efficient and have a specific interpretation in nonlinear filtering settings. We recall that in this situation the fitness functions are related to the observation process. Roughly speaking the selection transition evaluates the population structure and allocates reproductive opportunities in such a way that these particles which better match with the current observation are given more chance to "reproduce". This stabilizes the particles around certain values of the real signal in accordance with its noisy observations.
  It often appears that a single observation data is not really sufficient to distinguish in a clearly manner the relative fitness of individuals. For instance this may occurs in high noise environment. In this sense the IPS with periodic selections allows particles to learn the observation process between the selection dates in order to produce more effective selections.

2.3.2 Conditional Mutations

The IPS with conditional mutations is defined in a natural way as the IPS approximating model associated to the measure valued process (23) presented in section 2.1.1, p. 22. To clarify the presentation it is convenient here to change the time parameter in the fitness functions $\{g_n \ ; \ n \geq 1\}$ so that we will write g_n instead of g_{n+1}. Using these notations (23) is defined by the recursion

$$\forall\, n \geq 1, \qquad \widehat{\eta}_n \;=\; \widehat{\Phi}_n(\widehat{\eta}_{n-1}) \tag{77}$$

where $\widehat{\Phi}_n(\eta) = \widehat{\Psi}_n(\eta)\widehat{K}_n$ and for any $f \in \mathcal{B}_b(E)$,

$$\widehat{\Psi}_n(\eta)(f) \stackrel{\text{def.}}{=} \frac{\eta(\widehat{g}_n f)}{\eta(\widehat{g}_n)}, \qquad \widehat{K}_n f \stackrel{\text{def.}}{=} \frac{K_n(g_n f)}{K_n(g_n)}, \qquad \widehat{g}_n \stackrel{\text{def.}}{=} K_n(g_n).$$

As noticed in section 2.1.1 this model has the same form as (8). It is then clear that all the results developed in section 2.1.2 and section 2.2 can be translated in these settings.

In contrast to (8) we also note that the prediction transitions here depends on the fitness functions and therefore the corresponding IPS approximating model will involve mutation transitions which also depend on these functions. More precisely, let $\widehat{\zeta}_n = \left(\widehat{\zeta}_n^1, \ldots, \widehat{\zeta}_n^N\right) \in E^N$ be the N-IPS associated with (77) and defined as in (13) by the transition probability kernels

$$P\left(\widehat{\zeta}_n \in dx \,|\, \widehat{\zeta}_{n-1} = z\right) = \prod_{p=1}^{N} \widehat{\Phi}_n(m(z))(dx^p)$$

where $dx \stackrel{\text{def}}{=} dx^1 \times \cdots \times dx^N$ is an infinitesimal neighborhood of the point $x = (x^1, \ldots, x^N) \in E^N$, $z = (z^1, \ldots, z^N) \in E^N$. As before we see that the above transition involves two mechanisms

$$\widehat{\zeta}_{n-1} \xrightarrow{\text{Selection}} \widetilde{\zeta}_n \xrightarrow{\text{Mutation}} \widehat{\zeta}_n$$

which can also be modelled as follows

$$P(\tilde{\zeta}_n \in dx \,|\, \widehat{\zeta}_{n-1} = z) = \prod_{p=1}^{N}\sum_{i=1}^{N} \frac{\widehat{g}_n(z^i)}{\sum_{j=1}^{N} \widehat{g}_n(z^j)}\, \delta_{z^i}(dx^p)$$

$$P(\widehat{\zeta}_n \in dz \,|\, \tilde{\zeta}_n = x) = \prod_{p=1}^{N} \widehat{K}_n(x^p, dz^p)$$

As we already noticed in section 2.1.1 the fitness functions $\{\widehat{g}_n \; ; \; n \geq 0\}$ and the transitions $\{\widehat{K}_n \; ; \; n \geq 1\}$ involve integrations over the whole state space E so that another level of approximation is in general needed.

Nevertheless let us work out an example in which these two objects have a simple form

Example 2.35 *Let us suppose that $E = \mathbb{R}$ and the fitness functions $\{g_n \; ; \; n \geq 1\}$ and the transitions $\{K_n \; ; \; n \geq 1\}$ are given by*

$$g_n(x) = e^{-\frac{1}{2\,r_n}(y_n - c_n.x)^2} \qquad K_n(x, dz) = \frac{1}{\sqrt{2\pi}\; q_n}\; e^{-\frac{1}{2\,q_n}(z - a_n(x))^2}$$

where $a_n : \mathbb{R} \to \mathbb{R}$, $q_n > 0$ for any $n \geq 1$ and $r_n > 0$, $c_n, y_n \in \mathbb{R}$, for any $n \geq 0$. In this situation one gets easily

$$\widehat{K}_n(x, dz) = \frac{1}{\sqrt{2\,\pi\,|s_n|}} \exp\left(-\frac{1}{2\,|s_n|}\left(z - \left[a_n(x) + s_n\, c_n\, r_n^{-1}\,(y_n - c_n\, a_n(x))\right]\right)^2\right)$$

$$\text{and } \widehat{g}_n(x) = \frac{1}{\sqrt{2\,\pi\,|q_n|\,|r_n|/|s_n|}} \exp\left(-\frac{1}{2\,|q_n|\,|r_n|/|s_n|}\,(y_n - c_n\, a_n(x))^2\right)$$

with $s_n = (q_n^{-1} + c_n\, r_n^{-1}\, c_n)^{-1}$.

One idea to approximate the transition $\widehat{\zeta}_{n-1} \to \widehat{\zeta}_n$ is to introduce an auxiliary branching mechanism.

$$\widehat{\zeta}_{n-1} \xrightarrow{\text{Branching}} \zeta_n \xrightarrow{\text{Selection/Mutation}} \widehat{\zeta}_n$$

The branching transition is defined inductively as follows.
At each time $(n-1)$ each particle $\widehat{\zeta}^i_{n-1}$ branches independently of each other into M-auxiliary particles with common law $K_n\left(\widehat{\zeta}^i_{n-1}, \cdot\right)$, that is for any $1 \leq i \leq N$,

$$\widehat{\zeta}^i_{n-1} \xrightarrow{\text{Branching}} \zeta^{(i)}_n \stackrel{\text{def.}}{=} (\zeta^{i,1}_n, \ldots, \zeta^{i,M}_n)$$

where $(\zeta^{i,1}_n, \ldots, \zeta^{i,M}_n)$ are (conditionally) M-independent particles with common law $K_n\left(\widehat{\zeta}^i_{n-1}, \cdot\right)$.
At the end of this branching step the system consists in $N \times M$ particles

$$\zeta_n \stackrel{\text{def.}}{=} (\zeta^{(1)}_n, \ldots, \zeta^{(N)}_n) \in \underbrace{E^M \times \ldots \times E^M}_{N-\text{times}}$$

If the parameter M is sufficiently large then, in some sense, the empirical measures associated with each sub-group of M particles is an approximating measure of $K_n\left(\widehat{\zeta}_{n-1}^i,.\right)$, that is

$$\forall\, f\in\mathcal{B}_b(E),\qquad K_n^{(M)}(f)(\zeta_n^{(i)}) \underset{M\to\infty}{\longrightarrow} K_n(f)(\widehat{\zeta}_{n-1}^i) \tag{78}$$

where $K_n^{(M)}$ is the transition probability kernel from E^M into E given for any $x\in E^M$ and $f\in\mathcal{B}_b(E)$ by the formula

$$K_n^{(M)}(x,.) \overset{\text{def.}}{=} \frac{1}{M}\sum_{i=1}^M \delta_{x^i} \qquad\text{and}\qquad (K_n^{(M)}f)(x)=\int_E f(z)\,K_n^{(M)}(x,dz)$$

Using the above notations we also have, in some sense, that

$$\widehat{g}_n^{(M)}(\zeta_n^{(i)}) \overset{\text{def.}}{=} K_n^{(M)}(g_n)(\zeta_n^{(i)}) \underset{M\to\infty}{\longrightarrow} K_n(g_n)(\widehat{\zeta}_{n-1}^i)=\widehat{g}_n(\widehat{\zeta}_{n-1}^i) \tag{79}$$

and for any $f\in\mathcal{B}_b(E)$

$$\widehat{K}_n^{(M)}(f)(\zeta_n^{(i)}) \overset{\text{def.}}{=} \frac{K_n^{(M)}(g_n\,f)\left(\zeta_n^{(i)}\right)}{K_n^{(M)}(g_n)(\zeta_n^{(i)})} \underset{M\to\infty}{\longrightarrow} \widehat{K}_n(f)(\widehat{\zeta}_{n-1}^i) \tag{80}$$

Finally if we combine (79) and (80) one gets an M-approximation of the desired transition

$$\sum_{i=1}^N \frac{\widehat{g}_n^{(M)}(\zeta_n^{(i)})}{\sum_{j=1}^N \widehat{g}_n^{(M)}(\zeta_n^{(i)})}\ \widehat{K}_n^{(M)}(\zeta_n^{(i)},.) \underset{M\to\infty}{\longrightarrow} \widehat{\Phi}_n\left(\frac{1}{N}\sum_{i=1}^N \delta_{\widehat{\zeta}_{n-1}^i}\right) \tag{81}$$

The next particle system $\widehat{\zeta}_n=\left(\widehat{\zeta}_n^1,\ldots,\widehat{\zeta}_n^N\right)$ simply consists in N conditionally independent particles with common law the left hand side of (81).

Our new BIPS is now defined by the following Markov Model

$$\widehat{\zeta}_{n-1} \xrightarrow{\text{Branching}} \zeta_n=(\zeta_n^{(1)},\ldots,\zeta_n^{(N)}) \xrightarrow{\text{Selection}} \widetilde{\zeta}_n=\left(\widetilde{\zeta}_n^{(1)},\ldots,\widetilde{\zeta}_n^{(N)}\right) \xrightarrow{\text{Mutation}} \widehat{\zeta}_n$$

with the following transitions

- **Branchings**: The branching transition

$$\widehat{\zeta}_{n-1}\in E^N \longrightarrow \zeta_n=(\zeta_n^{(1)},\ldots,\zeta_n^{(N)})\in(E^M)^N$$

is defined by

$$P\left(\zeta_n\in dx^{(1)}\times\ldots dx^{(N)}\,\middle|\,\widehat{\zeta}_{n-1}=z\right)=\prod_{i=1}^N (K_n(z^i,\cdot))^{\otimes M}(dx^{(i)})$$

where $dx^{(i)}=dx^{i,1}\times\ldots\times dx^{i,M}$ is an infinitesimal neighborhood of the point $x^{(i)}=(x^{i,1},\ldots,x^{i,M})\in E^M$, $z=(z^1,\ldots,z^N)\in E^N$.

- **Selection**: The selection transition

$$\zeta_n = (\zeta_n^{(1)}, \ldots, \zeta_n^{(N)}) \in (E^M)^N \longrightarrow \tilde{\zeta}_n = \left(\tilde{\zeta}_n^{(1)}, \ldots, \tilde{\zeta}_n^{(N)}\right) \in (E^M)^N$$

is defined by

$$P\left(\tilde{\zeta}_n \in dy^{(1)} \times \ldots \times dy^{(N)} \,|\zeta_n = (x^{(1)}, \ldots, x^{(N)})\right)$$
$$= \prod_{i=1}^{N} \sum_{p=1}^{N} \frac{\widehat{g}_n^{(M)}(x^{(p)})}{\sum_{q=1}^{N} \widehat{g}_n^{(M)}(x^{(q)})} \, \delta_{x^{(p)}}(dy^{(i)})$$

- **Mutation:** The mutation transition

$$\tilde{\zeta}_n = \left(\tilde{\zeta}_n^{(1)}, \ldots, \tilde{\zeta}_n^{(N)}\right) \in (E^M)^N \longrightarrow \widehat{\zeta}_n \in E^N$$

is defined by

$$P\left(\widehat{\zeta}_n \in dz \middle| \tilde{\zeta}_n = (y^{(1)}, \ldots, y^{(N)})\right) = \prod_{i=1}^{N} \widehat{K}_n^{(M)}\left(y^{(i)}, dz^i\right)$$

This algorithm has been introduced in [33]. In this work exponential rates of convergence and $\mathbb{L}^p$-mean error bounds are discussed. The LDP associated with such branching strategy and comparisons with the rates presented in section 2.2.5 are described in [35].

2.3.3 Branching Selections

Roughly speaking, the selection transition is intended to improve the quality of the system by given individuals of "higher quality" to be copied into the next generation. In other words, selection focuses the evolution of the system on promising regions in the state space by allocating reproductive opportunities in such a way that those particles which have a higher fitness are given more chances to give an offspring than those which have a poorer fitness. They are number of ways to approximate the updating transitions but they are all based on the same natural idea. Namely, how to approximate an updated empirical measure of the following form

$$\Psi_n\left(\frac{1}{N}\sum_{i=1}^{N} \delta_{x^i}\right) = \sum_{i=1}^{N} \frac{g_n(x^i)}{\sum_{j=1}^{N} g_n(x^j)} \, \delta_{x^i} \tag{82}$$

by a new probability measure with atoms of size integers multiples of $1/N$?

In the generic IPS approximating model (13) this approximation is done by sampling N-independent random variables $\{\widehat{x}^i \,;\, 1 \leq i \leq N\}$ with common law (82) and the corresponding approximating measure is given by

$$\frac{1}{N}\sum_{i=1}^{N} \delta_{\widehat{x}^i} = \sum_{i=1}^{N} \frac{M^i}{N} \, \delta_{x^i}$$

where

$$(M^1,\ldots,M^N) \stackrel{\text{def.}}{=} \text{Multinomial}\left(N, \frac{g_n(x^1)}{\sum_{j=1}^N g_n(x^j)},\ldots,\frac{g_n(x^N)}{\sum_{j=1}^N g_n(x^j)}\right)$$

Using these notations the random and $\mathbb{N}$-valued random variables can be regarded as random number of offsprings created at the positions $(x^1,\ldots,x^N)$.

The above question is strongly related to weighted bootstrap and genetic algorithms theory (see for instance [10] and references therein). In this connection the above multinomial approximating strategy can be viewed as a weighted Efron bootstrap.

It is well known that sampling according to a multinomial may be "time consuming" mainly because it requires a sorting of the population. As in classical bootstrap literature the other idea consists in using independent random numbers $(M^1,\ldots,M^N)$ distributed according a suitably chosen branching law. In what follows we present an abstract BIPS approximating model which enables a unified description of several classes of branching laws that can be used in practice including Bernoulli, binomial and Poisson distributions.

Abstract BIPS Model

The abstract BIPS model will be a two step Markov chain

$$(N_n,\xi_n) \xrightarrow{\textit{Branching}} (\widehat{N}_n,\widehat{\xi}_n) \xrightarrow{\textit{Mutation}} (N_{n+1},\xi_{n+1}) \tag{83}$$

with product state space

$$\mathcal{E} = \bigcup_{\alpha\in\mathbb{N}} (\{\alpha\}\times E^\alpha)$$

with the convention $E^\alpha = \{\triangle\}$ a cemetery if $\alpha = 0$. We will note

$$\mathcal{F} = \{F_n, \widehat{F}_n \; : \; n \geq 0\}$$

the canonical filtration associated to (83) so that

$$F_n \subset \widehat{F}_n \subset F_{n+1}$$

The points of the set E^α, $\alpha \geq 0$ are called particle systems and are mostly denoted by the letters x and z. The parameter $\alpha \in \mathbb{N}$ represents the size of the system. The initial number of particles $N_0 \in \mathbb{N}$ is a fixed non-random number which represents the precision parameter of the BIPS algorithm.

The evolution in time of the BIPS is defined inductively as follows.

- At the time $n = 0$:
  The initial particle system $\xi_0 = (\xi_0^1,\ldots,\xi_0^{N_0})$ consists in N_0 independent and identically distributed particles with common law ε_0.

- Evolution in time:
  At the time n, the particle system ξ_n consists in N_n particles.
  If $N_n = 0$ the particle system died and we let $\widehat{N}_n = 0$ and $N_{n+1} = 0$.
  Otherwise the branching correction is defined as follows

1. **Branching Correction:**
   When $N_n > 0$ we associate to $\xi_n = (\xi_n^1, \ldots, \xi_n^{N_n}) \in E^{N_n}$ the weight vector $W_n = (W_n^1, \ldots, W_n^{N_n}) \in \mathbb{R}^{N_n}$ given by

$$\sum_{i=1}^{N_n} W_n^i \, \delta_{\xi_n^i} = \Psi_{n+1}(m(\xi_n)) \qquad \text{where} \quad m(\xi_n) = \frac{1}{N_n} \sum_{i=1}^{N_n} \delta_{\xi_n^i}$$

   Then, each particle ξ_n^i, $1 \leq i \leq N_n$, branches into a random number of offsprings M_n^i, $1 \leq i \leq N_n$ and the mechanism is chosen so that for any $f \in \mathcal{B}_b(E)$

$$\mathbb{E}\left(\sum_{i=1}^{N_n} M_n^i f(\xi_n^i) \, | F_n \right) = N_n \Psi_{n+1}(m(\xi_n)) f \tag{84}$$

   and there exists a finite constant $C < \infty$ so that

$$\mathbb{E}\left(\left[\sum_{i=1}^{N_n} M_n^i f(\xi_n^i) - N_n \Psi_{n+1}(m(\xi_n)) f \right]^2 |F_n \right) \leq C \, N_n \|f\|^2 \tag{85}$$

   At the end of this stage the particle system $\widehat{\xi}_n$ consists in $\widehat{N}_n = \sum_{i=1}^{N_n} M_n^i$ particles denoted by

$$\widehat{\xi}_n^i = \xi_n^k \qquad 1 \leq k \leq N_n \qquad \sum_{l=1}^{k-1} M_n^l + 1 \leq i \leq \sum_{l=1}^{k-1} M_n^l + M_n^k \tag{86}$$

2. **Mutation transition:**
   If $\widehat{N}_n = 0$ the particle system dies and $N_{n+1} = 0$.
   Otherwise, each particle moves independently of each other starting off from the parent particle branching site $\widehat{\xi}_n^i$ with law $K_{n+1}(\widehat{\xi}_n^i, dx)$, $1 \leq i \leq \widehat{N}_n$, for any $1 \leq i \leq \widehat{N}_n$. During this transition the total number of particles doesn't change ($N_{n+1} = \widehat{N}_n$) and the mechanism can be summarized as follows, for any $\alpha \geq 0$ and $z \in E^\alpha$

$$P\left(\xi_{n+1} \in dx | \widehat{\xi}_n = z, \widehat{N}_n = \alpha \right) = \prod_{i=1}^{\alpha} K_{n+1}(z^i, dx^i)$$

   where $dx = dx^1 \times \ldots \times dx^\alpha$ is an infinitesimal neighborhood of $x \in E^\alpha$ with the conventions $dx = \{\triangle\}$ and $\prod_{i=1}^{\alpha} = 1$ if $\alpha = 0$.

The approximation of the flow of distributions $\{\eta_n \, ; \, n \geq 0\}$ by the particle density profiles

$$\eta_n^N \stackrel{\text{def.}}{=} \frac{1}{N_0} \sum_{i=1}^{N_n} \delta_{\xi_n^i}$$

is guaranteed by the following theorem.

Theorem 2.36 *If the branching selection law satisfy (84) and (85) then, the total mass process $N = (N_n)_{n\geq 0}$ is a non-negative integer valued martingale with respect to the filtration $F = (F_n)_{n\geq 0}$ with the following properties*

$$\forall\, n \geq 0, \qquad \mathbb{E}\left(\sup_{0\leq k\leq n}\left(\frac{N_k}{N_0}-1\right)^2\right) \leq \frac{C\,n}{N_0} \qquad \text{and} \qquad P(N_n = 0) \leq \frac{C\,n}{N_0} \tag{87}$$

In addition, for any $n \geq 0$ and $f \in \mathcal{B}_b(E)$, $\|f\| \leq 1$ we have that

$$\mathbb{E}\left[\left(\eta_n^N(f) - \eta_n(f)\right)^2\right] \leq \frac{B_n}{N_0}$$

for some finite constant B_n which only depends on the time parameter n.

Branching Selections

Here we present several examples of branching laws which satisfy conditions (84) and (85). The first one is known as the *Remainder Stochastic Sampling* in genetic algorithms literature. It has been presented for the first time in [6, 7]. From a pure practical point of view this sampling technique seems to be the more efficient since it is extremely time saving and if the BIPS model is only based on this branching selection scheme then the size of the system remains constant.

1) Remainder Stochastic Sampling

In what follows we denote by $[a]$ (resp. $\{a\} = a - [a]$) the integer part (resp. the fractional part) of $a \in \mathbb{R}$.
At each time $n \geq 0$, each particle ξ_n^i branches directly into a fixed number of offsprings

$$\forall\, 1 \leq i \leq N_n, \qquad \overline{M}_n^i \;\; df \;\; [N_n W_n^i]$$

so that the intermediate population consists in $\overline{N}_n \stackrel{\text{def.}}{=} \sum_{i=1}^{N_n} \overline{M}_n^i$ particles. To prevent extinction and to keep the size of the system fixed it is convenient to introduce in this population $\tilde{N}_n$ additional particles with

$$\tilde{N}_n \stackrel{\text{def.}}{=} N_n - \overline{N}_n = \sum_{i=1}^{N_n}\{N_n W_n^i\}$$

One natural way to do this is to introduce the additional sequence of branching numbers

$$\left(\tilde{M}_n^1, \ldots, \tilde{M}_n^{N_n}\right) \stackrel{\text{def.}}{=} \text{Multinomial}\left(\tilde{N}_n, \frac{\{N_n W_n^1\}}{\sum_{j=1}^{N_n}\{N_n W_n^j\}}, \ldots, \frac{\{N_n W_n^{N_n}\}}{\sum_{j=1}^{N_n}\{N_n W_n^j\}}\right) \tag{88}$$

More precisely, if each particle ξ_n^i again produces a number of $\tilde{M}_n^i$ additional offsprings, $1 \leq i \leq N_n$, then the total size of the system is kept constant.

At the end of this stage, the particle system $\widehat{\xi}_n$ again consists in N_n particles denoted by

$$\begin{array}{rclll} \widehat{\xi}_n^i & = & \xi_n^k & 1 \leq k \leq N_n & \sum_{l=1}^{k-1} \overline{M}_n^l + 1 \leq i \leq \sum_{l=1}^{k-1} \overline{M}_n^l + \overline{M}_n^k \\ \widehat{\xi}_n^{\overline{N}_n+i} & = & \xi_n^k & 1 \leq k \leq N_n & \sum_{l=1}^{k-1} \tilde{M}_n^l + 1 \leq i \leq \sum_{l=1}^{k-1} \tilde{M}_n^l + \tilde{M}_n^k \end{array}$$

The multinomial random numbers (88) can also be defined as follows

$$\tilde{M}_n^k = \mathrm{Card}\left\{1 \leq j \leq \tilde{N}_n \ ; \ \tilde{\xi}_n^j = \xi^k\right\} \qquad 1 \leq k \leq N_n$$

where $(\tilde{\xi}_n^1, \ldots, \tilde{\xi}_n^{\tilde{N}_n})$ are $\tilde{N}_n$ independent random variables with common law

$$\sum_{i=1}^{N_n} \frac{\{N_n W_n^i\}}{\sum_{j=1}^{N_n} \{N_n W_n^j\}} \, \delta_{\xi_n^i}$$

It is easily checked that (84) is satisfied and (85) holds for $C = 1$.

Let us now present some classical examples of independent branching numbers that satisfy the non bias condition (84) and the $\mathbb{L}^2$-condition (85).

2) Bernoulli branching numbers:

The Bernoulli branching numbers were introduced in [23, 21] and further developed in [25].

They are defined as a sequence $M_n = (M_n^i,\ 1 \leq i \leq N_n)$ of conditionally independent random numbers with respect to F_n with distribution given for any $1 \leq i \leq N_n$ by

$$P(M_n^i = k | F_n) = \begin{cases} \{N_n W_n^i\} & \text{if } k = [N_n W_n^i] + 1 \\ 1 - \{N_n W_n^i\} & \text{if } k = [N_n W_n^i] \end{cases}$$

In addition it can be seen from the relation $\sum_{i=1}^{N_n} (N_n W_n^i) = N_n$ that at least one particle has one offspring (cf. [23] for more details). Therefore using the above branching correction the particle system never dies.

It is also worth observing that the Bernoulli branching numbers are defined as in the *Remainder Stochastic Sampling* by replacing the multinomial remainder branching law (88) by a sequence of N_n independent Bernoulli random variables $\left(\tilde{M}_n^1, \ldots, \tilde{M}_n^{N_n}\right)$ given by

$$P(\tilde{M}_i^{N_n} = 1 | F_n) = 1 - P(\tilde{M}_i^{N_n} = 0 | F_n) = \{N_n W_n^i\}$$

3) Poisson branching numbers:

The Poisson branching numbers are defined as a sequence $M_n = (M_n^i,\ 1 \le i \le N_n)$ of conditionally independent random numbers with respect to F_n with distribution given for any $1 \le i \le N_n$ by

$$\forall\, k \ge 0, \qquad P(M_n^i = k | F_n) \;=\; \exp\left(-N_n W_n^i\right)\ \frac{(N_n W_n^i)^k}{k!}$$

4) Binomial branching numbers:

These numbers are defined as a sequence $M_n = (M_n^i,\ 1 \le i \le N_n)$ of conditionally independent random numbers with respect to F_n with distribution given for any $1 \le i \le N_n$ by

$$\forall\, 0 \le k \le N_n, \qquad P(M_n^i = k | F_n) \;=\; \binom{N_n}{k}\ (W_n^i)^k\ (1 - W_n^i)^{N_n - k}$$

All of these models are described in full details in [25]. In particular it is shown that the BIPS with multinomial branching laws arises by conditioning a BIPS with Poisson branching laws. The law of large numbers and large deviations for the BIPS model with Bernoulli branching laws are studied in [25] and [22].

The convergence analysis of these BIPS approximating schemes is still in progress. They are many open problems such as that of finding the analog of the Donsker and Glivenko-Cantelli Theorems as well as the study of their long time behavior. This last question is maybe the most important one. The main difficulty here is that the total size process $\{N_n\ ;\ n \ge 0\}$ is an F-martingale with predictable quadratic variation

$$A_n = N_0^2 + \sum_{p=1}^{n} \mathbb{E}\left(|N_p - N_{p-1}|^2 / F_{p-1}\right) = N_0^2 + \sum_{p=0}^{n-1} \sum_{i=1}^{N_p} \mathbb{E}\left((M_p^i - N_p W_p^i)^2 / F_p\right)$$

and therefore a uniform convergence result presented in section 2.2.2 will take place only if

$$\sup_{n \ge 0} E\left(A_n^2\right) = \sum_{p=1}^{\infty} \mathbb{E}\left(|N_p - N_{p-1}|^2\right) < \infty$$

The following simple example shows that even for the minimum variance Bernoulli branching law one cannot expects to obtain the analog of the uniform convergence results as those presented in section 2.2.2. Let us assume that the state space $E = \{0,1\}$, the fitness functions $\{g_n\ ;\ n \ge 1\}$ and the transition kernels $\{K_n\ ;\ n \ge 1\}$ are time homogeneous and given by

$$g(1) = 3g(0) > 0 \qquad K(x, dz) = \nu(dz) \stackrel{\text{def}}{=} \frac{1}{2}\delta_0(dz) + \frac{1}{2}\delta_1(dz)$$

In this case one can check that for sufficiently large N_0

$$\mathbb{E}\left((\eta_n^N(1) - \eta_n(1))^2\right) \ge \frac{n}{5N_0} \xrightarrow[n \to \infty]{} \infty$$

In contrast to the situation described above in this simple case the IPS approximating model (13) will simply consist at each time in N_0 i.i.d. particles with common law ν and

$$\forall\, n \geq 0, \qquad \mathbb{E}\left((\eta_n^N(f) - \eta_n(f))^2\right) \;\leq\; \frac{1}{N_0}$$

for any bounded test function such that $\|f\| \leq 1$.

2.3.4 Regularized Mutations

The regularization of the mutation transition discussed in this section has been presented in [59]. Hereafter we briefly indicate why it is sometimes necessary to add a regularization step and how the previous analysis applies to this situation.

In nonlinear filtering settings the mutation transitions $\{K_n\ ;\ n \geq 1\}$ are given by the problem at hand. More precisely they are the transitions of the un-known signal process. If $E = \mathbb{R}^d$, it may happen that some coordinates of the signal are deterministic. For instance if the signal is purely deterministic the IPS approximating scheme (13) does not work since after a finite number of steps we would get a single deterministic path.

As noticed in [41] the standard regularization technique used in practice consists in adding a "small noise" in the deterministic parts of the signal. When $E = \mathbb{R}^d$, $d \geq 1$, the introduction of such a "small noise" in the signal dynamics structure consists in replacing the transitions $\{K_n\ ;\ n \geq 1\}$ by the regularized ones

$$K_n^{(\alpha)} = R^{(\alpha)} K_n \qquad \text{or} \qquad K_n^{(\alpha)} = K_n R^{(\alpha)}$$

where $R^{(\alpha)}$ is a new transition probability kernel on E and defined for any $f \in \mathcal{B}_b(E)$ by

$$R^{(\alpha)}(f)(x) \stackrel{\text{def.}}{=} \int \frac{1}{\alpha^d}\,\theta\left(\frac{y-x}{\alpha}\right)\,f(y)\,dy$$

where $\theta : \mathbb{R}^d \to (0,\infty)$ is a Borel bounded function such that

$$\int \theta(y)\,dy = 1 \qquad \text{and} \qquad \sigma \stackrel{\text{def.}}{=} \int |y|\,\theta(y)\,dy < \infty$$

The regularized limiting measure valued system is now defined by

$$\eta_n^{(\alpha)} = \Phi_n^{(\alpha)}\left(\eta_{n-1}^{(\alpha)}\right) \qquad n \geq 1 \tag{89}$$

with $\eta_0^{(\alpha)} = \eta_0$ and where $\Phi_n^{(\alpha)} : \mathbf{M}_1(E) \to \mathbf{M}_1(E)$ is defined as in (8) by replacing the transitions $\{K_n\ ;\ n \geq 1\}$ by $\{K_n^{(\alpha)}\ ;\ n \geq 1\}$.

Let us denote by

$$\xi_n^{(\alpha)} = \left(\xi_n^{(\alpha),1}, \ldots, \xi_n^{(\alpha),N}\right)$$

the N-IPS approximating model associated with (89) and by $\eta_n^{(\alpha),N}$ the empirical measure associated with the system $\xi_n^{(\alpha)}$ and defined as usual by

$$\eta_n^{(\alpha),N} = \frac{1}{N} \sum_{i=1}^{N} \delta_{\xi_n^{(\alpha),i}}$$

It is transparent from the above construction that the convergence results presented in section 2.2.2 can be applied. For instance for any bounded test function $f \in \mathcal{B}_b(E)$, $\|f\| \leq 1$, and $p \geq 1$ and $n \geq 0$

$$\mathbb{E}\left(\left|\eta_n^{(\alpha),N}(f) - \eta_n^{(\alpha)}(f)\right|^p\right)^{\frac{1}{p}} \leq \frac{B_p}{\sqrt{N}}\,(n+1)\,a_{0,n}^2$$

where B_p is a universal constant which only depends on the parameter p.
Of course we still have to check that the flow of distributions $\{\eta_n^{(\alpha)}\ ;\ n \geq 1\}$ is close to the desired flow $\{\eta_n\ ;\ n \geq 1\}$ as α is close to zero. To this end we introduce some additional notations.

We denote by Lip_1 the set of globally Lipschitz functions with Lipschitz norm less than 1 that is

$$|f(x) - f(y)| \leq |x - y|$$

and $\|f\| \leq 1$. We will also use the following assumption

($\mathcal{R}$) ***For any time*** $n \geq 1$***, there exist some constants*** $C_n^1, C_n^2 < \infty$ ***such that***

$$g_n \in C_n^1 \cdot \mathrm{Lip}_1 \qquad \text{and} \qquad K_n(\mathrm{Lip}_1) \subset C_n^2 \cdot \mathrm{Lip}_1.$$

Lemma 2.37 *Under* ($\mathcal{R}$)*, for any* $n \geq 1$ *there exists some constant* $C_n < \infty$ *such that for any* $f \in \mathrm{Lip}_1$

$$\left|\eta_n^{(\alpha)}(f) - \eta_n(f)\right| \leq C_n\,\alpha\,\sigma \tag{90}$$

Proof: For any $f \in \mathrm{Lip}_1$ and $x \in \mathbb{R}^d$ we clearly have

$$\left|R^{(\alpha)}(f)(x) - f(x)\right| \leq \int |f(x + \alpha\, y) - f(x)|\ \theta(y)\, dy \leq \alpha\ \sigma$$

and therefore

$$\sup_{f \in \mathrm{Lip}_1} \left\|R^{(\alpha)}(f) - f\right\| \leq \alpha\ \sigma$$

Under our assumptions this implies that for any $\eta \in \mathbf{M}_1(E)$

$$\sup_{f \in \mathrm{Lip}_1} \left|\Phi_n^{(\alpha)}(\eta)(f) - \Phi_n(\eta)(f)\right| \leq \alpha\ \sigma\ A_n$$

for some constant $A_n < \infty$. Let us prove (90) by induction on the parameter n. For $n = 0$ the result is trivial with $C_0 = 0$ so we assume that it holds for $(n-1)$. Using Lemma 2.2 we have the decomposition

$$\begin{aligned}
&\Phi_n(\eta_{n-1}^{(\alpha)})(f) - \Phi_n(\eta_{n-1})(f) \\
&= \frac{1}{\eta_{n-1}(g_n)} \\
&\quad \times \left[(\eta_{n-1}^{(\alpha)}(g_n K_n(f)) - \eta_{n-1}(g_n K_n(f))) + \Phi_n(\eta_{n-1}^{(\alpha)})(f)\ (\eta_{n-1}(g_n) - \eta_{n-1}^{(\alpha)}(g_n))\right]
\end{aligned}$$

for any $f \in \mathrm{Lip}_1$. There is no loss of generality to assume that $C_n^{(1)} \geq a_n$ and that $C_n^{(2)} \geq 1$. Thus one gets

$$\begin{aligned}
&\Phi_n(\eta_{n-1}^{(\alpha)})(f) - \Phi_n(\eta_{n-1})(f) \\
&= \frac{C_n^{(1)} + a_n C_n^{(2)}}{\eta_{n-1}(g_n)} \left[(\eta_{n-1}^{(\alpha)}(f_1) - \eta_{n-1}(f_1))\right] \\
&\qquad + \Phi_n(\eta_{n-1}^{(\alpha)})(f)\ \frac{C_n^{(1)}}{\eta_{n-1}(g_n)} \left[(\eta_{n-1}(f_2) - \eta_{n-1}^{(\alpha)}(f_2))\right]
\end{aligned}$$

with

$$f_1 = \frac{1}{C_n^{(1)} + a_n C_n^{(2)}}\ g_n K_n(f) \qquad \text{and} \qquad f_2 = \frac{1}{C_n^{(1)}}\ g_n$$

so that $f_1, f_2 \in \mathrm{Lip}_1$. Using the induction hypothesis we arrive at

$$\sup_{f \in \mathrm{Lip}_1} \left\| \Phi_n(\eta_{n-1}^{(\alpha)})(f) - \Phi_n(\eta_{n-1})(f) \right\| \leq \frac{2(C_n^{(1)} + a_n C_n^{(2)})}{\eta_{n-1}(g_n)}\ C_{n-1}\ \sigma\ \alpha$$

Therefore if we combine the above results one gets finally

$$\sup_{f \in \mathrm{Lip}_1} |\eta_n^{(\alpha)}(f) - \eta_n(f)| \ \leq\ C_n\ \sigma\ \alpha$$

with

$$C_n = A_n + \frac{2(C_n^{(1)} + a_n C_n^{(2)})}{\eta_{n-1}(g_n)}\ C_{n-1}.$$

■

A direct consequence of the above lemma is the following estimate

$$\mathbb{E}\left(|\eta_n^{(\alpha),N}(f) - \eta_n(f)|^p\right)^{\frac{1}{p}} \leq \frac{B_p}{\sqrt{N}}\ (n+1)\ a_{0,n}^2 + C_n\ \sigma\ \alpha$$

The approximation of $\{\eta_n\ ;\ n \geq 0\}$ by the regularized IPS is now guaranteed by the following proposition.

Proposition 2.38 *Assume that the condition $(\mathcal{R})$ is satisfied. Then for any $n \geq 0$ and $p \geq 1$ there exists some constant $C_{p,n} < \infty$ such that*

$$\alpha(N) = 1/\sqrt{N} \Longrightarrow \sup_{f \in \mathrm{Lip}_1} \mathbb{E}\left(|\eta_n^{(\alpha),N}(f) - \eta_n(f)|^p\right)^{\frac{1}{p}} \leq \frac{C_{p,n}}{\sqrt{N}}$$

3 The Continuous Time Case

We will try here to retranscribe some results obtained in previous parts for discrete time settings to continuous time models. Generally speaking, the same behaviors are expected, but the technicalities are more involved, even only for the definitions. Furthermore, this case has been less thoroughly investigated, and it seems that a genuine continuous time interacting particle system approximation has only recently been introduced in [45].

This last paper will be our reference for this part and we will keep working in the same spirit, but in the details our approach will be different in order to obtain numerous improvements and to prove new results: weak propagation of chaos valid for all bounded measurable functions and related upper bounds in terms of the supremum norm, as well as uniform convergence results with respect to time and central limit theorem and exponential upper bounds for the fluctuations.

Heuristically the main difference between the two time models is that for discrete time, in the selection step all the particles are allowed to change, whereas for continuous time, only one particle may change at a (random) selection time, but the length of the interval between two selection times is of order $1/N$ (N being as above the number of particles). So in some mean sense, in one unit time interval, "every particle should have had a good chance to change".
This is a first weak link between discrete and continuous time. But, even if this may not be clear at first reading, there are stronger relations between the formal treatment of the discrete and continuous times, and in our way to trying to understand the subject, they have influenced each other. In order to point out these connections, we have tried to use the same notations in both set-ups.

To finish this opening introduction, we must precise that the main results obtained here (the weak propagation of chaos and to some extend the central limit theorem) can also be deduced from the approach of Graham and Méléard [65], valid for a more general set-up, except they have put more restrictions on the state space, which is assumed to be $\mathbb{R}^d$, for some $d \geq 1$ (but perhaps this point is unessential).
Nevertheless, we have preferred to introduce another method, may be more immediate (e.g. without any reference to Boltzmann trees or Sobolev imbeddings ...), because we have taken into account that our models are simpler, since we are not in the need of considering the broader situation of [65].
More precisely, in our case, we have a nice a priori explicit expression (3) and (4) for the (deterministic) limiting objects, making them appear as a ratio of linear terms with respect to η_0, which is hidden in $\mathbb{E}$ as the initial distribution. This structure is more tractable than the information one would get by merely considering the nonlinear equation of evolution looking like (11) satisfied by the family $(\eta_t)_{t\geq 0}$. So we can make use of some associated nonnegative Feynman-Kac semigroups to obtain without difficulty the desired results.

3.1 Hypotheses on the Limiting Process

To accomplish the program described above it is first convenient to define more precisely the objects arising in formula (4). Our next objective is to introduce several kinds of assumptions needed in the sequel and to prove some preliminary results. As before, the metric space (E,r) is supposed to be Polish, and we denote by $\mathbf{B}(E)$ the σ-field of Borel subsets of E.

3.1.1 Definitions and Weak Regularity Assumption

Here we introduce the basic definitions and the weak hypothesis under which we will work. It is already largely weaker than the one considered in [45], and we believe that maybe it can even be removed (we have been making some recent progress in this direction, but at the expense of readability, considering tensorized empirical measures ...), but at least this weak assumption make clear the regularity problem one is to encounter when following the approach presented in this paper.

The simplest object to be explained is the family $(U_t)_{t\geq 0}$ of non-negative functions. We will assume that the mapping

$$U \,:\, \mathbb{R}_+ \times E \ni (t,x) \;\mapsto\; U_t(x) \in \mathbb{R}_+$$

is $\mathbf{B}(\mathbb{R}_+)\otimes\mathbf{B}(E)$-measurable (where $\mathbf{B}(\mathbb{R}_+)$ is the usual σ-field of the Borel subsets of $\mathbb{R}_+$) and locally bounded in the sense that for all $T>0$, its restriction to $[0,T]\times E$ is bounded.

Next we need to define the E-valued time-inhomogeneous Markov process X arising in the right hand side of (4). In our settings the more efficient and convenient way seems to be in terms of a martingale problem (cf [55] for a general reference), since our method will be largely based on properties of martingales (as it was already true for discrete time, so the set-up we are now presenting is quite a natural generalization of the previous one):

For $t \geq 0$, let $D([t,+\infty[,E)$ be the set of all càdlàg paths from $[t,+\infty[$ to E. We denote by $(X_s)_{s\geq t}$ the process of canonical coordinates on $D([t,+\infty[,E)$, which generates on this space the σ-algebra $\mathcal{D}_{t,+\infty}=\sigma(X_s\,:\,s\geq t)$. We will also use as customary the notation $\mathcal{D}_{t,s}=\sigma(X_u\,:\,t\leq u\leq s)$, for $0\leq t\leq s$.

Let $\mathcal{A}$ be a dense sub-algebra of $\mathcal{C}_b(E)$ which is supposed to contain $\mathbf{I}$ (that is the function taking everywhere the value 1).

A linear operator L_0 from the domain $\mathcal{A}$ to $\mathcal{C}_b(E)$ will be called a pregenerator, if for all $x\in E$, there exists a probability $\mathbb{P}_x$ on $(D([0,+\infty[,E),\mathcal{D}_{0,+\infty})$ such that

- $X_0 \circ \mathbb{P}_x = \delta_x$, the Dirac mass in x, and
- for all $\varphi\in\mathcal{A}$, the process

$$\left(\varphi(X_s)-\varphi(X_0)-\int_0^s L_0(\varphi)(X_u)\,du\right)_{s\geq 0}$$

is a $(\mathcal{D}_{0,s})_{s\geq 0}$-martingale under $\mathbb{P}_x$.

Let $(L_t)_{t\geq 0}$ be a measurable family of pregenerators: for each $t\geq 0$, $L_t\,:\,\mathcal{A}\to\mathcal{C}_b(E)$ is a pregenerator, and for each $\varphi\in\mathcal{A}$ fixed,

$$\mathbb{R}_+\times E\ni(t,x)\;\mapsto\;L_t(\varphi)(x)$$

is $\mathcal{B}(\mathbb{R}_+) \otimes \mathcal{E}$-measurable. For the sake of simplicity, we will furthermore impose that the above function is locally bounded.

Our first hypothesis is

(H1) ***For all*** $(t,x) \in \mathbb{R}_+ \times E$***, there exists a unique probability*** $\mathbb{P}_{t,x}$ ***on*** $(D([t,+\infty[,E), \mathcal{D}_{t,+\infty})$ ***such that***

- $X_t \circ \mathbb{P}_{t,x} = \delta_x$***, and***
- ***for all*** $\varphi \in \mathcal{A}$***, the process***

$$\left(\varphi(X_s) - \varphi(X_t) - \int_t^s L_u(\varphi)(X_u)\, du\right)_{s \geq t}$$

is a $(\mathcal{D}_{t,s})_{s\geq t}$***-martingale under*** $\mathbb{P}_{t,x}$***.***
Furthermore, it is assumed that for all $A \in \mathcal{D}_{0,+\infty}$***, the mapping***

$$\mathbb{R}_+ \times E \ni (t,x) \quad \mapsto \quad \mathbb{P}_{t,x}(\theta_t(A))$$

is measurable (where $(\theta_t)_{t\geq 0}$ ***denotes the family of usual time shifts on*** $D([0,+\infty[,E)$***).***

Combining the measurability and the uniqueness assumption (H1), one can check that

$$\left((X_t)_{t\geq 0}, (\mathbb{P}_{t,x})_{(t,x)\in\mathbb{R}_+\times E}\right) \tag{91}$$

is a strong Markov process (this uniqueness condition will only be needed here, so it can be removed if we rather suppose that the previous object is a Markov process). One can also define a probability $\mathbb{P}_{\eta_0}$ on $(D([0,+\infty[,E), \mathcal{D}_{0,+\infty})$, for any distribution $\eta_0 \in M_1(E)$, by

$$\forall\, A \in \mathcal{D}_{0,+\infty}, \qquad \mathbb{P}_{\eta_0}(A) \;=\; \int_E \mathbb{P}_{0,x}(A)\,\eta_0(dx)$$

which is easily seen to be the unique solution to the martingale problem associated to $(L_t)_{t\geq 0}$ whose initial law is η_0 (from now on, $\mathbb{E}_{\eta_0}$ will stand for the expectation relative to $\mathbb{P}_{\eta_0}$, the probability $\eta_0 \in \mathbf{M}_1(E)$ being fixed).

The previous martingale problem can be extended to a time-space version as follows:

Let $\mathbf{A}$ be the set of absolutely continuous functions $g : \mathbb{R}_+ \to \mathbb{R}$ with bounded derivative in the sense that there exists a bounded measurable function $g' : \mathbb{R}_+ \to \mathbb{R}$, such that for all $t \geq 0$,

$$g(t) \;=\; g(0) + \int_0^t g'(s)\, ds$$

On $\mathbf{A} \otimes \mathcal{A}$, we define the operator L given on functions of the form $f = g \otimes \varphi$, with $g \in \mathbf{A}$ and $\varphi \in \mathcal{A}$, by

$$\forall\, t \geq 0, \forall\, x \in E, \qquad L(f)(t,x) \;=\; g'(t)\varphi(x) + g(t)L_t(\varphi)(x) \tag{92}$$

Then we have the standard result:

Lemma 3.1 *Let $(t,x) \in \mathbb{R}_+ \times E$ be fixed. Under $\mathbb{P}_{t,x}$, for each $f \in \mathbf{A} \otimes \mathcal{A}$, the process $(M_s(f))_{s \geq t}$ defined by*

$$\forall\, s \geq t, \qquad M_s(f) \;=\; f(s,X_s) - f(t,X_t) - \int_t^s L(f)(u,X_u)\, du$$

is a square integrable martingale and its increasing process has the form

$$\forall\, s \geq t, \qquad \langle M(f) \rangle_s \;=\; \int_t^s \Gamma(f,f)(u,X_u)\, du$$

where Γ is the "carré du champ" bilinear operator associated to the pregenerator L and defined by

$$\forall\, f,g \in \mathbf{A} \otimes \mathcal{A}, \qquad \Gamma(f,g) \;=\; L(fg) - f\, L(g) - g\, L(f) \tag{93}$$

We can consider, for $s \geq 0$, the "carré du champ" bilinear operator Γ_s associated to the pregenerator L_s, which is naturally defined by

$$\forall\, \phi, \varphi \in \mathcal{A}, \qquad \Gamma_s(\phi,\varphi) \;=\; L_s(\phi\varphi) - \phi\, L_s(\varphi) - \varphi\, L_s(\phi)$$

We easily see that for all $f,g \in \mathbf{A} \otimes \mathcal{A}$,

$$\forall\, (s,x) \in \mathbb{R}_+ \times E, \qquad \Gamma(f,g)(s,x) \;=\; \Gamma_s\left(f(s,.), g(s,.)\right)(x)$$

In Lemma 3.1, the fact that $\mathbf{A} \otimes \mathcal{A}$ is an algebra is crucial in order to describe the increasing process. But the domain $\mathbf{A} \otimes \mathcal{A}$ is rather too small for our purposes. We extend it in the following way: for $T > 0$ fixed, we denote by $\mathcal{B}_b([0,T] \times E)$ the set of all measurable bounded functions $f : [0,T] \times E \to \mathbb{R}$ and by $\mathcal{B}_T$ the vector space of applications $f \in \mathcal{B}_b([0,T] \times E)$ for which there exists a function $\bar{L}(f) \in \mathcal{B}_b([0,T] \times E)$ such that the process $(M_t(f))_{0 \leq t \leq T}$ defined by

$$\forall\, 0 \leq t \leq T, \qquad M_t(f) \;=\; f(t,X_t) - f(0,X_0) - \int_0^t \bar{L}(f)(s,X_s)\, ds$$

is a $\mathbb{P}_{\eta_0}$-martingale.

In this article and unless otherwise stated, all martingales will be implicitly assumed to be càdlàg (a.s.). Note that those coming from (H1) or from Lemma 3.1 are automatically càdlàg. Let us furthermore introduce $\mathcal{A}_T$ the subset of function $f \in \mathcal{B}_T$ for which there exists a non-negative mapping $\bar{\Gamma}(f,f) \in \mathcal{B}_b([0,T] \times E)$ such that the increasing process associated with $(M_t(f))_{0 \leq t \leq T}$ has the form

$$\forall\, 0 \leq t \leq T, \qquad \langle M(f) \rangle_t \;=\; \int_0^t \bar{\Gamma}(f,f)(s,X_s)\, ds$$

Remark that $\bar{L}(f)$ and $\bar{\Gamma}(f,f)$ may not be uniquely defined by $f \in \mathcal{A}_T$ (but we will keep abusing of these notations), nevertheless this is not really important, since for martingale problems, one can consider multi-valued operators, cf [55].

Next we will need some regularity conditions on the function U and the family of probability measures $(\mathbb{P}_{t,x})_{(t,x)\in\mathbb{R}_+\times E}$. These are expressed in the following assumption (in the subsection 3.1.2, we will give separate and stronger hypotheses on U and $(\mathbb{P}_{t,x})_{(t,x)\in\mathbb{R}_+\times E}$ which insure that (H2) is satisfied):

(H2) ***For all*** $T > 0$ ***and*** $\varphi \in \mathcal{A}$ ***fixed, the application***

$$F_{T,\varphi} : [0,T]\times E \ni (t,x) \ \mapsto \ \mathbb{E}_{t,x}\left[\exp\left(\int_t^T U_s(X_s)\,ds\right)\varphi(X_T)\right] \tag{94}$$

belongs to $\mathcal{A}_T$.

We notice that $F_{T,\varphi}$ satisfies almost surely the first assumption hidden in (H2), that is $F_{T,\varphi} \in \mathcal{B}_T$. To see this claim let us denote for any $T > 0$ and $\varphi \in \mathcal{A}$ fixed,

$$\forall\, 0 \le t \le T, \qquad N_t(T,\varphi) \ = \ F_{T,\varphi}(t, X_t)$$

The Markov property of X implies that the process $(\widetilde{M}_t)_{0\le t\le T}$ defined by

$$\begin{aligned}\forall\, 0 \le t \le T, \qquad \widetilde{M}_t \ &= \ \exp\left(\int_0^t U_s(X_s)\,ds\right) N_t(T,\varphi) \\ &= \ \mathbb{E}_{\eta_0}\left[\exp\left(\int_0^T U_s(X_s)\,ds\right)\varphi(X_T)\middle| \mathcal{D}_{0,t}\right]\end{aligned}$$

is a martingale. So we have

$$N_t(T,\varphi) \ = \ \exp\left(-\int_0^t U_s(X_s)\,ds\right)\widetilde{M}_t \tag{95}$$

$$\begin{aligned} &= \ N_0(T,\varphi) + \int_0^t \exp\left(-\int_0^s U_u(X_u)\,du\right)\, d\widetilde{M}_s \\ &\qquad - \int_0^t U_s(X_s) N_s(T,\varphi)\,ds \end{aligned} \tag{96}$$

from which it follows that

$$(\widehat{M}_t)_{0\le t\le T} \ \stackrel{\text{def.}}{=} \ \left(N_t(T,\varphi) - N_0(T,\varphi) + \int_0^t U_s(X_s)N_s(T,\varphi)\,ds\right)_{0\le t\le T}$$

is a martingale. Note that it is not necessarily càdlàg, and this could be annoying for some calculations afterwards (in (95) there was already a little difficulty, nevertheless it is possible to consider there a càdlàg modification, and end up with the fact that $(\widehat{M}_t)_{0\le t\le T}$ is a (contingently non-càdlàg) martingale). If (H2) is fulfilled, a classical uniqueness argument for semimartingale expansion will show that $(\widehat{M}_t)_{0\le t\le T}$ is almost surely càdlàg. Thus, we can take

$$\forall\, 0 \le t \le T, \forall\, x \in E, \qquad \bar{L}(F_{T,\varphi})(t,x) \ = \ -U_t(x)F_{T,\varphi}(t,x)$$

In addition if the mapping $F_{T,\varphi}$ defined in (94) belongs to $\mathbf{A}_T \otimes \mathcal{A}$, where $\mathbf{A}_T$ is the set of restrictions to $[0,T]$ of functions from $\mathbf{A}$, then we conclude that one can choose

$$\bar{\Gamma}(F_{T,\varphi}, F_{T,\varphi}) \ = \ \Gamma(F_{T,\varphi}, F_{T,\varphi})$$

(where the action of Γ on $(\mathbf{A}_T \otimes \mathcal{A})^2$ is defined similarly as the one on $(\mathbf{A} \otimes \mathcal{A})^2$).

But our setting is too weak to insure this property, leading us to make the assumption (H2), which will serve as an ersatz.

One way to check this condition is to follow the procedure given below:

Lemma 3.2 *Let $f \in \mathcal{B}_T$ such that there exists a sequence $(f_n)_{n\geq 0}$ of elements of $\mathbf{A}_T \otimes \mathcal{A}$ satisfying, as n and m tend to infinity, $f_n \rightharpoonup f$ and*

$$\Gamma(f_n - f_m, f_n - f_m) \rightharpoonup 0,$$

where $\rightharpoonup$ stands for the bounded pointwise convergence on $[0,T] \times E$, and

$$\sup_{n\in\mathbb{N},\, 0\leq t\leq T} \|L(f_n)(t,\cdot)\| < +\infty$$

Then we have that $f \in \mathcal{A}_T$.

Proof: By dominated convergence, we obtain that

$$\lim_{n,m\to\infty} \mathbb{E}_{\eta_0}[\langle M(f_n - f_m)\rangle_T] \;=\; 0$$

in other words

$$\lim_{n,m\to\infty} \mathbb{E}_{\eta_0}[(M_T(f_n) - M_T(f_m))^2] \;=\; 0$$

It is now quite standard to deduce from this Cauchy convergence that there exists a martingale $(M_t)_{0\leq t\leq T}$ (that we can choose càdlàg) such that

$$\lim_{n\to\infty} \mathbb{E}_{\eta_0}[\sup_{0\leq t\leq T} (M_t(f_n) - M_t)^2] \;=\; 0$$

Since for any $0 \leq t \leq T$ $f_n(t, X_t) - f_n(0, X_0)$ converges in $\mathbb{L}^2(\mathbb{P}_{\eta_0})$ to $f(t, X_t) - f(0, X_0)$ as $n \to \infty$ one concludes that

$$\int_0^t L(f_n)(s, X_s)\, ds - \int_0^t \bar{L}(f)(s, X_s)\, ds$$

converges in $\mathbb{L}^2(\mathbb{P}_{\eta_0})$ to

$$M_t(f) - M_t \;=\; f(t, X_t) - f(0, X_0) - \int_0^t \bar{L}(f)(s, X_s)\, ds - M_t$$

This shows that the latter process (for $0 \leq t \leq T$) is predictable, as a limit of predictable processes, and we already know that it is also a martingale. Furthermore, from our hypotheses, there exists a finite constant $C_T > 0$, such that for all $n \in \mathbb{N}$ and all $0 \leq t_1 \leq t_2 \leq T$,

$$\left| \int_{t_1}^{t_2} L(f_n)(s, X_s)\, ds - \int_{t_1}^{t_2} \bar{L}(f)(s, X_s)\, ds \right| \;\leq\; C_T(t_2 - t_1)$$

so in the end (considering a subsequence), $(M_t(f) - M_t)_{t\geq 0}$ will also be of bounded variations. It is now well known that up to an evanescent set, it is the null process, that is $(M_t)_{0\leq t\leq T} = (M_t(f))_{0\leq t\leq T}$.

On the other hand since L_s is a pregenerator, for any $s \geq 0$, then Γ_s satisfies

$$\forall\, \varphi \in \mathcal{A}, \forall\, x \in E, \qquad \Gamma_s(\varphi,\varphi)(x) \;\geq\; 0$$

(see (98) below).

This implies that

$$\forall\, n,m \geq 0, \forall\, (s,x) \in [0,T] \times E,$$

$$\left|\sqrt{\Gamma(f_n,f_n)(s,x)} - \sqrt{\Gamma(f_m,f_m)(s,x)}\right| \;\leq\; \sqrt{\Gamma(f_n - f_m, f_n - f_m)(s,x)} \tag{97}$$

Therefore there exists a function $\bar{\Gamma}(f,f) \,:\, [0,T] \times E \to \mathbb{R}_+$ such that

$$\Gamma(f_n,f_n) \rightharpoonup \bar{\Gamma}(f,f)$$

as n tends to infinity.

Again using convergence theorems (but now in $\mathbb{L}^1(\mathbb{P}_{\eta_0})$) one obtains that

$$\left(M_t^2(f) - \int_0^t \bar{\Gamma}(f,f)(s,X_s)\,ds\right)_{0\leq t\leq T}$$

is a martingale. ■

Remark 3.3:

(a) In the hypothesis $f \in \mathcal{B}_T$ of the previous lemma, we don't really need to assume that $(M_t(f))_{0\leq t\leq T}$ is càdlàg, since the equality $(M_t(f))_{0\leq t\leq T} = (M_t)_{0\leq t\leq T}$ insures this property.

(b) The result is also true if instead of assuming that $f_n \in \mathbf{A}_T \otimes \mathcal{A}$, we rather suppose that f_n, $f_n - f_m \in \mathcal{A}_T$, for $n,m \geq 0$, and that (97) is satisfied with Γ replaced by $\bar{\Gamma}$ (the other hypotheses remaining the same, for instance $\left\|\bar{L}(f_n)(t,\,\cdot\,)\right\|$ is uniformly bounded in $0 \leq t \leq T$ and $n \in \mathbb{N}$).

Example 3.4

Assume that E is $\mathbb{R}^n$ for some $n \in \mathbb{N}^*$ and $\mathcal{A}$ contains at least all C^2 functions with compact support and that the L_t, $t \geq 0$, are second order differential operators (without 0-order terms) with locally bounded coefficients (in time but uniformly in space).

Then Lemma 3.2 shows that $\mathcal{A}_T$ contains at least all $C_b^{1,2}$ functions.

This approach is a little more flexible that the one using only stochastic calculus (to see that at least in case there is only continuous martingales (as in a Brownian filtration), $\mathcal{A}_T$ is stable by composition with C^2 functions, in the sense that if f_1, $f_2, \cdots, f_p$ belongs to $\mathcal{A}_T$ and if $F \in C^2(\mathbb{R}^p)$, then $F(f_1, f_2, \cdots, f_p)$ belongs to $\mathcal{A}_T$).

Using a classical embedding theorem the same results are also true for smooth separable manifolds.

Let us observe that if $f \in \mathcal{A}_T$, then the process given by

$$\left(f^2(t,X_t) - f^2(0,X_0) - \int_0^t \left[\bar{\Gamma}(f,f)(s,X_s) + 2f(s,X_s)\bar{L}(f)(s,X_s)\right]\, ds\right)_{0\leq t\leq T}$$

is also a martingale.

To see this claim we use the following decomposition (for $0 \leq t \leq T$)

$$\begin{aligned}
&\left(f(t,X_t) - f(0,X_0) - \int_0^t \bar{L}(f)(s,X_s)\, ds\right)^2 \\
&= f^2(t,X_t) - f^2(0,X_0) + \left(\int_0^t \bar{L}(f)(s,X_s)\, ds\right)^2 \\
&\quad -2f(t,X_t)\int_0^t \bar{L}(f)(s,X_s)\, ds + 2f(0,X_0)M_t(f) \\
&= f^2(t,X_t) - f^2(0,X_0) - 2\int_0^t f(s,X_s)\bar{L}(f)(s,X_s)\, ds + 2f(0,X_0)M_t(f) \\
&\quad -2\int_0^t \left(\int_0^s \bar{L}(f)(u,X_u)\, du\right) dM_s(f)
\end{aligned}$$

where the last equality comes from an integration by parts (this is also the proof of the second part of Lemma 3.1). In other words this means that if $f \in \mathcal{A}_T$, then $f^2 \in \mathcal{B}_T$, and we can take $\bar{L}(f^2) = \bar{\Gamma}(f,f) + 2f\bar{L}(f)$.

If we look at $(\mathcal{B}_T, \bar{L})$ as an (contingently multi-valued) linear operator, then $\mathbb{P}_{\eta_0}$ is also the solution to the time-space martingale problem associated with the initial condition η_0 and to $(\mathcal{B}_T, \bar{L})$.

Finally, let us mention other properties of a pregenerator L_0 which will be useful latter on. If Γ_0 and $(\mathbb{P}_x)_{x\in E}$ are respectively the "carré du champs" and solutions of the martingale problems associated with L_0, then we have for all $\varphi \in \mathcal{A}$ and $x \in E$,

$$\Gamma_0(\varphi,\varphi)(x) \;=\; \lim_{t\to 0_+} \frac{\mathbb{P}_x[(\varphi(X_t) - \varphi(x))^2]}{t} \tag{98}$$

It follows that if $\phi, \varphi \in \mathcal{A}$ and $f \in \mathbf{A}_T \otimes \mathcal{A}$, then we have that

$$\|\Gamma_0(\phi\varphi,\phi\varphi)\| \;\leq\; 2(\|\phi\|^2\, \|\Gamma_0(\varphi,\varphi)\| + \|\varphi\|^2\, \|\Gamma_0(\phi,\phi)\|)$$

and for all $x \in E$,

$$\Gamma_0\left(\int_0^T f(s,\cdot)\, ds, \int_0^T f(s,\cdot)\, ds\right)(x) \;\leq\; T\int_0^T \Gamma_0(f(s,\cdot), f(s,\cdot))(x)\, ds$$

3.1.2 Strong Regularity Assumptions

The hypotheses introduced below are not strictly necessary for the results presented in the following sections, but they are often the shortest way to check condition (H2), at least in abstract setting (ie not in the cases of example 3.4), so maybe this section should be skipped at a first reading. More precisely, they allow for an understanding of the latter hypothesis, by separating the roles of U and several aspects of regularity for the semigroup associated to X.

To make a clear differentiation, each of them has been associated to a hypothesis below. Thus the strong regularity assumption consists in the set of the following conditions $(\mathrm{H4})_T$, $(\mathrm{H7})_T$, $(\mathrm{H8})_T$, $(\mathrm{H9})_T$ and $(\mathrm{H10})_T$, the other ones are only intermediate steps. Furthermore the considerations presented here are more or less standard in the so-called semigroup approach to the theory of Markov processes, and they will enable us to give more tractable expressions for some norms arising in the study of exponential bounds presented in section 3.3.3.

The time inhomogeneous semigroup associated with the Markov process (91) is the family $(P_{s,t})_{0\leq s\leq t}$ of operators on $\mathcal{B}_\mathrm{b}(E)$ defined by

$$\forall\, 0 \leq s \leq t,\, \forall\, \varphi \in \mathcal{B}_\mathrm{b}(E),\, \forall\, x \in E, \qquad P_{s,t}(\varphi)(x) \;=\; \mathbb{E}_{s,x}[\varphi(X_t)] \tag{99}$$

For $T > 0$ which will be supposed fixed in this subsection, we consider the following set of strong hypotheses:

$(\mathrm{H3})_T$ ***There exists a constant*** $C_{1,T} \geq 1$ ***such that for all*** $0 \leq t \leq T$***,***

$$C_{1,T}^{-1}\Gamma_0 \;\leq\; \Gamma_t \;\leq\; C_{1,T}\Gamma_0$$

in the sense that for all $\varphi \in \mathcal{A}$ ***and all*** $x \in E$***,***

$$C_{1,T}^{-1}\Gamma_0(\varphi,\varphi)(x) \leq \Gamma_t(\varphi,\varphi)(x) \leq C_{1,T}\Gamma_0(\varphi,\varphi)(x).$$

Let us define a norm $|\!|\!| \cdot |\!|\!|$ on $\mathcal{A}$ by

$$\forall\, \varphi \in \mathcal{A}, \qquad |\!|\!|\varphi|\!|\!| \;=\; \sqrt{\|\varphi\|^2 + \|\Gamma_0(\varphi,\varphi)\|}$$

$(\mathrm{H4})_T$ ***For all*** $0 \leq t \leq T$***, we have*** $U_t \in \mathcal{A}$***, and the application***

$$[0,T] \ni t \;\mapsto\; U_t \in \mathcal{A}$$

is continuous with respect to $|\!|\!| \cdot |\!|\!|$***.***

$(\mathrm{H5})_T$ ***For all*** $0 \leq s \leq t \leq T$***,*** $\mathcal{A}$ ***is stable by*** $P_{s,t}$***, and for*** $\varphi \in \mathcal{A}$ ***and*** $0 \leq t \leq T$ ***fixed, the mappings***

$$\begin{aligned}
[0,t] \ni s \;&\mapsto\; P_{s,t}(\varphi) \in \mathcal{A} \\
[t,T] \ni s \;&\mapsto\; P_{t,s}(\varphi) \in \mathcal{A}
\end{aligned}$$

are continuous with respect to $|\!|\!| \cdot |\!|\!|$***.***

$(\mathrm{H6})_T$ ***For all*** $0 \leq s \leq t \leq T$***,*** $\mathcal{A}$ ***is stable by*** $P_{s,t}$***, and there exists a constant*** $C_{2,T} \geq 1$ ***such that for all*** $0 \leq s \leq t \leq T$ ***and all*** $\varphi \in \mathcal{A}$***,***

$$\|\Gamma_0(P_{s,t}(\varphi), P_{s,t}(\varphi))\| \;\leq\; C_{2,T}\,\|\Gamma_0(\varphi,\varphi)\|$$

(H7)$_T$ ***For all*** $0 \le s \le t \le T$***,*** $\mathcal{A}$ ***is stable by*** $P_{s,t}$***, and for all*** $\varphi \in \mathcal{A}$***, there is a finite constant*** $C_{3,T}(\varphi) > 0$ ***such that for all*** $0 \le s, t \le T$***,***

$$\|L_s(P_{t,T}(\varphi))\| \le C_{3,T}(\varphi)$$

(H8)$_T$ ***For all*** $0 \le s \le t \le T$***,*** $\mathcal{A}$ ***is stable by*** $P_{s,t}$***, and we have in the*** $\|\!|\cdot|\!\|$ ***sense, on*** $\mathcal{A}$***,***

$$\begin{cases} \frac{d}{ds} P_{s,t} & = & -L_s P_{s,t} \\ \frac{d}{dt} P_{s,t} & = & P_{s,t} L_t \end{cases}$$

(H9)$_T$ ***For all*** $\varphi \in \mathcal{A}$***,*** $[0,T] \ni t \mapsto \Gamma_t(\varphi,\varphi)$ ***is differentiable (in the sense of the norm*** $\|\cdot\|$***), and there exists a finite constant*** $C_{4,T} \ge 0$ ***such that for all*** $0 \le t \le T$ ***and all*** $\varphi \in \mathcal{A}$***,***

$$\left\| \frac{\partial_t \Gamma_t(\varphi,\varphi)}{\Gamma_t(\varphi,\varphi)} \right\| \le C_{4,T}$$

(where ∂_t ***stands for*** d/dt***).***

If $\mathcal{A}$ is stable by L_t, for a given $t \ge 0$, we can define for $\phi, \varphi \in \mathcal{A}$,

$$\Gamma_{2,t}(\phi,\varphi) = \frac{1}{2}(L_t(\Gamma_t(\phi,\varphi)) - \Gamma_t(L_t(\phi),\varphi) - \Gamma_t(\phi, L_t(\varphi)))$$

We are now in position to introduce our last assumption:

(H10)$_T$ ***For all*** $0 \le t \le T$***,*** $\mathcal{A}$ ***is stable by*** L_t***, and there exists a constant*** $R_T \in \mathbb{R}$ ***such that for all*** $\varphi \in \mathcal{A}$ ***and*** $x \in E$***,***

$$\Gamma_{2,t}(\varphi,\varphi)(x) \ge R_T \Gamma_t(\varphi,\varphi)(x)$$

(R_T ***will then denote the best constant possible verifying this property, i.e. the largest one).***

Before studying several links between these hypotheses and (H2)$_T$ (which corresponds to (H2) for a $T > 0$ fixed), let us introduce $B(T,\mathcal{A})$ the set of $f \in \mathcal{B}_b([0,T]\times E)$ such that for all $0 \le t \le T$, $f(t,\cdot) \in \mathcal{A}$ and such that

$$\|\!|f|\!\|_{[0,T]} \stackrel{\text{def.}}{=} \sup_{0 \le t \le T} \|\!|f(t,\cdot)|\!\| < +\infty$$

This quantity is a norm on $B(T,\mathcal{A})$, and we will note $\|\!|f|\!\|_t$ the semi-norm $\|\!|f(t,\cdot)|\!\|$, for any $0 \le t \le T$ and $f \in B(T,\mathcal{A})$.

In much the same way let $\|f\|_t = \|f(t,\cdot)\|$ and denote

$$\|f\|_{[0,T]} = \sup_{0 \le t \le T} \|f\|_t$$

for $f \in \mathcal{B}_b([0,T] \times E)$.

Our first remark is:

Lemma 3.5 *Under* $(H3)_T$, $(H5)_T$ *and* $(H7)_T$, *for all* $\varphi \in \mathcal{A}$, *the mapping*

$$G_{T,\varphi} : [0,T] \times E \ni (t,x) \mapsto P_{t,T}(\varphi)(x)$$

belongs to $\mathcal{A}_T$ *(the continuity in the first variable is only necessary in* $(H5)_T$*).*

Proof: For each $n \in \mathbb{N}$ we introduce the functions $f_n \in \mathbf{A}_T \otimes \mathcal{A}$ defined by $\forall\, 0 \leq t \leq T, \forall\, x \in E$,

$$f_n(t,x) \;=\; \left(1 + k - \frac{(1+n)t}{T}\right) P_{\frac{kT}{1+n},T}(\varphi)(x) + \left(\frac{(1+n)t}{T} - k\right) P_{\frac{(1+k)T}{1+n},T}(\varphi)(x)$$

where $k = \lfloor (n+1)t/T \rfloor$. Clearly, under $(\mathrm{H5})_T$ we have that $G_{T,\varphi} \in B(T,\mathcal{A})$ and

$$\lim_{n\to\infty} |||G_{T,\varphi} - f_n|||_{[0,T]} \;=\; 0$$

By Lemma 3.2 and condition $(\mathrm{H3})_T$, to prove the announced result, it remains to show that

$$\sup_{n\in\mathbf{N}} \|L(f_n)\|_{[0,T]} \;<\; +\infty$$

To this end we first notice that for any $0 \leq t \leq T$ and $x \in E$

$$\begin{aligned} L(f_n)(t,x) \;=\;& \frac{1+n}{T}\left(P_{\frac{(1+k)T}{1+n},T}(\varphi)(x) - P_{\frac{kT}{1+n},T}(\varphi)(x)\right) \\ &+ \left(1 + k - \frac{(1+n)t}{T}\right) L_t[P_{\frac{kT}{1+n},T}(\varphi)](x) \\ &+ \left(\frac{(1+n)t}{T} - k\right) L_t[P_{\frac{(1+k)T}{1+n},T}(\varphi)](x) \end{aligned}$$

where k is defined as above. Therefore, in view of $(\mathrm{H7})_T$, to see the previous affirmation, we only need to check that

$$\sup_{n\in\mathbf{N},\, 0\leq k\leq n} \frac{1+n}{T}\left\|P_{\frac{(1+k)T}{1+n},T}(\varphi) - P_{\frac{kT}{1+n},T}(\varphi)\right\| \;<\; +\infty$$

To see this claim it is sufficient to write, for all $x \in E$,

$$\begin{aligned} &P_{\frac{(1+k)T}{1+n},T}(\varphi)(x) - P_{\frac{kT}{1+n},T}(\varphi)(x) \\ &= \mathbb{E}_{\frac{kT}{1+n},x}\left[P_{\frac{(1+k)T}{1+n},T}(\varphi)(x) - P_{\frac{(1+k)T}{1+n},T}(\varphi)(X_{\frac{(1+k)T}{1+n}})\right] \\ &= -\mathbb{E}_{\frac{kT}{1+n},x}\left[\int_{\frac{kT}{1+n}}^{\frac{(k+1)T}{1+n}} L_s(P_{\frac{(1+k)T}{1+n},T}(\varphi))(X_s)\, ds\right] \end{aligned}$$

■

We consider next the collection $(Q_{s,t})_{0\leq s\leq t}$ of linear operator on $\mathcal{B}_\mathrm{b}(E)$ defined for any $0 \leq s \leq t$, $\varphi \in \mathcal{B}_\mathrm{b}(E)$, $x \in E$, as follows

$$Q_{s,t}(\varphi)(x) \;=\; \mathbb{E}_{s,x}\left[\exp\left(\int_s^t U_u(X_u)\, du\right) \varphi(X_t)\right]$$

It is easily seen that $(Q_{s,t})_{0\leq s\leq t}$ is a well defined time inhomogeneous semigroup of non-negative operators on $\mathcal{B}_b(E)$ (but in general non-Markovian, except in the trivial case where $U \equiv 0$, ie $(Q_{s,t})_{0\leq s\leq t} = (P_{s,t})_{0\leq s\leq t}$).

To see that Lemma 3.5 is also true for $(Q_{s,t})_{0\leq s\leq t}$, under the additional conditions $(\mathrm{H4})_T$ and $(\mathrm{H6})_T$, let us work out a relation between $(P_{s,t})_{0\leq s\leq t}$ and $(Q_{s,t})_{0\leq s\leq t}$:

Taking $t = T$ in (95) and integrating the above equality with respect to $\mathbb{P}_{0,x}$, we obtain (in fact for all $\varphi \in \mathcal{B}_b(E)$),

$$Q_{0,T}(\varphi)(x) \;=\; P_{0,T}(\varphi)(x) + \int_0^T P_{0,s}(U_s Q_{s,T}(\varphi))(x)\,ds$$

More generally and in the same way one can prove that for any $0 \leq t \leq T$,

$$Q_{t,T}(\varphi) \;=\; P_{t,T}(\varphi) + \int_t^T P_{t,s}(U_s Q_{s,T}(\varphi))\,ds$$

This identity leads us to consider, for $\varphi \in \mathcal{A}$ fixed, the application Z defined on $\mathcal{B}_b([0,T]\times E)$ by

$\forall\, f \in \mathcal{B}_b([0,T]\times E),\ \forall\, 0 \leq t \leq T,\ \forall\, x \in E,$

$$Z(f)(t,x) \;=\; P_{t,T}(\varphi)(x) + \int_t^T P_{t,s}(U_s f(s,\,\cdot\,))(x)\,ds$$

Under $(\mathrm{H4})_T$ there exists a constant $C_T^{(1)} > 0$ such that

$\forall\, f_1, f_2 \in \mathcal{B}_b([0,T]\times E),\ \forall\, 0 \leq t \leq T,$

$$\|Z(f_1) - Z(f_2)\|_t \;\leq\; C_T^{(1)} \int_t^T \|f_1 - f_2\|_s\,ds$$

Let $Z^n = Z \circ \ldots \circ Z$ (n-times) be the n-step iterate of Z. It is standard to check that

$$\forall\, f_1, f_2 \in \mathcal{B}_b([0,T]\times E), \qquad \|Z^n(f_1) - Z^n(f_2)\|_{[0,T]} \;\leq\; \frac{(C_T^{(1)}T)^n}{n\,!}\,\|f_1 - f_2\|_{[0,T]}$$

This implies that $F_{T,\varphi} = Q_{\cdot,T}(\varphi)(\,\cdot\,)$ is the unique solution of the equation $Z(f) = f$, with $f \in \mathcal{B}_b([0,T]\times E)$.

Another useful remark is that elementary calculations show that for any $f \in \mathcal{B}_b([0,T]\times E)$

$$\left(Z(f)(t,X_t) - Z(f)(0,X_0) - \int_0^t P_{s,T}(U_s f(s,\,\cdot\,))(X_s)\,ds\right)_{0\leq t\leq T}$$

is a martingale (the càdlàg property comes from $(\mathrm{H5})_T$). This means that one can take

$$\forall\, 0 \leq t \leq T,\ \forall\, x \in E, \qquad \bar{L}(Z(f))(t,x) \;=\; P_{t,T}(U_t f(t,\,\cdot\,))(x) \tag{100}$$

Proposition 3.6 *Under conditions $(H3)_T$, $(H4)_T$, $(H5)_T$, $(H6)_T$ and $(H7)_T$, assumption $(H2)_T$ holds, and there exists a finite constant $C_T^{(2)} > 0$, such that for all $\varphi \in \mathcal{A}$,*

$$|\!|\!|\bar{F}_{T,\varphi}|\!|\!|_{[0,T]} \;\leq\; C_T^{(2)}\,|\!|\!|\varphi|\!|\!|$$

Proof: Let $\bar{B}(T,\mathcal{A})$ denote the Banach space which is the $|\!|\!| \cdot |\!|\!|_{[0,T]}$-completion of $\mathbf{A}_T \otimes \mathcal{A}$. Note that if $\bar{f} \in \bar{B}(T,\mathcal{A})$, then we can naturally associate to it numbers $|\!|\!|\bar{f}|\!|\!|_t$, for $0 \leq t \leq T$, and as above $|\!|\!|\bar{f}|\!|\!| = \sup_{0\leq t\leq T} |\!|\!|\bar{f}|\!|\!|_t$. In the same way, as the norm $|\!|\!| \cdot |\!|\!|_{[0,T]}$ dominates $\|\cdot\|_{[0,T]}$, to each $\bar{f} \in \bar{B}(T,\mathcal{A})$ we associate a measurable bounded function $f : [0,T] \times E \to \mathbb{R}$ (we will occasionally abuse notation, saying that $f \in \bar{B}(T,\mathcal{A})$, and also note that we can associate to $\bar{f}$ another measurable bounded function which could be in an obvious way written $\Gamma_0(\bar{f},\bar{f})$). In view of the Lemma 3.2 and $(\mathrm{H3})_T$, if it wasn't for the boundedness condition on the $(L(f_n))_{n\in\mathbf{N}}$, such a function f would have a good chance to belong to $\mathcal{A}_T$. We will use this procedure here to show the belonging of $F_{T,\varphi}$ to $\mathcal{A}_T$, but rather taking into account the remark after Lemma 3.2 and (100).

From now on we consider the restriction of Z to $\mathbf{A}_T\otimes\mathcal{A}$. We slight abuse notations and still denote Z these restrictions. Using approximations techniques (as the one presented in the proof of Lemma 3.5), the remarks at the end of section 3.1.1 and the hypothesis $(\mathrm{H6})_T$, it is easily seen that $Z(f) \in \bar{B}(T,\mathcal{A})$, for $f \in \mathbf{A}_T \otimes \mathcal{A}$. More precisely, Z can be extended to $\bar{B}(T,\mathcal{A})$ and one can find a finite constant $C_T^{(3)} > 0$ depending on $C_{2,T}$ and $|\!|\!|U|\!|\!|_{[0,T]}$ and such that for any $f_1,\ f_2 \in \bar{B}(T,\mathcal{A})$ and $0 \leq t \leq T$

$$|\!|\!|Z(f_1) - Z(f_2)|\!|\!|_t^2 \;\leq\; C_T^{(3)} \int_t^T |\!|\!|f_1 - f_2|\!|\!|_s^2\, ds$$

Then proceeding as above, it appears that for all $n \geq 1$,

$$\forall\ f_1, f_2 \in \bar{B}(T,\mathcal{A}), \qquad |\!|\!|Z^n(f_1) - Z^n(f_2)|\!|\!|_{[0,T]}^2 \;\leq\; \frac{(C_T^{(3)}T)^n}{n\,!}|\!|\!|f_1 - f_2|\!|\!|_{[0,T]}^2$$

As a result there is a unique solution denoted by $\bar{F}_{T,\varphi}$ to the equation

$$Z(\bar{f}) = \bar{f}$$

with $\bar{f} \in \bar{B}(T,\mathcal{A})$. Its corresponding bounded measurable function is clearly $F_{T,\varphi}$. Furthermore, $\bar{F}_{T,\varphi}$ is classically shown to be the limit in $\bar{B}(T,\mathcal{A})$ of $Z^n(0)$, for n large. By Lemma 3.2, to be convinced that $F_{T,\varphi} \in \mathcal{A}_T$, it remains to prove that

$$\sup_{n\in\mathbf{N}} \|L(Z^n(0))\|_{[0,T]} \;<\; +\infty$$

but the remark before the proposition shows that this is a consequence of

$$\sup_{n\in\mathbf{N}} \|Z^n(0)\|_{[0,T]} \;<\; +\infty$$

which itself follows from the considerations above the proposition.
The last part of this proposition is now clear from the previous approximations.

■

Remark 3.7: In practice it may be important to know the dependence of $C_T^{(2)}$ on U. Using the above proof we first observe that $C_T^{(3)} = C_T^{(4)}|\!|\!|U|\!|\!|_{[0,T]}^2$, where $C_T^{(4)} > 0$ is a finite constant which does not depend on U.

Therefore for any $n \in \mathbb{N}^*$ and $f_1, f_2 \in \bar{B}(T, \mathcal{A})$ we have that

$$\|Z^n(f_1) - Z^n(f_2)\|_{[0,T]} \leq \sqrt{\frac{(C_T^{(4)}T)^n}{n!}} \|U\|_{[0,T]}^n \|f_1 - f_2\|_{[0,T]}$$
$$\leq \frac{\sqrt{C(C_T^{(4)}T)^n}}{\lfloor n/2 \rfloor !} \|U\|_{[0,T]}^n \|f_1 - f_2\|_{[0,T]}$$

where we have use the Stirling formula to find a finite constant $C > 0$ such that for all $n \geq 1$,

$$\frac{1}{n!} \leq C \frac{1}{(\lfloor n/2 \rfloor !)^2}$$

Writing

$$\|\bar{F}_{T,\varphi}\|_{[0,T]} \leq \sum_{n \geq 0} \|Z^{n+1}(0) - Z^n(0)\|_{[0,T]}$$
$$\leq \|P_{\cdot,T}(\varphi)(\cdot)\|_{[0,T]} \sum_{n \geq 0} \frac{\sqrt{C(C_T^{(4)}T)^n}}{\lfloor n/2 \rfloor !} \|U\|_{[0,T]}^n$$

makes it clear that there exist two finite constants $C_T^{(5)}$, $C_T^{(6)} > 0$ such that

$$C_T^{(2)} \leq C_T^{(5)} \exp(C_T^{(6)} \|U\|_{[0,T]})$$

In the end, we have

Proposition 3.8 *The conditions (H8)$_T$, (H9)$_T$ and (H10)$_T$ implies (H3)$_T$, (H5)$_T$ and (H6)$_T$*

For a proof of this result and more discussions on the curvature hypothesis (H10)$_T$, see [45].

Remarks 3.9:
a) We believe that the hypothesis (H7)$_T$ is not natural here, and we would have preferred to work only with closures related to functions and their carrés du champs. Nevertheless, note that if there is a finite constant $C_{5,T} > 0$ such that for all $\varphi \in \mathcal{A}$ and all $0 \leq s, t \leq T$,

$$\|L_t(\varphi) - L_s(\varphi)\| \leq C_{5,T} \|\varphi\|$$

(as it is the case in our applications to nonlinear filtering, cf. section 4.2), and if the right hand side of the first equation in (H8)$_T$ is $\|\cdot\|$-continuous, then (H7)$_T$ is satisfied.

b) In view of the generality of our setting, the reader may wondering why we have not only considered the time-homogeneous case, by adding the time as a coordinate of the Markov process. The corresponding generator on the state space $\mathbb{R}_+ \times E$ is L

acting on $\mathbf{A} \otimes \mathcal{A}$. But in general U does not belong to this domain, since typically in our applications to nonlinear filtering U will not be differentiable with respect to time (only a regularity of Hölder exponent less than 1/2 can be expected). So the hypothesis $(H4)_T$ will not be verified in this context, giving one reason for which it seems interesting to us to separate the role of time.
Nevertheless, note that a way to get round this particular difficulty would be to complete $\mathcal{A}$ (or $\mathbf{A} \otimes \mathcal{A}$ in the homogeneous case) with respect to $\|\!|\ \|\!|$.

3.1.3 Asymptotic Stability

In this section we take up the study of the asymptotic stability of the limiting process $\eta = \{\eta_t \ ; \ t \geq 0\}$ which was begun in section 2.1.2 in discrete time settings. Before getting into the details we first need to make a few general observations and give some definitions. We retain notations of section 2.1.2 and we denote $\alpha(H)$ and $\beta(H)$ the contraction and Dobrushin ergodic coefficients associated with a given transition probability kernel H on E and given by (25) and (24).

Next we denote $\Phi = \{\Phi_{s,t} \ ; \ s \leq t\}$ the nonlinear semigroup in distribution space associated with the dynamics structure of η, namely

$$\forall\, s \leq t, \qquad \eta_t \ = \ \Phi_{s,t}(\eta_s) \tag{101}$$

One can check that each mapping $\Phi_{s,t}$, $s \leq t$, has the following handy form given for any $\eta \in \mathbf{M}_1(E)$ and for any bounded Borel function f on E by

$$\Phi_{s,t}(\eta)(f) = \frac{\eta(H_{s,t}(f))}{\eta(H_{s,t}(1))}$$

where

$$H_{s,t}(f)(x) \stackrel{\text{def.}}{=} \mathbb{E}_{s,x}\left(f(X_t)\, Z_{s,t} \right) \qquad \text{and} \qquad Z_{s,t} \stackrel{\text{def.}}{=} \exp\left(\int_s^t U_r(X_r)\, dr \right)$$

As in the discrete time case there is a simple trick which allows to connect the nonlinear semigroup Φ with a family of linear semigroups.

Lemma 3.10 *For any $s \leq t$ and $\mu \in \mathbf{M}_1(E)$ we have the following decomposition*

$$\Phi_{s,t}(\mu) = \Psi_{s,t}(\mu) H^{(t)}_{s,t}$$

where the mapping $\Psi_{s,t} : \mathbf{M}_1(E) \to \mathbf{M}_1(E)$ is defined by

$$\Psi_{s,t}(\mu)(f) = \frac{\mu(\ g_{s,t}\ f\ )}{\mu(\ g_{s,t}\ )} \qquad \text{where} \qquad g_{s,t} \stackrel{\text{def.}}{=} H_{s,t}1$$

and for any $t \geq 0$, $H^{(t)} = \{H^{(t)}_{s,r} \ ; \ s \leq r \leq t\}$ is a linear semigroup defined for any $f \in \mathcal{B}_b(E)$ and $\mu \in \mathbf{M}_1(E)$ and $s \leq r \leq t$ by

$$\mu H^{(t)}_{s,r} f = \int_E \mu(dx)\ H^{(t)}_{s,r} f(x) \qquad \text{and} \qquad H^{(t)}_{s,r} f = \frac{H_{s,r}(g_{r,t}\ f)}{H_{s,r}(g_{r,t})}$$

Remark 3.11: By construction and using the Markov property of X it is easy to see that the transition kernels $H^{(t)}_{s,r}(x,dz)$, $s \leq r \leq t$ may likewise be defined for any bounded Borel function f by setting

$$\forall\, x \in E, \qquad H^{(t)}_{s,r} f(x) \;=\; \frac{\mathbb{E}_{s,x}\left(f(X_r)\, Z_{s,t}\right)}{\mathbb{E}_{s,x}\left(Z_{s,t}\right)}$$

As in discrete time settings the asymptotic stability properties of Φ can now be characterized in terms of the contraction coefficient of the linear semigroups $\{H^{(t)}\,;\, t \geq 0\}$.

Lemma 3.12 *For any $s \leq t$ we have that*

$$\beta\left(H^{(t)}_{s,t}\right) = \sup_{\mu,\nu} \|\Phi_{s,t}(\mu) - \Phi_{s,t}(\nu)\|_{\mathrm{tv}}$$

where the supremum is taken over all distributions $\mu, \nu \in \mathbf{M}_1(E)$.

On the basis of the definition of H it is now easy to establish that for any $f \in \mathcal{B}_b(E)$, $x \in E$ and any $s \leq r \leq t$

$$H^{(t)}_{s,r} f(x) = \frac{\mathbb{E}_{s,x}\left(\, f(X_r)\, g_{r,t}(X_r)\, \mathbb{E}_{s,x}\left(Z_{s,r}\,|X_r\,\right)\right)}{\mathbb{E}_{s,x}\left(\, g_{r,t}(X_r)\, \mathbb{E}_{s,x}\left(Z_{s,r}\,|X_r\,\right)\right)}$$

Recalling the definition of the ergodic coefficient $\alpha(K)$ of a given transition probability kernel $K(x,dz)$ on E the above formula yields that for any $s \leq r \leq t$

$$\alpha\left(S^{(t)}_{s,r}\right)\ \exp -\left(\int_s^r \mathrm{osc}(U_\tau)\, d\tau\right) \leq \alpha\left(H^{(t)}_{s,r}\right) \leq \alpha\left(S^{(t)}_{s,r}\right)\ \exp\left(\int_s^r \mathrm{osc}(U_\tau)\, d\tau\right)$$

where

- For any $t \geq 0$, $\mathrm{osc}(U_t) \stackrel{\text{def.}}{=} \sup\left\{U_t(y) - U_t(x)\,;\, (x,y) \in E^2\right\}$.
- For any $t \geq 0$, $S^{(t)} = \{S^{(t)}_{s,r}\,;\, s \leq r\}$ is a collection of transition probability functions on E and defined for any bounded Borel function f by

$$S^{(t)}_{s,r} f = \frac{P_{s,r}(g_{r,t}\, f)}{P_{s,r}(g_{r,t})}$$

Using the semigroup property of $H^{(t)}$, by definition of the ergodic and the contraction coefficients $\alpha(K)$ and $\beta(K)$ of a transition probability kernel K on E one proves the following result.

Proposition 3.13 *For any $v > 0$ and $s \leq r \leq t$ we have*

$$\beta\left(H^{(t)}_{s,r}\right) \leq \prod_{m \in I_v(s,r)} \left(1 - \alpha\left(S^{(t)}_{m,m+v}\right)\ \exp -\left(\int_m^{m+v} \mathrm{osc}(U_r)\, dr\right)\right)$$

where for any $v > 0$ and $s \leq r \leq t$, $I_v(s,r)$ is the subset of $\mathbb{R}_+$ defined by $I_v(s,r) \stackrel{\text{def.}}{=} \{s + pv\,;\, 0 \leq p < [(r-s)/v]\}$ with $[a]$ the integer part of $a \in \mathbb{R}$.

If U is chosen so that

$$\mathrm{osc}^\star(U) \stackrel{\mathrm{def.}}{=} \int_0^\infty \mathrm{osc}(U_t)\,dt \;<\; +\infty \tag{102}$$

then by definition of $S^{(t)}$ one gets the inequalities

$$\forall\, s \leq r \leq t, \qquad \alpha(P_{s,r})\; e^{-\mathrm{osc}^\star(U)} \;\leq\; \alpha(S^{(t)}_{s,r}) \;\leq\; \alpha(P_{s,r})\; e^{\mathrm{osc}^\star(U)}$$

from which one concludes that

Theorem 3.14 *If the function U satisfies (102) then for any $v > 0$ the following implications hold*

$$\sum_{n\geq 1} \alpha\left(P_{(n-1)v,nv}\right) = \infty \implies \lim_{t\to\infty} \beta\left(H^{(t)}_{0,t}\right) = 0$$

$$\lim_{n\to\infty} \frac{1}{n}\sum_{k=1}^{n} \alpha\left(P_{(k-1)v,kv}\right) \stackrel{\mathrm{def.}}{=} \overline{\alpha}(v) \Longrightarrow \limsup_{t\to\infty} \frac{1}{t}\log\beta\left(H^{(t)}_{0,t}\right) \leq -\frac{\overline{\alpha}(v)}{v}\; e^{-2\mathrm{osc}^\star(U)}$$

In addition, if $\inf_{|t-s|=v} \alpha\left(P_{s,t}\right) \stackrel{\mathrm{def.}}{=} \tilde{\alpha}(v)$ for some $v > 0$ then for any $p \geq 1$ and $T \geq p\,.\,v$ we have that

$$\sup_{t\geq 0}\sup_{\mu,\nu} \frac{1}{T}\log\|\Phi_{t,t+T}(\mu) - \Phi_{t,t+T}(\nu)\|_{\mathrm{tv}} \leq -\frac{\tilde{\alpha}(v)}{q\,.\,v}\; e^{-2\mathrm{osc}^\star(U)}$$

Next we examine an additional sufficient condition for the asymptotic stability of Φ in terms of the mixing properties of $P = \{P_{s,t}\ ;\ s \leq t\}$. Assume that the semigroup P satisfies the following condition.

$(\mathcal{P})$ ***There exists some*** $v > 0$ ***such that for any*** $t \geq 0$

$$\forall\, x \in E, \qquad \epsilon_t^{1/2}(v) \;\leq\; \frac{dP_{t,t+v}(x,\bullet)}{d\mu_{t,v}} \;\leq\; \epsilon_t^{-1/2}(v)$$

for some positive constant $\epsilon_t(v) > 0$ ***and some reference probability measure*** $\mu_{t,v} \in \mathbf{M}_1(E)$.

The main simplification due to condition $(\mathcal{P})$ is the following: for any non-negative test function f we clearly have for any $0 \leq s \leq s+v \leq t$

$$\epsilon_s(v)\;\Psi_{s+v,t}(\mu)(f) \leq S^{(t)}_{s,s+v}(f) \leq \epsilon_s^{-1}(v)\;\Psi_{s+v,t}(\mu)(f)$$

This implies that $\alpha\left(S^{(t)}_{s,s+v}\right) \geq \epsilon_s(v)$ from which one can prove the following theorem.

Theorem 3.15 *Assume that the semigroup P satisfies condition $(\mathcal{P})$ for some constants $v > 0$, $\epsilon_t(v) > 0$ and some reference probability measure $\mu_{t,v} \in \mathbf{M}_1(E)$. If the function U is such that*

$$\|\mathrm{osc}(U)\| \stackrel{\mathrm{def.}}{=} \sup_{t\geq 0}\mathrm{osc}(U_t) < \infty$$

then for any $v > 0$ the following implications hold

$$\sum_{n\geq 0} \epsilon_{nv}(v) = \infty \implies \lim_{t\to\infty} \beta\left(H_{0,t}^{(t)}\right) = 0$$

$$\lim_{n\to\infty} \frac{1}{n}\sum_{k=0}^{n-1} \epsilon_{kv}(v) \stackrel{\text{def.}}{=} \overline{\epsilon}(v) \implies \limsup_{t\to\infty} \frac{1}{t}\log\beta\left(H_{0,t}^{(t)}\right) \leq -\frac{\overline{\epsilon}(v)}{v}\, e^{-v\,.\,\|\text{osc}(U)\|}$$

In addition, if $\inf_{t\geq 0} \epsilon_t(v) \stackrel{\text{def.}}{=} \tilde{\epsilon}(v)$ *for some $v > 0$ then for any $p \geq 1$ and $T \geq p\,.\,v$ we have that*

$$\sup_{t\geq 0}\sup_{\mu,\nu} \frac{1}{T}\log\|\Phi_{t,t+T}(\mu) - \Phi_{t,t+T}(\nu)\|_{\text{tv}} \leq -\frac{\tilde{\epsilon}(v)}{q\,.\,v}\, e^{-v\,.\,\|\text{osc}(U)\|}$$

for any $p, q \geq 1$ such that $1/p + 1/q = 1$.

In time homogeneous settings several examples of semigroups P satisfying condition $(\mathcal{P})$ can be found in [9] and in [27]. For instance if X is a sufficiently regular diffusion on a compact manifold then $(\mathcal{P})$ holds with

$$\epsilon_t(v) = \epsilon(v) = A\ \exp -(B/v) \tag{103}$$

for some constants $0 < A, B < \infty$ and for the uniform Riemannian measure $\mu_{t,v} = \mu$ on the manifold. To illustrate our result let us examine the situation in which U is also time-homogeneous, that is $U_t = U$.

Corollary 3.16 *Assume that U is time-homogeneous and the semigroup of X satisfies condition $(\mathcal{P})$ with $\epsilon_t(v) = \epsilon(v)$ given by (103). Then for any $p \geq 1$ and $v > 0$ and $T \geq p\,.\,v$ we have that*

$$\sup_{\mu,\nu\in\mathbf{M}_1(E)} \|\Phi_{t,t+T}(\mu) - \Phi_{t,t+T}(\nu)\|_{\text{tv}} \leq \exp -(\gamma\,.\,T)$$

with

$$\gamma \geq \frac{A}{q\ v}\ \exp -\left(\frac{B}{v} + v\,.\,\text{osc}(U)\right) \qquad \text{and} \qquad \frac{1}{p} + \frac{1}{q} = 1$$

The best bound in term of the constants A, B and osc(U) is obtained for

$$v = v^\star \stackrel{\text{def.}}{=} \frac{2B}{1 + \sqrt{1 + 4B\ \text{osc}(U)}}$$

The above asymptotic stability study can be extended in a simple way, but this is outside the scope of these notes, to study stability properties of the nonlinear filtering equation and its robust version. The interested reader is recommended to consult [43].

3.2 The Interacting Particle System Model

The purpose of this section is to design an IPS

$$(\xi_t)_{t\geq 0} = (\xi_t^1, \xi_t^2, \cdots, \xi_t^N)_{t\geq 0}$$

taking values in E^N, where $N \geq 1$ is the number of particles and such that for any $t \geq 0$, the empirical measures given by

$$\eta_t^N(\,\cdot\,) \;=\; \frac{1}{N}\sum_{i=1}^{N} \delta_{\xi_t^i}(\,\cdot\,)$$

are good approximations of η_t, $t \geq 0$, for N large enough.

As announced in the introduction, the Markov process $(\xi_t)_{t\geq 0}$ will also be defined by a martingale problem. At this point it is convenient to give a more detailed description of its pregenerators which was already presented in (15). At time $t \geq 0$, the pregenerator $\mathcal{L}_t^{(N)}$ of the IPS will be of genetic type in the sense that it is defined as the sum of two pregenerators, namely

$$\mathcal{L}_t^{(N)} \;=\; \widetilde{\mathcal{L}}_t^{(N)} + \widehat{\mathcal{L}}_t^{(N)}$$

where

• The first pregenerator $\widetilde{\mathcal{L}}_t^{(N)}$ is called the mutation pregenerator, it denotes the pregenerator at time t coming from N-independent processes having the same evolution as X and it is given on $\mathcal{A}^{\otimes N}$ by

$$\forall\, \phi \in \mathcal{A}^{\otimes N}, \qquad \widetilde{\mathcal{L}}_t^{(N)}(\phi) \;=\; \sum_{i=1}^{N} L_t^{(i)}(\phi)$$

where $L_t^{(i)}$ denotes the action of L_t on the i-th variable x_i, that is

$$L_t^{(i)} = \mathrm{Id} \otimes \cdots \otimes \underbrace{L_t}_{i-\mathrm{th}} \otimes \cdots \otimes \mathrm{Id},$$

where Id is the identity operator.

• The second one is denoted by $\widehat{\mathcal{L}}_t^{(N)}$ and it is called the selection pregenerator. It is defined as the jump type generator defined for any $\phi \in \mathcal{A}^{\otimes N}$ and $x = (x_1, \ldots, x_N) \in E^N$ by

$$\widehat{\mathcal{L}}_t^{(N)}(\phi)(x) \;=\; \frac{1}{N}\sum_{i=1}^{N}\sum_{j=1}^{N} \left(\phi(x^{i,j}) - \phi(x)\right) U_t(x_j)$$

where for $1 \leq i, j \leq N$ and $x = (x_1, \cdots, x_N) \in E^N$, $x^{i,j}$ is the element of E^N given by

$$\forall\, 1 \leq k \leq N, \qquad x_k^{i,j} \;=\; \begin{cases} x_k & , \text{ if } k \neq i \\ x_j & , \text{ if } k = i \end{cases}$$

(this is meaningful for all functions $\phi \in \mathcal{B}_{\mathrm{b}}(E^N)$, and in fact $\widehat{\mathcal{L}}_t^{(N)}$ is a bounded generator on $\mathcal{B}_{\mathrm{b}}(E^N)$).

Heuristically the motion of the particles $(\xi_t)_{t\geq 0}$ is decomposed into the two following rules.
Between the jumps due to interaction between particles, each particle evolves independently from the others and randomly according to the time-inhomogeneous semigroup of X.
At some random times, say τ, we introduce a competitive interaction between the particles, during this stage a given particle ξ_τ^i will be replaced by a new particle ξ_τ^j, $1 \leq j \leq N$, with a probability proportional to its "adaptation" $U_\tau(\xi_\tau^j)$.

This mechanism is similar to the one of a Moran IPS, except in the form of the intensity of replacing ξ_τ^i by ξ_τ^j (which should be symmetric in these variables) and in the total jump rate which is here "proportional" to N (instead of N^2). In fact that renormalization shows that our Moran's type IPS can also be regarded as a Nanbu's type interacting particle approximating model for a simple generalized spatially homogeneous Boltzmann equation.

There is no real difficulty to construct a probability $\mathbb{P}$ (on $D([0,+\infty[, E^N))$ which is solution to the martingale problem associated with the initial distribution $\eta_0^{\otimes N}$ and to the time-inhomogeneous family of pregenerators $(\mathcal{L}_t^{(N)})_{t\geq 0}$. More precisely, we proceed in two steps: we first consider the product process on E^N corresponding to the family of pregenerators $(\widetilde{\mathcal{L}}_t^{(N)})_{t\geq 0}$ and to the initial distribution $\eta_0^{\otimes N}$. This is quite immediate and the N coordinates are independent and have the same law as X starting from η_0 (cf for instance the Theorem 10.1 p. 253 of [55]). Then, for all $t \geq 0$, $\mathcal{L}_t^{(N)}$ is just seen as a bounded perturbation of $\widetilde{\mathcal{L}}_t^{(N)}$ by $\widehat{\mathcal{L}}_t^{(N)}$. So we can apply general results about this kind of martingale problems, see the Proposition 10.2 p. 256 of [55].

From now on, $\mathbb{E}$ will designate the expectation relative to the process $\xi = \{\xi_t\ ;\ t \geq 0\}$ under $\mathbb{P}$.

For more information about how to construct and simulate this process, see section 3 of [45].

In fact, the law $\mathbb{P}$ of the particle system ξ satisfies a more extended martingale problem, since the proofs presented by Ethier and Kurtz [55] enable us to transpose the whole pregenerator $(\mathcal{B}_T, \bar{L})$ considered in the section 3.1.1:

Let us denote by $\mathcal{B}_{T,N}$ the vector sub-space of $\mathcal{B}_b([0,T] \times E^N)$ generated by the functions $f : [0,T] \times E^N \to \mathbb{R}$ such that there exist $f_1, \cdots, f_N \in \mathcal{B}_T$ for which we have

$$\forall\, t \in [0,T],\, \forall\, x = (x_1, \cdots, x_N) \in E^N, \qquad f(t,x) \;=\; \prod_{1\leq i\leq N} f_i(t,x_i)$$

If such a function f is given, we define for all $(t,x) \in [0,T] \times E$,

$$\widetilde{\mathcal{L}}^{(N)}(f)(t,x)$$

$$= \sum_{1\leq i\leq N} f_1(t,x_1)\cdots f_{i-1}(t,x_{i-1})\bar{L}(f_i)(t,x_i) f_{i+1}(t,x_{i+1})\cdots f_N(t,x_N)$$

and then we extend linearly this (contingently multi-valued) operator $\widetilde{\mathcal{L}}^{(N)}$ on $\mathcal{B}_{T,N}$.

We also need to consider the pregenerator $\widehat{\mathcal{L}}^{(N)}$ acting on $\mathcal{B}_b([0,T]\times E^N)$ in the following way:

$$\forall\, f\in\mathcal{B}_b([0,T]\times E^N), \forall\,(t,x)\in[0,T]\times E,$$
$$\widehat{\mathcal{L}}^{(N)}(f)(t,x) \;=\; \widehat{\mathcal{L}}_t^{(N)}(f(t,\cdot))(x)$$

and next we introduce on $\mathcal{B}_{T,N}$,

$$\mathcal{L}^{(N)} \;=\; \widetilde{\mathcal{L}}^{(N)}+\widehat{\mathcal{L}}^{(N)}$$

This pregenerator coincides naturally with $\partial_t+\mathcal{L}_t^{(N)}$ on $\mathbf{A}_T\otimes\mathcal{A}^{\otimes N}\subset\mathcal{B}_{T,N}$.

Lemma 3.17 *Under $\mathbb{P}$, for all $T>0$ and all $f\in\mathcal{B}_{T,N}$, the process $(M_t^{(N)}(f))_{0\le t\le T}$ defined by*

$$\forall\, 0\le t\le T,\qquad M_t^{(N)}(f) \;=\; f(t,\xi_t)-f(0,\xi_0)-\int_0^t\mathcal{L}^{(N)}(f)(s,\xi_s)\,ds$$

is a bounded martingale.

Now let $\mathcal{A}_{T,N}$ be the set of all $f\in\mathcal{B}_{T,N}$ for which $f^2\in\mathcal{B}_{T,N}$. With some obvious notations for such functions we can define

$$\Gamma^{(N)}(f,f) \;\stackrel{\text{def.}}{=}\; \mathcal{L}^{(N)}(f^2)-2f\mathcal{L}^{(N)}(f)$$
$$= \;\widetilde{\Gamma}^{(N)}(f,f)+\widehat{\Gamma}^{(N)}(f,f)$$

Then an easy calculation we have already met several times shows that

Lemma 3.18 *For any $f\in\mathcal{A}_{T,N}$, the increasing process associated to the martingale $(M_t^{(N)}(f))_{0\le t\le T}$ is given by the formula*

$$\forall\, 0\le t\le T,\qquad \langle M^{(N)}(f)\rangle_t \;=\; \int_0^t\Gamma^{(N)}(f,f)(s,\xi_s)\,ds$$

Next we will apply this result to some special functions, for which $x\in E^N$ is seen only through its empirical measure $m^{(N)}(x)=\frac{1}{N}\sum_{1\le i\le N}\delta_{x_i}$, and more precisely to mappings f of the following type

$$\forall\,(t,x)\in[0,T]\times E^N,\qquad f(t,x)=m^{(N)}(x)(\widetilde{f}(t,\cdot))$$

where $\widetilde{f}\in\mathcal{A}_T$, since clearly such a function f belongs to $\mathcal{A}_{T,N}$.

3.3 Asymptotic Behavior

An important role will be played here by the unnormalized Feynman-Kac stochastic flow defined in (18). In all this section, the finite horizon $T>0$ and the initial condition η_0 are fixed.

3.3.1 Weak Propagation of Chaos

Our objective is to prove the following result, which somehow gives a justification for the interacting particle system just introduced:

Theorem 3.19 *Under the assumption $(H2)_T$, there exists a finite constant $C_T > 0$, such that for all $\varphi \in \mathcal{B}_b(E)$ and all $0 \le t \le T$,*

$$\mathbb{E}[|\eta_t^N(\varphi) - \eta_t(\varphi)|] \;\le\; C_T \frac{\|\varphi\|}{\sqrt{N}}$$

The basic idea behind the proof of this result is quite simple: it consists in finding a martingale indexed by the interval $[0, T]$ whose terminal value at time T is precisely $\gamma_T^N(\varphi)$, for any given $\varphi \in \mathcal{A}$. The reader may be wondering why this quantity and not directly $\eta_T^N(\varphi)$. The reason for this choice is that for the unnormalized Feynman-Kac stochastic flow, we can take advantage of the underlying linear structure of the limiting dynamical system $(\gamma_t)_{t\ge 0}$ (ie of the relation $\gamma_t(\cdot) = \gamma_s(Q_{s,t}(\cdot))$ valid for all $t \ge s \ge 0$). As we will see, there is then a straight way to find such a martingale, via the martingale problem satisfied by the law of the particle system, but naturally we have to use the semigroup $(Q_{s,t})_{0\le s\le t}$ considered above. All the calculations would be immediate if one can use the heuristic formula

$$\partial_s Q_{s,t} \;=\; -L_s Q_{s,t} - U_s Q_{s,t}$$

(in the usual regular cases, this is satisfied, and maybe that at a first reading one should concentrate on these situations ...). To treat the little difficulties associated to the general case, we consider this equation in the sense of the martingale problem. This point of view led us to introduce the weak regularity condition (H2), insuring that some functions constructed through the semigroup $(Q_{s,t})_{0\le s\le t}$ are in the domain of an extended pregenerator.

Another important aspect of the martingales we will consider is that they are quite small, in the sense that their increasing process will be of order $1/N$. Then taking into account some results about iid random variables in order to estimate the initial approximation at time 0, the expected convergence will follow easily for the unnormalized Feynman-Kac formulae (and there will be no problem with the renormalization, so we will also end up with the convergence of the empirical measure toward the normalized Feynman-Kac formulae).

Furthermore, this approach gives at once the central limit theorem and exponential bounds for the fluctuations.

So from now on, we will look for nice martingales, through calculations of the action of the pregenerators on convenient functions, and $(H2)_T$ will be assumed fulfilled.

Lemma 3.20 *For any $\varphi \in \mathcal{A}$, the process*

$$\left(B_t^N(\varphi)\right)_{0\le t\le T}$$

$$\stackrel{\text{def.}}{=} \left(\eta_t^N(Q_{t,T}(\varphi)) - \eta_0^N(Q_{0,T}(\varphi)) + \int_0^t \eta_s^N(U_s)\eta_s^N(Q_{s,T}(\varphi))\, ds\right)_{0\le t\le T}$$

is a martingale and its increasing process is given by

$$\langle B^N(\varphi)\rangle_t \;=\; \frac{1}{N}\int_0^t G(s,T,\eta_s^N,\varphi)\,ds$$

where for any $0 \le s \le T$, $m \in \mathbf{M}_1(E)$ *and* $\varphi \in \mathcal{A}$,

$$G(s,T,m,\varphi)$$

$$= m\left[\bar{\Gamma}(F_{T,\varphi},F_{T,\varphi})(t,\cdot)\right] + m\left[(Q_{s,T}(\varphi) - m[Q_{s,T}(\varphi)])^2(U_s + m[U_s])\right]$$

Proof: Applying Lemma 3.17 and Lemma 3.18 to the function

$$f\,:\,[0,T]\times E^N \ni (t,x) \mapsto m^{(N)}(x)(Q_{t,T}(\varphi)) \tag{104}$$

it appears easily that

$$\begin{aligned}\widetilde{\mathcal{L}}^{(N)}(f)(t,x) &= -m^{(N)}(x)[U_tQ_{t,T}(\varphi)]\\ \widehat{\mathcal{L}}^{(N)}(f)(t,x) &= m^{(N)}(x)[U_tQ_{t,T}(\varphi)] - m^{(N)}(x)[U_t]m^{(N)}(x)[Q_{t,T}(\varphi)]\end{aligned}$$

and the proof of the first assertion is now complete.
For the second one, calculations are a little more tedious, but in the end, we get

$$N\widetilde{\Gamma}^{(N)}(f)(t,x) = m^{(N)}(x)[\bar{\Gamma}(F_{T,\varphi},F_{T,\varphi})(t,\cdot)]$$

and

$$N\widehat{\Gamma}^{(N)}(f)(t,x) = m^{(N)}(x)\left[(Q_{s,T}(\varphi) - m^{(N)}(x)[Q_{s,T}(\varphi)])^2(U_s + m^{(N)}(x)[U_s])\right]$$

■

In view of the above lemma to find a good upper bound of $N\mathbb{E}[\langle B^N(\varphi)\rangle_T]$ it is tempting to introduce the norm on $\mathcal{A}$ defined for any $\varphi \in \mathcal{A}$ by

$$|||\varphi|||_{[0,T]} \;=\; \sqrt{\int_0^T \|F_{T,\varphi}(t,\cdot)\|^2 + \left\|\bar{\Gamma}(F_{T,\varphi},F_{T,\varphi})(t,\cdot)\right\|\,dt} \tag{105}$$

(more rigorously, the infimum over all possible choices of $\bar{\Gamma}(F_{T,\varphi},F_{T,\varphi})$ of this quantities). Using the considerations of section 3.1.2 we see that this norm is clearly related to $|||\cdot|||$

Next result shows that the above choice is in fact far from being optimal.

Lemma 3.21 *There exists a finite constant* $\bar{C}_T > 0$, *such that for all* $\varphi \in \mathcal{A}$,

$$\mathbb{E}[\langle B^N(\varphi)\rangle_T] \;\le\; \bar{C}_T\frac{\|\varphi\|^2}{N}$$

Proof: Let us write that for the function f defined in (104),
$\forall\,(t,x)\in[0,T]\times E^N$,

$$\begin{aligned}N\widetilde{\Gamma}^{(N)}(f,f)(t,x) &= m^{(N)}(x)[\bar{L}(Q^2_{\cdot,T}(\varphi))(t,\cdot) - 2Q_{t,T}(\varphi)(\cdot)\bar{L}(Q_{\cdot,T}(\varphi))(t,\cdot)]\\ &= m^{(N)}(x)[\bar{L}(Q^2_{\cdot,T}(\varphi))(t,\cdot)] + 2m^{(N)}(x)[U_tQ^2_{t,T}(\varphi)]\end{aligned}$$

If we define

$$g\,:\,[0,T]\times E^N\ni(t,x)\mapsto m^{(N)}(x)[Q^2_{t,T}(\varphi)]$$

then

$$\widetilde{\mathcal{L}}^{(N)}(g)(t,x)\;=\;m^{(N)}(x)[\bar{L}(Q^2_{\cdot,T}(\varphi))(t,\cdot)]$$

therefore

$$\begin{aligned}N\widetilde{\Gamma}^{(N)}(f,f)(t,x)&=\widetilde{\mathcal{L}}^{(N)}(g)(t,x)+2m^{(N)}(x)[U_tQ^2_{t,T}(\varphi)]\\&=\mathcal{L}^{(N)}(g)(t,x)-\widehat{\mathcal{L}}^{(N)}(g)(t,x)+2m^{(N)}(x)[U_tQ^2_{t,T}(\varphi)]\end{aligned}$$

and

$$\begin{aligned}&N\widetilde{\Gamma}^{(N)}(f,f)(t,x)\\&=\mathcal{L}^{(N)}(g)(t,x)+m^{(N)}(x)[U_tQ^2_{t,T}(\varphi)]+m^{(N)}(x)[U_t]m^{(N)}(x)[Q^2_{t,T}(\varphi)]\end{aligned}$$

and finally one gets

$$\begin{aligned}&N\Gamma^{(N)}(f,f)(t,x)\\&=\;\mathcal{L}^{(N)}(g)(t,x)+2m^{(N)}(x)[U_tQ^2_{t,T}(\varphi)]\\&\quad-2m^{(N)}(x)[U_tQ_{t,T}(\varphi)]m^{(N)}(x)[Q_{t,T}(\varphi)]+2m^{(N)}(x)[U_t]m^{(N)}(x)[Q^2_{t,T}(\varphi)]\end{aligned}$$

This implies that

$$\begin{aligned}&N\mathbb{E}[\langle B^N(\varphi)\rangle_T]\\&=\;\mathbb{E}\left[\int_0^T\Gamma^{(N)}(f,f)(t,\xi_t)\,dt\right]\\&=\;\mathbb{E}\Big[\eta^N_T[Q^2_{T,T}(\varphi)]-\eta^N_0[Q^2_{0,T}(\varphi)]+\\&\quad\int_0^T 2\eta^N_t[U_tQ^2_{t,T}(\varphi)]-2\eta^N_t[U_tQ_{t,T}(\varphi)]\eta^N_t[Q_{t,T}(\varphi)]+2\eta^N_t[U_t]\eta^N_t[Q^2_{t,T}(\varphi)]\,dt\Big]\end{aligned}$$

and the desired upper bound is now clear.

■

As announced, it is preferable to first obtain a result similar to Theorem 3.19 for γ^N.

Proposition 3.22 *There exists a finite constant $\widetilde{C}_T>0$, such that for any $\varphi\in\mathcal{A}$ and $0\le t\le T$,*

$$\mathbb{E}[|\gamma^N_t(\varphi)-\gamma_t(\varphi)|]\;\le\;\widetilde{C}_T\frac{\|\varphi\|}{\sqrt{N}}\tag{106}$$

Proof: By construction, for any $\varphi\in\mathcal{A}$, we have that

$$\gamma^N_t(\varphi)\;=\;\exp\left(\int_0^t\eta^N_s(U_s)\,ds\right)\eta^N_t(\varphi)$$

By Lemma 3.20 it follows that for $0 \leq t \leq T$,

$$\gamma_t^N(Q_{t,T}(\varphi)) = \gamma_0^N(Q_{0,T}(\varphi)) + \widetilde{B}_t^N(\varphi) \tag{107}$$

where

$$\widetilde{B}_t^N(\varphi) = \int_0^t \exp\left(\int_0^s \eta_u^N(U_u)\,du\right) dB_s^N(\varphi)$$

is a martingale. Its increasing process is clearly given for any $0 \leq t \leq T$ by

$$\begin{aligned}\langle \widetilde{B}^N(\varphi)\rangle_t &= \int_0^t \exp\left(2\int_0^s \eta_u^N(U_u)\,du\right) d\langle B^N(\varphi)\rangle_s \\ &= \frac{1}{N}\int_0^t \exp\left(2\int_0^s \eta_u^N(U_u)\,du\right) G(s,T,\eta_s^N,Q_{s,T}(\varphi))\,ds \\ &\leq \frac{\widehat{C}_T}{N}\langle B(\varphi)\rangle_T\end{aligned}$$

for a finite constant $\widehat{C}_T > 0$ which does not depend on φ (but it depends on U, through $\|U\|_{[0,T]}$).
Recalling that $\gamma_T(\varphi) = \gamma_0(Q_{0,T}(\varphi))$ one concludes that

$$\gamma_T^N(\varphi) - \gamma_T(\varphi) = \gamma_0^N(Q_{0,T}(\varphi)) - \gamma_0(Q_{0,T}(\varphi)) + \widetilde{B}_T^N(\varphi) \tag{108}$$

from which one gets that

$$\mathbb{E}[(\gamma_T^N(\varphi) - \gamma_T(\varphi))^2] = \mathbb{E}[(\gamma_0^N(Q_{0,T}(\varphi)) - \gamma_0(Q_{0,T}(\varphi)))^2] + \mathbb{E}[(\widetilde{B}_T^N)^2(\varphi)]$$

Now the result follows, via a Cauchy-Schwarz inequality, from

$$\mathbb{E}[(\widetilde{B}_T^N)^2(\varphi)] = \mathbb{E}[\langle \widetilde{B}^N(\varphi)\rangle_T]$$

and from the classical equality for iid variables:

$$\begin{aligned}\mathbb{E}[(\gamma_0^N(Q_{0,T}(\varphi)) - \gamma_0(Q_{0,T}(\varphi)))^2] &= \mathbb{E}[(\eta_0^N(Q_{0,T}(\varphi)) - \eta_0(Q_{0,T}(\varphi)))^2] \\ &= \frac{\eta_0[(Q_{0,T}(\varphi) - \eta_0(Q_{0,T}(\varphi)))^2]}{N} \\ &\leq \exp(2T\,\|U\|_{[0,T]})\frac{\|\varphi\|^2}{N}\end{aligned}$$

■

Proof of Theorem 3.19: First let us prove that the upper bound of the previous proposition is true for any $\varphi \in \mathcal{B}_b(E)$. By density of $\mathcal{A}$ in $\mathcal{C}_b(E)$, that is at least clear for all $\varphi \in \mathcal{C}_b(E)$. Next let $\varphi \in \mathcal{B}_b(E)$ be fixed. We have

$$\mathbb{E}[|\gamma_t^N(\varphi) - \gamma_t(\varphi)|] = \sup_{H\in\mathbb{L}^\infty(\mathbb{P}),\|H\|_{\mathbb{L}^\infty(\mathbb{P})}\leq 1} \mathbb{E}[H(\gamma_t^N(\varphi) - \gamma_t(\varphi))]$$

Let $H \in \mathbb{L}^\infty(\mathbb{P})$ be given with $\|H\|_{\mathbb{L}^\infty(\mathbb{P})} \leq 1$. Then the following formula

$$\forall\, \widetilde{\varphi} \in \mathcal{B}_b(E), \qquad m_H(\widetilde{\varphi}) \stackrel{\text{def.}}{=} \mathbb{E}[H(\gamma_t^N(\widetilde{\varphi}) - \gamma_t(\widetilde{\varphi}))]$$

defines a measure $m_H \in \mathbf{M}(E)$.
But a classical approximation result says that there exists a sequence $(\varphi_n)_{n\in\mathbb{N}}$ of elements of $\mathcal{C}_b(E)$, with $\|\varphi_n\| \leq \|\varphi\|$ for all $n \in \mathbb{N}$, such that

$$\lim_{n\to\infty} m_H(\varphi_n) \;=\; m_H(\varphi)$$

Putting together these facts, (106) is satisfied for all $\varphi \in \mathcal{B}_b(E)$.
Now it is enough to write that for all $\varphi \in \mathcal{A}$,

$$\eta_T^N(\varphi) - \eta_t(\varphi) = \frac{1}{\gamma_T(\mathbf{1})}\left[\gamma_T^N(\varphi) - \gamma_T(\varphi) + \eta_T^N(\varphi)(\gamma_T(\mathbf{1}) - \gamma_T^N(\mathbf{1}))\right] \tag{109}$$

■

Theorem 3.19 can be improved. As a first step in this direction, let us come back to the martingale $(B_t^N(\varphi))_{0\leq t\leq T}$, for a given $\varphi \in \mathcal{A}$, and study its jumps:

Lemma 3.23 *There exists a finite constant $C_T^{(7)} > 0$ such that $\mathbb{P}$-a.s., we have*

$$\sup_{0\leq t\leq T} \left|\triangle B_t^N(\varphi)\right| \;\leq\; C_T^{(7)}\frac{\|\varphi\|}{N}$$

Proof: From the definition of $(B_t^N(\varphi))_{0\leq t\leq T}$, we see that this upper bound would be clear if we could show that $\mathbb{P}$-a.s., the coordinates of the $\mathbb{R}^N$-valued process

$$\left(\left(Q_{t,T}(\varphi)(\xi_t^i)\right)_{1\leq i\leq N}\right)_{0\leq t\leq T}$$

never jump together.

From the construction of $\mathbb{P}$ (cf [45] for more details), it is sufficient to prove this property for the process $\widetilde{\xi}$ which is just the product of N independent copies of X (starting from the distribution η_0). So it is in fact enough to convince oneself that the jumps of the martingale $(N_t(T,\varphi))_{0\leq t\leq T} = (F_{T,\varphi}(t,X_t))_{0\leq t\leq T}$ (under $\mathbb{P}_{\eta_0}$) are totally inaccessible, but this is a consequence of the continuity of its increasing process.

■

A similar statement holds for $(\widetilde{B}_t^N(\varphi))_{0\leq t\leq T}$.

Lemma 3.24 *For any $p > 0$, there exists a finite constant $C_{T,p}^{(8)} > 0$, which does not depend on φ and such that*

$$\sup_{0\leq t\leq T} \mathbb{E}\left[\left|\widetilde{B}_t^N(\varphi)\right|^p\right]^{1/p} \;\leq\; C_{T,p}^{(8)}\frac{\|\varphi\|}{\sqrt{N}}$$

Proof: By Hölder inequality, we only need to prove this result for $p = 2q$, with $q \in \mathbb{N}^*$, and we will proceed by an induction on q. We have already shown the case $q = 1$, so let $q \in \mathbb{N}\setminus\{0,1\}$ be given. Applying the Itô's formula for the mapping $\mathbb{R} \ni x \mapsto x^{2q}$, we get (writing $\widetilde{B} \stackrel{\text{def.}}{=} \widetilde{B}^N(\varphi)$ for simplicity) for $0 \leq t \leq T$,

$$\widetilde{B}_t^{2q} = 2q\int_0^t \widetilde{B}_s^{2q-1}\,d\widetilde{B}_s + q(2q-1)\int_0^t \widetilde{B}_s^{2q-2}\,d\langle\widetilde{B}^c\rangle_s$$
$$+\sum_{0\leq s\leq t} \widetilde{B}_s^{2q} - \widetilde{B}_{s-}^{2q} - 2q\widetilde{B}_{s-}^{2q-1}\triangle\widetilde{B}_s$$

where $\widetilde{B}^c$ is the continuous martingale part of $\widetilde{B}$.
But quite obviously, there exists a finite constant $C_q > 0$ depending only on $q \geq 2$ and such that for any $0 \leq s \leq T$,

$$\widetilde{B}_s^{2q} - \widetilde{B}_{s-}^{2q} - 2q\widetilde{B}_{s-}^{2q-1}\Delta\widetilde{B}_s \;\leq\; C_q(\widetilde{B}_s^{2q-2} + (\Delta\widetilde{B}_s)^{2q-2})\Delta B_s^2$$

Using the previous lemma and the general fact that

$$\left(\langle\widetilde{B}\rangle_t - \langle\widetilde{B}^c\rangle_t - \sum_{0\leq s\leq t} \Delta B_s^2\right)_{0\leq t\leq T}$$

is a martingale, one can find a finite constant $C_{T,p} > 0$ such that

$$\mathbb{E}[(\widetilde{B}_t)^{2q}] \;\leq\; C_{T,p}\mathbb{E}\left[\int_0^t \left(\widetilde{B}_s^{2q-2} + \left(\frac{\|\varphi\|}{N}\right)^{2q-2}\right) d\langle\widetilde{B}\rangle_s\right]$$

Now the desired result follows from the bounds on $\langle\widetilde{B}\rangle$ we have already met in the proof of Proposition 3.22 and from the induction hypothesis (after an appropriate application of Fatou's Lemma).

■

A consequence of these preliminary results is that

Proposition 3.25 *For all $p \geq 1$, there exists a finite constant $C_{p,T} > 0$ such that for all $\varphi \in \mathcal{B}_b(E)$ and $0 \leq t \leq T$*

$$\mathbb{E}[|\eta_t^N(\varphi) - \eta_t(\varphi)|^p]^{1/p} \leq C_{p,T}\frac{\|\varphi\|}{\sqrt{N}}$$

Proof: Using the same arguments as in the proofs of Proposition 3.22 and Theorem 3.19, this inequality is a consequence of previous lemma and standard Marcinkiewicz-Zygmund's inequality, ensuring that there is a finite constant $C_p > 0$ such that

$$\mathbb{E}[|\eta_0^N(Q_{0,T}(\varphi)) - \eta_0(Q_{0,T}(\varphi))|^p]^{1/p} \;\leq\; \frac{C_p\,\|Q_{0,T}(\varphi)\|}{\sqrt{N}}$$

■

Using for instance the previous proposition with $p = 4$ (note that for this case, we could have rather take into account the Lemma 3.34 p. 382 of [68], instead of Lemma 3.24), we can apply the Borel-Cantelli Lemma to see that almost surely, for a $\varphi \in \mathcal{B}_b(E)$ fixed,

$$\lim_{N\to\infty} \eta_t^N(\varphi) \;=\; \eta_t(\varphi)$$

In fact the Proposition 3.25 can be quantitatively improved for $p = 1$, if we put the absolute value outside the expectation:

Proposition 3.26 *For $T > 0$ given, there exists a constant $\bar{C}_T \geq 0$ such that for all $\varphi \in \mathcal{B}_b(E)$ and all $N \geq 1$, we have*

$$\left|\mathbb{E}[\eta_T^N(\varphi)] - \eta_T(\varphi)\right| \;\leq\; \bar{C}_T\frac{\|\varphi\|}{N}$$

Proof: Using the equation (108) valid for all $\varphi \in \mathcal{A}$, we get

$$\mathbb{E}[\gamma_T^N(\varphi)] \;=\; \gamma_T(\varphi)$$

But clearly this equality is then true for all $\varphi \in \mathcal{B}_b(E)$, via usual arguments (ie γ_T^N is an estimator without biais of γ_T).

Then taking into account (109), we see that

$$\begin{aligned}
\left|\mathbb{E}[\eta_T^N(\varphi)] - \eta_T(\varphi)\right| &= \left|\mathbb{E}\left[\frac{1}{\gamma_T(\mathbf{1})}\eta_T^N(\varphi)(\gamma_T^N(\mathbf{1}) - \gamma_T(\mathbf{1}))\right]\right| \\
&= \left|\mathbb{E}\left[\frac{1}{\gamma_T(\mathbf{1})}(\eta_T^N(\varphi) - \eta_T(\varphi))(\gamma_T^N(\mathbf{1}) - \gamma_T(\mathbf{1}))\right]\right| \\
&\leq \frac{1}{\gamma_T(\mathbf{1})}\sqrt{\mathbb{E}[(\eta_T^N(\varphi) - \eta_T(\varphi))^2]}\sqrt{\mathbb{E}[(\gamma_T^N(\mathbf{1}) - \gamma_T(\mathbf{1}))^2]} \\
&\leq \frac{\bar{C}_T}{N}
\end{aligned}$$

for a constant $\bar{C}_T > 0$, according to the Proposition 3.25 and its proof.

■

The proportionality to $1/N$ of the upper bound we have just obtained is related to the strong propagation of chaos. This question will not be thoroughly investigated here, but let us give some immediate remarks about it.

First we recall the definition of this property: let $\bar{X} = (\bar{X}_t)_{t\geq 0}$ be the time inhomogeneous Markovian process taking values in E, whose initial law is η_0 and whose family of pregenerators is $(\mathcal{L}_{t,\eta_t})_{t\geq 0}$, where $(\eta_t)_{t\geq 0}$ is defined by (3) and (4) (starting from the same η_0) and where the pregenerator $\mathcal{L}_{t,\eta}$, for $t \geq 0$ and $\eta \in \mathbf{M}_1(E)$, is given by (12).

In the generalized Boltzmann equations literature, $\bar{X}$ is called the nonlinear process (or sometimes the tarjet process) because at any time $t \geq 0$, the law of $\bar{X}_t$ is η_t, but at this instant its pregenerator also uses in its definition the probability η_t. So the evolution depends on the time marginal and this is the nonlinear aspect of $\bar{X}$.

For $T > 0$ fixed, let $\bar{\mathbb{P}}_{\eta_0,[0,T]}$ denote the law of $(\bar{X}_t)_{0\leq t\leq T}$. Then the strong propagation of chaos can be expressed as the existence of a constant $C_T^{(9)} \geq 0$ such that for all $N \geq 1$ and $1 \leq k \leq N$, we are assured of

$$\left\|\mathbb{P}_{\eta_0,[0,T]}^{(N,1,\cdots,k)} - \bar{\mathbb{P}}_{\eta_0,[0,T]}^{\otimes k}\right\|_{\mathrm{tv}} \;\leq\; \frac{C_T^{(9)}k^2}{N}$$

where $\|\cdot\|_{\mathrm{tv}}$ stands for the total variation norm and where $\mathbb{P}_{\eta_0,[0,T]}^{(N,1,\cdots,k)}$ denotes the law over the time interval $[0,T]$ of the kth first particles $(\xi_t^1, \xi_t^2, \cdots, \xi_t^k)_{0\leq t\leq T}$ (cf Graham and Méléard [65]).

Using on one hand a generalization of the approach presented here, but for the tensorized empirical measures alluded to in the beginning of section 3.1.1, and on the other hand two straightforward coupling arguments, it is possible to prove such a behavior, except we have not yet been able to get the right dependence in k. Nevertheless, note that as a direct consequence of Proposition 3.26, we have the simpler bound

$$\left\| \mathbb{P}^{(N,1)}_{\eta_0,T} - \bar{\mathbb{P}}_{\eta_0,T} \right\|_{\text{tv}} \leq \frac{\bar{C}_T}{N}$$

where $\mathbb{P}^{(N,1)}_{\eta_0,T}$ (resp. $\bar{\mathbb{P}}_{\eta_0,T}$) denote the law of ξ^1_T (resp. $\bar{X}_T$). This comes from the identity $\bar{\mathbb{P}}_{\eta_0,T} = \eta_T$ and the fact that the distribution of ξ^1_T is also $\mathbb{E}[\eta^N_T]$, by exchangeability of the particles.

Furthermore, if we assume that the semigroup Φ is exponentially asymptotically stable, in the sense (111) given below, then we obtain an uniform in time result: for some constants $\bar{C} \geq 0$ and $0 < \alpha \leq 1$,

$$\sup_{t \geq 0} \left\| \mathbb{P}^{(N,1)}_{\eta_0,t} - \bar{\mathbb{P}}_{\eta_0,t} \right\|_{\text{tv}} \leq \frac{\bar{C}}{N^\alpha}$$

More generally, to get uniform upper bounds with respect to the time parameter (under additional stability assumptions) it is important to control the dependence of the constant $C_{p,T}$ arising in Proposition 3.25. A more cautious study shows that in fact there exists a universal constant $A_p > 0$ depending on the parameter p and an additional finite constant $B > 0$ (which do not depend on p) such that

$$C_{p,T} \leq A_p \exp\left(B \int_0^T (1 + \|U_s\|) \, ds \right) \tag{110}$$

We end this section with a uniform convergence result with respect to the time parameter.

Theorem 3.27 *Assume that the semigroup Φ associated with the dynamics structure of η and defined in (101) is asymptotically stable in the sense that*

$$\lim_{T\to\infty} \sup_{\mu,\nu \in \mathbf{M}_1(E)} \sup_{t\geq 0} \|\Phi_{t,t+T}(\mu) - \Phi_{t,t+T}(\nu)\|_{\text{tv}} = 0$$

If the function U satisfies $\|U\|^\star \stackrel{\text{def.}}{=} \sup_{t\geq 0} \|U_t\| < \infty$ then for any bounded Borel function φ we have the following uniform convergence result

$$\lim_{N\to\infty} \sup_{t\geq 0} \mathbb{E}\left(|\eta^N_t \varphi - \eta_t \varphi| \right) = 0$$

In addition, assume that the semigroup Φ is exponentially asymptotically stable in the sense that there exist some positive constant $\gamma > 0$ and $T_0 \geq 0$ such that for any $\mu, \nu \in \mathbf{M}_1(E)$ and $T \geq T_0$

$$\sup_{t\geq 0} \|\Phi_{t,t+T}(\mu) - \Phi_{t,t+T}(\nu)\|_{\text{tv}} \leq e^{-\gamma T} \tag{111}$$

Then for any $p \geq 1$ and for any Borel function φ, $\|\varphi\| \leq 1$, we have the following uniform $\mathbb{L}^p$ error bound

$$\sup_{t\geq 0} \mathbb{E}\left(|\eta_t^N(\varphi) - \eta_t(\varphi)|^p\right)^{\frac{1}{p}} \leq \frac{A'_p\, e^{\gamma'}}{N^{\frac{\alpha}{2}}}$$

for any $N \geq 1$ such that

$$T(N) \stackrel{\text{def}}{=} \frac{1}{2}\frac{\log N}{\gamma+\gamma'} \geq T_0$$

where A'_p is a universal constant which only depends on $p \geq 1$ and α and γ' are given by

$$\alpha = \frac{\gamma}{\gamma+\gamma'} \quad \text{and} \quad \gamma' = B(1+\|U\|^*)$$

and B is the finite constant arising in (110).

Proof: To prove this theorem we follow the same line of arguments as in the proof of Theorem 2.11. Since most of the computations are similar to those made in discrete time settings the proof will be only sketched.
The only point we have to check is that for any φ, $\|\varphi\| \leq 1$, $p \geq 1$, $T \geq T_0$ and $t \geq 0$

$$\mathbb{E}\left(\left|\eta_{t+T}^N(\varphi) - \Phi_{t,t+T}(\eta_t^N)(\varphi)\right|^p\right)^{\frac{1}{p}} \leq A_p \frac{\exp(\gamma' T)}{\sqrt{N}} \quad \text{with} \quad \gamma' = B(1+\|U\|^*)$$

Subtracting the equalities (107) at times $t = T$ and $t = s$ (and taking into account the definition of $\widetilde{B}$ given below (107)), we get that for any $s \leq T$ and $\varphi \in \mathcal{A}$,

$$\gamma_T^N(\varphi) = \gamma_s^N(Q_{s,T}(\varphi)) + \int_s^T \exp\left(\int_0^\tau \eta_u^N(U_u)\,du\right) dB_\tau^N(\varphi)$$

Since by definition we have

$$\gamma_s^N(1) = \exp\left(\int_0^s \eta_u^N(U_u)\,du\right)\eta_s^N(1) = \exp\left(\int_0^s \eta_u^N(U_u)\,du\right)$$

it appears by construction that

$$\frac{\gamma_s^N(Q_{s,T}(\varphi))}{\gamma_s^N(1)} = \eta_s^N(Q_{s,T}(\varphi))$$

so the above decomposition yields that

$$\eta_{s,T}^N(\varphi) \stackrel{\text{def.}}{=} \frac{\gamma_T^N(\varphi)}{\gamma_s^N(1)} = \eta_s^N(Q_{s,T}(\varphi)) + \int_s^T \exp\left(\int_s^\tau \eta_u^N(U_u)\,du\right) dB_\tau^N$$

By the same reasoning as in Lemma 3.24 one can check that for any $p \geq 1$, there exists a universal constant $A_p > 0$ depending on the parameter p and an additional finite constant $B > 0$ (which do not depend on p) such that

$$\mathbb{E}\left(\left|\eta_{s,T}^N(\varphi) - \eta_s^N(Q_{s,T}(\varphi))\right|^p\right)^{\frac{1}{p}} \leq A_p \frac{e^{B(1+\|U\|^*)(T-s)}}{\sqrt{N}}$$

Instead of (109) we now use the decomposition

$$\eta_T^N(\varphi) - \Phi_{s,T}(\eta_s^N)(\varphi)$$

$$= \frac{1}{\eta_s^N(Q_{s,T}(1))} \left((\eta_{s,T}^N(\varphi) - \eta_s^N(Q_{s,T}(\varphi))) + \eta_T^N(\varphi) \left(\eta_s^N(Q_{s,T}(1)) - \eta_{s,T}^N(1) \right) \right)$$

to prove that

$$\mathbb{E}\left(\left| \eta_T^N(\varphi) - \Phi_{s,T}(\eta_s^N)(\varphi) \right|^p \right)^{\frac{1}{p}} \leq A_p \, \frac{e^{B(1+\|U\|^*)\,(T-s)}}{\sqrt{N}}$$

The desired upper bound is now clear by replacing the pair parameter (s,T) by $(t, t+T)$. ■

3.3.2 Central Limit Theorem

We now turn to fluctuactions associated with the weak propagation of chaos. We proceed in much the same way as in discrete time settings. Let us introduce the "normalized" semigroup $(\bar{Q}_{s,t})_{0\leq s \leq t}$ defined by

$$\forall\, 0 \leq s \leq t,\ \forall\, \varphi \in \mathcal{A}, \qquad \bar{Q}_{s,t}(\varphi) = \frac{Q_{s,t}(\varphi)}{\eta_s(Q_{s,t}(\mathbf{1}))}$$

To see that it is in fact a semigroup, we first notice that

$$\begin{aligned} \forall\, 0 \leq s \leq t \leq T, \qquad \eta_s(Q_{s,t}(\mathbf{1})) &= \mathbb{E}_{s,\eta_s}\left[\exp\left(\int_s^t U_u(X_u)\, du \right) \right] \\ &= \exp\left(\int_s^t \eta_u(U_u)\, du \right) \end{aligned}$$

where the latter equality correspond to the basic identity

$$\gamma_t(1) = \exp\left(\int_0^t \eta_s(U_s)\, ds\right)$$

proved at the end of section 1.3, but for the shifted dynamical system $(\eta_{s+t})_{t\geq 0}$ instead of $(\eta_t)_{t\geq 0}$.

Consequently the above semigroup can be rewritten as follows

$$\forall\, 0 \leq s \leq t \leq T,\ \forall\, x \in E,\ \forall\, \varphi \in \mathcal{A},$$

$$\bar{Q}_{s,t}(\varphi)(x) = \mathbb{E}_{s,x}\left[\varphi(X_t) \exp\left(\int_s^t U_u(X_u) - \eta_u(U_u)\, du \right) \right]$$

Therefore it is clear that $(\bar{Q}_{s,t})_{0\leq s\leq t\leq T}$ is defined as $(Q_{s,t})_{0\leq s\leq t\leq T}$, by replacing the mapping U by the mapping

$$\widetilde{U} : \mathbb{R}_+ \times E \ni (t,x) \mapsto U_t(x) - \eta_t(U_t)$$

Theorem 3.28 *For $\varphi \in \mathcal{A}$, define*

$$W_T^N(\varphi) = \sqrt{N}(\eta_T^N(\varphi) - \eta_T(\varphi))$$

and let us denote as in (61), for $0 \leq t \leq T$,

$$\varphi_{t,T} = \bar{Q}_{t,T}(\varphi - \eta_T(\varphi))$$

Then under $(H2)_T$, the family $(W_T^N(\varphi))_{\varphi\in\mathcal{A}}$ converges in law to a centered Gaussian field $(W_T(\varphi))_{\varphi\in\mathcal{A}}$ whose covariances are given by

$$\forall\, \phi, \varphi \in \mathcal{A}, \qquad \mathbb{E}[W_T(\phi)W_T(\varphi)] = \eta_0[\phi_{0,T}\varphi_{0,T}] + \int_0^T \bar{G}(s, T, \eta_s, \phi, \varphi)\, ds$$

where for all $0 \leq s \leq T$, all $m \in \mathbf{M}_1(E)$ and all $\phi, \varphi \in \mathcal{A}$,

$$\bar{G}(s, T, m, \phi, \varphi)$$

$$= m\left[\bar{\Gamma}(\phi_{\cdot,T}, \varphi_{\cdot,T})(s, \cdot)\right] + m\left[(\phi_{s,T} - m[\phi_{s,T}])(\varphi_{s,T} - m[\varphi_{s,T}])(U_s + m[U_s])\right]$$

First, we will only consider one function $\varphi \in \mathcal{A}$, for which we have the analogous result of Lemma 3.20, whose proof is quite identical:

Lemma 3.29 *For $\varphi \in \mathcal{A}$, the process*

$$(\bar{B}_t^N(\varphi))_{0\leq t\leq T} \stackrel{\text{def.}}{=} \left(\eta_t^N(\varphi_{t,T}) + \int_0^t (\eta_s^N(U_s) - \eta_s(U_s))\eta_s^N(\varphi_{s,T})\, ds\right)_{0\leq t\leq T}$$

is a martingale, whose initial value is

$$\bar{B}_0^N(\varphi) = \eta_0^N(\varphi_{0,T})$$

and whose increasing process is given by

$$\langle \bar{B}^N(\varphi)\rangle_t = \int_0^t \bar{G}(s, T, \eta_s^N, \varphi, \varphi)\, ds$$

We will examine separately each term arising in the above lemma. For the first one, we have

Lemma 3.30 *Under $(H2)_T$, the random variables*

$$\sqrt{N}\int_0^T \left|(\eta_s^N(U_s) - \eta_s(U_s))\eta_s^N(\varphi_{s,T})\right|\, ds$$

converge in probability to 0 as N tends to infinity.

Proof: It is enough to show that

$$\lim_{N\to\infty} \mathbb{E}\left[\sqrt{N}\int_0^T \left|(\eta_s^N(U_s) - \eta_s(U_s))\eta_s^N(\varphi_{s,T})\right|\, ds\right] = 0 \qquad (112)$$

To this end we use the following Cauchy-Schwarz upper bound of this quantity (before going to the limit for N large)

$$\sqrt{\mathbb{E}\left[\int_0^T (\eta_s^N(U_s) - \eta_s(U_s))^2\, ds\right]}\sqrt{N\mathbb{E}\left[\int_0^T (\eta_s^N(\varphi_{s,T}))^2\, ds\right]}$$

Since $\|U\|_{[0,T]} < +\infty$, using the last assertion of the previous lemma and dominated convergence, the first factor goes to zero as N tends to infinity.
For the second term, let us write that for $0 \leq s \leq T$,

$$\begin{aligned}
\eta_s^N[\varphi_{s,T}] &= \eta_s^N[\bar{Q}_{s,T}(\varphi)] - \eta_s^N[\bar{Q}_{s,T}(\mathbf{1})]\eta_T(\varphi) \\
&= \eta_s^N[\bar{Q}_{s,T}(\mathbf{1})]\left(\frac{\eta_s^N[\bar{Q}_{s,T}(\varphi)]}{\eta_s^N[\bar{Q}_{s,T}(\mathbf{1})]} - \eta_T(\varphi)\right) \\
&= \eta_s^N[\bar{Q}_{s,T}(\mathbf{1})]\left(\frac{\eta_s^N[Q_{s,T}(\varphi)]}{\eta_s^N[Q_{s,T}(\mathbf{1})]} - \frac{\eta_s[Q_{s,T}(\varphi)]}{\eta_s[Q_{s,T}(\mathbf{1})]}\right)
\end{aligned}$$

This yields that

$$\begin{aligned}
\eta_s^N[\varphi_{s,T}]^2 \leq\ & 2\eta_s^N[\bar{Q}_{s,T}(\mathbf{1})]^2\left(\frac{(\eta_s^N[Q_{s,T}(\varphi)] - \eta_s[Q_{s,T}(\varphi)])^2}{\eta_s^N[Q_{s,T}(\mathbf{1})]^2} + \right. \\
& \left. \frac{\eta_s[Q_{s,T}(\varphi)]^2}{\eta_s[Q_{s,T}(\mathbf{1})]^2\eta_s^N[Q_{s,T}(\mathbf{1})]^2}\left(\eta_s^N[Q_{s,T}(\mathbf{1})] - \eta_s[Q_{s,T}(\mathbf{1})]\right)^2\right)
\end{aligned}$$

But the proof of Theorem 3.19 shows in fact that for all $\varphi \in \mathcal{A}$

$$\sup_{0\leq s\leq T} N\mathbb{E}[(\eta_s^N[Q_{s,T}(\varphi)] - \eta_s[Q_{s,T}(\varphi)])^2] < +\infty$$

from which one concludes that (112) holds.

■

The important step in this section is the following

Proposition 3.31 *Under (H2)$_T$, the martingale $(\bar{B}_t^N(\varphi))_{0\leq t\leq T}$ converges in law (for the Skorokhod topology in $D([0,+\infty[,\mathbb{R})$) for N large toward a Gaussian centered martingale $(\bar{B}_t(\varphi))_{0\leq t\leq T}$ whose increasing process is the deterministic mapping*

$$[0,T] \ni t \mapsto \int_0^t \bar{G}(s,T,\eta_s,\varphi,\varphi)\, ds$$

and whose initial value $\bar{B}_0(\varphi)$ admits

$$\sigma(\varphi) \stackrel{def.}{=} \eta_0[(\bar{Q}_{0,T}(\varphi - \eta_T(\varphi)))^2]$$

for variance.

Proof: Conditionning with respect to the σ-algebra associated with time 0, it is not so difficult to realize that it is sufficient to prove the two following lemmas.

■

Lemma 3.32 *The random variables $\sqrt{N}\eta_0^N(\varphi_{0,T})$ converge in law for N large toward a centered Gaussian law of variance $\sigma(\varphi)$.*

Proof: It is just the usual central limit theorem for the independent variables

$$(\varphi_{0,T}(\xi_{i,0}^N))_{1\leq i\leq N},$$

where the $\xi_{i,0}^N$, $1 \leq i \leq N$, have the same law η_0. We would have noticed that $\eta_0(\varphi_{0,T}) = 0$.

■

Lemma 3.33 *Consider the process $(\check{B}_t^N(\varphi))_{0\leq t\leq T} = (\bar{B}_t^N(\varphi) - \bar{B}_0^N(\varphi))_{0\leq t\leq T}$, under the law $\widetilde{\mathbb{P}}$ which is constructed as $\mathbb{P}$, except that the initial distribution of ξ_0^N is a Dirac measure $\delta_{x_0^N}$ for some $x_0^N \in E^N$. Then $(\check{B}_t^N(\varphi))_{0\leq t\leq T}$ converge in law toward a Gaussian centered martingale $(\check{B}_t(\varphi))_{0\leq t\leq T}$ starting in 0 and whose increasing process is the deterministic mapping*

$$[0,T] \ni t \mapsto \int_0^t \bar{G}(s,T,\eta_s,\varphi,\varphi)\, ds$$

Proof: Since $(\mathrm{H2})_T$ is assumed to be satisfied for all $\eta_0 \in \mathbf{M}_1(E)$, $(\check{B}_t^N(\varphi))_{0\leq t\leq T}$ is again a martingale under $\widetilde{\mathbb{P}}$, and we also have that

$$\langle \check{B}^N(\varphi)\rangle_t = \int_0^t \bar{G}(s,T,\eta_s^N,\varphi,\varphi)\, ds$$

Thanks to Theorem 3.11 p. 432 of [68] it is now enough to prove that

$$\forall\, \epsilon > 0, \qquad \lim_{N\to+\infty} \widetilde{\mathbb{P}}\left[\sup_{0\leq s\leq T} |\Delta \check{B}_s^N(\varphi)| \geq \epsilon\right] = 0 \tag{113}$$

and

$$\forall\, 0 \leq t \leq T, \qquad \lim_{N\to+\infty} \langle \check{B}^N(\varphi)\rangle_t = \int_0^t G(s,T,\eta_s,\varphi)\, ds \tag{114}$$

where the last limit is understood in probability ($\widetilde{\mathbb{P}}$).
But (113) is proved in quite the same way as Lemma 3.23, and (114) comes from a dominated convergence theorem, using the weak propagation of chaos and the fact that for all $0 \leq s \leq T$, U_s, $\varphi_{s,T}$ and $\bar{\Gamma}(\varphi_{\cdot,T},\varphi_{\cdot,T})(s,\cdot) \in \mathcal{B}_b(E)$.

■

Putting together the previous calculations, we also see that the process

$$(\eta_t^N(\varphi_{t,T}))_{0\leq t\leq T}$$

converge to the same limit as the one presented in Proposition 3.31.

Now the Theorem 3.28 follows, by considering terminal values. More precisely, it remains to replace φ by $(\varphi_1,\varphi_2,\cdots,\varphi_p)$, where $\varphi_1,\varphi_2,\cdots,\varphi_p \in \mathcal{A}$, $p \geq 1$, and to use linear relations like $\bar{B}^N(\varphi_i + \varphi_j) = \bar{B}^N(\varphi_i) + \bar{B}^N(\varphi_j)$, $1 \leq i \neq j \leq p$, which implies for instance that for any $1 \leq i \neq j \leq p$

$$\langle \bar{B}^N(\varphi_i), \bar{B}^N(\varphi_j) \rangle = \frac{1}{4}(\langle \bar{B}^N(\varphi_i + \varphi_j) \rangle - \langle \bar{B}^N(\varphi_i - \varphi_j) \rangle)$$

The details are left to the reader.

The expression for the limit covariance can be simplified, in order to see that in fact it depends continuously on ϕ, φ with respect to the norm $\|\cdot\|$. Furthermore, it will confirm that it doesn't depend on the choice of $\bar{\Gamma}(\varphi_{\cdot,T}, \varphi_{\cdot,T})$.

Lemma 3.34 *More precisely, for any $\phi, \varphi \in \mathcal{A}$ we have*

$$\mathbb{E}[W_T(\phi) W_T(\varphi)] = \eta_T[(\varphi - \eta_T(\varphi))(\phi - \eta_T(\phi))] + 2 \int_0^T \eta_s[\varphi_{s,T} \phi_{s,T} U_s] \, ds$$

Proof: By a symmetrization procedure, it is sufficient to consider the case $\phi = \varphi$. Then we use calculations similar to those of Lemma 3.21.

To this end we first recall that

$$\begin{aligned} &\textstyle\int_0^T \eta_s[\bar{L}(\varphi^2_{\cdot,T})(s, \cdot)] \, ds \\ &= \int_0^T \mathbb{E}_{\eta_0} \left[\bar{L}(\varphi^2_{\cdot,T})(s, X_s) \exp\left(\int_0^s U_u(X_u) - \eta_u(U_u) \, du \right) \right] ds \end{aligned}$$

Writing for all $0 \leq s \leq T$,

$$\varphi^2_{s,T}(X_s) = \varphi^2_{0,T}(X_0) + \int_0^s \bar{L}(\varphi^2_{\cdot,T})(u, X_u) \, du + M_s^{(\varphi^2_{\cdot,T})}$$

with a certain martingale $(M_s^{(\varphi^2_{\cdot,T})})_{s \geq 0}$, we deduce that

$$\begin{aligned} &\int_0^T \mathbb{E}_{\eta_0} \left[\bar{L}(\varphi^2_{\cdot,T})(s, X_s) \exp\left(\int_0^s U_u(X_u) - \eta_u(U_u) \, du \right) \right] ds \\ &= \mathbb{E}_{\eta_0} \left[\varphi^2_{T,T}(X_T) \exp\left(\int_0^T U_s(X_s) - \eta_s(U_s) \, ds \right) \right] - \mathbb{E}_{\eta_0} \left[\varphi^2_{0,T}(X_0) \right] - \\ &\quad \mathbb{E}_{\eta_0} \left[\int_0^T \varphi^2_{s,T}(X_s)(U_s(X_s) - \eta_s(U_s)) \exp\left(\int_0^s U_u(X_u) - \eta_u(U_u) \, du \right) ds \right] \\ &= \eta_T[\varphi^2_{T,T}] - \eta_0[\varphi^2_{0,T}] - \int_0^T \eta_s[\varphi^2_{s,T}(U_s - \eta_s(U_s))] \, ds \end{aligned}$$

Therefore

$$\begin{aligned} &\int_0^T \eta_s[\bar{\Gamma}(\varphi_{\cdot,T}, \varphi_{\cdot,T})(s, \cdot)] \, ds \\ &= \int_0^T \eta_s[\bar{L}(\varphi^2_{\cdot,T})(s, \cdot) - 2\varphi_{s,T} \bar{L}(\varphi_{\cdot,T})(s, \cdot)] \, ds \\ &= \eta_T[\varphi^2_{T,T}] - \eta_0[\varphi^2_{0,T}] - \int_0^T \eta_s[\varphi^2_{s,T}(U_s - \eta_s(U_s)) + 2\varphi_{s,T} \bar{L}(\varphi_{\cdot,T})] \, ds \end{aligned}$$

But it appears from the definition of $\varphi_{s,T}$, that

$$\forall\ (s,x) \in [0,T] \times E^N,$$

$$\begin{aligned}\bar{L}(\varphi_{\cdot,T})(s,x) &= \exp\left(-\int_s^T \eta_u(U_u)\,du\right)\bar{L}(Q_{\cdot,T}(\varphi))(s,x) + \eta_s(U_s)\varphi_{s,T}(x) \\ &= -(U_s(x) - \eta_s(U_s))\varphi_{s,T}(x)\end{aligned}$$

so

$$\begin{aligned}&\int_0^T \eta_s[\bar{\Gamma}(\varphi_{\cdot,T},\varphi_{\cdot,T})(s,\cdot)]\,ds \\ &= \eta_T[\varphi_{T,T}^2] - \eta_0[\varphi_{0,T}^2] + \int_0^T \eta_s[\varphi_{s,T}^2(U_s - \eta_s(U_s))]\,ds\end{aligned}$$

and finally the lemma follows, taking into account that

$$\forall\ 0 \le s \le T, \qquad \eta_s(\varphi_{s,T}) = \eta_T(\varphi - \eta_T(\varphi)) = 0$$

■

Remark 3.35: This is a first step which could lead to the conclusion that the Theorem 3.28 is true more generally for the family $(W_T^N(\varphi))_{\varphi\in\mathcal{B}_b(E)}$. Note also that in the trivial case where $U \equiv 0$, we find the classical covariance for independent particles. The term $2\int_0^T \eta_s[\varphi_{s,T}^2 U_s]\,ds$ gives a measurement of the noise introduced by interactions.

3.3.3 Exponential Bounds

We can also take advantage of the martingales we have exhibited in the previous sections to obtain exponential bounds on deviations from the limit. Here we will use the strong regularity assumptions considered in section 3.1.2, because a.s. bounds on the increasing processes will be needed (and not only $\mathbb{L}^1$ estimations, as for the weak propagation of chaos).

Our starting point will be the following basic result from the general theory of martingales (cf for instance the Corollary 3.3 of [86] using calculations from the section 4.13 of [81]):

Proposition 3.36 *Let $(M_t)_{0\le t\le T}$ be a martingale starting from 0, ie $M_0 = 0$, and such that for a constant $a \ge 0$, we have a.s., $\sup_{0\le t\le T} |\triangle M_t| \le a$. Then M is locally square integrable and we are assured of the bounds*

$$\forall\ G \ge 0,\ \forall\ 0 \le \epsilon \le G/a, \qquad \mathbb{P}\left[\sup_{0\le t\le T} |M_t| \ge \epsilon,\ \langle M\rangle_T \ge G\right] \le 2\exp\left(\frac{-\epsilon^2}{4G}\right)$$

Now the procedure to get exponential upper bounds for the deviations $W_T^N(\varphi)$ introduced in Theorem 3.28 is quite standard.

First we have to estimate the "characteristics" of the martingale $\widetilde{B}(\varphi)$ defined in the proof of Proposition 3.22, for $\varphi \in \mathcal{A}$.

Lemma 3.37 *For $T > 0$ given, there exist two constants $C_T^{(10)}$, $C_T^{(11)} \geq 0$, such that for all $\varphi \in \mathcal{A}$, we have almost surely,*

$$\sup_{0\leq t\leq T} \left|\triangle \widetilde{B}_t^N(\varphi)\right| \leq \frac{C_T^{(10)} \|\varphi\|}{N}$$
$$\langle \widetilde{B}^N(\varphi)\rangle_T \leq \frac{C_T^{(11)} \vert\vert\vert\varphi\vert\vert\vert_{[0,T]}^2}{N}$$

where the norm $\vert\vert\vert \cdot \vert\vert\vert_{[0,T]}$ was defined by the formula (105) given page 96.

Proof: Using the following relation valid for all $0 \leq t \leq T$,

$$\widetilde{B}_t^N(\varphi) = \int_0^t \exp\left(\int_0^s \eta_u^N(U_u)\, du\right) dB_s^N(\varphi)$$

clearly it is enough to get these upper bounds with $\widetilde{B}^N(\varphi)$ replaced by the martingale $B^N(\varphi)$ considered in Lemma 3.20. But then the required estimations are deduced at once from Lemma 3.20 and Lemma 3.23.

∎

Taking into account a basic result on iid random variables, we end up with

Proposition 3.38 *There exists a constant $C_T^{(12)} \geq 0$ such that for all $\varphi \in \mathcal{A}$,*

$$\forall\, \epsilon > 0,\ \mathbb{P}\left[\sup_{0\leq t\leq T} \left|\gamma_t^N(Q_{t,T}(\varphi)) - \gamma_T(\varphi)\right| \geq \epsilon\right] \leq 4\exp\left(\frac{-NC_T^{(12)}\epsilon^2}{\|\varphi\|^2 \vee \vert\vert\vert\varphi\vert\vert\vert_{[0,T]}^2}\right)$$

Proof: Recall that $\gamma_T(\varphi) = \gamma_0(Q_{0,T}(\varphi))$, so we can write the decomposition

$$\begin{aligned}
&\gamma_t^N(Q_{t,T}(\varphi)) - \gamma_T(\varphi)\\
&= \gamma_t^N(Q_{t,T}(\varphi)) - \gamma_0^N(Q_{0,T}(\varphi)) + \gamma_0^N(Q_{0,T}(\varphi)) - \gamma_0(Q_{0,T}(\varphi))\\
&= \widetilde{B}_t^N(\varphi) + \gamma_0^N(Q_{0,T}(\varphi)) - \gamma_0(Q_{0,T}(\varphi))
\end{aligned}$$

It is then sufficient to prove that for all $\epsilon > 0$,

$$\mathbb{P}\left[\sup_{0\leq t\leq T} \left|\widetilde{B}_t^N(\varphi)\right| \geq \epsilon/2\right] \leq 2\exp\left(\frac{-NC_T^{(13)}\epsilon^2}{\|\varphi\|^2 \vee \vert\vert\vert\varphi\vert\vert\vert_{[0,T]}^2}\right)$$

$$\mathbb{P}\left[\left|\gamma_0^N(Q_{0,T}(\varphi)) - \gamma_0(Q_{0,T}(\varphi))\right| \geq \epsilon/2\right] \leq 2\exp\left(\frac{-NC_T^{(14)}\epsilon^2}{\|\varphi\|^2}\right)$$

for some constants $C_T^{(13)}$, $C_T^{(14)} \geq 0$.
For the first inequality, we note that obviously

$$\left|\widetilde{B}_T^N(\varphi)\right| \leq \exp(T\,\|U\|_{[0,T]})\,\|\varphi\|$$

so we only have to prove it for $0 < \epsilon \leq \exp(T \|U\|_{[0,T]}) \|\varphi\|$, but then, by using the estimates of the lemma above, it comes from the Proposition 3.36 applied with $M = \widetilde{B}(\varphi)$, $a = C_T^{(10)} \|\varphi\| / N$ and

$$G = [C_T^{(11)} \vee C_T^{(10)} \exp(T \|U\|_{[0,T]})](\|\varphi\|^2 \vee |||\varphi|||_{[0,T]}^2)/N.$$

The second bound is just a consequence of the Hoeffding inequality for iid random variables (cf [50]), which states that for all $\epsilon > 0$,

$$\mathbb{P}\left[\left|\gamma_0^N(Q_{0,T}(\varphi)) - \gamma_0(Q_{0,T}(\varphi))\right| \geq \epsilon/2\right] \;\leq\; 2\exp\left(\frac{-N\epsilon^2}{8\|Q_{0,T}(\varphi)\|^2}\right)$$

so we can take $C_T^{(14)} = \exp(-2T\|U\|_{[0,T]})/8$.

■

Now the conclusion follows easily:

Theorem 3.39 *There exists a constant $C_T^{(15)} > 0$ such that*

$$\forall \varphi \in \mathcal{A}, \forall \epsilon > 0, \qquad \mathbb{P}[|W_T^N(\varphi)| \geq \epsilon] \;\leq\; 4\exp\left(\frac{-C_T^{(15)}\epsilon^2}{\|\varphi\|^2 \vee |||\varphi|||_{[0,T]}^2}\right)$$

where we recall that the fluctuation are given by $W_T^N(\varphi) = \sqrt{N}(\eta_T^N(\varphi) - \eta_T(\varphi))$, for $\varphi \in \mathcal{A}$.

Proof: We deduce this result from the usual decomposition (108) and the inequality of the latter proposition, considered at time $t = T$. If we use the full strength of the uniformity over the interval $[0,T]$, then for any $\varphi \in \mathcal{A}$ and for any $\epsilon > 0$ we rather end up with

$$\mathbb{P}[\sup_{0\leq t\leq T} \left|\Phi_{t,T}(\eta_t^N)(\varphi) - \eta_T(\varphi)\right| \geq \epsilon] \;\leq\; 4\exp\left(\frac{-NC_T^{(15)}\epsilon^2}{\|\varphi\|^2 \vee |||\varphi|||_{[0,T]}^2}\right)$$

■

Remarks 3.40:

a) The norm $|||\cdot|||_{[0,T]}$ does not seem an easy object to manipulate, but the considerations of section 3.1.2 enables us to replace it by more convenient ones, for instance, under hypotheses $(H4)_T$, $(H7)_T$, $(H8)_T$, $(H9)_T$ and $(H10)_T$, it appears that for a constant $C_T^{(16)} > 0$ depending only on $T \geq 0$, we have

$$|||\cdot|||_{[0,T]} \;\leq\; C_T^{(16)}\|\cdot\|$$

b) Theorem 3.39 is a first step in the direction of a $\mathbb{L}^p$ Glivenko-Cantelli result, since it shows that for $T \geq 0$ fixed, the process

$$(\eta_T^N(\varphi) - \eta_T(\varphi))_{\varphi\in\mathcal{A}}$$

indexed by $\mathcal{A}$ is sub-Gaussian with respect to the norm $\|\cdot\| \vee |||\cdot|||_{[0,T]}$.

So if for a class of functions $\mathcal{F} \subset \mathcal{A}$, we have enough information about the packing and covering numbers of $\mathcal{F}$ with respect to $\|\cdot\| \vee |||\cdot|||_{[0,T]}$ (or more conveniently, with respect to $|||\cdot|||$, under the appropriate hypotheses), then we could conclude to results similar to those presented in section 2.2.3.

c) We also notice that the Proposition 3.25 could classically be deduced from the previous theorem (cf for instance [41]), except that we end up with the norm $\|\cdot\| \vee |||\cdot|||_{[0,T]}$ in the rhs of the inequality given there, instead of $\|\cdot\|$. This leads to the question of whether the Theorem 3.39 would not be satisfied with that norm.

4 Applications to Non Linear Filtering

4.1 Introduction

The object of this section is to apply the results obtained in previous sections to nonlinear filtering problems. We will study continuous time as well as discrete time filtering problems. For a detailed discussion of the filtering problem the reader is referred to the pioneering paper of Stratonovich [104] and to the more rigorous studies of Shiryaev [99] and Kallianpur-Striebel [72]. More recent developments can be found in Ocone [89] and Pardoux [90].

In continuous time settings the desired conditional distributions can be regarded as a Markov process taking values in the space of all probability measures. The corresponding evolution equation is usually called the Kushner-Stratonovitch equation. The most important measure of complexity is the infinite dimensionality of the state space of this equation.

In the first section 4.2 we formulate the continuous time nonlinear filtering problem in such a way that the results of section 3 can be applied. In section 4.3 we present an alternative approach to approximate a continuous time filtering problem. This approach is based on a commonly used time discretization procedure (see for instance [82, 74, 76, 58, 90] and [92, 94, 93, 108]).
We shall see that the resulting discrete time model has the same form as in (8) and it also characterizes the evolution in time of the optimal filter for a suitably defined discrete time filtering problem.
The fundamental difference between the Moran's type IPS and the genetic type IPS associated with this additional level of discretization lies in the fact that in the Moran IPS competitive interactions occur randomly. The resulting scheme is therefore a genuine continuous time and particle approximating model of the nonlinear filtering equation.

In section 4.4 we briefly describe the discrete time filtering problem. It will be transparent from this formulation that the desired flow of distributions have the same form as the one considered in this work (see (8) section 1.3). We will also remark that the fitness functions $\{g_n \,;\, n \geq 1\}$ and therefore the constants $\{a_n \,;\, n \geq 1\}$ defined in condition $(\mathcal{G})$ depend on the observation process so that the analysis given in previous sections will also lead to quenched results.

For instance the covariance function in Donsker Theorem and the rates functions in LDP will now depend on the observation record.

One natural question that one may ask if whether the averaged version of the stability results of section 2.1.2 and the $\mathbb{L}^p$-error bounds given in section 2.2.2 hold. In many practical situations the functions $\{a_n \; ; \; n \geq 1\}$ have a rather complicated form and it is difficult to obtain an averaged version of some results such as the exponential rates given in Theorem 2.15. Nevertheless we will see in section 4.4.2 that the averaged version of the stability results given in section 2.1.2 as well as the averaged version of the $\mathbb{L}^p$-uniform bounds given in section 2.2.2 hold for a large class of nonlinear sensors.

4.2 Continuous Time Filtering Problems

4.2.1 Description of the Models

The aim of this section is to formulate some classical nonlinear filtering problems in such a way that the previous particles interpretations can be naturally applied to them. Here is the heuristic model: let a signal process $S = \{S_t \; ; \; t \geq 0\}$ be given, it is assumed to be a time-homogeneous Markov process with càdlàg paths taking values in the Polish space E. We suppose that this signal is seen through a $\mathbb{R}^d$-valued noisy observation process $Y = \{Y_t \; ; \; t \geq 0\}$ defined by

$$\forall\, t \geq 0, \qquad Y_t \;=\; \int_0^t h(S_s)\, ds + V_t$$

where $V = \{V_t \; ; \; t \geq 0\}$ is a d-vector standard Wiener process independent of S, and h maps somewhat smoothly the signal state space E into $\mathbb{R}^d$.

The traditional filtering problem is concerned with estimating the conditional distribution of S_t given the observational information that is available at time t, $\mathcal{Y}_{[0,t]} = \sigma(Y_s \; ; \; 0 \leq s \leq t)$, ie to evaluate for all $f \in \mathcal{C}_b(E)$,

$$\pi_t(f) \;\stackrel{\text{def.}}{=}\; \mathbb{E}\left(f(S_t) \,\middle|\, \mathcal{Y}_{[0,t]}\right)$$

More precisely, we will make the assumption that there exist an algebra $\mathcal{A} \subset \mathcal{C}_b(E)$ and a pregenerator $L_0 : \mathcal{A} \to \mathcal{C}_b(E)$ such that for any initial distribution $\eta_0 \in \mathbf{M}_1(E)$, there is a unique solution $\widetilde{\mathbb{P}}_{\eta_0}$ to the martingales problem (we refer to the section 3.1.1 for more details) associated with η_0 and L_0 on $\Omega_1 = D(\mathbb{R}_+, E)$, the space of all càdlàg paths from $\mathbb{R}_+$ to E, endowed with its natural σ-algebra. Then $S \stackrel{\text{def.}}{=} (S_t)_{t\geq 0}$ will denote the canonical coordinate process on Ω_1, and from now on, the initial distribution η_0 will be supposed fixed, and we will consider the Markov process S under $\widetilde{\mathbb{P}}_{\eta_0}$.

Let $h = (h_i)_{1\leq i\leq d}$ be the map from E to $\mathbb{R}^d$ alluded to above, we make the hypothesis that for all $1 \leq i \leq d$, $h_i \in \mathcal{A}$.

Let us also introduce the canonical probability space associated with the observation process: $\Omega_2 = C(\mathbb{R}_+, \mathbb{R}^d)$, the set of all continuous functions from $\mathbb{R}_+$ to $\mathbb{R}^d$, and $Y = (Y_t)_{t\geq 0}$ is the coordinate process on Ω_2.

Let us denote $\Omega = \Omega_1 \times \Omega_2$ and $\widetilde{\mathbb{P}}$ the probability on its usual Borelian σ-field such that its marginal on Ω_1 is $\widetilde{\mathbb{P}}_{\eta_0}$ and such that

$$V = (V_t)_{t\geq 0} \stackrel{\text{def.}}{=} \left(Y_t - \int_0^t h(S_s)\, ds\right)_{t\geq 0}$$

is a d-vector standard Brownian motion, as previously mentionned.

In practice, this probability $\widetilde{\mathbb{P}}$ is usually constructed via Girsanov's Theorem from another reference probability measure $\widehat{\mathbb{P}}$ on Ω, under which S and Y are independent, S has law $\widetilde{\mathbb{P}}_{\eta_0}$ and Y is a d-vector standard Brownian motion. For $t \geq 0$, let

$$\mathcal{F}_t = \sigma((S_s, Y_s)\ ;\ 0 \leq s \leq t)$$

be the σ-algebra of events up to time t, the probabilities $\widetilde{\mathbb{P}}$ and $\widehat{\mathbb{P}}$ are in fact equivalent on $\mathcal{F}_t$, and their density is given by

$$\frac{d\widetilde{\mathbb{P}}}{d\widehat{\mathbb{P}}}|\mathcal{F}_t = Z_t(S,Y) \stackrel{\text{def.}}{=} \exp\left(\int_0^t h^\star(S_s)dY_s - \frac{1}{2}\int_0^t h^\star(S_s)h(S_s)ds\right)$$

where we have used standard matrix notations, for instance

$$\int_0^t h^\star(S_s)\, dY_s = \sum_{1\leq i\leq d}\int_0^t h_i(S_s)\, dY_{i,s}$$

Under our assumptions ([56, 90]) one can prove that

$$\pi_t(f) = \frac{\widehat{\mathbb{E}}\left(f(S_t)Z_t(S,Y)\,|\mathcal{Y}_{[0,t]}\right)}{\widehat{\mathbb{E}}\left(Z_t(S,Y)\,|\mathcal{Y}_{[0,t]}\right)} = \frac{\int_{\Omega_1} f(\theta_t)\; Z_t(\theta,Y)\; \widetilde{\mathbb{P}}_{\eta_0}(d\theta)}{\int_{\Omega_1} Z_t(\theta,Y)\; \widetilde{\mathbb{P}}_{\eta_0}(d\theta)} \tag{115}$$

Using Itô's integration by part formula, in the differential sense we have that

$$h^\star(S_s)dY_s \;=\; d(h^\star(S_s)Y_s) - Y_s^\star\, L_0(h)(S_s)ds - Y_s^\star\, dM_s^{(h)}$$

where $L(h) = (L(h_i))_{1\leq i\leq d} : E \to \mathbb{R}^d$, and where $M^{(h)} = (M^{(h_i)})_{1\leq i\leq d}$ is a d-vector square integrable continuous martingale (relative to the natural filtration $(\mathcal{F}_t)_{t\geq 0}$) with cross-variation processes given by

$$\forall\, 1 \leq i,j \leq d, \qquad \langle M^{(h_i)}, M^{(h_j)}\rangle_t \;=\; \int_0^t \Gamma_0(h_i,h_j)(S_s)\, ds$$

(as usual Γ_0 is the carré du champ associated with the pregenerator L_0). This yields the decomposition

$$\begin{aligned}&\ln Z_t(S,Y)\\ &= h^\star(S_t)Y_t - \int_0^t Y_s^\star\, L_0(h)(S_s)\, ds - \int_0^t Y_s^\star\, dM_s^h - \frac{1}{2}\int_0^t h^\star(S_s)h(S_s)ds\end{aligned}$$

Before going further, let us make the following remark:
As in section 3.1.1, we denote L the operator acting on $\mathbf{A}\otimes\mathcal{A}$ by formula (92), where L_t is replaced by L_0. We also consider $\mathcal{B}_\infty$ the vector space of functions $f \in \mathcal{B}_b(\mathbb{R}_+ \times E)$

for which there exists a function $\bar{L}(f) \in \mathcal{B}_b(\mathbb{R}_+ \times E)$ such that under $\widetilde{\mathbb{P}}_{\eta_0}$, the process $(M_t(f))_{0\leq t\leq T}$ defined by

$$\forall\, 0 \leq t \leq T, \qquad M_t(f) \;=\; f(t, X_t) - f(0, X_0) - \int_0^t \bar{L}(f)(s, X_s)\, ds$$

is a martingale.

We have the following stability property:

Lemma 4.1 *If $f \in \mathbf{A} \otimes \mathcal{A}$, then $\exp(f) \in \mathcal{B}_\infty$.*

Proof: Using the formula $\exp(f) = \sum_{n\geq 0} f^n/n\,!$ and an approximation technique as the one presented in Lemma 3.2, it is enough to show that for all $T > 0$ given, we have as $n, m \to \infty$ (and on $[0, T] \times E$)

$$\Gamma\left(\sum_{n\leq p\leq m} f^p/p!, \sum_{n\leq p\leq m} f^p/p!\right) \rightharpoonup 0$$

$$L\left(\sum_{n\leq p\leq m} f^p/p!\right) \rightharpoonup 0$$

But these convergence results are easy to obtain, because of the general bounds

$$\sqrt{\Gamma\left(\sum_{n\leq p\leq m} f^p/p!, \sum_{n\leq p\leq m} f^p/p!\right)} \;\leq\; \sum_{n\leq p\leq m} \sqrt{\Gamma\left(f^p/p!, f^p/p!\right)}$$

$$\forall\, p \geq 1, \qquad \|\Gamma(f^p, f^p)\|_{[0,T]} \;\leq\; p^2 \, \|f\|_{[0,T]}^{2p-2} \, \|\Gamma(f,f)\|_{[0,T]}$$

and of the upper bound on $L(f^n)$ which can be deduced by induction from

$$\begin{aligned}
\forall\, n \geq 1, \qquad & \left\|L(f^{n+1})\right\|_{[0,T]} \\
\leq\; & \|f^n\|_{[0,T]} \|L(f)\|_{[0,T]} + \|f\|_{[0,T]} \|L(f^n)\|_{[0,T]} + \|\Gamma(f^n, f)\|_{[0,T]} \\
\leq\; & \|f^n\|_{[0,T]} \|L(f)\|_{[0,T]} + \|f\|_{[0,T]} \|L(f^n)\|_{[0,T]} \\
& + \sqrt{\|\Gamma(f,f)\|_{[0,T]}}\sqrt{\|\Gamma(f^n, f^n)\|_{[0,T]}}
\end{aligned}$$

This ends the proof of the lemma. ∎

Let $y \in C(\mathbb{R}_+, \mathbb{R}^d)$ be given. For $t \geq 0$ fixed, the map

$$h_t \,:\, E \ni x \;\mapsto\; \sum_{1\leq i\leq d} y_{i,t} h_i(x)$$

belongs to $\mathcal{A}$, so in the same way as above, we can define a function

$$\bar{L}_0(\exp(-h_t)) \in \mathcal{B}_b(E).$$

We easily realize that the application

$$\mathbb{R}_+ \times E \ni (t, x) \;\mapsto\; \bar{L}_0(\exp(-h_t))(x)$$

is continuous as a locally uniform limit of continuous functions.

Then we can consider

$$\breve{Z}_t(S,y) \stackrel{\text{def.}}{=} \exp\left(-\int_0^t y_s^\star \, dM_s^{(h)} - \int_0^t \exp(h_s(S_s))\bar{L}_0(\exp(-h_s))(S_s) + L_0(h_s)(S_s)\, ds\right)$$

Note that if a sequence $(y_n)_{n\geq 0}$ of elements of $C(\mathbb{R}_+,\mathbb{R}^d)$ converges uniformly on compact subsets of $\mathbb{R}_+$ toward y, then uniformly for t belonging to compact subsets, $\breve{Z}_t(S,y_n)$ converges in probability towards $\breve{Z}_t(S,y)$.

The interest of this quantity is that

Lemma 4.2 *Let a function $y \in C(\mathbb{R}_+,\mathbb{R}^d)$ be fixed. Under $\widetilde{\mathbb{P}}_{\eta_0}$, the process $(\breve{Z}_t(S,y))_{t\geq 0}$ is a martingale.*

Proof: We will first consider the case $y \in C^1(\mathbb{R}_+,\mathbb{R}^d)$ (even if this situation a.s. never occur for Y).

Under this additional assumption, the mapping

$$\bar{h} \,:\, \mathbb{R}_+ \times E \ni (t,x) \;\mapsto\; h_t(x)$$

clearly belongs to $\mathbf{A}\otimes\mathcal{A}$, and it appears that if $M^{(\bar{h})}$ is the martingale such that for all $t\geq 0$,

$$\bar{h}(t,S_t) \;=\; \bar{h}(0,S_0) + \int_0^t L(\bar{h})(s,S_s)\, ds + M_t^{(\bar{h})}$$

then in fact it is given by

$$\forall\, t\geq 0, \qquad M_t^{(\bar{h})} \;=\; \int_0^t y_s^\star \, dM_s^{(h)}$$

Furthermore, from Lemma 4.1, for any $t\geq 0$ we can write that

$$\exp(-\bar{h}(t,S_t)) \;=\; \exp(-\bar{h}(0,S_0)) + \int_0^t \bar{L}(\exp(-\bar{h}))(s,S_s)\, ds + M_t^{(\exp(-\bar{h}))}$$

for a certain martingale $M^{(\exp(-\bar{h}))}$.

On the other hand it appears without difficulty that for any $(s,x)\in\mathbb{R}_+\times E$

$$\begin{aligned}&\exp(h_s(x))\bar{L}_0(\exp(-h_s))(x) + L_0(h_s)(x)\\ &= \exp(\bar{h}(s,x))\bar{L}(\exp(-\bar{h}))(s,x) + L(\bar{h})(s,x)\end{aligned}$$

(ie the time derivatives cancel). As a result, for any $t\geq 0$ we have that

$$\breve{Z}_t(S,y) = \exp\left(-\bar{h}(t,S_t) + \bar{h}(0,S_0) + \int_0^t \exp(\bar{h}(s,S_s))\bar{L}(\exp(-\bar{h}))(s,S_s)\, ds\right)$$

This may also be written in differential form

$$\begin{aligned}
&d\breve{Z}_t(S,y)\\
&= \breve{Z}_{t-}(S,y)\Big(\exp(\bar{h}(t-,S_{t-}))\bar{L}(\exp(-\bar{h}))(t-,S_{t-})\,dt\\
&+\exp(\bar{h}(t-,S_{t-}))\,dM_t^{(\exp(-\bar{h}))} - \exp(\bar{h}(t-,S_{t-}))\bar{L}(\exp(-\bar{h}))(t-,S_{t-})\,dt\Big)\\
&= \breve{Z}_{t-}(S,y)\exp(\bar{h}(t-,S_{t-}))\,dM_t^{(\exp(-\bar{h}))}
\end{aligned}$$

It follows that $(\breve{Z}_t(S,y))_{t\geq 0}$ is a martingale. Indeed it is more precisely the Doléans-Dade exponential of the martingale

$$\left(\int_0^t \exp(\bar{h}(s-,S_{s-}))\,dM_s^{(\exp(-\bar{h}))}\right)_{t\geq 0}$$

Thus, we see that for all $0\leq s\leq t$ and any random variables H_s which are measurable with respect to $\sigma(S_u\,;\,0\leq u\leq s)$, we have

$$\widetilde{\mathbb{E}}_{\eta_0}[H_s(\breve{Z}_t(S,y)-\breve{Z}_s(S,y))] \;=\; 0$$

We end the proof by noting that the left hand side is continuous with respect to $y\in C(\mathbb{R}_+,\mathbb{R}^d)$, if this set is endowed with the uniform convergence on compact subsets of $\mathbb{R}_+$.

■

Let us write, for $y\in C(\mathbb{R}_+,\mathbb{R}^d)$,

$$\forall\, t\geq 0, \qquad \ln Z_t(S,y) \;=\; h^\star(S_t)y_t + \int_0^t V(S_s,y_s)\,ds + \ln\breve{Z}_t(S,y)$$

where for all $(x,y)\in E\times\mathbb{R}^d$,

$$V(x,y) \;=\; \exp(y^\star h(x))\bar{L}_0[\exp(-y^\star h(\,\cdot\,))](x) - \frac{1}{2}h^\star(x)h(x)$$

Together with (115) this decomposition implies that

$$\pi_t(f) = \frac{\displaystyle\int_{\Omega_1} f(\theta_t)\,e^{h^\star(\theta_t)Y_t+\int_0^t V(\theta_s,Y_s)\,ds}\,\mathbb{P}_{\eta_0}^{[Y]}(d\theta)}{\displaystyle\int_{\Omega_1} e^{h^\star(\theta_t)Y_t+\int_0^t V(\theta_s,Y_s)\,ds}\,\mathbb{P}_{\eta_0}^{[Y]}(d\theta)}$$

where, for any $y\in C(\mathbb{R}_+,\mathbb{R}^d)$, $\mathbb{P}_{\eta_0}^{[y]}$ is the probability measure on Ω_1 defined by its restrictions to $\mathcal{F}_t^{(1)}=\sigma(S_s,\,0\leq s\leq t)$:

$$\frac{d\mathbb{P}_{\eta_0}^{[y]}}{d\widetilde{\mathbb{P}}_{\eta_0}}|\mathcal{F}_t^{(1)} = \breve{Z}_t(S,y)$$

Here is a more tractable caracterisation of $\mathbb{P}^{[y]}_{\eta_0}$:

Proposition 4.3 *Fix a mapping $y \in C(\mathbb{R}_+, \mathbb{R}^d)$, and consider for $t \geq 0$ an operator L_t given on $\mathcal{A}$ by*

$$\forall\, \varphi \in \mathcal{A}, \qquad L_t(\varphi) \;=\; L_0(\varphi) + \exp(h_t)\bar{\Gamma}_0(\exp(-h_t), \varphi)$$

Then $(L_t)_{t\geq 0}$ is a measurable family of pregenerators, and $\mathbb{P}^{[y]}_{\eta_0}$ is the unique solution to the martingale problem associated with the initial condition η_0 and to this family.

We would have noticed there is no real difficulty in defining $\bar{\Gamma}(\exp(-h_t), \varphi)$ for $\varphi \in \mathcal{A}$.
Proof: We shall verify that for all $\varphi \in \mathcal{A}$, all $0 \leq s \leq t$ and all random variable H_s which is $\sigma(S_u\,;\, 0 \leq u \leq s)$-measurable, we have

$$\mathbb{E}^{[y]}_{\eta_0}\left[H_s\left(\varphi(S_t) - \varphi(S_s) - \int_s^t L_u(\varphi)(X_u)\,du\right)\right] \;=\; 0$$

i.e.

$$\widetilde{\mathbb{E}}_{\eta_0}\left[\check{Z}_t(S,y)H_s\,(N_t - N_s)\right] \;=\; 0$$

where for all $t \geq 0$,

$$N_t \;=\; \varphi(S_t) - \varphi(S_0) - \int_0^t L_u(\varphi)(X_u)\,du$$

So as in Lemma 4.2, by continuity, we can assume that $y \in C^1(\mathbb{R}_+, \mathbb{R}^d)$.
With this assumption enforced, we have

$$\begin{aligned}
&d(\check{Z}_t(S,y)N_t)\\
&= \check{Z}_{t-}(S,y)((L_0(\varphi) - L_t(\varphi))dt + dM_t^{(\varphi)}) + N_{t-}d\check{Z}_t(S,y) + d\langle M^{(\varphi)}, \check{Z}(S,y)\rangle_t\\
&= \check{Z}_{t-}(S,y)[(L_0(\varphi) - L_t(\varphi))dt + \exp(\bar{h}(t,S_t))d\langle M^{(\varphi)}, M^{(\exp(-\bar{h}))}\rangle_t]\\
&\qquad +\check{Z}_{t-}(S,y)dM_t^{(\varphi)} + N_{t-}d\check{Z}_t(S,y)\\
&= \check{Z}_{t-}(S,y)[L_0(\varphi) - L_t(\varphi) + \exp(\bar{h}(t,S_t))\bar{\Gamma}(\exp(-\bar{h})), \varphi)(t,S_t)]\,dt\\
&\qquad +\check{Z}_{t-}(S,y)dM_t^{(\varphi)} + N_{t-}d\check{Z}_t(S,y)
\end{aligned}$$

Since

$$\widetilde{\mathbb{E}}_{\eta_0}\left[\check{Z}_t(S,y)H_s\,(N_t - N_s)\right] \;=\; \widetilde{\mathbb{E}}_{\eta_0}\left[H_s\left(\check{Z}_t(S,y)N_t - \check{Z}_s(S,y)N_s\right)\right]$$

we need to check that $(\check{Z}_t(S,y)N_t)_{t\geq 0}$ is a martingale. To this end it is enough to see that for any $(t,x) \in \mathbb{R}_+ \times E$

$$L_0(\varphi)(x) - L_t(\varphi)(x) + \exp(\bar{h}(t,x))\bar{\Gamma}(\exp(-\bar{h}), \varphi)(t,x) \;=\; 0$$

Next, since for any $(t, x) \in \mathbb{R}_+ \times E$

$$\exp(\bar{h}(t,x))\bar{\Gamma}(\exp(-\bar{h}),\varphi)(t,x) = \exp(h_t(x))\bar{\Gamma}_0(\exp(-h_t),\varphi)(x)$$

L_t is well defined.

The fact that, for any $t \geq 0$ L_t is a pregenerator comes from the previous considerations, taking $y \in C(\mathbb{R}_+, \mathbb{R}^d)$ defined by

$$\forall\, s \geq 0, \qquad y_s = y_t$$

And the uniqueness property comes from the one of $\widetilde{\mathbb{P}}_{\eta_0}$, since if $\check{\mathbb{P}}$ is a solution to the martingale problem associated to $(L_t)_{t\geq 0}$, then it can be shown that the probability $\check{\mathbb{P}}$ defined on Ω_1 by

$$\forall\, t \geq 0, \qquad \frac{d\check{\mathbb{P}}}{d\check{\mathbb{P}}}|\mathcal{F}_t^{(1)} = \frac{1}{\check{Z}_t(S,y)}$$

is solution to the time-homogeneous martingale problem associated to L_0 (all initial conditions being η_0).

■

The above formulation of the optimal filter can be regarded as a path-wise filter

$$\begin{array}{rccc} \pi_t: & \mathcal{C}([0,T]) & \longrightarrow & \mathbf{M}_1(E) \\ & y & \mapsto & \pi_{y,t} \end{array}$$

where the probability $\pi_{y,t}$ is given by

$$\forall\, f \in \mathcal{B}_b(E), \qquad \pi_{y,t}(f) = \frac{\displaystyle\int_{\Omega_1} f(\theta_t)\, e^{h^\star(\theta_t)y_t + \int_0^t V(\theta_s,y_s)\,ds}\, \mathbb{P}_{\eta_0}^{[y]}(d\theta)}{\displaystyle\int_{\Omega_1} e^{h^\star(\theta_t)y_t + \int_0^t V(\theta_s,y_s)\,ds}\, \mathbb{P}_{\eta_0}^{[y]}(d\theta)}$$

This gives a description of the optimal filter in terms of Feynman-Kac formulae as those presented in (3) and (4) in the introduction. Namely,

$$\forall\, f \in \mathcal{B}_b(E), \qquad \pi_{y,t}(f) = \frac{\displaystyle\int_E f(x)\, e^{h^\star(x)y_t}\, \eta_{y,t}(dx)}{\displaystyle\int_E e^{h^\star(x)y_t}\, \eta_{y,t}(dx)}$$

where

$$\forall\, f \in \mathcal{B}_b(E), \qquad \eta_{y,t}(f) \stackrel{\text{def.}}{=} \frac{\gamma_{y,t}(f)}{\gamma_{y,t}(1)}$$

and

$$\forall\, f \in \mathcal{B}_b(E), \qquad \gamma_{y,t}(f) = \mathbb{E}_{\eta_0}^{[y]}\left[f(X_t)\, e^{\int_0^t V(X_s,y_s)\,ds}\right]$$

In contrast to (115) we notice that the previous formulations do not involve stochastic integrations and therefore it is well defined for all observation paths and not only on a set of probability measure 1. This formulation is necessary to study the robustness of the optimal filter (that is the continuity of the filter with respect to the observation process), and is also essential to construct robust approximations of the optimal filter, as our interacting particles scheme.

Remarks 4.4:

(a) Note that the condition (H1) is automatically verified for family of pregenerators $(L_t)_{t\geq 0}$ constructed as in Proposition 4.3.

(b) The change of probability presented in that proposition is rather well known in case of diffusions (cf for instance [90] and [96]), ie when the trajectories of S are continuous.

Then we can suppose that $\mathcal{A}$ is stable by composition with C^∞ functions and we have

$$\forall\, F \in C^\infty(\mathbb{R}), \forall\, \phi, \varphi \in \mathcal{A},$$

$$\begin{aligned} L_0(F(\varphi)) &= F'(\varphi)L_0(\varphi) + \frac{F''(\varphi)}{2}\Gamma_0(\varphi,\varphi) \\ \Gamma_0(F(\varphi),\phi) &= F'(\varphi)\Gamma_0(\varphi,\phi) \end{aligned}$$

So in the previous expressions, we can replace

$$\begin{aligned} \exp(h_t)\Gamma_0(\exp(-h_t))(x) \quad &\text{by} \quad -\Gamma_0(h_t,\varphi) \\ \exp(y^\star h(x))\bar{L}_0(\exp(-y^\star h(\,\cdot\,)))(x) \quad &\text{by} \quad \frac{1}{2}\, y^\star\, \Gamma_0(h,h)(x)y - y^\star\, L_0(h)(x) \end{aligned}$$

where $\Gamma_0(h,h)(x)$ denote the matrix $(\Gamma_0(h_i,h_j)(x))_{1\leq i,j\leq d}$.

4.2.2 Examples

Here are some classical examples that can be handled in our framework. The map h and the function $y \in C(\mathbb{R}_+, \mathbb{R}^d)$ will be given as before.

• Bounded Generators

The simplest example a pregenerator L_0 is that of a bounded generator. Namely, let $L_0 : E \times \mathcal{E} \to \mathbb{R}$ be a signed kernel such that

- for any $x \in E$, $L_0(x, .\cap (E \setminus \{x\}) \in \mathbf{M}(E)$ and $L_0(x, E) = 0$
- for any $A \in \mathcal{E}$, $E \ni x \mapsto L_0(x, A) \in \mathbb{R}$ is a measurable function
- there exists a constant $0 \leq M < \infty$ such that

$$\forall\, x \in E, \qquad L_0(x, E \setminus \{x\}) \;\leq\; M$$

Then for any function $f \in \mathcal{B}_b(E)$, we define

$$\forall\, x \in E, \qquad L_0(f)(x) \;=\; \int f(y)\, L_0(x,dy)$$

We can take here $\mathcal{A} = \mathcal{B}_b(E)$. We calculate that the carré du champ is given for any $\phi, \varphi \in \mathcal{A}$ and $x \in E$ by

$$\Gamma_0(\phi,\varphi)(x) \;=\; \int L_0(x,dy)\,(\phi(y)-\phi(x))(\varphi(y)-\varphi(x))$$

so it appears that for any $t \geq 0$, $\varphi \in \mathcal{A}$ and $x \in E$

$$L_t(\varphi)(x) = \int L_0(x, dy) \exp(h_t(x) - h_t(y)) (\varphi(y) - \varphi(x))$$

They are again jump generators, and the rate of transition from x to y at time $t \geq 0$ has just been multiplied by $\exp(h_t(x) - h_t(y))$.
For this kind of generators, all our hypotheses are trivially satisfied.

• **Riemannian Diffusions**

Let E be a compact Riemannian manifold. As usual, $\langle \cdot, \cdot \rangle$, $\nabla \cdot$ and $\triangle \cdot$ will denote the scalar product, the gradient and the Laplacian associated with this structure. Let $\mathcal{A}$ be the algebra of smooth functions, i.e. $\mathcal{A} = C^\infty(E)$. Suppose that we are given a vector field b, we denote

$$\begin{aligned} L_0 : \mathcal{A} &\to \mathcal{A} \\ \varphi &\mapsto \frac{\triangle\varphi}{2} + \langle b, \nabla\varphi \rangle \end{aligned}$$

It is immediate to realize that in this example the carré du champ does not depend on b and satisfy

$$\forall\, f, g \in \mathcal{A}, \qquad \Gamma_0(f, f) = \langle \nabla f, \nabla g \rangle$$

(by the way, this equality gave the name "carré du champs").

The existence and uniqueness assumption for the associated martingale problem is well known to be fulfilled. We calculate that for $t \geq 0$, L_t is obtained from L_0 by a change of drift:

$$\forall\, \varphi \in \mathcal{A}, \qquad L_t(\varphi) = \frac{\triangle\varphi}{2} + \langle b - \nabla h_t, \nabla\varphi \rangle$$

This example is also a typical one where all the assumptions of section 3.1.2 are verified.

• **Euclidean Diffusions**

Except for the compactness of the state space, these processes are similar to those of the previous example.

So here $E = \mathbb{R}^n$, $n \geq 1$, and let for $x \in E$,

$$\sigma(x) = (\sigma^{i,j}(x))_{1 \leq i,j \leq n} \quad \text{and} \quad b(x) = (b^i(x))_{1 \leq i \leq n}$$

be respectively a symmetric nonnegative definite matrix and a n-vector. We suppose they are uniformly Lipschitz in their dependence on $x \in \mathbb{R}^n$.

Then denoting $a = \sigma^2$, let us consider on $\mathcal{A} \stackrel{\text{def.}}{=} C_b^2(\mathbb{R}^n)$ the pregenerator

$$\forall\, \varphi \in \mathcal{A}, \forall\, x \in \mathbb{R}^n, \quad L_0(\varphi)(x) = \sum_{1 \leq i,j \leq n} \frac{a^{i,j}}{2}(x) \partial_{i,j}\varphi(x) + \sum_{1 \leq i \leq n} b^i(x) \partial_i \varphi(x)$$

It is a classical result that the associated martingale problems are well-posed (for further details about this problem, see [105]), and more precisely, for all $x \in \mathbb{R}^n$, $\widetilde{\mathbb{P}}_{\delta_x}$ is the law of the (unique strong) solution of the stochastic differential equation

$$\begin{cases} S_0 &= x \\ dS_t &= \sigma(S_t)\, dB_t + b(S_t)\, dt \qquad ; t \geq 0 \end{cases}$$

where $(B_t)_{t\geq 0}$ is a standard n-vector Brownian motion.

Here the carré du champ is given by

$$\forall\, \phi, \varphi \in \mathcal{A}, \qquad \Gamma_0(\phi, \varphi) \;=\; \sum_{1\leq i,j\leq n} a^{i,j}(x)\partial_i\phi(x)\partial_j\varphi(x)$$

so we find that

$\forall\, t \geq 0, \forall\, \varphi \in \mathcal{A}, \forall\, x \in \mathbb{R}^n,$

$$L_t(\varphi)(x) \;=\; \sum_{1\leq i,j\leq n} \frac{a^{i,j}}{2}(x)\partial_{i,j}\varphi(x) + \sum_{1\leq i\leq n} (b^i(x) - \sum_{1\leq j\leq n} a^{i,j}(x)\partial_j h_t(x))\partial_i\varphi(x)$$

Let us make the hypothesis that a is uniformly elliptic: there exists a constant $\epsilon > 0$ such that for all $x \in \mathbb{R}^n$,

$$\forall\, z = (z_i)_{1\leq i\leq n} \in \mathbb{R}^n, \qquad \sum_{1\leq i,j\leq n} a^{i,j}(t,x)z_i z_j \;\geq\; \epsilon \sum_{1\leq i\leq n} z_i^2$$

Under the extra assumption that a and b are C_b^∞, we see that all the requirement of section 3.1.2 are met (rather taking there $\mathcal{A} = C_b^\infty(\mathbb{R}^n)$). But considering the parabolic equation satisfied by $F_{T,\varphi}$, it appears that (H2) will be verified under much less regularity for the coefficients a and b.

4.3 Time Discretization of Continuous Time Filtering Problems

In this section we discuss a time discretization approximating model for the non linear filtering problem associated to the previous Euclidean diffusion signal. To clarify the presentation all processes considered in this section will be indexed on the compact interval $[0,1] \subset \mathbb{R}_+$.

The basic model for the continuous time filtering problem considered here consists in an $\mathbb{R}^p \times \mathbb{R}^q$-valued Markov process $\{(X_t, Y_t) \;:\; t \in [0,1]\}$, strong solution on a probability space (Ω, F, P) of the Itô's type stochastic differential equations

$$\begin{cases} dX_t &= a(X_t)dt + b(X_t)d\beta_t \\ dY_t &= h(X_t)dt + dV_t \end{cases}$$

where

1. $a : \mathbb{R}^p \to \mathbb{R}^p$, $b : \mathbb{R}^p \to \mathbb{R}^{p\times m}$ and $h : \mathbb{R}^p \to \mathbb{R}^q$ are bounded and Lipschitz continuous functions.

2. $\{(\beta_t, V_t) : t \in [0,1]\}$ is a $(\mathbb{R}^m \times \mathbb{R}^q)$-valued standard Brownian motion.

3. $Y_0 = 0$ and X_0 is a random variable independent of $\{(\beta_t, V_t) : t \in [0,1]\}$ with law ν so that $E(|X_0|^2) < \infty$.

The classical filtering problem is to find the conditional distribution of the signal X at time t with respect to the observations Y up to time t, that is

$$\forall\, f \in \mathcal{B}_b(\mathbb{R}^p), \qquad \pi_t f = \mathbb{E}(f(X_t)|\mathcal{Y}_{[0,t]}) \tag{116}$$

equation where $\mathcal{Y}_{[0,t]}$ is the filtration generated by the observations Y up to time t.

The first step in this direction consists in obtaining a more tractable description of the conditional expectations (116).
Introducing $Z_t > 0$ such that

$$\forall t \in [0,1], \qquad \log Z_t = \int_0^t h^\star(X_s)\, dY_s - \frac{1}{2}\int_0^t |h(X_s)|^2\, ds$$

it is well known that the original probability measure P is equivalent to a so called *reference probability measure* P_0 given by

$$P = Z_1\, P_0.$$

In addition, under P_0, $\{(\beta_t, Y_t) : t \in [0,1]\}$ is a $(\mathbb{R}^m \times \mathbb{R}^q)$-valued standard Brownian motion and, X_0 is a random variable with law ν, independent of (β, Y).
The following well known result gives a functional integral representation for the conditional expectations (116), which is known as the Kallianpur-Striebel formula:

$$\forall\, f \in \mathcal{B}_b(\mathbb{R}^p), \forall\, t \in [0,1], \quad \pi_t f \quad = \quad \frac{\mathbb{E}_0(f(X_t)\, Z_t|\mathcal{Y}_{[0,t]})}{\mathbb{E}_0(Z_t|\mathcal{Y}_{[0,t]})} = \frac{\mathbb{E}_0^Y(f(X_t)\, Z_t|\mathcal{Y}_{[0,t]})}{\mathbb{E}_0^Y(Z_t|\mathcal{Y}_{[0,t]})} \tag{117}$$

We use $\mathbb{E}_0^Y(.)$ to denote the integration with respect to the Brownian paths $\{\beta_t : t \in [0,1]\}$ and the variable X_0.

In this section a program for the numerical solving of (117) by using a discrete time IPS scheme is embarked on. As announced in the introduction such IPS approach is obtained by first approximating the original model by a discrete time and measure valued process. The treatment that follows is standard in nonlinear filtering literature and it is essentially contained in [75, 76] and [92].

The former discrete time approximating model of (117) is obtained by first introducing a time discretization scheme of the basic model. To this end we introduce a sequence a meshes $\{(t_0, \ldots, t_M) : M \geq 1\}$ given by

$$t_n = \frac{n}{M} \qquad n \in \{0, \ldots, M\}.$$

To obtain a computationally feasible solution we will also use the following natural assumptions:

- For any $M \geq 1$ there exists a transition probability kernel $P^{(M)}$ such that

$$\sup_{t\in[0,1]} \mathbb{E}\left(|X_{t_n} - X_{t_n}^{(M)}|^2\right) \leq \frac{K}{M}, \qquad K < \infty$$

  where $\{X_{t_n}^{(M)} \ : \ n = 0, \ldots, M\}$ is the time homogeneous Markov chain with transition probability kernel $P^{(M)}$ and such that $X_0^{(M)} = X_0$.

- We can exactly simulate random variables according to the law $P^{(M)}(x, .)$ for any $x \in \mathbb{R}^p$.

It is worth noting that an example of an approximating Markov chain

$$\{X_{t_n}^{(M)} \ : \ n = 0, \ldots, M\}$$

satisfying these assumptions is given by the classical Euler scheme

$$X_{t_n}^{(M)} = X_{t_{n-1}}^{(M)} + a(X_{t_{n-1}}^{(M)})(t_n - t_{n-1}) + c(X_{t_{n-1}}^{(M)})\,(\beta_{t_n} - \beta_{t_{n-1}}) \qquad n = 1, \ldots, M \quad (118)$$

with $X_0^{(M)} = X_0$. This scheme is the crudest of the discretization scheme that can be used in our settings. Other time discretization schemes for diffusive signals are described in full detail in [91] and [108]. In view of (117) the optimal filters $\{\pi_{t_n} : n = 0, \ldots, M\}$ can be written as

$$\pi_{t_n} f = \frac{\mathbb{E}_0^Y\left(Z_{t_{n-1}}\,(H_{t_n} f)(X_{t_{n-1}})\right)}{\mathbb{E}_0^Y\left(Z_{t_{n-1}}\,(H_{t_n} 1)(X_{t_{n-1}})\right)}$$

where H_{t_n} is the finite transition measure on $\mathbb{R}^p$ given by

$$\begin{aligned} H_{t_n} f(x) &\stackrel{\text{def}}{=} \int H_{t_n}(x, dz)\, f(z) = E_0^Y\left(f(X_{t_n})\, g_{t_n}(X, Y) / X_{t_{n-1}} = x\right) \\ \log g_{t_n}(X, Y) &= \int_{t_{n-1}}^{t_n} h^\star(X_s)\, dY_s - \frac{1}{2}\int_{t_{n-1}}^{t_n} |h(X_s)|^2\, ds. \end{aligned} \quad (119)$$

If, for any transition measure H and any probability measure π on $\mathbb{R}^p$ we denote by πH the finite measure so that for any bounded continuous function $f \in \mathcal{B}_b(\mathbb{R}^p)$,

$$\pi H(f) = \int \pi(dx)\, (Hf)(x),$$

then, given the observations, the dynamics structure of the conditional distributions $\{\pi_{t_n} \ : \ n = 0, \ldots, M\}$ is defined by the recursion

$$\pi_{t_n}(f) = \frac{\pi_{t_{n-1}} H_{t_n}(f)}{\pi_{t_{n-1}} H_{t_n}(1)}, \qquad n = 1, \ldots, M \qquad \text{with} \quad \pi_{t_0} = \nu$$

To approximate the stochastic integrals (119) it is convenient to note that, in a sense to be given,

$$\log g_{t_n}(X,Y) \underset{\Delta t_n \sim 0}{\sim} h^\star(X_{t_{n-1}})\,\Delta Y_{t_n} - \frac{1}{2}|h(X_{t_{n-1}})|^2\,\Delta t_n$$

with $\Delta t_n = t_n - t_{n-1}$. In this connection, a first step to obtain a computationally feasible solution consists in replacing H_{t_n} by the approximating multiplication operator

$$(H^M_{t_n} f)(x) = g^M_{t_n}(\Delta Y_{t_n}, x)\;(P^{(M)} f)(x)$$

where $g^M_{t_n}(\Delta Y_{t_n}, .) : \mathbb{R}^p \to \mathbb{R}_+$ is the positive and continuous function given by

$$g^M_{t_n}(\Delta Y_{t_n}, x) = \exp\left(h^\star(x)\,\Delta Y_{t_n} - \frac{1}{2M}|h(x)|^2\right) \quad \text{with} \quad \Delta Y_{t_n} = Y_{t_n} - Y_{t_{n-1}}$$

Remark 4.5: The choice of the approximating function $g^M_{t_n}$ given above is not unique. We can also use the functions $\tilde{g}^M_{t_n}$ given by

$$\tilde{g}^M_{t_n}(\Delta Y_{t_n}, x) = 1 + h^\star(x)\Delta Y_{t_n} + \frac{1}{2}|h(x)|^2\left(|\Delta Y_{t_n}|^2 - \frac{1}{M}\right)$$

The function h being bounded we can choose M large enough so that

$$\forall\, x \in \mathbb{R}^p, \qquad \|h\| < \sqrt{M} \qquad \text{and} \qquad g^M_{t_n}(\Delta Y_{t_n}, x) > 0$$

From now on we denote by $\{\pi^M_{t_n} \;:\; n = 0, \ldots, M\}$ the solution of the resulting approximating discrete time model

$$\left\{\begin{array}{lcll} \pi^M_{t_n} & = & \Phi^M_n(\Delta Y_{t_n}, \pi^M_{t_{n-1}}), & n = 1, \ldots, M \\ \pi^M_0 & = & \nu & \end{array}\right. \tag{120}$$

where $\Phi^M_n(y,\pi) = \Psi^M_n(y,\pi) P^{(M)}$ and

$$\forall\, f \in \mathcal{B}_b(E), \qquad \Psi^M_{t_n}(y,\pi) f = \frac{\displaystyle\int f(x)\, g^M_{t_n}(y,x)\,\pi(dx)}{\displaystyle\int g^M_{t_n}(y,x)\,\pi(dx)}$$

for any $y \in \mathbb{R}^q$ and $\pi \in \mathbf{M}_1(\mathbb{R}^p)$.

Elementary manipulations show that the solution of the latter system is also given by the formula

$$\pi^M_{t_n} f = \frac{\displaystyle\int f(x_n) \prod_{m=1}^n g^M_{t_m}(\Delta Y_{t_m}, x_{m-1}) \prod_{m=1}^n P^{(M)}(x_{m-1}, dx_m)\,\nu(dx_0)}{\displaystyle\int \prod_{m=1}^n g^M_{t_m}(\Delta Y_{t_m}, x_{m-1}) \prod_{m=1}^n P^{(M)}(x_{m-1}, dx_m)\,\nu(dx_0)} \tag{121}$$

This gives a description of a discrete time approximating model for the optimal filter in terms of Feynman-Kac formulae for measure valued systems as those presented in section 1.1 and section 1.3.1.

The error bound caused by the discretization of the time interval $[0, 1]$ and the approximation of the signal semigroup is well understood (see for instance Proposition 5.2 p. 31 [75], Theorem 2 in [92], Theorem 4.2 in [76] and also Theorem 4.1 in [82]). More precisely if for any $n = 0, \ldots, M-1$ and $t \in [t_n, t_{n+1})$ we denote by $\pi_t^M \stackrel{\text{def.}}{=} \pi_{t_n}^M$ we have the well known result.

Theorem 4.6 ([76]) *Let f be a bounded test function on $\mathbb{R}^p$ satisfying the Lipschitz condition*

$$|f(x) - f(z)| \leq k(f)\, |x - z|.$$

Then

$$\sup_{t\in[0,1)} \mathbb{E}\left(|\pi_t f - \pi_t^M f|\right) \leq \frac{C}{\sqrt{M}} (\|f\| + k(f)) \tag{122}$$

where C is some finite constant.

We shall see in section 4.4 that the discrete time approximating model (120) can be regarded as the optimal filter of a suitably defined discrete time filtering problem. In most of the applications we have in mind the whole path of observation process $\{Y_t\ ;\ t \in [0, 1]\}$ is not completely known.
Instead of that the acquisition of the observation data is made at regularly spaced times. In this specific situation the approximating model (120) and the sampled observation record $\{\Delta Y_{t_n}\ ;\ n = 1, \ldots, M\}$ give a natural framework for formulating this filtering problem and for applying the BIPS approaches developed in previous sections.

By $\{\xi_{t_n}\ ;\ n \geq 0\}$ we denote the N interacting particle scheme associated with the limiting system (120) and defined as in (13) by replacing the functions $\{\Phi_n\ ;\ n \geq 1\}$ by the functions $\{\Phi_n^{(M)}(\Delta Y_{t_n}, .)\ ;\ n \geq 1\}$.
The results of section 2.2 can be used to study the convergence of the random measures

$$\pi_{t_n}^{M,N} \stackrel{\text{def.}}{=} \frac{1}{N} \sum_{i=1}^{N} \delta_{\xi_{t_n}^i}$$

to the flow of distributions $\{\pi_{t_n}^M\ ;\ n \geq 0\}$ as $N \to \infty$.

An immediate question is to know how the discrete time N-particle scheme and the M-discretization time scheme combine? This study is still in progress. The only known result in this direction has been obtained in [25].

For any $n = 0, \ldots, M-1$ and $t \in [t_n, t_{n+1})$ we denote by $\pi_t^{M,N}$, the empirical measures associated with the system ξ_{t_n}, namely

$$\pi_t^{M,N} = \frac{1}{N} \sum_{i=1}^{N} \delta_{\xi_{t_n}^i}.$$

Theorem 4.7 *For any bounded Lipschitz test function f such that*

$$|f(x) - f(z)| \leq k(f)\ |x - z|$$

we have that

$$\sup_{t\in[0,1)} \mathbb{E}\left(|\pi_t f - \pi_t^{M,N} f|\right) \leq \frac{C_1}{\sqrt{M}}\ (\|f\| + k(f)) + C_2\ \sqrt{\frac{M}{N}}\ \|f\| \tag{123}$$

where C_1 is the finite constant appeared in Theorem 4.6 and $C_2 = 2\sqrt{2}e^{12\|h\|^2}$. In addition, if $p = q = 1$ and a, b, f, h are four times continuously differentiable with bounded derivatives then we have

$$\sup_{t\in[0,1)} \mathbb{E}\left(|\pi_t f - \pi_t^{M,N} f|\right) \leq \mathrm{Cte}\left(\frac{1}{M} + \sqrt{\frac{M}{N}}\right). \tag{124}$$

4.4 Discrete Time Filtering Problems

4.4.1 Description of the Models

The discrete time filtering problem consists in a signal process $X = (X_n\ ;\ n \geq 0)$ taking values in a Polish space E and an "observation" process $Y = (Y_n\ ;\ n \geq 1)$ taking values in $\mathbb{R}^d$ for some $d \geq 1$. We assume that the transition probability kernels $\{K_n\ ;\ n \geq 1\}$ are Feller and the initial value X_0 of the signal is an E-valued random variable with law $\eta_0 \in \mathbf{M}_1(E)$. The observation process has the form

$$\forall n \geq 1, \qquad Y_n = h_n(X_{n-1}) + V_n$$

where $h_n : E \to \mathbb{R}^d$ are bounded continuous and $(V_n\ ;\ n \geq 0)$ are independent random variables with positive continuous density $(\varphi_n\ ;\ n \geq 0)$ with respect to Lebesgue measure on $\mathbb{R}^d$. It is furthermore assumed that the observation noise $(V_n\ ;\ n \geq 0)$ and the signal $(X_n\ ;\ n \geq 0)$ are independent.

The filtering problem can be summarized as to find the conditional distributions

$$\forall\, f \in \mathcal{C}_b(E), \forall\, n \geq 1, \qquad \eta_n(f) \ = \ \mathbb{E}(f(X_n)/Y_1, \ldots, Y_n)$$

A version of η_n is given by a Feynman-Kac formula as the one presented in (1), namely

$$\eta_n(f) = \frac{\displaystyle\int f(x_n) \prod_{m=1}^{n} \varphi_m\left(Y_m - h_m(x_{m-1})\right) \prod_{m=1}^{n} K_m(x_{m-1}, dx_m)\ \eta_0(dx_0)}{\displaystyle\int \prod_{m=1}^{n} \varphi_m\left(Y_m - h_m(x_{m-1})\right) \prod_{m=1}^{n} K_m(x_{m-1}, dx_m)\ \eta_0(dx_0)} \tag{125}$$

It is transparent from this formulation that the discrete time approximating model (120) given in section 4.3 can be regarded as the optimal filter associated with a discrete time nonlinear filtering problem.

Given the observations $\{y_n\ ;\ n \geq 1\}$ the flow of distributions $\{\eta_n\ ;\ n \geq 0\}$ is again solution of a $\mathbf{M}_1(E)$-valued dynamical system of the form (8), that is

$$\forall\, n \geq 1,\ \forall\, \eta_0 \in \mathbf{M}_1(E), \qquad \eta_n = \Phi_n(y_n, \eta_{n-1}) \tag{126}$$

where for any $y \in \mathbb{R}^d$, $\Phi_n(y, .) : \mathbf{M}_1(E) \to \mathbf{M}_1(E)$ is the continuous function given by

$$\forall\, \eta \in \mathbf{M}_1(E), \qquad \Phi_n(y, \eta) \stackrel{\text{def.}}{=} \Psi_n(y_n, \eta)\, K_n$$

and $\Psi_n(y, .) : \mathbf{M}_1(E) \to \mathbf{M}_1(E)$ is the continuous function given by

$$\forall\, \eta \in \mathbf{M}_1(E),\ \forall\, f \in \mathcal{C}_b(E), \qquad \Psi_n(y, \eta)(f) \ = \ \frac{\int f(x)\, \varphi_n(y - h_n(x))\, \eta(dx)}{\int \varphi_n(y - h_n(z))\, \eta(dz)}$$

In this formulation the flow of distributions $\{\eta_n\ ;\ n \geq 0\}$ is parameterized by a given observation record $\{y_n\ :\ n \geq 1\}$ and it is solution of the measure valued dynamical system having the form (8) so that the IPS and BIPS approaches introduced in section 1.3.2 and section 2.3 can be applied.

4.4.2 Averaged Results

Our next objective is to present averaged versions of stability results given in section 2.1.2 and the averaged version of Theorem 2.11.

The only difficulty in directly applying the results of the end of section 2.1.2 stems from the fact that in our setting the fitness functions are random in the observation parameter. Instead of $(\mathcal{G})$ we will use the following assumption

$(\mathcal{G}')$ ***For any time*** $n \geq 1$***, there exist a positive function***

$$a_n : \mathbb{R}^d \to [1, \infty)$$

and a nondecreasing function $\theta : \mathbb{R} \to \mathbb{R}$ ***such that***

$$\forall\, x \in E,\ \forall\, y \in \mathbb{R}^d, \qquad \frac{1}{a_n(y)} \leq \frac{\varphi_n(y - h_n(x))}{\varphi_n(y)} \leq a_n(y) \tag{127}$$

and

$$|\log a_n(y + u) - \log a_n(y)| \leq \theta(\|u\|)$$

Theorem 4.8 *Assume that* $(\mathcal{G}')$ *holds with* $\sup_{n\geq 1} \|h_n\| < \infty$*. For any* $\mu \in \mathbf{M}_1(E)$ *we write* $\{\eta_n^\mu\ ;\ n \geq 0\}$ *the solution of the nonlinear filtering equation (126) starting at* μ*.*
If $(\mathcal{K})_2$ *holds then for any* $\mu, \nu \in \mathbf{M}_1(E)$ *we have the following implication*

$$\sup_{n\geq 1} \mathbb{E}(\log a_n(V_n)) < \infty \Longrightarrow \lim_{n\to\infty} \mathbb{E}(\|\eta_n^\mu - \eta_n^\nu\|_{\text{tv}}) = 0$$

If $(\mathcal{K})_3$ *holds then we also have for any* $\mu, \nu \in \mathbf{M}_1(E)$

$$\sum_{n\geq 1} \mathbb{E}(a_n^{-2}(V_n)) = \infty \ \Longrightarrow\ \lim_{n\to\infty} \mathbb{E}(\|\eta_n^\mu - \eta_n^\nu\|_{\text{tv}}) = 0$$

$$\lim_{n\to\infty} \frac{1}{n} \sum_{p=1}^{n} \mathbb{E}(a_p^{-2}(V_p)) > 0 \ \Longrightarrow\ \limsup_{n\to\infty} \frac{1}{n} \log \mathbb{E}(\|\eta_n^\mu - \eta_n^\nu\|_{\text{tv}}) < 0$$

The averaged version of Theorem 2.11 can be stated as follows

Theorem 4.9 *Let $\mathcal{F}$ be a countable collection of functions f such that $\|f\| \leq 1$ and satisfying the entropy condition $I(\mathcal{F}) < \infty$. Assume that $(\mathcal{G})'$ holds and the following conditions are met*

$$\sup_{n\geq 1} \log \mathbb{E}(a_n^2(V_n))^{1/2} \stackrel{\text{def}}{=} L < \infty, \qquad \sup_{n\geq 1} \|h_n\| \stackrel{\text{def}}{=} M < \infty$$

Assume moreover that the nonlinear filtering equation (126) is asymptotically stable in the sense that,

$$\lim_{T\to\infty} \sup_{\mu,\nu\in \mathbf{M}_1(E)} \sup_{p\geq 0} \mathbb{E}\left(\|\Phi_{p,p+T}(\mu) - \Phi_{p,p+T}(\nu)\|_{\mathcal{F}} \,|Y_1,\ldots,Y_p\right) = 0$$

then we have the following uniform convergence with respect to time

$$\lim_{N\to\infty} \sup_{n\geq 0} \mathbb{E}\left(\left\|\eta_n^N - \eta_n\right\|_{\mathcal{F}}\right) = 0 \tag{128}$$

In addition, if the evolution equation (126) is exponentially asymptotically stable in the sense that there exists some positive constant $\gamma > 0$ such that for any $\mu, \nu \in \mathbf{M}_1(E)$ and $T \geq 0$

$$\sup_{p\geq 0} \mathbb{E}\left(\|\Phi_{p,p+T}(\mu) - \Phi_{p,p+T}(\nu)\|_{\mathcal{F}} \,|Y_1,\ldots,Y_p\right) \leq e^{-\gamma T}$$

then we have for any $p \geq 1$, the uniform $\mathbb{L}^p$-error bound given by

$$\sup_{n\geq 0} \mathbb{E}\left(\left\|\eta_n^N - \eta_n\right\|_{\mathcal{F}}^p\right)^{\frac{1}{p}} \leq \frac{C_p\, e^{\gamma'}}{N^{\frac{\alpha}{2}}} \;\; I(\mathcal{F})$$

where C_p is a universal constant which only depends on $p \geq 1$ and α and γ' are given by

$$\alpha = \frac{\gamma}{\gamma + \gamma'} \quad \text{and} \quad \gamma' = 1 + 2\,(L + \theta(M))$$

Remark 4.10: We now present a class of discrete time nonlinear filtering problems for which the BIPS approaches developed in this work do not apply.
Let us assume that the pair process (X, Y) takes values in $\mathbb{R}^p \times \mathbb{R}^{p'}$ and evolves according to the following Itô's differential equations

$$\begin{cases} dX_t &= a(X_t, Y_t)\, dt + b(X_t, Y_t)\, dW_t \\ dY_t &= a'(X_t, Y_t)\, dt + b'(X_t, Y_t)\, dW'_t \qquad Y_0 = 0 \end{cases}$$

where a, b, a', b' are known functions suitably defined and (W, W') is a $p+p'$-dimensional Wiener process. Suppose moreover that the acquisition of the observations is only made at times $n \in \mathbb{N}$ and we want to compute for all reasonable functions $f : \mathbb{R}^p \to \mathbb{R}$

$$\pi_n f = \mathbb{E}(f(X_n)\,|Y_1,\ldots,Y_n)$$

This problem is clearly a discrete time nonlinear filtering problem but the BIPS approaches developed in previous sections do not apply.

A novel BIPS strategy has been proposed in [41] to solve this problem. In contrast to the latter this new particle scheme consists in N-pair particles and the mutation transition is related to the continuous semigroup of the pair process $\{(X_t, Y_t)\ ;\ t \geq 0\}$.

Exponential rates of convergence and L^1-mean error estimates are given in [41] and central limit theorems for the particle density profiles are presented in [40, 39] but many questions such as large deviations, fluctuations on path space as well as uniform convergence results with respect to time remain unsolved.

5 Appendix and Index of Notations

Since we have tried to use similar notations for discrete and continuous time, we hope that the following separations of the indexes for both settings will be convenient for the reader.

Index of Symbols (discrete time)

Index of Conditions for Asymptotic Theorems (discrete time)

Index of Symbols (continuous time)

Index of Conditions for Asymptotic Theorems (continuous time)

Examples

Despite our best efforts, the variety of approaches to study asymptotic stability properties and limit theorems for the IPS approximating models inevitably require a set of specific assumptions which might be confusing at a first reading. We therefore gather together in this section a short discussion on the conditions we have defined and a succinct series of implications. As a guide to their usage we also analyze these assumptions in two academic examples, namely the Gaussian and bi-exponential transitions $\{K_n \; ; \; n \geq 1\}$.

It is also worth observing immediatly that the limit theorems for the particle density profiles as the number of particles tends to infinity only depend on a boundedness condition $(\mathcal{G})$ defined on page 21. Furthermore in nonlinear filtering settings the fitness functions $\{g_n \; ; \; n \geq 1\}$ also depend on the observation process and the appropriate condition corresponding to $(\mathcal{G})$ is the assumption $(\mathcal{G})'$ defined on page 129. It is clear that the situation becomes more involved when dispensing with the assumption $(\mathcal{G})$ or $(\mathcal{G})'$. It turns out that several results can be proved without these conditions, see for instance Proposition 2.9 and its proof on page 35. Furthermore these boundedness conditions are not really restrictive and they are commonly used in nonlinear filtering literature. We conclude this section with a short collection of fitness functions and observation processes satisfying the former conditions.

Section 2.1.2 is concerned with the asymptotic stability properties of the limiting measure valued systems $\{\eta_n \; ; \; n \geq 0\}$ and $\{\widehat{\eta}_n \; ; \; n \geq 0\}$. Under appropriate mixing conditions on the transition probability kernels $\{K_n \; ; \; n \geq 1\}$ it is proven that the resulting limiting systems forget exponentially fast their initial conditions. The mixing type conditions we employed are $(\mathcal{K})_\epsilon$, $(\mathcal{K})_1$, $(\mathcal{K})_2$, $(\mathcal{K})_3$ and $(\mathcal{K}\mathcal{G})$. Recalling their description given respectively on pages 20, 24, 26 and 27 it is easy to establish the following implications

$$\begin{array}{ccccc} & & (\mathcal{K})_3 & & \\ & & \Downarrow & & \\ (\mathcal{K})_\epsilon & \Rightarrow \; (\mathcal{K})_1 & \Rightarrow \; (\mathcal{K})_2 & \qquad & (\mathcal{K})_2 + (\mathcal{G}) \Rightarrow (\mathcal{K}\mathcal{G}) \end{array}$$

As we said previously, the limit theorems for the IPS approximating schemes presented in section 2.2 only require the assumption $(\mathcal{G})$ on the fitness functions $\{g_n \; ; \; n \geq 1\}$ except for the central limit theorem and large deviations principles on path space (see for instance section 2.2.1 as well as section 2.2.4 and section 2.2.5). The weakest and preliminary condition $(\mathcal{K})_0$ employed in the study the convergence of the empirical measures on path space can be regarded as a mixing type condition. Recalling its description on page 32 we clearly have

$$(\mathcal{K})_2 \Longrightarrow (\mathcal{K})_0$$

Condition $(\mathcal{K})_0$ is also a natural and appropriate condition for using a reference product measure and express the law of the IPS as a simple mean field Gibbs measure on path space.

The only additional restriction we place in the study of the central limit theorem on path space is the exponential moment condition $(\mathcal{TCL})$ given on page 48. As we shall see in the further development this condition holds for many typical examples of signal transitions. It is also worth noting that

$$(\mathcal{K})_1 \Longrightarrow (\mathcal{TCL}) \Longrightarrow (\mathcal{K})_0$$

In section 2.2.5 we analyze large deviations for the IPS approximating models for general and abstract functions $\{\Phi_n \; ; \; n \geq 1\}$. We have presented a number of simple criterion only involving these one step functions. When applied to Feynman-Kac type systems we have seen that the large deviation principles on path space only require the assumptions $(\mathcal{K})_1'$ and $(\mathcal{K})_1''$ defined on page 54 and page 56. It is also easy to see that

$$(\mathcal{K})_1' \Longrightarrow (\mathcal{K})_1'' \Longrightarrow (\mathcal{K})_0$$

In view of the previous remarks the fluctuations and deviations on path space rely on the existence of a reference probability measure satisfying $(\mathcal{K})_0$. One way to remove this assumption is to study the particle density profiles.

The analysis of fluctuations of the particle density profiles is based on the dynamics structure of the limiting system given in section 2.1.1 and on limit theorems for stochastic processes. These results only depend on the boundedness condition $(\mathcal{G})$.

The same condition is used in proving large deviations for the particle density profiles. More precisely we have employed condition $(\mathcal{G})$ to check an exponential tightness condition $(\mathcal{ET})$ (cf. page 57)

$$(\mathcal{G}) \Longrightarrow (\mathcal{ET})$$

Gaussian Transitions

Let us now investigate the chain of assumptions $(\mathcal{K})_1$, $(\mathcal{K})_2$, $(\mathcal{K})_3$, $(\mathcal{K})_1'$, $(\mathcal{K})_1''$, and $(\mathcal{TCL})$ through the following Gaussian example.

Suppose that $E = \mathbb{R}^m$, $m \geq 1$ and K_n, $n \geq 1$ are given by

$$K_n(x, dz) = \frac{1}{((2\pi)^m |Q_n|)^{1/2}} \exp\left(-\frac{1}{2}(z - b_n(x))' \, Q_n^{-1} \, (z - b_n(x))\right)$$

where Q is a $m \times m$ symmetric nonnegative matrix and $b_n : \mathbb{R}^m \to \mathbb{R}^m$ is a bounded continuous function. It is not difficult to check that $(\mathcal{K})_1'$, is satisfied with

$$\lambda_n(dz) = \frac{1}{((2\pi)^m |Q_n|)^{1/2}} \exp\left(-\frac{1}{2} z' \, Q_n^{-1} \, z\right) dz.$$

Indeed, we then find out that

$$\log \frac{dK_n(x,.)}{d\lambda_n}(z) = \text{const.} - b_n(x)'Q_n^{-1}z$$

which insures the Lipschitz property as well as the growth property with

$$\forall\, z \in \mathbb{R}, \qquad \varphi(z) \;=\; \frac{1}{2}\|b_n\|^2\|Q_n^{-1}\| + \|Q_n^{-1}\|\;\|b_n\|\;|z|$$

From the previous observation it is also not difficult to check that condition $(\mathcal{TCL})$ is also satisfied.

Let us discuss conditions $(\mathcal{K})_1$ and $(\mathcal{K})_3$ in time homogeneous settings (that is $K_n = K$) and when $E = \mathbb{R}$ and $Q_n = 1$ and $b_n = b$.
If $b : \mathbb{R} \to \mathbb{R}$ is only a bounded function, then $(\mathcal{K})_1$ and $(\mathcal{K})_2$ do not hold. For instance let us suppose that $b : \mathbb{R} \to \mathbb{R}$ is a bounded $\mathbf{B}(E)$-measurable function such that $b(0) = 0$ and $b(1) = -1$. Then, hypothesis $(\mathcal{K})_1$ is not satisfied. Suppose K satisfies $(\mathcal{K})_1$. Clearly there exists an absolutely continuous probability measure with density p such that

$$\forall\, x, z \in \mathbb{R}, \qquad c^{-1}\, p(z) \;\leq\; e^{-\frac{1}{2}(z-b(x))^2} \;\leq\; c\, p(z)$$

for some positive constant c. Using the fact that $b(1) = -1$ we obtain

$$\lim_{z\to\infty} p(z)e^{\frac{z^2}{2}} = 0$$

On the other hand $b(0) = 0$ implies $p(z)e^{\frac{z^2}{2}} \geq c^{-1}$ which is absurd.

Now we examine condition $(\mathcal{K})_3$. First we note that for any $|z| \leq M$ where $M \geq 0$ is chosen so that $\|b\| \leq M$ we have

$$\forall\, x \in E, \qquad \epsilon \;\leq\; \frac{dK(x,.)}{d\lambda}(z) \;\leq\; \frac{1}{\epsilon}$$

where

$$\lambda(dz) = \frac{1}{\sqrt{2\pi}}\,\exp -\frac{z^2}{2}\,dz \qquad \text{and} \qquad \log \epsilon = -2M^2$$

In other words and roughly speaking $(\mathcal{K})_1$ is satisfied on the compact set

$$\{z \in \mathbb{R}\;;\; |z| \leq M\}.$$

Let us assume that the drift function b satisfies

$$\forall\, |x| \geq M, \qquad b(x) \;=\; b(\text{sign}(x)M)$$

In this situation it is not difficult to check that $(\mathcal{K})_3$ holds with

$$A = [-M, M] \qquad B_1 = (-\infty, -M) \quad B_2 = (M, +\infty)$$

and

$$\begin{aligned}\lambda_1 &= \delta_{-M} K, & \lambda_2 = \delta_{-M} K \\ \gamma_1(dz) &= \frac{1}{\sqrt{2\pi}} \exp -\frac{1}{2}(z-M)^2 \, dz, & \gamma_2(dz) = \frac{1}{\sqrt{2\pi}} \exp -\frac{1}{2}(z+M)^2 dz.\end{aligned}$$

Let us examine a Gaussian situation where $(\mathcal{K})_1''$ is not met. Again we suppose that $E = \mathbb{R}$ and

$$K_n(x, dz) = \sqrt{\frac{\epsilon_n(x)}{2\pi}} \exp\left(-\frac{1}{2}\epsilon_n(x) \, z^2\right) dz$$

where $\epsilon_n : \mathbb{R} \to \mathbb{R}$ is a continuous function such that

$$\forall \, x \in \mathbb{R}, \qquad \epsilon_n(x) > 0 \qquad \text{and} \qquad \lim_{|x| \to \infty} \epsilon_n(x) = 0.$$

It is not difficult to see that K_n is Feller. On the other hand, let us assume that K_n satisfies $(\mathcal{K})_1''$ for some function φ. Since $K_n(x,.)$ is absolutely continuous with respect to Lebesgue measure for any $x \in \mathbb{R}$, the probability measure λ_n described in $(\mathcal{K})_1''$ is absolutely continuous with respect to Lebesgue measure. Therefore, there exists a probability density p_n such that

$$\forall \, x, z \in \mathbb{R}, \qquad e^{-\varphi(z)} \, p_n(z) \;\leq\; \sqrt{\epsilon_n(x)} \, \exp\left(-\frac{1}{2}\epsilon_n(x) \, z^2\right) \;\leq\; e^{\varphi(z)} \, p_n(z).$$

Letting $|x| \to \infty$ one gets $e^{-\varphi(z)} p_n(z) = 0$ for any $z \in \mathbb{R}$ which is absurd since we also assumed $\int e^{\varphi^{1+\epsilon}(z)} p_n(z) dz < \infty$.

Our study is not restricted to nonlinear filtering problem with Gaussian transitions K_n or with observations corrupted by Gaussian perturbations. We now present another kind of densities that can be handled in our framework.

Bi-exponential Transitions

Suppose $E = \mathbb{R}$ and K_n, $n \geq 1$, are given by

$$K_n(x, dz) = \frac{1}{2} \, \alpha_n \exp\left(-\alpha_n |z - b_n(x)|\right) dz, \qquad \alpha_n > 0, \quad b_n \in \mathcal{C}_b(\mathbb{R})$$

This corresponds to the situation where the signal process X is given by

$$X_n = b_n(X_{n-1}) + W_n \qquad n \geq 1$$

where $(W_n)_{n \geq 1}$ is a sequence of real valued and independent random variables with bilateral exponential densities. Note that K_n may be written

$$K_n(x, dz) = \frac{1}{2} \, \alpha_n \exp\left(\alpha_n (|z| - |z - b_n(x)|)\right) \, \lambda_n(dz)$$

with

$$\lambda_n(dz) = \frac{1}{2} \, \alpha_n \exp\left(-\alpha_n |z|\right) dz$$

It follows that $(\mathcal{K})_1''$ holds since $|\log \frac{K_n(x,.)}{d\lambda_n}(z)|$ has Lipschitz norm $2\alpha_n + \alpha_n \|b_n\|$.

It is also clear that $(\mathcal{K})_1$ and $(\mathcal{TCL})$ hold since we have in this situation

$$\exp\left(-\alpha_n \|b_n\|\right) \leq \frac{K_n(x,.)}{d\lambda_n}(z) \leq \exp\left(\alpha_n \|b_n\|\right)$$

Conditions $(\mathcal{G})$ and $(\mathcal{G})'$

Next we examine condition $(\mathcal{G})'$.
As a typical example of nonlinear filtering problem assume the functions $h_n : E \to \mathbb{R}^d$, $n \geq 1$, are bounded continuous and the densities φ_n given by

$$\varphi_n(v) = \frac{1}{((2\pi)^d |R_n|)^{1/2}} \exp\left(-\frac{1}{2} v' R_n^{-1} v\right)$$

where R_n is a $d \times d$ symmetric positive matrix. This correspond to the situation where the observations are given by

$$\forall\, n \geq 1, \qquad Y_n \;=\; h_n(X_{n-1}) + V_n \tag{129}$$

where $(V_n)_{n\geq 1}$ is a sequence of $\mathbb{R}^d$-valued and independent random variables with Gaussian densities.
After some easy manipulations one gets that $(\mathcal{G})'$ holds with

$$\log a_n(y) = \frac{1}{2} \|R_n^{-1}\| \, \|h_n\|^2 + \|R_n^{-1}\| \, \|h_n\| \, |y|$$

where $\|R_n^{-1}\|$ is the spectral radius of R_n^{-1}. In addition we have

$$|\log a_n(y+u) - \log a_n(y)| \leq L_n \, |u| \qquad \text{with} \quad L_n = \|R_n^{-1}\| \, \|h_n\|$$

It is therefore not difficult to check that the assumptions of Theorem 4.9 is satisfied when

$$\sup_{n\geq 1} (\|h_n\|, \|R_n^{-1}\|) < \infty$$

In this situation it is also clear that the conditions of Theorem 4.8 and Theorem 4.9 are met. To see this claim it suffices to note that Jensen's inequality yields that

$$\log E(a_n^{-2}(V_n)) \geq -\|R_n^{-1}\| \, \|h_n\|^2 - 2\|R_n^{-1}\| \, \|h_n\| \, E(|V_n|)$$

and we also have

$$E(\log a_n(V_n)) = \frac{1}{2} \|R_n^{-1}\| \, \|h_n\|^2 + \|R_n^{-1}\| \, \|h_n\| \, E(|V_n|)$$

Our result is not restricted to Gaussian noise sources. For instance, let us assume that $d = 1$ and φ_n is a bilateral exponential density

$$\varphi_n(v) = \frac{\alpha_n}{2} \exp -(\alpha_n |v|) \qquad \alpha_n > 0$$

In this case one gets that $(\mathcal{G})'$ holds with

$$\log a_n(y) = \alpha_n \, \|h_n\|$$

which is independent of the observation parameter y. One concludes easily that the conditions of Theorem 4.8 and Theorem 4.9 are satisfied as soon as

$$\sup_{n\geq 0} \{\alpha_n, \|h_n\|\} < \infty$$

On the other hand if $\sum_{n\geq 0} \alpha_n \|h_n\| < \infty$ then condition $(\mathcal{KG})$ is satisfied and Theorem 2.7 can be used to study the asymptotic stability of the nonlinear filtering equation for any strongly ergodic signal process (cf. remark 2.8).
We end this section with an example of Cauchy noise sources. Suppose that $d = 1$ and φ_n is the density given by

$$\varphi_n(v) = \frac{\theta_n}{\pi\,(v^2 + \theta_n^2)} \qquad \theta_n > 0$$

In this situation one can check that

$$\frac{y^2 + \theta_n^2}{y^2 + \theta_n^2 + \|h_n\|^2 + 2|y|\,\|h_n\|} \leq \frac{\varphi_n(y - h_n(x))}{\varphi_n(y)} \leq 1 + \left(\frac{y}{\theta_n}\right)^2$$

Thus, $(\mathcal{G})'$ holds with

$$a_n(y) = 1 + \left(\left(\frac{y}{\theta_n}\right)^2 \vee \frac{(|y| + \|h_n\|)^2}{y^2 + \theta_n^2}\right)$$

References

[1] J. Abela, D. Abramson, A. De Silval, M. Krishnamoorthy, and G. Mills. Computing optimal schedules for landing aircraft. Technical report, Department of Computer Systems Eng. R.M.I.T., Melbourne, May 1993.

[2] J.-M. Alliot, D. Delahaye, J.-L. Farges, and M. Schoenauer. Genetic algorithms for automatic regrouping of air traffic control sectors. In J. R. McDonnell, R. G. Reynolds, and D. B. Fogel, editors, *Proceedings of the 4th Annual Conference on Evolutionary Programming*, pages 657–672. MIT Press, March 1995.

[3] G. Ben Arous and M. Brunaud. Methode de laplace : étude variationnelle des fluctuations de diffusions de type “champ moyen”. *Stochastics*, 31-32:79–144, 1990.

[4] R. Atar and O. Zeitouni. Exponential stability for nonlinear filtering. *Annales de l'Institut Henri Poincaré*, 33(6):697–725, 1997.

[5] R. Atar and O. Zeitouni. Lyapunov exponents for finite state space nonlinear filtering. *Society for Industrial and Applied Mathematics. Journal on Control and Optimization*, 35(1):36–55, January 1997.

[6] J. Baker. Adaptative selection methods for genetic algorithms. In J. Grefenstette, editor, *Proc. International Conf. on Genetic Algorithms and their Applications*. L. Erlbaum Associates, 1985.

[7] J. Baker. Reducing bias and inefficiency in the selection algorithm. In J. Grefenstette, editor, *Proc. of the Second International Conf. on Genetic Algorithms and their Applications*. L. Erlbaum Associates, 1987.

[8] A. Bakirtzis, S. Kazarlis, and V. Petridis. A genetic algorithm solution to the economic dispatch problem. http://www.dai.ed.ac.uk/groups/evalg/eag_local_copies_of_papers.body.html.

[9] D. Bakry. L'hypercontractivité et son utilisation en théorie des semigroupes. In P. Bernard, editor, *Lectures on Probability Theory. Ecole d'Eté de Probabilités de Saint-Flour XXII-1992*, Lecture Notes in Mathematics 1581. Springer-Verlag, 1994.

[10] P. Barbe and P. Bertail. *The Weighted Bootstrap.* Lecture Notes in Statistics 98. Springer-Verlag, 1995.

[11] E. Beadle and P. Djuric. A fast weighted Bayesian bootstrap filter for nonlinear model state estimation. *Institute of Electrical and Electronics Engineers. Transactions on Aerospace and Electronic Systems*, AES-33:338–343, January 1997.

[12] B.E. Benes. Exact finite-dimensional filters for certain diffusions with nonlinear drift. *Stochastics*, 5:65–92, 1981.

[13] E. Bolthausen. Laplace approximation for sums of independent random vectors i. *Probability Theory and Related Fields*, 72:305–318, 1986.

[14] C.L. Bridges and D.E. Goldberg. An analysis of reproduction and crossover in a binary-coded genetic algorithm. In J.J. Grefenstette, editor, *Proc. of the Second International Conf. on Genetic Algorithms and their Applications.* L. Erlbaum Associates, 1987.

[15] R.S. Bucy. *Lectures on discrete time filtering, Signal Processing and Digital Filtering.* Springer-Verlag, 1994.

[16] A. Budhiraja and D. Ocone. Exponential stability of discrete time filters for bounded observation noise. *Systems and Control Letters*, 30:185–193, 1997.

[17] Z. Cai, F. Le Gland, and H. Zhang. An adaptative local grid refinement method for nonlinear filtering. Technical report, INRIA, October 1995.

[18] H. Carvalho, P. Del Moral, A. Monin, and G. Salut. Optimal nonlinear filtering in gps/ins integration. *Institute of Electrical and Electronics Engineers. Transactions on Aerospace and Electronic Systems*, 33(3):835–850, July 1997.

[19] R. Cerf. *Une théorie asymptotique des algorithmes génétiques.* Thèse de doctorat, Université Montpellier II, March 1994.

[20] M. Chaleyat-Maurel and D. Michel. Des résultats de non existence de filtres de dimension finie. *Comptes Rendus de l'Académie des Sciences de Paris. Série I. Mathématique*, 296, 1983.

[21] D. Crisan, J. Gaines, and T.J. Lyons. A particle approximation of the solution of the kushner-stratonovitch equation. *Society for Industrial and Applied Mathematics. Journal on Applied Mathematics*, 58(5):1568–1590, 1998.

[22] D. Crisan and M. Grunwald. Large deviation comparison of branching algorithms versus resampling algorithm. Preprint, 1998.

[23] D. Crisan and T.J. Lyons. Nonlinear filtering and measure valued processes. *Probability Theory and Related Fields*, 109:217–244, 1997.

[24] D. Crisan, P. Del Moral, and T.J. Lyons. Interacting particle systems approximations of the Kushner-Stratonovitch equation. *Advances in Applied Probability*, 31(3), September 1999.

[25] D. Crisan, P. Del Moral, and T.J. Lyons. Non linear filtering using branching and interacting particle systems. *Markov Processes and Related Fields*, 5(3):293–319, 1999.

[26] G. Da Prato, M. Furhman, and P. Malliavin. Asymptotic ergodicity for the Zakai filtering equation. *Comptes Rendus de l'Académie des Sciences de Paris. Série I. Mathématique*, 321(5):613–616, 1995.

[27] E.B. Davies. *Heat Kernels and Spectral Theory.* Cambridge University Press, 1989.

[28] M. Davis. New approach to filtering for nonlinear systems. *Institute of Electrical and Electronics Engineers. Proceedings*, 128(5):166–172, 1981. Part D.

[29] D. Dawson. Measure-valued Markov processes. In P.L. Hennequin, editor, *Lectures on Probability Theory. Ecole d'Eté de Probabilités de Saint-Flour XXI-1991*, Lecture Notes in Mathematics 1541. Springer-Verlag, 1993.

[30] P. Del Moral. Non-linear filtering: interacting particle resolution. *Markov Processes and Related Fields*, 2(4):555–581, 1996.

[31] P. Del Moral. Filtrage non linéaire par systèmes de particules en interaction. *Comptes Rendus de l'Académie des Sciences de Paris. Série I. Mathématique*, 325:653–658, 1997.

[32] P. Del Moral. Maslov optimization theory: optimality versus randomness. In V.N. Kolokoltsov and V.P. Maslov, editors, *Idempotency Analysis and its Applications*, Mathematics and its Applications 401, pages 243–302. Kluwer Academic Publishers, Dordrecht/Boston/London, 1997.

[33] P. Del Moral. Measure valued processes and interacting particle systems. Application to non linear filtering problems. *The Annals of Applied Probability*, 8(2):438–495, 1998.

[34] P. Del Moral. A uniform convergence theorem for the numerical solving of non linear filtering problems. *Journal of Applied Probability*, 35:873–884, 1998.

[35] P. Del Moral and A. Guionnet. Large deviations for interacting particle systems. Applications to non linear filtering problems. *Stochastic Processes and their Applications*, 78:69–95, 1998.

[36] P. Del Moral and A. Guionnet. On the stability of measure valued processes. Applications to non linear filtering and interacting particle systems. Publications du Laboratoire de Statistique et Probabilités, no 03-98, Université Paul Sabatier, 1998.

[37] P. Del Moral and A. Guionnet. A central limit theorem for non linear filtering using interacting particle systems. *The Annals of Applied Probability*, 9(2):275–297, 1999.

[38] P. Del Moral and A. Guionnet. On the stability of measure valued processes with applications to filtering. *Comptes Rendus de l'Académie des Sciences de Paris. Série I. Mathématique*, 329:429–434, 1999.

[39] P. Del Moral and J. Jacod. Interacting particle filtering with discrete observations. Publications du Laboratoire de Statistiques et Probabilités, no 11-99, 1999.

[40] P. Del Moral and J. Jacod. The monte-carlo method for filtering with discrete time observations. central limit theorems. Publications du Laboratoire de Probabilités, no 515, 1999.

[41] P. Del Moral, J. Jacod, and P. Protter. The Monte Carlo method for filtering with discrete time observations. Publications du Laboratoire de Probabilités, no 453, June 1998.

[42] P. Del Moral and M. Ledoux. Convergence of empirical processes for interacting particle systems with applications to nonlinear filtering. to appear in Journal of Theoretical Probability, January 2000.

[43] P. Del Moral and L. Miclo. Asymptotic stability of non linear semigroup of Feynman-Kac type. Préprint, publications du Laboratoire de Statistique et Probabilités, no 04-99, 1999.

[44] P. Del Moral and L. Miclo. On the convergence and the applications of the generalized simulated annealing. *SIAM Journal on Control and Optimization*, 37(4):1222–1250, 1999.

[45] P. Del Moral and L. Miclo. A Moran particle system approximation of Feynman-Kac formulae. to appear in Stochastic Processes and their Applications, 2000.

[46] P. Del Moral, J.C. Noyer, and G. Salut. Résolution particulaire et traitement non-linéaire du signal : application radar/sonar. In *Traitement du signal*, September 1995.

[47] P. Del Moral, G. Rigal, and G. Salut. Estimation et commande optimale non linéaire. Technical Report 2, LAAS/CNRS, March 1992. Contract D.R.E.T.-DIGILOG.

[48] B. Delyon and O. Zeitouni. Liapunov exponents for filtering problems. In M.H.A. Davis and R.J. Elliot, editors, *Applied Stochastic Analysis*, pages 511–521. Springer-Verlag, 1991.

[49] A. Dembo and O. Zeitouni. *Large Deviations Techniques and Application.* Jones and Bartlett, 1993.

[50] A.N. Van der Vaart and J.A. Wellner. *Weak Convergence and Empirical Processes with Applications to Statistics.* Springer Series in Statistics. Springer, 1996.

[51] J.-D. Deuschel and D.W. Stroock. *Large Deviations.* Pure and applied mathematics 137. Academic Press, 1989.

[52] P. Diaconis and B. Efron. Méthodes de calculs statistiques intensifs sur ordinateurs. *Pour la Science*, 1983. translation of the American Scientist.

[53] R.L. Dobrushin. Central limit theorem for nonstationnary Markov chains, i,ii. *Theory of Probability and its Applications*, 1(1 and 4):66–80 and 330–385, 1956.

[54] E.B. Dynkin and A. Mandelbaum. Symmetric statistics, Poisson processes and multiple Wiener integrals. *The Annals of Statistics*, 11:739–745, 1983.

[55] S. Ethier and T. Kurtz. *Markov Processes, Characterization and Convergence.* Wiley series in probability and mathematical statistics. John Wiley and Sons, New York, 1986.

[56] M. Fujisaki, G. Kallianpur, and H. Kunita. Stochastic differential equations for the non linear filtering problem. *Osaka J. Math.*, 1:19–40, 1972.

[57] F. Le Gland. Monte-Carlo methods in nonlinear filtering. In *Proceedings of the 23rd IEEE Conference on Decision and Control*, pages 31–32, Las Vegas, December 1984.

[58] F. Le Gland. High order time discretization of nonlinear filtering equations. In *28th IEEE CDC*, pages 2601–2606, Tampa, 1989.

[59] F. Le Gland, C. Musso, and N. Oudjane. An analysis of regularized interacting particle methods for nonlinear filtering. In *Proceedings of the 3rd IEEE European Workshop on Computer-Intensive Methods in Control and Signal Processing*, Prague, September 1998.

[60] D.E. Goldberg. Genetic algorithms and rule learning in dynamic control systems. In *Proceedings of the First International Conference on Genetic Algorithms*, pages 8–15, Hillsdale, NJ, 1985. L. Erlbaum Associates.

[61] D.E. Goldberg. Simple genetic algorithms and the minimal deceptive problem. In L. Davis, editor, *Genetic Algorithms and Simulated Annealing.* Pitman, 1987.

[62] D.E. Goldberg. *Genetic Algorithms in Search, Optimization and Machine Learning.* Addison-Wesley, Reading, MA., 1989.

[63] D.E. Goldberg and P. Segrest. Finite Markov chain analysis of genetic algorithms. In J.J. Grefenstette, editor, *Proc. of the 2nd Int. Conf. on Genetic Algorithms.* L. Erlbaum Associates, 1987.

[64] N.J. Gordon, D.J. Salmon, and A.F.M. Smith. Novel approach to nonlinear/non-Gaussian Bayesian state estimation. *IEE Proceedings F*, 140:107–113, 1993.

[65] C. Graham and S. Méléard. Stochastic particle approximations for generalized Boltzmann models and convergence estimates. *The Annals of Probability*, 25(1):115–132, 1997.

[66] A. Guionnet. About precise Laplace's method; applications to fluctuations for mean field interacting particles. Preprint, 1997.

[67] J.H. Holland. *Adaptation in Natural and Artificial Systems.* University of Michigan Press, Ann Arbor, 1975.

[68] J. Jacod and A.N. Shiryaev. *Limit Theorems for Stochastic Processes.* A Series of Comprehensive Studies in Mathematics 288. Springer-Verlag, 1987.

[69] J.M. Johnson and Y. Rahmat-Samii. Genetic algorithms in electromagnetics. In *IEEE Antennas and Propagation Society International Symposium Digest*, volume 2, pages 1480–1483, 1996.

[70] F. Jouve, L. Kallel, and M. Schoenauer. Mechanical inclusions identification by evolutionary computation. *European Journal of Finite Elements*, 5(5-6):619–648, 1996.

[71] F. Jouve, L. Kallel, and M. Schoenauer. Identification of mechanical inclusions. In D. Dagsgupta and Z. Michalewicz, editors, *Evolutionary Computation in Engeneering*, pages 477–494. Springer Verlag, 1997.

[72] G. Kallianpur and C. Striebel. Stochastic differential equations occuring in the estimation of continuous parameter stochastic processes. Tech. Rep. 103, Department of statistics, Univ. of Minnesota, September 1967.

[73] G. Kitagawa. Monte-Carlo filter and smoother for non-Gaussian nonlinear state space models. *Journal on Computational and Graphical Statistics*, 5(1):1–25, 1996.

[74] H. Korezlioglu. Computation of filters by sampling and quantization. Technical Report 208, Center for Stochastic Processes, University of North Carolina, 1987.

[75] H. Korezlioglu and G. Maziotto. Modelization and filtering of discrete systems and discrete approximation of continuous systems. In *Modélisation et Optimisation des Systèmes, VI Conférence INRIA*, Nice, 1983.

[76] H. Korezlioglu and W.J. Runggaldier. Filtering for nonlinear systems driven by nonwhite noises: an approximating scheme. *Stochastics and Stochastics Reports*, 44(1-2):65–102, 1993.

[77] H. Kunita. Asymptotic behavior of nonlinear filtering errors of Markov processes. *Journal of Multivariate Analysis*, 1(4):365–393, 1971.

[78] H. Kunita. Ergodic properties nonlinear filtering processes. In K.C. Alexander and J.C. Watkins, editors, *Spatial Stochastic Processes.* Birkhaüser Boston, Boston, MA, 1991.

[79] S. Kusuoda and Y . Tamura. Gibbs measures for mean field potentials. *J. Fac. Sci. Univ. Tokyo, Sect. IA, Math*, 31, 1984.

[80] J.W. Kwiatkowski. Algorithms for index tracking. Technical report, Department of Business Studies, The University of Edinburgh, 1991.

[81] R.S. Liptser and A.N. Shiryayev. *Theory of Martingales.* Dordrecht: Kluwer Academic Publishers, 1989.

[82] G.B. Di Masi, M. Pratelli, and W.G. Runggaldier. An approximation for the nonlinear filtering problem with error bounds. *Stochastics*, 14(4):247–271, 1985.

[83] S. Méléard. Asymptotic behaviour of some interacting particle systems; McKean-Vlasov and Boltzmann models. In D. Talay and L. Tubaro, editors, *Probabilistic Models for Nonlinear Partial Differential Equations, Montecatini Terme, 1995*, Lecture Notes in Mathematics 1627. Springer-Verlag, 1996.

[84] C. Musso and N. Oudjane. Regularization schemes for branching particle systems as a numerical solving method of the nonlinear filtering problem. In *Proceedings of the Irish Signals Systems Conference, Dublin*, June 1998.

[85] C. Musso and N. Oudjane. Regularized particle schemes applied to the tracking problem. In *International Radar Symposium, Munich, Proceedings*, September 1998.

[86] Y. Nishiyama. Some central limit theorems for l^∞-valued semimartingales and their applications. *Probability Theory and Related Fields*, 108:459–494, 1997.

[87] A. Nix and M.D. Vose. Modelling genetic algorithms with Markov chains. *Annals of Mathematics and Artificial Intelligence*, 5:79–88, 1991.

[88] D. Ocone and E Pardoux. Asymptotic stability of the optimal filter with respect to its initial condition. *Society for Industrial and Applied Mathematics. Journal on Control and Optimization*, 34:226–243, 1996.

[89] D.L. Ocone. *Topics in nonlinear filtering theory.* Phd thesis, MIT, Cambridge, 1980.

[90] E. Pardoux. Filtrage non linéaire et équations aux dérivés partielles stochastiques associées. In P.L. Hennequin, editor, *Ecole d'Eté de Probabilités de Saint-Flour XIX-1989*, Lecture Notes in Mathematics 1464. Springer-Verlag, 1991.

[91] E. Pardoux and D. Talay. Approximation and simulation of solutions of stochastic differential equations. *Acta Applicandae Mathematicae*, 3(1):23–47, 1985.

[92] J. Picard. Approximation of nonlinear filtering problems and order of convergence. In *Filtering and Control of Random Processes*, Lecture Notes Control and Inf. Sc. 61. Springer, 1984.

[93] J. Picard. An estimate of the error in time discretization of nonlinear filtering problems. In C.I. Byrnes and A. Lindquist, editors, *Proceedings of the 7th MTNS — Theory and Applications of nonlinear Control Systems, Stockholm, 1985*, pages 401–412, Amsterdam, 1986. North–Holland Pub.

[94] J. Picard. Nonlinear filtering of one-dimensional diffusions in the case of a high signal-to-noise ratio. *Society for Industrial and Applied Mathematics. Journal on Applied Mathematics*, 16:1098–1125, 1986.

[95] S.T. Rachev. *Probability Metrics and the Stability of Stochastic Models.* Wiley, New York, 1991.

[96] D. Revuz and M. Yor. *Continuous Martingales and Brownian Motion.* Springer-Verlag, 1991.

[97] J. Shapcott. Index tracking: genetic algorithms for investment portfolio selection. Technical Report SS92-24, EPCC, September 1992.

[98] T. Shiga and H. Tanaka. Central limit theorem for a system of markovian particles with mean field interaction. *Zeitschrift für Wahrscheinlichkeitstheorie verwandte Gebiete*, 69, 1985. 439–459.

[99] A.N. Shiryaev. On stochastic equations in the theory of conditional Markov processes. *Theor. Prob. Appl.*, 11:179–184, 1966.

[100] A.N. Shiryaev. *Probability.* Number 95 in Graduate Texts in Mathematics. Springer-Verlag, New-York, second edition, 1996.

[101] B. Simon. *Trace ideals and their applications.* London Mathematical Society Lecture Notes Series 35. Cambridge University Press, 1977.

[102] L. Stettner. On invariant measures of filtering processes. In K. Helmes and N. Kohlmann, editors, *Stochastic Differential Systems, Proc. 4th Bad Honnef Conf.*, Lecture Notes in Control and Inform. Sci., pages 279–292, 1989.

[103] L. Stettner. Invariant measures of pair state/approximate filtering process. In *Colloq. Math. LXII*, pages 347–352, 1991.

[104] R.L. Stratonovich. Conditional Markov processes. *Theor. Prob. Appl.*, 5:156–178, 1960.

[105] D. Stroock and S. R. S. Varadhan. *Multidimensional Diffusion Processes.* Springer-Verlag, 1979.

[106] D. W. Stroock. *An Introduction to the Theory of Large Deviations.* Universitext. Springer-Verlag, New-York, 1984.

[107] M. Talagrand. Sharper bounds for Gaussian and empirical processes. *The Annals of Probability*, 22:28–76, 1994.

[108] D. Talay. Efficient numerical schemes for the approximation of expectations of functionals of the solution of s.d.e. and applications. In *Filtering and Control of random processes (Paris, 1983)*, Lecture Notes in Control and Inform. Sci. 61, pages 294–313, Berlin-New York, 1984. Springer.

[109] H. Tanaka. Limit theorems for certain diffusion processes. In *Proceedings of the Taniguchi Symp., Katata, 1982*, pages 469–488, Tokyo, 1984. Kinokuniya.

[110] D. Treyer, D.S. Weile, and E. Michielsen. The application of novel genetic algorithms to electromagnetic problems. In *Applied Computational Electromagnetics, Symposium Digest*, volume 2, pages 1382–1386, Monterey, CA, March 1997.

[111] M. D. Vose. *The Simple Genetic Algorithm, Foundations and Theory.* The MIT Press Books, August 1999.

[112] D. Williams. *Probability with Martingales.* Cambridge Mathematical Textbooks, 1992.

Laboratoire de Statistiques et Probabilités
CNRS UMR C5583
Université Paul Sabatier
118, route de Narbonne
31062 Toulouse cedex, France
E-mail: delmoral@cict.fr
E-mail: miclo@cict.fr

EXPONENTIAL INEQUALITIES FOR BESSEL PROCESSES

Nathalie Eisenbaum

Laboratoire de Probabilités, Tour 56, 3ème étage, 4 Place Jussieu 75252 Paris Cedex 05.

Abstract : Let $R_d^*(t)$ be the supremum at time t of a Bessel process with dimension d. For T a stopping time, Burkholder has compared the expectationsof $\left(\frac{R_d^*(T)}{\sqrt{d}}\right)^p$ and $(\sqrt{T})^p$ for p>0. Replacing the function x^p by exponential functions, we obtain some variant of his results.

I - Introduction and notations

Let $(B_s)_{s\geq 0}$ be a linear Brownian motion starting from zero. Let R_d be a positive process such that R_d^2 is a solution of the equation :

$$X_t = 2\int_0^t \sqrt{|X_s|}\,dB_s + dt \ , \ d>0 \ ;$$

i.e. : R_d is a Bessel process of dimension d starting from 0.
We set : $R_d^*(t) = \sup_{0\leq s\leq t} R_d(s)$.
Let T be a stopping time with respect to the natural filtration of B.

Burkholder [B] established for any p>0, that $\left(E\left(\frac{R_d^*(T)}{\sqrt{d}}\right)^p / E(\sqrt{T})^p\right)_{d\in\mathbb{N}^*}$ converges to 1, uniformly in T, as d tends to ∞.

Refinements of this convergence have since been proved by Davis [D]. Making use of Poincaré's Lemma, Yor ([Y],p.55) could prove adequate modifications ofthese results for other times than stopping times.

A consequence of Burkholder's result is that $\left(E\left(\frac{R_d^*(T)}{d^{1/2}}\right)^p / E(T^{p/2})\right)_{d\in\mathbb{N}^*}$ is uniformly bounded in d and T. In this paper we consider the following question : What happens if the moderate function x^p is replaced by an exponential function ? For example, is there a function F such that the sequence $\left(E\left(\exp\left(\frac{R_d^*(T)}{d^{1/2}}\right)\right) / E(F(\sqrt{T}))\right)_{d\geq 0}$ is uniformly bounded in d and T, or even converging as d tends to ∞.

We will see that the answer is affirmative for the uniform boundedness question and negative for the convergence question.

II - Exponential inequalities for Bessel processes

Theorem 1 :

(i) There exist two strictly positive constants c *and* β *such that for any* $\lambda>0$ *and any* d *in* $\mathbb{N}^*$ *, we have for any stopping time* T :

$$E\left(\exp\left\{\lambda\left(\frac{R_d^*(T)}{d^{1/2}}\right)\right\} \right) \leq c\, E\left(\exp\left\{ \frac{\beta^2}{2} \lambda^2 T \right\} \right)$$

Moreover β *can be taken equal to* $2\sqrt{e}$.

(ii) For any p *in* (0,2) *there exist two strictly positive constants* b_p *and* β_p *such that for any* $\lambda>0$ *and any* d *in* $\mathbb{N}^*$, *we have for any stopping time* T :

$$E\left(\exp\left\{\lambda\left(\frac{R_d^*(T)}{d^{1/2}}\right)^p\right\} \right) \leq b_p\, E\left(\exp\left\{ \beta_p\, \lambda^{\frac{2}{2-p}}\, T^{\frac{p}{2-p}} \right\} \right)$$

Moreover β_p *can be taken equal to* $(\frac{1}{p} - \frac{1}{2})\, p^{\frac{2}{2-p}}\, (4e)^{\frac{p}{2-p}}$.

The proof of Theorem 1 is based on the following result.

Theorem 2 *There exists a strictly positive constant* c *such that for every stopping time* T, *every* d *in* $\mathbb{N}^*$ *and every* p>0, *we have :*

$$(1)_p \qquad E(R_d^*(T))^p \leq c\, (2\sqrt{e})^p\, d^{p/2}\, E((B_1^*)^p)\, E(T^{p/2})$$

Proof of Theorem 2 : Jacka and Yor have proved (see [J-Y] section 4) that there exists a constant a_p such that for all stopping times T with respect to the natural filtration of B and every d in $\mathbb{N}^*$:

$$E(R_d^*(T))^p \leq a_p\, d^{p/2}\, E(T^{p/2})$$

with $a_p \leq 2e\ 2^p\ (p + \frac{1}{2})^{p/2}$ when $p \geq 2$.

Thus, we are looking for $\beta > 0$ such that there exists $c > 0$ with :

$$c.\beta^p\ E(B_1^*)^p \geq a_p$$

Since for any $p>0$: $E(|B_1|^{2p}) = \dfrac{2^p}{\pi^{1/2}} \Gamma(p + \frac{1}{2})$,

Stirling's asymptotic formula gives the following equivalency :

$$E(|B_1|^{2p}) \underset{p \to \infty}{\sim} 2^{p+\frac{1}{2}} (p + \frac{1}{2})^p \exp\{-(p + \frac{1}{2})\}$$

Hence, for any $\beta \geq 2\sqrt{e}$, there exists a constant $c>0$ such that for every $p>0$

$$c.\beta^p\ E(B_1^*)^p \geq a_p .$$ □

We note that $(1)_p$ can be rewritten as follows :

$(2)_p$ $$E(\ R_d^*(T))^p \leq c\ (2\sqrt{e})^p\ \ d^{p/2}\ E\Big(\ (\tilde{B}_1^*)^p\ T^{p/2}\Big)$$

where $\tilde{B}$ is an independent copy of B.

Summing the inequalities $(2)_{np}$, n running through $\mathbb{N}$ and p being a fixed value in (0,2), we obtain the following result :

$$E\Big(\exp\{\lambda\ (R_d^*(T)^p\}\Big) \leq c\ E\Big(\ \exp\{\lambda\ (2\sqrt{e})^p\ d^{p/2}\ (\tilde{B}_1^*)^p\ T^{p/2}\}\Big)$$

We then use the following majorizations already established in [D-E] :

$$E(\exp(\lambda B_1^*)) \leq 4\ \exp(\frac{\lambda^2}{2}) \qquad \text{and}$$

$$E(\exp\{\lambda(B_1^*)^p\}) \leq b_p\ \exp\{\ (\frac{1}{p} - \frac{1}{2})(\lambda p)^{\frac{2}{2-p}}\ \}$$

where b_p is a strictly positive constant

to obtain Theorem 1.

We can write similar relations for exponential functions vanishing at zero. As an example, we have the following theorem .

Theorem 3 : *There exist two strictly positive constants* c *and* β *such that* *for any* $\lambda>0$ *and any* d *in* $\mathbb{N}^*$, *we have for any stopping time* T :

$$E\Big(\cosh\{\lambda(\frac{R_d^*(T)}{d^{1/2}})\} - 1 \Big) \le c\, E\Big(\exp \{ \frac{\beta^2}{2} \lambda^2 T \} - 1 \Big)$$

Moreover: $\beta \le 2\sqrt{e}$.

<u>Proof</u> : By summing the inequalities $(2)_{2n}$, n running through $\mathbb{N}^*$, we obtain :

$$E\Big(\cosh\{\lambda(\frac{R_d^*(T)}{d^{1/2}})\} - 1 \Big) \le c\, E\Big(\cosh\{\lambda\beta\, \tilde{B}_1^* \,.T^{1/2}\} - 1 \Big)$$

We then note that : $E\Big(\cosh\{\lambda\, B_1^*\} - 1 \Big) \le 2\, (\exp(\frac{\lambda^2}{2}) - 1)$.

<u>Remark</u> : In view of the results of Burkholder [B] and Davis [D] it is natural to look for a function F such that, for example,

$E(\exp(\frac{R_d^*(T)}{d^{1/2}}))/ E(F(T^{1/2}))$ would converge to 1 , uniformly in T, when d tends to infinity.

Assuming such a function F exists, we would obtain that for a given $\varepsilon>0$, if d is big enough, for every stopping time T :

$$(*) \qquad (1-\varepsilon)\, E(F(T^{1/2})) \le E(\exp(\frac{R_d^*(T)}{d^{1/2}})).$$

We would also have : $\lim_{d\to\infty} E(\exp(\frac{R_d^*(t)}{d^{1/2}})) = F(t^{1/2})$, for $t>0$.

But we know that for every p in $\mathbb{N}^*$, $E\left(\frac{R_d^*(t)}{d^{1/2}}\right)^p$ converges to $t^{p/2}$.

Consequently for every n in $\mathbb{N}^*$:

$$\lim_{d\to\infty} E(\exp(\frac{R_d^*(t)}{d^{1/2}})) \ge \lim_{d\to\infty} \sum_{p=0}^{n} \frac{1}{p!}\, E\left(\frac{R_d^*(t)}{d^{1/2}}\right)^p = \sum_{p=0}^{n} \frac{t^{p/2}}{p!} .$$

We finally obtain : $F(t^{1/2}) \ge \exp(t^{1/2})$, for every $t>0$.

Hence the inequality (*) implies :

$$(1-\varepsilon)\, E(\exp(T^{1/2})) \le E(\exp(\frac{R_d^*(T)}{d^{1/2}})).$$

In [J-Y] Jacka and Yor have proved that such an inequality can not hold when d=1. Since their argument is exclusively based on the scaling property of the Brownian motion, we can easily extend their result to any d>1.

In conclusion, there is no function F verifying such hypothesis.

Moreover, we see thanks to the same kind of argument, that there is not even a function F such that $E\left(\exp\left(\frac{R_d^*(T)}{d^{1/2}}\right)\right)/\ E(F(T^{1/2}))$ would be uniformly minorized by a strictly positive constant.

References

[B] D.L. BURKHOLDER "Exit times of Brownian motion, and Hardy spaces" *Advances in Math.26, 182-205 (1977).*

[D] B. DAVIS "On stopping times for n- dimensional Brownian motion" *Annals of Proba.6 ,651-659,(1978).*

[D-E] V.H. DE LA PENA and N.EISENBAUM "Exponential Burkholder Davis Gundy inequalities" *Bull.Lond.Math.Soc.29 ,239-242 (1996).*

[J-Y] S.D. JACKA and M. YOR "Inequalities for non-moderate functions of a pair of stochastic processes".*Proc.London Math.Soc. (3) 67,649-672(1993).*

[Y] M. YOR "Some aspects of Brownian motion - II - Some recent martingale problems" *Lecture in Math.(Zürich) Birkhäuser(1997).*

On Sums of iid Random Variables Indexed by N Parameters★

By D. Khoshnevisan
The University of Utah

Summary. Motivated by the works of J.L. Doob and R. Cairoli, we discuss reverse N-parameter inequalities for sums of i.i.d. random variables indexed by N parameters. As a corollary, we derive Smythe's law of large numbers.

1. Introduction

For any integer $N \geqslant 1$, let us consider $\mathbb{Z}_+^N \triangleq \{1, 2, \cdots\}^N$ and endow it with the following partial order: for all $\mathbf{n}, \mathbf{m} \in \mathbb{Z}_+^N$,

$$\mathbf{n} \preccurlyeq \mathbf{m} \quad \Longleftrightarrow \quad n_i \leqslant m_i, \quad \text{for all } 1 \leqslant i \leqslant N.$$

Suppose $\{X, X(\mathbf{k});\ \mathbf{k} \in \mathbb{Z}_+^N\}$ is a sequence of independent, identically distributed random variables, indexed by $\mathbb{Z}_+^N$. The corresponding random walk S is given by:

$$S(\mathbf{n}) \triangleq \sum_{\mathbf{k} \preccurlyeq \mathbf{n}} X(\mathbf{k}), \qquad \mathbf{n} \in \mathbb{Z}_+^N.$$

According to Cairoli and Dalang [CD], for all $p > 1$,

$$\begin{aligned} \mathbb{E} \sup_{\mathbf{n}} \left| \frac{S(\mathbf{n})}{\langle \mathbf{n} \rangle} \right| < \infty \quad &\Longleftrightarrow \quad \mathbb{E}\big[|X| \big(\log_+ |X|\big)^N\big] < \infty, \\ \mathbb{E} \sup_{\mathbf{n}} \left| \frac{S(\mathbf{n})}{\langle \mathbf{n} \rangle} \right|^p < \infty \quad &\Longleftrightarrow \quad \mathbb{E}|X|^p < \infty. \end{aligned} \tag{1.1}$$

Here and throughout, for all $x > 0$,

$$\log_+ x \triangleq \begin{cases} \ln(x), & \text{if } x > e \\ 1, & \text{if } 0 < x \leqslant e \end{cases},$$

★ Research partially supported by NSA and NSF

and for all $\mathbf{n} \in \mathbb{Z}_+^N$, $\langle \mathbf{n} \rangle \triangleq \prod_{j=1}^N n_j$. When $N = 1$, this is classical. In this case, J.L. Doob has given a more probabilistic interpretation of this fact by observing that $S(n)/n$ is a reverse martingale; cf. Chung [Ch] for this and more. The goal of this note is to show how a quantitative version of the method of Doob can be carried out, even when $N > 1$. Our approach involves projection arguments which are reminiscent of some old ideas of R. Cairoli; see Cairoli [Ca], Cairoli and Dalang [CD] and Walsh [W].

Perhaps the best way to explain the proposed approach is by demonstrating the following result which may be of independent interest. For related results and a wealth of further references, see [CD], Shorack and Smythe [S1] and Smythe [S2].

Theorem 1. *For all $p > 1$,*

$$\mathbb{E} \sup_{\mathbf{n} \in \mathbb{Z}_+^N} \left| \frac{S(\mathbf{n})}{\langle \mathbf{n} \rangle} \right|^p \leqslant \left(\frac{p}{p-1} \right)^{Np} \mathbb{E}|X|^p. \tag{1.2}$$

Moreover, the corresponding L^1 norm has the following bound:

$$\mathbb{E} \sup_{\mathbf{n} \in \mathbb{Z}_+^N} \left| \frac{S(\mathbf{n})}{\langle \mathbf{n} \rangle} \right| \leqslant \left(\frac{e}{e-1} \right)^N \left\{ N + \mathbb{E}\left[|X| \left(\log_+ |X| \right)^N \right] \right\}. \tag{1.3}$$

Theorem 1 implies the "hard" half of both displays in eq. (1.1). The easy half is obtained upon observing that for all $p \geqslant 1$,

$$\mathbb{E} \sup_{\mathbf{n}} \left| \frac{S(\mathbf{n})}{\langle \mathbf{n} \rangle} \right|^p \geqslant 2^{-p} \mathbb{E} \sup_{\mathbf{n}} \left| \frac{X(\mathbf{n})}{\langle \mathbf{n} \rangle} \right|^p,$$

and directly calculating the above.

An enhanced version of Theorem 1 is stated and proved in Section 2. There, we also demonstrate how to use Theorem 1 together with Banach space arguments to obtain the law of large numbers for $S(\mathbf{n})$ due to Smythe [S2].

2. Proof of Theorem 1

I will prove (1.3) of Theorem 1. Eq. (1.2) follows along similar lines. In fact, it turns out to be alot simpler to prove more. Define for all $p \geqslant 0$,

$$\Psi_p(x) \triangleq x \left(\log_+ x \right)^p, \qquad x > 0.$$

I propose to prove the following extension of Theorem 1:

Theorem 1-bis. *For all $p \geqslant 0$,*

$$\mathbb{E} \sup_{\mathbf{n} \in \mathbb{Z}_+^N} \Psi_p \left(\frac{S(\mathbf{n})}{\langle \mathbf{n} \rangle} \right) \leqslant (p+1)^N \left(\frac{e}{e-1} \right)^N \left\{ N + \mathbb{E} \Psi_{p+N} (|X|) \right\}.$$

Setting $p \equiv 0$ in Theorem 1-bis, we arrive at Theorem 1.

Let us recall the following elementary fact:

Lemma 2.1. *Suppose $\{M_n; n \geqslant 1\}$ is a reverse martingale. Then for all $p > 1$,*

$$\mathbb{E} \sup_{n \geqslant 1} |M_n|^p \leqslant \Big(\frac{p}{p-1}\Big)^p \mathbb{E}|M_1|^p. \tag{2.1}$$

For any $p \geqslant 0$,

$$\mathbb{E} \sup_{n \geqslant 1} \Psi_p(|M_n|) \leqslant (p+1)\Big(\frac{e}{e-1}\Big)\Big\{1 + \mathbb{E}\Psi_{p+1}(|X|)\Big\}. \tag{2.2}$$

Proof. Eq. (2.1) follows from integration by parts and the maximal inequality of Doob. Likewise, one shows that

$$\mathbb{E} \sup_{n \geqslant 1} \Psi_p(|M_n|) \leqslant \Big(\frac{e}{e-1}\Big)\Big\{1 + \mathbb{E}\Big[\Psi_p(|M_1|) \ln_+ \Psi_p(|M_1|)\Big]\Big\}.$$

For all $x > 0$, $\ln_+ \Psi_p(x) \leqslant \ln_+ x + p \ln_+ \ln_+ x$. Eq. (2.2) follows easily. ◇

Now, each $\mathbf{n} \in \mathbb{Z}_+^N$ can be thought of as $\mathbf{n} = (\widehat{\mathbf{n}}, n_N)$, where $\widehat{\mathbf{n}}$ is defined by $\widehat{\mathbf{n}} \triangleq (n_1, \cdots, n_{N-1}) \in \mathbb{Z}_+^{N-1}$. For all $\mathbf{n} \in \mathbb{Z}_+^N$ and all $1 \leqslant j \leqslant n_N$, define

$$Y(\widehat{\mathbf{n}}, j) \triangleq \frac{1}{\prod_{j=1}^{N-1} n_j} \sum_{i_1=1}^{n_1} \cdots \sum_{i_{N-1}=1}^{n_{N-1}} X(\widehat{\mathbf{i}}, j).$$

Clearly,

$$\frac{S(\mathbf{n})}{\langle \mathbf{n} \rangle} = \frac{1}{n_N} \sum_{j=1}^{n_N} Y(\widehat{\mathbf{n}}, j), \qquad \mathbf{n} \in \mathbb{Z}_+^N. \tag{2.3}$$

Let

$$\mathcal{R}(k) \triangleq \sigma\{X(\mathbf{m}); m_N > k\} \vee \sigma\{S(\mathbf{m}); m_N = k\}, \qquad k \geqslant 1,$$

where $\sigma\{\cdots\}$ represents the ($\mathbb{P}$–completed) σ–field generated by $\{\cdots\}$.

Lemma 2.2. *$\{\mathcal{R}(k); r \geqslant 1\}$ is a reverse filtration indexed by $\mathbb{Z}_+^1$.*

Proof. This means that $\mathcal{R}(k) \supset \mathcal{R}(k+1)$ — a simple fact. ◇

Lemma 2.3. *For all $\mathbf{n} \in \mathbb{Z}_+^N$,*

$$\frac{S(\mathbf{n})}{\langle \mathbf{n} \rangle} = \mathbb{E}\big[Y(\widehat{\mathbf{n}}, 1) \mid \mathcal{R}(n_N)\big].$$

Assuming Lemma 2.3 for the moment, let us prove Theorem 1.

Proof of Theorem 1-bis. Without loss of generality, we can and will assume that

$$\mathbb{E}\Psi_{p+N}(|X|) < \infty. \tag{2.4}$$

Otherwise, there is nothing to prove. When $N = 1$, the result follows immediately from Lemma 2.1. Our proof proceeds by induction over N. Suppose Theorem 1-bis holds for all sums of iid random variables indexed by $\mathbb{Z}_+^{N-1}$ whose incremental distribution is the same as that of X. We will prove it holds for N. By Lemma 2.3,

$$\mathbb{E} \sup_{\mathbf{n}\in\mathbb{Z}_+^N} \Psi_p\Big(\frac{S(\mathbf{n})}{\langle\mathbf{n}\rangle}\Big) \leqslant \mathbb{E} \sup_{k\geqslant 1} \Psi_p\Big(\mathbb{E}\big[W \mid \mathcal{R}(k)\big]\Big),$$

where

$$W \triangleq \sup_{n_1,\cdots,n_{N-1}\geqslant 1} \big|Y(\widehat{\mathbf{n}},1)\big|.$$

However, $\{Y(\widehat{\mathbf{n}},1); \widehat{\mathbf{n}} \in \mathbb{Z}_+^{N-1}\}$ is the average of a random walk indexed by $\mathbb{Z}_+^{N-1}$ with the same increments as S. Therefore, by the induction assumption,

$$\mathbb{E}\Psi_p(W) \leqslant (p+1)^{N-1}\Big(\frac{e}{e-1}\Big)^{N-1}\Big\{N-1+\mathbb{E}\Psi_{p+N}(|X|)\Big\}. \tag{2.5}$$

In particular, $\mathbb{E}W < \infty$. Together with with Lemma 2.1's eq. (2.2), this implies that $M_k \triangleq \mathbb{E}\big[W \mid \mathcal{R}(k)\big]$ is a reverse martingale, By eq. (2.2) of Lemma 2.1,

$$\mathbb{E}\Big[\sup_{\mathbf{n}\in\mathbb{Z}_+^N} \Psi_p\Big(\frac{S(\mathbf{n})}{\langle\mathbf{n}\rangle}\Big)\Big] \leqslant (p+1)\Big(\frac{e}{e-1}\Big)\Big\{1+\mathbb{E}\big[\Psi_p(W)\big]\Big\}.$$

Note that $(p+1)e(e-1)^{-1} \geqslant 1$. Therefore, applying (2.5) to this inequality, we obtain Theorem 1-bis. ◇

Proof of Lemma 2.3. Recall (2.3). It remains to show that for $1 \leqslant j \leqslant n_N$,

$$\mathbb{E}\big[Y(\widehat{\mathbf{n}},j) \mid \mathcal{R}(n_N)\big] = \mathbb{E}\big[Y(\widehat{\mathbf{n}},1) \mid \mathcal{R}(n_N)\big]. \tag{2.6}$$

To this end, we observe that $\{Y(\widehat{\mathbf{n}},j); 1 \leqslant j \leqslant n_N\}$ is a sequence of iid random variables. By exchangeability,

$$\mathbb{E}\big[Y(\widehat{\mathbf{n}},j) \mid \mathcal{B}(\mathbf{n})\big] = \mathbb{E}\big[Y(\widehat{\mathbf{n}},1) \mid \mathcal{B}(\mathbf{n})\big], \tag{2.7}$$

where for all $\mathbf{n} \in \mathbb{Z}^N$,,

$$\mathcal{B}(\mathbf{n}) \triangleq \sigma\big\{S(\mathbf{k}); \mathbf{k} \in \mathbb{Z}_+^N \text{ with } k_N = n_N \text{ and } k_j \leqslant n_j, \text{ for all } 1 \leqslant j \leqslant N-1\big\}.$$

Let $\mathcal{C}_0(n_N)$ denote the sigma-field generated by $\{X(\mathbf{k});\ k_N > n_N\}$ and define

$$\mathcal{C}(n_N) \triangleq \mathcal{C}_0(n_N) \vee \sigma\big\{X(\mathbf{k});\ k_N = n_N \text{ and for some } 1 \leqslant j \leqslant N-1,\ k_j > n_j\big\}.$$

It is easy to see that $\mathcal{B}(\mathbf{n})$ is independent of $\mathcal{C}(n_N)$ and

$$\mathcal{R}(n_N) = \mathcal{C}(n_N) \vee \mathcal{B}(\mathbf{n}). \tag{2.8}$$

Eq. (2.6) follows from (2.7), (2.8) and the elementary fact that the collection $\{Y(\widehat{\mathbf{n}}, j); 1 \leqslant j \leqslant n_N\}$ is independent of $\mathcal{C}(n_N)$. ◇

Open Problem.* Motivated by the proof of Theorem 1-bis — and in the notation of that proof — consider:

$$T(n_N)(\widehat{\mathbf{n}}) \triangleq \frac{1}{n_N} \sum_{j=1}^{n_N} Y(\widehat{\mathbf{n}}, j).$$

It is easy to see that $T(n_N)$ is a reverse martingale which takes its values in the space of all sequences indexed by $\mathbb{Z}_+^{N-1}$. For all $\mathbf{n} \in \mathbb{Z}_+^N$ and any two reals $a < b$, define $U_{a,b}(n_N)(\widehat{\mathbf{n}})$ to be the total number of upcrossings of the interval $[a, b]$ before time n_N of the (real valued) reverse martingale $k \mapsto T(k)(\widehat{\mathbf{n}})$. Is it true that there exist constants C_1 and C_2 (which depend **only** on N) such that

$$\mathbb{E}\big[\sup_{\widehat{\mathbf{n}} \in \mathbb{Z}_+^{N-1}} U_{a,b}(n_N)(\widehat{\mathbf{n}})\big] \leqslant C_1 \frac{\mathbb{E}\big[\sup_{\widehat{\mathbf{n}} \in \mathbb{Z}_+^{N-1}} |T(1)(\widehat{\mathbf{n}}) - a|\big]}{(b-a)^{C_2}}? \tag{2.9}$$

Note that when $N = 1$, the supremum is vacuous. In this case, the above holds with $C_1 = C_2 = 1$ and is DOOB's upcrossing inequality for the reversed martingale T. If it holds, (2.9) and Theorem 1 together imply SMYTHE's strong law of large numbers; cf. [S2]. The main part of the aforementioned result is the following:

Theorem 2. ([S2]) *Suppose*

$$\mathbb{E}\big[|X|\big(\log_+ |X|\big)^{N-1}\big] < \infty \quad \textit{and} \quad \mathbb{E}X = 0. \tag{2.10}$$

Then almost surely,

$$\lim_{\langle \mathbf{n} \rangle \to \infty} \frac{S(\mathbf{n})}{\langle \mathbf{n} \rangle} = 0.$$

Remark. Classical arguments show that condition (2.10) is necessary as well.

Proof. I will first prove Theorem 2 for $N = 2$. Let c_0 denote the collection of all bounded functions $a : \mathbb{Z}_+^1 \mapsto \mathbb{R}$ such that $\lim_{k \to \infty} |a(k)| = 0$. Topologize c_0 with the supremum norm: $\|a\| \triangleq \sup_k |a(k)|$. Then, c_0 is a separable Banach space. Let

$$\xi_j(k) \triangleq \frac{1}{k} \sum_{i=1}^{k} X(i, j).$$

---

* **Added Note.** Since this article was accepted for publication, we have found the answer to the open problem above to be affirmative.

Note that ξ_j are i.i.d. random functions from $\mathbb{Z}^1_+$ to $\mathbb{R}$. By Theorem 1, for all $j \geqslant 1$, $\mathbb{E}\|\xi_j\| \leqslant e^2(e-1)^{-2}\{2 + \mathbb{E}[|X| \log_+ |X|]\} < \infty$. By the classical strong law of large numbers, $\xi_1, \xi_2, \cdots$ are i.i.d. elements of c_0. The most elementary law of large numbers on Banach spaces will show that as elements of c_0, almost surely,

$$\lim_{n\to\infty} \frac{1}{n} \sum_{j=1}^{n} \xi_j = 0.$$

See LEDOUX AND TALAGRAND [LT; Corollary 7.10] for this and much more. In other words, almost surely

$$\lim_{n_1\to\infty} \frac{1}{n_1} \sum_{i_1=1}^{n_1} X(i_1, i_2) = 0,$$

uniformly over all $i_2 \geqslant 1$. Plainly, this implies the desired result and much more when $N = 2$. The general case follows by inductive reasoning; the details are omitted. ◊

REFERENCES.

[Ca] R. CAIROLI, (1970). Une ingálité pour martingales á indices multiples et ses applications, *Sém. de Prob.* IV, 1–27, Lecture Notes in Math., **124**, Springer, New York

[CD] R. CAIROLI AND R.C. DALANG, (1996). *Sequential Stochastic Optimization*, Wiley, New York

[Ch] K.L. CHUNG, (1974). *A Course in Probability Theory*, Second Ed., Academic Press, New York

[LT] M. LEDÓUX AND M. TALAGRAND, (1991). *Probability in Banach Spaces*, Springer, New York

[S1] G.R. SHORACK AND R.T. SMYTHE, (1976). Inequalities for $\max|S_{\mathbf{k}}|/b_{\mathbf{k}}$ where $\mathbf{k} \in N^r$, *Proc. Amer. Math. Soc.*, **54**, 331–336

[S2] R.T. SMYTHE, (1973). Strong law of large numbers for r–dimensional arrays of random variables, *Ann. Prob.*, **1**(1), 164–170

[W] J.B. WALSH, (1986). Martingales with a multidimensional parameter and stochastic integrals in the plane, *Lectures in Probability and Statistics*, 329–491, Lecture Notes in Math. **1215**, Springer, New York

Department of Mathematics
Salt Lake City, UT. 84112
davar@math.utah.edu

SERIES OF ITERATED QUANTUM STOCHASTIC INTEGRALS[1]

Stéphane ATTAL[2] & Robin L. HUDSON[3]

[2]Institut Fourier
Université de Grenoble I
BP 74
38402 St Martin d'Hères Cedex, France

[3]Mathematics Department
Nottingham Trent University
Burton street
Nottingham NG1 4BU, United Kingdom

Abstract

We consider series of iterated non-commutative stochastic integrals of scalar operators on the boson Fock space. We give a sufficient condition for these series to converge and to define a reasonable operator. An application of this criterion gives a condition for the convergence of some formal series of generalized integrator processes such as considered in [CEH].

1 Introduction

On the multiple boson Fock space $\Phi = \Gamma(L^2(\mathbb{R}^+; \mathbb{C}^N))$ the quantum stochastic calculus ([HP1], [Me1], [Par]) gives the definition of stochastic integrals of the form $\int_0^t H_j^i(s)\, dA_j^i(s)$ where A_0^i, A_i^0, A_j^i, $i, j \in \{1, \ldots, N\}$ are the creation, annihilation and conservation processes respectively; the H_j^i being adapted processes of operators on Φ. These quantum stochastic integrals can be seen as non-commutative extensions of the usual stochastic integrals with respect to the Brownian motion (for example). In classical stochastic calculus one considers chaotic expansion of random variables that is, series of iterated stochastic integrals of scalar processes. It is also useful in many problems of quantum stochastic calculus to consider series of iterated non-commutative stochastic integrals that is, operators of the form

$$T_t = \lambda I + \sum_{n=1}^{\infty} \sum_{\varepsilon \in E^n} \int_{0<t_1<\ldots<t_n<t} h^{\varepsilon}_{t_1 \ldots t_n}\, dA^{\varepsilon_1}_{t_1} \ldots dA^{\varepsilon_n}_{t_n} \tag{1}$$

where I is the identity operator on Φ and λ is a complex number; where $E = \{0, 1, \ldots, N\}^2 \setminus \{(0,0)\}$ and for each $\eta = (\eta^1, \eta^2) \in E$ the symbol A_t^η denotes the operator $A^{\eta^1}_{\eta^2}(t)$; where $\{h^{\varepsilon}_{t_1 \ldots t_n}; n \in \mathbb{N}^*, 0 < t_1 < \ldots < t_n, \varepsilon \in E^n\}$ are scalar operators; and finally $\int_{0<t_1<\ldots<t_n<t} h^{\varepsilon}_{t_1 \ldots t_n}\, dA^{\varepsilon_1}_{t_1} \ldots dA^{\varepsilon_n}_{t_n}$ denotes the iterated non-commutative stochastic integral

$$\int_0^t \Big[\int_0^{t_n} \Big[\cdots \int_0^{t_2} h^{\varepsilon}_{t_1 \ldots t_n}\, dA^{\varepsilon_1}_{t_1}\Big]\, dA^{\varepsilon_2}_{t_2}\Big] \ldots dA^{\varepsilon_{n-1}}_{t_{n-1}}\Big] dA^{\varepsilon_n}_{t_n}.$$

We propose here to give a sufficient condition on the family $\{h^{\varepsilon}_{t_1 \ldots t_n}\}$ for the expression (1) to define a reasonable operator on Φ, that is for each iterated stochastic integral to be well-defined and for the series to converge weakly.

---

[1] Work supported by EU HCM Contract CHRX-CT93-0094

Let us first present a useful "short notation": the *Guichardet space* notation. Let $\mathcal{I}$ be the set $\{1,\ldots,N\}$. Let $\mathcal{P}$ be the set of finite subsets of $\mathbb{R}^+$, then $\mathcal{P} = \cup_{n\in\mathbb{N}}\mathcal{P}_n$ where $\mathcal{P}_0 = \{\emptyset\}$ and $\mathcal{P}_n$ is the set of n-element subsets of $\mathbb{R}^+$, for $n \geq 1$. Each set $\mathcal{P}_n$ can be identified with the increasing simplex $\{(t_1,\ldots,t_n) \in \mathbb{R}^n; 0 < t_1 < \ldots < t_n\}$, so it is equipped with the restriction of the Lebesgue measure on $\mathbb{R}^n$. Thus the set $\mathcal{P}$ can be equipped with a measure space structure whose element of volume is simply denoted $d\sigma$, $\sigma \in \mathcal{P}$ (in the following, elements of $\mathcal{P}$ will always be denoted by small greek letters $\alpha,\beta,\gamma,\sigma,\tau,\ldots$). It can be easily seen that the Fock space $\Phi = \Gamma(L^2(\mathbb{R}^+;\mathbb{C}^N))$ is isomorphic to the space $L^2(\mathcal{P}^\mathcal{I})$ (see [Me1], p. 103-104), called the *Guichardet space.* Thus a vector f of Φ is determined by the family of complex numbers $f(\sigma)$, $\sigma = (\sigma_i)_{i\in\mathcal{I}} \in \mathcal{P}^\mathcal{I}$ which satisfies $\int_{\mathcal{P}^\mathcal{I}} |f(\sigma)|^2\, d\sigma < \infty$ where $d\sigma$ denotes $\prod_{i\in\mathcal{I}} d\sigma_i$. The family $\{f(\sigma), \sigma \in \mathcal{P}^\mathcal{I}\}$ is called the *chaotic expansion* of f.

In the following the symbol $+$ used for elements of $\mathcal{P}$ denotes the union of *disjoints* elements of $\mathcal{P}$.

Recall a fundamental property of integrals over Guichardet space, known as the $\sum\!\!\!\!\int$-Lemma.

$\sum\!\!\!\!\int$-Lemma (see [L-P])– *Let φ be a positive (resp. integrable) measurable function on $\mathcal{P}^n$, then the function*

$$\alpha \mapsto \sum_{\alpha_1+\ldots+\alpha_n=\alpha} \varphi(\alpha_1,\ldots,\alpha_n)$$

is a positive (resp. integrable) measurable function on $\mathcal{P}$ and one has

$$\int_\mathcal{P}\cdots\int_\mathcal{P} \varphi(\alpha_1,\ldots,\alpha_n)\, d\alpha_1\ldots d\alpha_n = \int_\mathcal{P} \sum_{\alpha_1+\ldots+\alpha_n=\alpha} \varphi(\alpha_1,\ldots,\alpha_n)\, d\alpha. \quad \blacksquare$$

Let $L^2_{lb}(\mathbb{R}^+;\mathbb{C}^N)$ be the space of locally bounded elements of $L^2(\mathbb{R}^+;\mathbb{C}^N)$. For f in $L^2_{lb}(\mathbb{R}^+;\mathbb{C}^N)$, one puts $f_{t]} = f\mathbb{1}_{[0,t]}$, $f_{[t} = f\mathbb{1}_{[t,+\infty[}$ and denotes by $\pi(f)$ the associated *coherent vector*, that is the element of Φ whose chaotic expansion is given by $[\pi(f)](\sigma) = \prod_{i\in\mathcal{I}}\prod_{s\in\sigma_i} f_i(s)$ (where as usual the empty product is equal to 1), where the f_i's are the coordinates of f. The space of finite linear combinations of coherent vectors is denoted $\mathcal{E}_{lb}$. It is a dense subspace of Φ.

Let $\Phi_{t]}$ be the space $\Gamma(L^2([0,t];\mathbb{C}^N))$ and $\Phi_{[t}$ the space $\Gamma(L^2([t,\infty[;\mathbb{C}^N))$, then one has the *continuous tensor product structure* $\Phi \simeq \Phi_{t]} \otimes \Phi_{[t}$, in which we have $\pi(f) = \pi(f_{t]}) \otimes \pi(f_{[t})$ (cf [Me2]). Actually, in the rest of the article, the spaces Φ and $\Phi_{t]} \otimes \Phi_{[t}$ are not distinguished. The tensor product symbol is even omited: $\pi(f) = \pi(f_{t]})\pi(f_{[t})$.

Recall that an *adapted process* of operators (in the sense of [HP1]) is a family $(H_t)_{t\geq 0}$ of operators from Φ to Φ, defined on $\mathcal{E}_{lb}$ and such that:

i) the mapping $t \mapsto H_t\,\pi(f)$ is strongly measurable for all f;

ii) $H_t\,\pi(f_{t]}) \in \Phi_{t]}$ for all t, all f;

iii) $H_t\,\pi(f) = [H_t\pi(f_{t]})]\pi(f_{[t})$ for all t, all f.

Recall that if an adapted process of operators $(H_t)_{t\geq 0}$ is such that for all $f \in L^2_{lb}(\mathbb{R}^+;\mathbb{C}^N)$

$$\int_0^t \|H_s\,\pi(f)\|^2\, ds < \infty \quad \text{for all } t \geq 0, \tag{2}$$

then for every $\eta = (\eta^1, \eta^2) \in E$ the process $\left(\int_0^t H_s\, dA_s^\eta\right)_{s\geq 0}$ is well-defined as an adapted process of operators on $\mathcal{E}_{lb}$ given by

$$<\pi(g)\,,\int_0^t H_s\, dA_s^\eta\, \pi(f)> = \int_0^t \overline{g}_{\eta^1}(s) f_{\eta^2}(s) <\pi(g)\,,\, H_s\,\pi(f)> ds \qquad (3)$$

where $g_0 = f_0 \equiv 1$.

2 Definition of the iterated integrals

Lemma 1 – *Let $(H_t)_{t\geq 0}$ be an adapted process of operators satisfying (2). Then, for every $T > 0$, every $f \in L^2_{lb}(\mathbb{R}^+; \mathbb{C}^N)$ there exist two constants $C, C' \geq 0$ such that for all $0 \leq t < T$, all $\eta \in E$ one has*

$$\Big|\Big|\int_0^t H_s\, dA_s^\eta\, \pi(f)\Big|\Big|^2 \leq C e^{C't} \int_0^t \|H_s\,\pi(f)\|^2\, ds.$$

Proof

One can find many proofs of this kind of estimate. This particular one is taken from [At2] p. 93. One can find an analogous one in [Par] p. 188. ■

Lemma 2 – *Let $t \in \mathbb{R}^+$. If h is a function on $\mathcal{P}_n$ such that*

$$\int_{0<t_1<\ldots<t_n<t} |h^\varepsilon_{t_1\ldots t_n}|^2\, dt_1 \ldots dt_n < \infty$$

then the iterated non-commutative stochastic integrals

$$T_t = \int_{0<t_1<\ldots<t_n<t} h(t_1, \ldots, t_n)\, dA^{\varepsilon_1}_{t_1} \ldots dA^{\varepsilon_n}_{t_n}$$

are well-defined on $\mathcal{E}_{lb}$ for every $\varepsilon = (\varepsilon_1, \ldots, \varepsilon_n) \in E^n$.

If h satisfies

$$\int_{0<t_1<\ldots<t_n<\infty} e^{Ct_n} |h^\varepsilon_{t_1\ldots t_n}|^2\, dt_1 \ldots dt_n < \infty$$

for all $C \in \mathbb{R}^+$ then the iterated integral

$$T = \int_{0<t_1<\ldots<t_n<\infty} h(t_1, \ldots, t_n)\, dA^{\varepsilon_1}_{t_1} \ldots dA^{\varepsilon_n}_{t_n}$$

is well-defined on $\mathcal{E}_{lb}$ for every $\varepsilon = (\varepsilon_1, \ldots, \varepsilon_n) \in E^n$.

Proof

¿From (2) it is sufficient to have for every $f \in L^2_{lb}(\mathbb{R}^+; \mathbb{C}^N)$

$$\int_0^t \Big|\Big|\Big(\int_{0<t_1<\ldots<t_{n-1}<t_n} h(t_1, \ldots, t_{n-1}, t_n)\, dA^{\varepsilon_1}_{t_1} \ldots dA^{\varepsilon_{n-1}}_{t_{n-1}}\Big)\pi(f)\Big|\Big|^2 dt_n < \infty.$$

By successive applications of Lemma 1, one gets that this quantity is indeed dominated by

$$C^{n-1} e^{(n-1)C't} \int_{0<t_1<\ldots<t_n<t} \|h(t_1, \ldots, t_n)\pi(f)\|^2\, dt_1 \ldots dt_n$$

$$= C^{n-1} e^{(n-1)C't} \|\pi(f)\|^2 \int_{0<t_1<\ldots<t_n<t} |h(t_1, \ldots, t_n)|^2\, dt_1 \ldots dt_n.$$

Which is finite if h is locally square integrable.

The case $t = +\infty$ is easy to get in the same way. ∎

3 Correspondance between non-commutative chaotic expansions and Maassen-Meyer kernels

We now consider operators of the form (1) and call their representation as series of iterated non-commutative stochastic integrals, the *non-commutative chaotic expansion* of the operator. Of course, for the moment, we have not given a sense to the series; so let us consider operators of the form (1) such that the

$$\sum_{\varepsilon\in E^n} \int_{0<t_1<\ldots<t_n<t} h^{\varepsilon}_{t_1\ldots t_n}\, dA^{\varepsilon_1}_{t_1}\ldots dA^{\varepsilon_n}_{t_n},$$

$n \in \mathbb{N}$, do not vanish only for a finite number of n. We are going to show that the operators of the form (1) have a *Maassen-Meyer kernel*. The reader does not need to know the theory of Maassen-Meyer kernels, we just use them as a useful language for our computations. All that is needed in this note is going to be defined. Note that what we here call Maassen-Meyer kernels are in fact Dermoune's extension ([Der]) to multiple Fock space of Maassen-Meyer kernels ([Maa], [Me2]).

Recall that $\mathcal{I} = \{1,\ldots N\}$. Let $\mathcal{M} = \mathcal{I}^2$. We consider in the following elements of $\mathcal{P}^{\mathcal{I}}$ and of $\mathcal{P}^{\mathcal{M}}$. An element $\alpha \in \mathcal{P}^{\mathcal{I}}$ is then a "vector" $(\alpha_i)_{i\in\mathcal{I}}$, an element $\beta \in \mathcal{P}^{\mathcal{M}}$ is written as a "matrix" $(\beta^j_i)_{i,j\in\mathcal{I}}$. We also underline ($\underline{\beta}$) the elements of $\mathcal{P}^{\mathcal{M}}$ in order to distinguish them from the elements of $\mathcal{P}^{\mathcal{I}}$. Thus when one integrates with respect to $\underline{\beta} \in \mathcal{P}^{\mathcal{M}}$, the symbol $d\underline{\beta}$ actually means $\prod_{i\in\mathcal{I}}\prod_{j\in\mathcal{I}} d\beta^i_j$.

An operator T from Φ to Φ is said to have a *Maassen-Meyer kernel* on a domain $\mathcal{D}$ if there exists a measurable mapping, also denoted T, from $\mathcal{P}^{\mathcal{I}} \times \mathcal{P}^{\mathcal{M}} \times \mathcal{P}^{\mathcal{I}}$ to $\mathbb{C}$ such that $\mathcal{D}$ is included in $\operatorname{Dom} T$ and for every $f \in \mathcal{D}$ the chaotic expansion of Tf is given by

$$[Tf](\alpha) = \int_{\mu\in\mathcal{P}^{\mathcal{I}}} \sum_{\forall i,\, \rho_i+\Sigma_j\sigma^j_i+\tau_i=\alpha_i} T(\rho,\underline{\sigma},\mu) f\big((\mu_i + \Sigma_j \sigma^i_j + \tau_i)_i\big)\, d\mu.$$

Let us make precise some more notations. For $(\rho,\underline{\sigma},\tau) \in \mathcal{P}^{\mathcal{I}} \times \mathcal{P}^{\mathcal{M}} \times \mathcal{P}^{\mathcal{I}}$ we denote by $\operatorname{ord}(\rho,\underline{\sigma},\tau)$ the set $\cup_{i\in\mathcal{I}}(\rho_i \cup \cup_j \sigma^j_i \cup \tau_i)$ but *ordered* in the increasing direction. We denote by $\varepsilon(\rho,\underline{\sigma},\tau)$ the element of E^n, where $n = \#\operatorname{ord}(\rho,\underline{\sigma},\tau)$, such that

$$\varepsilon(\rho,\underline{\sigma},\tau)_k = \begin{cases} (i,0) & \text{if the } k\text{-th smallest element of } \operatorname{ord}(\rho,\underline{\sigma},\tau) \text{ is in } \rho_i \\ (j,i) & \text{if the } k\text{-th smallest element of } \operatorname{ord}(\rho,\underline{\sigma},\tau) \text{ is in } \sigma^j_i \\ (0,i) & \text{if the } k\text{-th smallest element of } \operatorname{ord}(\rho,\underline{\sigma},\tau) \text{ is in } \tau_i. \end{cases}$$

Conversely, for every $\varepsilon \in E^n$, every $(i,j) \in E$, let $\varepsilon(i,j) = \{k \in \{1,\ldots n\};\, \varepsilon_k = (i,j)\}$. For given $\varepsilon \subset E^n$ and $\sigma = \{t_1,\ldots,t_n\} \in \mathcal{P}_n$, let σ^+ be the element of $\mathcal{P}^{\mathcal{I}}$ defined by $\sigma^+_i = \{t_k\,;\, k \in \varepsilon(i,0)\}$; let $\underline{\sigma}$ be the element of $\mathcal{P}^{\mathcal{M}}$ defined by $(\underline{\sigma})^i_j = \{t_k\,;\, k \in \varepsilon(i,j)\}$; let σ^- be the element of $\mathcal{P}^{\mathcal{I}}$ defined by $\sigma^-_i = \{t_k\,;\, k \in \varepsilon(0,i)\}$. This way, we have defined from σ an element $(\sigma^+,\underline{\sigma},\sigma^-)$ of $\mathcal{P}^{\mathcal{I}} \times \mathcal{P}^{\mathcal{M}} \times \mathcal{P}^{\mathcal{I}}$ such that $\varepsilon(\sigma^+,\underline{\sigma},\sigma^-) = \varepsilon$.

Proposition 3 – *Let $t \in \mathbb{R}^+ \cup \{+\infty\}$. Let T_t be an operator on Φ which admits a non-commutative chaotic expansion*

$$T_t = \lambda I + \sum_{n=1}^{\infty}\sum_{\varepsilon\in E^n} \int_{0<t_1<\ldots<t_n<t} h^{\varepsilon}_{t_1\ldots t_n}\, dA^{\varepsilon_1}_{t_1}\ldots dA^{\varepsilon_n}_{t_n}$$

which is well-defined on $\mathcal{E}_{lb}$ in so far as the corresponding conditions of Lemma 2 are satisfied and that the sum over n is finite. Then T_t admits a Maassen-Meyer kernel described by

$$\begin{cases} T_t(\emptyset,\underline{\emptyset},\emptyset) & = \lambda \\ T_t(\rho,\underline{\sigma},\tau) & = h^{\varepsilon(\rho,\underline{\sigma},\tau)}_{\mathrm{ord}(\rho,\underline{\sigma},\tau)}\mathbb{1}_{[0,t]}(\rho,\underline{\sigma},\tau) \end{cases}$$

where $\mathbb{1}_{[0,t]}(\rho,\underline{\sigma},\tau)$ is the indicator function of "$\rho_i, \sigma_i^j, \tau_i \subset [0,t]$, for all $i,j \in \mathcal{I}$".

Conversely, let T be an operator on Φ having a Maassen-Meyer kernel representation given by the kernel $(\rho,\underline{\sigma},\tau)\mapsto T(\rho,\underline{\sigma},\tau)$. Then T admits a non-commutative chaotic expansion

$$T = \lambda I + \sum_{n=1}^{\infty}\sum_{\varepsilon\in E^n}\int_{0<t_1<\ldots<t_n<\infty} h^{\varepsilon}_{t_1\ldots t_n}\, dA^{\varepsilon_1}_{t_1}\ldots dA^{\varepsilon_n}_{t_n}$$

where, for $\sigma = \{t_1,\ldots t_n\}$, $h^\varepsilon_\sigma = T(\sigma^+,\underline{\sigma},\sigma^-)$.

Proof

By (3) one has

$$<\pi(g)\,,\,T_t\,\pi(f)> = \lambda<\pi(g)\,,\,\pi(f)> + \sum_{n=1}^{\infty}\sum_{\varepsilon\in E^n}\int_{0<t_1<\ldots<t_n<t} \overline{g}_{\varepsilon_1^1}(t_1)\ldots\overline{g}_{\varepsilon_n^1}(t_n)$$
$$\times f_{\varepsilon_1^2}(t_1)\ldots f_{\varepsilon_n^2}(t_n)\, h^\varepsilon_{t_1\ldots t_n}\, dt_1\ldots dt_n <\pi(g)\,,\,\pi(f)>.$$

For every $\varepsilon\in E^n$, every $(i,j)\in E$, recall that $\varepsilon(i,j) = \{k\in\{1,\ldots n\}; \varepsilon_k = (i,j)\}$. Then

$$<\pi(g)\,,\,T_t\,\pi(f)> = \lambda<\pi(g)\,,\,\pi(f)> + \sum_{n=1}^{\infty}\sum_{\varepsilon\in E^n}\int_{0<t_1<\ldots<t_n<t}\prod_{i\in\mathcal{I}}\prod_{k\in\varepsilon(i,0)}\overline{g}_i(t_k)$$
$$\times\prod_{i\in\mathcal{I}}\prod_{j\in\mathcal{I}}\prod_{k\in\varepsilon(i,j)} f_j(t_k)\,\overline{g}_i(t_k)\prod_{i\in\mathcal{I}}\prod_{k\in\varepsilon(0,i)} f_i(t_k)$$
$$\times\, h^\varepsilon_{t_1\ldots t_n}\, dt_1\ldots dt_n <\pi(g)\,,\,\pi(f)>.$$

But, for a given $\alpha = \{0<t_1<\ldots<t_n<t\}\in\mathcal{P}_n$, every $\varepsilon\in E^n$ defines a partition, $\alpha = \sum_i\alpha_i$ and $\rho_i + \sum_j\sigma_i^j + \tau_i = \alpha_i$, of α simply by taking $\rho_i = \alpha_i^+$, $\sigma_j^i = \underline{\alpha}_j^i$, $\tau_i = \alpha_i^-$. Conversely, let $(\rho,\underline{\sigma},\tau)\in\mathcal{P}^{\mathcal{I}}\times\mathcal{P}^{\mathcal{M}}\times\mathcal{P}^{\mathcal{I}}$, then $\varepsilon(\rho,\underline{\sigma},\tau)$ is an element of E^n which corresponds to the partition $\rho_i + \sum_j\sigma_i^j + \tau_i = \alpha_i$ of $\mathrm{ord}(\rho,\underline{\sigma},\tau)$. So

$$<\pi(g)\,,\,T_t\,\pi(f)> = \lambda<\pi(g)\,,\,\pi(f)>$$
$$+\sum_{n=1}^{\infty}\int_{\alpha\subset\mathcal{P}_n}\mathbb{1}_{[0,t]}(\alpha)\sum_{\Sigma_i\alpha_i=\alpha}\ \sum_{\forall i,\,\rho_i+\Sigma_j\sigma_i^j+\tau_i=\alpha_i}\ \prod_{i\in\mathcal{I}}\prod_{s\in\rho_i}\overline{g}_i(s)$$
$$\times\prod_{i\in\mathcal{I}}\prod_{j\in\mathcal{I}}\prod_{s\in\sigma_i^j} f_j(s)\,\overline{g}_i(s)\prod_{i\in\mathcal{I}}\prod_{s\in\tau_i} f_i(s)$$
$$\times\, h_\alpha^{\varepsilon(\rho,\underline{\sigma},\tau)}\, d\alpha <\pi(g)\,,\,\pi(f)>$$
$$= \int_{\alpha\in\mathcal{P}}\mathbb{1}_{[0,t]}(\alpha)\sum_{\Sigma_i\alpha_i=\alpha}\ \sum_{\forall i,\,\rho_i+\Sigma_j\sigma_i^j+\tau_i=\alpha_i}\overline{[\pi(g)]}((\rho_i+\Sigma_j\sigma_i^j)_i)$$
$$\times\,[\pi(f)]((\Sigma_j\sigma_j^i+\tau_i)_i)\,h_\alpha^{\varepsilon(\rho,\underline{\sigma},\tau)}\, d\alpha<\pi(g)\,,\,\pi(f)>$$

by putting $h_{\emptyset}^{\varepsilon(\emptyset,\underline{\emptyset},\emptyset)} = \lambda$. By the $\not\!\!\int$-Lemma, this gives

$$<\pi(g)\,,\,T_t\,\pi(f)> = \int_{\rho\in\mathcal{P}^{\mathcal{I}}}\int_{\underline{\sigma}\in\mathcal{P}^{\mathcal{M}}}\int_{\tau\in\mathcal{P}^{\mathcal{I}}} \mathbb{1}_{[0,t]}(\rho,\underline{\sigma},\tau)\,\overline{[\pi(g)]}\big((\rho_i+\Sigma_j\sigma_i^j)_i\big)$$
$$\times\,[\pi(f)]\big((\Sigma_j\sigma_j^i+\tau_i)_i\big)\;h_{\mathrm{ord}(\rho,\underline{\sigma},\tau)}^{\varepsilon(\rho,\underline{\sigma},\tau)}\;d\rho\,d\underline{\sigma}\,d\tau <\pi(g)\,,\,\pi(f)>.$$

But $<\pi(g)\,,\,\pi(f)> = \int_{\mathcal{P}^{\mathcal{I}}}\overline{[\pi(g)]}(\mu)[\pi(f)](\mu)\,d\mu$, therefore

$$<\pi(g)\,,\,T_t\,\pi(f)> = \int_{\rho\in\mathcal{P}^{\mathcal{I}}}\int_{\underline{\sigma}\in\mathcal{P}^{\mathcal{M}}}\int_{\tau\in\mathcal{P}^{\mathcal{I}}}\int_{\mu\in\mathcal{P}^{\mathcal{I}}} \mathbb{1}_{[0,t]}(\rho,\underline{\sigma},\tau)$$
$$\times\,\overline{[\pi(g)]}\big((\rho_i+\Sigma_j\sigma_i^j+\mu_i)_i\big)[\pi(f)]\big((\mu_i+\Sigma_j\sigma_j^i+\tau_i)_i\big)$$
$$\times\,h_{\mathrm{ord}(\rho,\underline{\sigma},\tau)}^{\varepsilon(\rho,\underline{\sigma},\tau)}\,d\rho\,d\underline{\sigma}\,d\tau\,d\mu.$$

Now, in order to get coherent notations, we exchange μ and τ in the previous identity. This gives

$$<\pi(g)\,,\,T_t\,\pi(f)> = \int_{\rho\in\mathcal{P}^{\mathcal{I}}}\int_{\underline{\sigma}\in\mathcal{P}^{\mathcal{M}}}\int_{\tau\in\mathcal{P}^{\mathcal{I}}}\int_{\mu\in\mathcal{P}^{\mathcal{I}}} \mathbb{1}_{[0,t]}(\rho,\underline{\sigma},\mu)$$
$$\times\,\overline{[\pi(g)]}\big((\rho_i+\Sigma_j\sigma_i^j+\tau_i)_i\big)\,h_{\mathrm{ord}(\rho,\underline{\sigma},\mu)}^{\varepsilon(\rho,\underline{\sigma},\mu)}$$
$$\times\,[\pi(f)]\big((\mu_i+\Sigma_j\sigma_j^i+\tau_i)_i\big)\,d\mu\,d\rho\,d\underline{\sigma}\,d\tau.$$

Using the $\not\!\!\int$-Lemma once again, we get

$$<\pi(g)\,,\,T_t\,\pi(f)> = \int_{\alpha\in\mathcal{P}^{\mathcal{I}}}\overline{[\pi(g)]}(\alpha)\int_{\mu\in\mathcal{P}^{\mathcal{I}}}\sum_{\forall i,\,\rho_i+\Sigma_j\sigma_i^j+\tau_i=\alpha_i}\mathbb{1}_{[0,t]}(\rho,\underline{\sigma},\mu)$$
$$\times\,h_{\mathrm{ord}(\rho,\underline{\sigma},\mu)}^{\varepsilon(\rho,\underline{\sigma},\mu)}\,[\pi(f)]\big((\mu_i+\Sigma_j\sigma_j^i+\tau_i)_i\big)\,d\mu\,d\sigma.$$

Let $\widetilde{T}_t$ be the Maassen-Meyer kernel given by the statement of the Proposition. We have proved that for every $f,g\in L^2_{lb}(\mathbb{R}^+)$ we have

$$<\pi(g)\,,\,T_t\,\pi(f)> = <\pi(g)\,,\,\widetilde{T}_t\,\pi(f)>.$$

So one concludes.

The converse is now easy. ∎

4 A criterion for the convergence of Maassen-Meyer kernels

In the previous section we have identified any finite series of iterated non-commutative stochastic integrals with a Maassen-Meyer kernel. Now, in the Fock space with multiplicity one (that is, $N = 1$) Belavkin and Lindsay [B-L] have given a criterion for a mapping $T : \mathcal{P}^3 \to \mathbb{C}$ to define a "reasonable operator" on Φ whose Maassen-Meyer kernel is T. We give an extension of their result to the case of any finite multiplicity.

Let $a \in (0,+\infty)$. Define

$$\Phi(a) = \{f\in L^0(\mathcal{P}^{\mathcal{I}});\int_{\mathcal{P}^{\mathcal{I}}} a^{\Sigma_i\#\sigma_i}\,|f(\sigma)|^2\,d\sigma<\infty\}.$$

Equipped with the scalar product $<g\,,\,f>_{(a)} = \int_{\mathcal{P}^{\mathcal{I}}} a^{\Sigma_i\#\sigma_i}\,\overline{g}(\sigma)f(\sigma)\,d\sigma$ the space $\Phi(a)$ is a Hilbert space, whose norm is denoted $||\cdot||_{(a)}$; it is a dense subspace of Φ for $a\geq 1$.

Let T be a measurable mapping from $\mathcal{P}^{\mathcal{I}} \times \mathcal{P}^{\mathcal{M}} \times \mathcal{P}^{\mathcal{I}}$ to $\mathbb{C}$. One identifies the mapping T with the operator T from $L^0(\mathcal{P}^{\mathcal{I}})$ into itself defined by

$$[Tf](\alpha) = \int_{\mu\in\mathcal{P}^{\mathcal{I}}} \sum_{\forall i,\, \rho_i+\Sigma_j\sigma_i^j+\tau_i=\alpha_i} T(\rho,\underline{\sigma},\mu) f\big((\mu_i + \Sigma_j\sigma_j^i + \tau_i)_i\big)\, d\mu. \qquad (4)$$

Let us consider the quantity

$$\int_{\mathcal{P}^{\mathcal{I}}} |\overline{g}(\alpha)\,[Tf](\alpha)|\, d\alpha.$$

It is dominated by

$$\int_{\mathcal{P}^{\mathcal{I}}}\int_{\mathcal{P}^{\mathcal{I}}} \sum_{\forall i,\, \rho_i+\Sigma_j\sigma_i^j+\tau_i=\alpha_i} \big|T(\rho,\underline{\sigma},\mu) f\big((\mu_i + \Sigma_j\sigma_j^i + \tau_i)_i\big)$$
$$\times g\big((\rho_i + \Sigma_j\sigma_i^j + \tau_i)_i\big)\big|\, d\mu\, d\alpha$$

$$= \int_{\mathcal{P}^{\mathcal{I}}}\int_{\mathcal{P}^{\mathcal{I}}}\int_{\mathcal{P}^{\mathcal{I}}}\int_{\mathcal{P}^{\mathcal{M}}} \big|T(\rho,\underline{\sigma},\mu) f\big((\mu_i + \Sigma_j\sigma_j^i + \tau_i)_i\big)$$
$$\times g\big((\rho_i + \Sigma_j\sigma_i^j + \tau_i)_i\big)\big|\, d\mu\, d\rho\, d\tau\, d\underline{\sigma}.$$

Let us change the notations and put $\beta_i^i = \sigma_i^i + \tau_i$ and $\beta_j^i = \sigma_j^i$ for $i \neq j$. The previous expression then becomes

$$\int_{\mathcal{P}^{\mathcal{I}}}\int_{\mathcal{P}^{\mathcal{I}}}\int_{\mathcal{P}^{\mathcal{M}}} \sum_{\forall i,\, \gamma_i\subset\beta_i^i} \big|T(\rho,\underline{\beta{:}\gamma},\mu) f\big((\mu_i + \Sigma_j\beta_j^i)_i\big) g\big((\rho_i + \Sigma_j\beta_i^j)_i\big)\big|\, d\mu\, d\rho\, d\underline{\beta}$$

where $\underline{\beta{:}\gamma}$ denotes the element $\underline{\delta}$ of $\mathcal{P}^{\mathcal{M}}$ such that $\delta_i^i = \gamma_i$ and $\delta_j^i = \beta_j^i$ for $i \neq j$.

Define the measurable mapping

$$T' : \mathcal{P}^{\mathcal{I}} \times \mathcal{P}^{\mathcal{M}} \times \mathcal{P}^{\mathcal{I}} \longrightarrow \mathbb{C}$$
$$(\rho,\underline{\beta},\tau) \mapsto \sum_{\forall i,\, \gamma_i\subset\beta_i^i} T(\rho,\underline{\beta{:}\gamma},\tau).$$

¿From now on, for $\alpha \in \mathcal{P}^{\mathcal{I}}$ we denote by $|\alpha|$ the quantity $\sum_i \#\alpha_i$, and for $\underline{\beta} \in \mathcal{P}^{\mathcal{M}}$ we denote by $|\underline{\beta}|$ the quantity $\sum_{i,j} \#\beta_j^i$.

For $a, b, c \in (0, +\infty)$, let

$$T'_{a,b,c}(\alpha,\underline{\beta},\gamma) = \frac{T'(\alpha,\underline{\beta},\gamma)}{\sqrt{a^{|\alpha|} b^{|\underline{\beta}|} c^{|\gamma|}}}$$

$$||T'||_{a,b,c} = \left(\int_{\mathcal{P}^{\mathcal{I}}}\int_{\mathcal{P}^{\mathcal{I}}} \sup_{\underline{\beta}\in\mathcal{P}^{\mathcal{M}}} |T'_{a,b,c}(\alpha,\underline{\beta},\gamma)|^2\, d\alpha\, d\gamma\right)^{1/2}.$$

The following estimate is inspired by [B-L].

Lemma 4 – *Let $p, a, q, c \in (0, +\infty)$ with $p > a$, $q > c$. Let $f, g \in L^0(\mathcal{P}^{\mathcal{I}})$ and T be a measurable map from $\mathcal{P}^{\mathcal{I}} \times \mathcal{P}^{\mathcal{M}} \times \mathcal{P}^{\mathcal{I}}$ to $\mathbb{C}$. Then one has*

$$\int_{\mathcal{P}^{\mathcal{I}}} |\overline{g}(\sigma)\,[Tf](\sigma)|\, d\sigma \leq ||g||_{(p)}\, ||f||_{(q)}\, ||T'||_{a,b,c}$$

where $b = \frac{\sqrt{(p-a)(q-c)}}{N}$.

Proof

Let $b = \frac{\sqrt{(p-a)(q-c)}}{N}$. One has

$$\int_{\mathcal{P}^{\mathcal{I}}} |\overline{g}(\sigma)\,[Tf](\sigma)|\, d\sigma$$

$$\le \int_{\mathcal{P}^{\mathcal{I}}}\int_{\mathcal{P}^{\mathcal{I}}}\int_{\mathcal{P}^{\mathcal{M}}} \left|T'(\rho,\underline{\beta},\mu) f\big((\mu_i+\Sigma_j\beta_j^i)_i\big) g\big((\rho_i+\Sigma_j\beta_i^j)_i\big)\right| d\mu\, d\rho\, d\underline{\beta}$$

$$\le \int_{\mathcal{P}^{\mathcal{I}}}\int_{\mathcal{P}^{\mathcal{I}}}\int_{\mathcal{P}^{\mathcal{M}}} \sqrt{a^{|\rho|}\Big(\frac{p-a}{N}\Big)^{|\underline{\beta}|}\Big(\frac{q-c}{N}\Big)^{|\underline{\beta}|} c^{|\mu|}}$$
$$\times \left|T'_{a,b,c}(\rho,\underline{\beta},\mu) f\big((\mu_i+\Sigma_j\beta_j^i)_i\big) g\big((\rho_i+\Sigma_j\beta_i^j)_i\big)\right| d\mu\, d\rho\, d\underline{\beta}$$

$$\le \Big(\int_{\mathcal{P}^{\mathcal{I}}}\int_{\mathcal{P}^{\mathcal{M}}} a^{|\rho|}\Big(\frac{p-a}{N}\Big)^{|\underline{\beta}|} |g((\rho_i+\Sigma_j\beta_i^j)_i)|^2\, d\rho\, d\underline{\beta}\Big)^{\frac12}$$
$$\times \int_{\mathcal{P}^{\mathcal{I}}} \sqrt{c^{|\mu|}}\Big(\int_{\mathcal{P}^{\mathcal{I}}}\int_{\mathcal{P}^{\mathcal{M}}} \Big(\frac{q-c}{N}\Big)^{|\underline{\beta}|}\left|T'_{a,b,c}(\rho,\underline{\beta},\mu)\right|^2$$
$$\times \left|f\big((\mu_i+\Sigma_j\beta_j^i)_i\big)\right|^2 d\rho\, d\underline{\beta}\Big)^{\frac12} d\mu$$

$$\le \Big(\int_{\mathcal{P}^{\mathcal{I}}}\int_{\mathcal{P}^{\mathcal{I}}} \sum_{\forall i,\,\Sigma_j\beta_i^j=\lambda_i} a^{|\rho|}\Big(\frac{p-a}{N}\Big)^{|\lambda|} |g((\rho_i+\lambda_i)_i)|^2\, d\rho\, d\lambda\Big)^{\frac12}$$
$$\times \int_{\mathcal{P}^{\mathcal{I}}} \sqrt{c^{|\mu|}}\Big(\int_{\mathcal{P}^{\mathcal{M}}} \Big(\frac{q-c}{N}\Big)^{|\underline{\beta}|}\left|f\big((\mu_i+\Sigma_j\beta_j^i)_i\big)\right|^2 d\underline{\beta}$$
$$\times \int_{\mathcal{P}^{\mathcal{I}}} \sup_{\underline{\beta}}\left|T'_{a,b,c}(\rho,\underline{\beta},\mu)\right|^2 d\rho\Big)^{\frac12} d\mu$$

$$\le \Big(\int_{\mathcal{P}^{\mathcal{I}}}\int_{\mathcal{P}^{\mathcal{I}}} N^{|\lambda|} a^{|\rho|}\Big(\frac{p-a}{N}\Big)^{|\lambda|} |g((\rho_i+\lambda_i)_i)|^2\, d\rho\, d\lambda\Big)^{\frac12}$$
$$\times \Big(\int_{\mathcal{P}^{\mathcal{I}}} c^{|\mu|}\int_{\mathcal{P}^{\mathcal{M}}} \Big(\frac{q-c}{N}\Big)^{|\underline{\beta}|}\left|f\big((\mu_i+\Sigma_j\beta_j^i)_i\big)\right|^2 d\underline{\beta}\, d\mu\Big)^{\frac12}$$
$$\times \Big(\int_{\mathcal{P}^{\mathcal{I}}}\int_{\mathcal{P}^{\mathcal{I}}} \sup_{\underline{\beta}}\left|T'_{a,b,c}(\rho,\underline{\beta},\mu)\right|^2 d\rho\, d\mu\Big)^{\frac12}$$

$$\le ||g||_{(p)}||f||_{(q)}||T'||_{a,b,c}. \qquad \blacksquare$$

Proposition 5 – *Let T be a measurable map from $\mathcal{P}^{\mathcal{I}}\times\mathcal{P}^{\mathcal{M}}\times\mathcal{P}^{\mathcal{I}}$ to $\mathbb{C}$ satisfying $||T'||_{a,b,c}<\infty$ for some $a,b,c\in(0,+\infty)$. Then, for every $e\in(0,+\infty)$, T defines a bounded operator from $\Phi(c+b\sqrt{N}/e)$ to $\Phi((a+b\sqrt{N}e)^{-1})$ with norm at most $||T'||_{a,b,c}$.*

Proof

By Lemma 4 we have proved that $|<g\,,\widetilde{T}f>|\le ||g||_{(p)}\,||f||_{(q)}\,||T'||_{a,b,c}$ for any p,q such that $p>a$, $q>c$ and $b=\frac{\sqrt{(p-a)(q-c)}}{N}$. Take $e\in(0,+\infty)$, put $b'=\sqrt{N}be$ and $b''=\sqrt{N}b/e$, put $p=a+b'$ and $q=c+b''$, we then have $b=\frac{\sqrt{(p-a)(q-c)}}{N}$. Take $f,g\in L^0(\mathcal{P}^{\mathcal{I}})$ and define $\widetilde{g}$ in $L^0(\mathcal{P}^{\mathcal{I}})$ by $\widetilde{g}(\sigma)=1/(a+b')^{|\sigma|}g(\sigma)$. Then applying Lemma 4 one gets

$$|<\widetilde{g}\,,Tf>|$$
$$=\Big|\int_{\mathcal{P}^{\mathcal{I}}}\frac{1}{(a+b')^{|\sigma|}}\overline{g}(\sigma)[Tf](\sigma)\,d\sigma\Big|$$
$$\le \Big(\int_{\mathcal{P}^{\mathcal{I}}}\Big(\frac{p}{(a+b')^2}\Big)^{|\sigma|}|g(\sigma)|^2\,d\sigma\Big)^{1/2}\Big(\int_{\mathcal{P}^{\mathcal{I}}} q^{|\sigma|}\,|f(\sigma)|^2\,d\sigma\Big)^{1/2}||T'||_{a,b,c}$$

$$= \left(\int_{\mathcal{P}^\mathcal{I}} \left(\frac{1}{a+b'}\right)^{|\sigma|} |g(\sigma)|^2 \, d\sigma\right)^{1/2} \left(\int_{\mathcal{P}^\mathcal{I}} (c+b'')^{|\sigma|} \, |f(\sigma)|^2 \, d\sigma\right)^{1/2} ||T'||_{a,b,c}.$$

So one concludes easily. ■

Now notice that if T is a measurable mapping from $\mathcal{P}^\mathcal{I} \times \mathcal{P}^\mathcal{M} \times \mathcal{P}^\mathcal{I}$ to $\mathbb{C}$, then the operator T on the Fock space whose Maassen-Meyer kernel is given by the mapping T satisfies identity (4) wherever it is meaningful. So by Proposition 5, one easily gets the following results.

Proposition 6 – *Let T be a measurable mapping from $\mathcal{P}^\mathcal{I} \times \mathcal{P}^\mathcal{M} \times \mathcal{P}^\mathcal{I}$ to $\mathbb{C}$ such that $||T'||_{a,b,c} < \infty$ for some $a, b, c \in (0, +\infty)$.*

i) *If $a < 1$ then the associated Maassen-Meyer kernel is well defined as an operator from Φ to Φ, with a dense domain containing $\Phi(c + N^2b^2/(1-a))$. This operator is bounded from $\Phi(c + N^2b^2/(1-a))$ to Φ with norm at most $||T'||_{a,b,c}$.*

ii) *If $a < 1$, $c < 1$ and $b = \frac{\sqrt{(1-a)(1-c)}}{N}$ then the Maassen-Meyer kernel T is a bounded operator on Φ, with same bound for the norm.* ■

Let us give a simple example to illustrate these estimates.

If H is a Hilbert-Schmidt operator on Φ, it is then a Hilbert-Schmidt operator on $L^2(\mathcal{P}^\mathcal{I})$, thus there exists a mapping φ from $\mathcal{P}^\mathcal{I} \times \mathcal{P}^\mathcal{I}$ to $\mathbb{C}$ such that

$$\int_{\mathcal{P}^\mathcal{I}} \int_{\mathcal{P}^\mathcal{I}} |\varphi(\alpha, \gamma)|^2 \, d\alpha \, d\gamma < \infty$$

and satisfying

$$[Hf](\sigma) = \int_{\mathcal{P}^\mathcal{I}} \varphi(\sigma, \mu) f(\mu) \, d\mu \tag{5}$$

for all $f \in \Phi$. Now consider the Maassen-Meyer kernel T defined by

$$T(\alpha, \underline{\beta}, \gamma) = \begin{cases} (-1)^{\Sigma_i \# \beta_i^i} \varphi(\alpha, \gamma) & \text{if } \beta_j^i = \emptyset \text{ for all } i \neq j \\ 0 & \text{otherwise.} \end{cases}$$

Applying (4) we get

$$\begin{aligned}
[Tf](\alpha) &= \int_{\mu \in \mathcal{P}^\mathcal{I}} \sum_{\forall i,\, \rho_i + \Sigma_j \sigma_i^j + \tau_i = \alpha_i} T(\rho, \underline{\sigma}, \mu) f\big((\mu_i + \Sigma_j \sigma_j^i + \tau_i)_i\big) \, d\mu \\
&= \int_{\mu \in \mathcal{P}^\mathcal{I}} \sum_{\forall i,\, \rho_i + \Sigma_j \sigma_i^j + \tau_i = \alpha_i} \mathbb{1}_{\forall i \neq j,\, \sigma_j^i = \emptyset} (-1)^{\Sigma_i \# \sigma_i^i} \varphi(\rho, \mu) \\
&\qquad \times f\big((\mu_i + \Sigma_j \sigma_j^i + \tau_i)_i\big) \, d\mu \\
&= \int_{\mu \in \mathcal{P}^\mathcal{I}} \sum_{\forall i,\, \rho_i + \sigma_i^i + \tau_i = \alpha_i} (-1)^{\Sigma_i \# \sigma_i^i} \varphi(\rho, \mu) f\big((\mu_i + \sigma_i^i + \tau_i)_i\big) \, d\mu \\
&= \int_{\mu \in \mathcal{P}^\mathcal{I}} \sum_{\forall i,\, \delta_i \subset \alpha_i} \varphi(\alpha \setminus \delta, \mu) f\big((\mu_i + \delta_i)_i\big) \sum_{\forall i,\, \sigma_i^i \subset \delta_i} (-1)^{\Sigma_i \# \sigma_i^i} \, d\mu \\
&= \int_{\mu \in \mathcal{P}^\mathcal{I}} \sum_{\forall i,\, \delta_i \subset \alpha_i} \varphi(\alpha \setminus \delta, \mu) f\big((\mu_i + \delta_i)_i\big) \mathbb{1}_{\delta = \emptyset} \, d\mu
\end{aligned}$$

(by Mœbius inversion formula)

$$= \int_{\mu \in \mathcal{P}\mathcal{I}} \varphi(\alpha,\mu) f(\mu)\, d\mu$$
$$= [Hf](\alpha).$$

Thus the Hilbert-Schmidt operator H admits $T(\alpha, \underline{\beta}, \gamma)$ as a Maassen-Meyer kernel. Let us apply our criterion to this kernel. We get

$$\int_{\mathcal{P}\mathcal{I}} \int_{\mathcal{P}\mathcal{I}} \sup_{\underline{\beta}} \frac{\left|\sum_{\forall i,\, C_i \subset \beta_i^i} T(\alpha, \underline{\beta}{:}\underline{C}, \gamma)\right|^2}{\sqrt{a^{|\alpha|} b^{|\underline{\beta}|} c^{|\gamma|}}}\, d\alpha\, d\gamma$$
$$= \int_{\mathcal{P}\mathcal{I}} \int_{\mathcal{P}\mathcal{I}} \sup_{\underline{\beta}} \frac{\sum_{\forall i,\, C_i \subset \beta_i^i} (-1)^{\sum_i \# C_i} 1\!\!1_{\forall i \neq j,\, \beta_j^i = \emptyset} \left|\varphi(\alpha,\gamma)\right|^2}{\sqrt{a^{|\alpha|} b^{|\underline{\beta}|} c^{|\gamma|}}}\, d\alpha\, d\gamma$$
$$= \int_{\mathcal{P}\mathcal{I}} \int_{\mathcal{P}\mathcal{I}} \frac{1\!\!1_{\forall i,j,\, \beta_j^i = \emptyset} \left|\varphi(\alpha,\gamma)\right|^2}{\sqrt{a^{|\alpha|} b^{|\underline{\beta}|} c^{|\gamma|}}}\, d\alpha\, d\gamma$$
$$= \int_{\mathcal{P}\mathcal{I}} \int_{\mathcal{P}\mathcal{I}} \frac{\left|\varphi(\alpha,\gamma)\right|^2}{\sqrt{a^{|\alpha|} c^{|\gamma|}}}\, d\alpha\, d\gamma.$$

This quantity is finite for $a = 1$, $c = 1$ and $b = 0$. Hence we recover that this kernel defines a bounded operator (Proposition 5).

5 Convergence of series of iterated non-commutative stochastic integrals

We are now able to give our final result, which gives a condition for a series of iterated non-commutative stochastic integrals of the form (1) to define a densely defined operator on Φ.

Theorem 7 – *Let $t \in \mathbb{R}^+ \cup \{+\infty\}$. For all $n \in \mathbb{N}$, all $\varepsilon \in E^n$, let h^ε be a function on $\mathcal{P}_n$ satisfying the condition of Lemma 2. Suppose that the functions h^ε satisfy*

$$||T'||^2_{a,b,c} = \int_{\mathcal{P}\mathcal{I}} \int_{\mathcal{P}\mathcal{I}} 1\!\!1_{[0,t]}(\alpha,\gamma) \sup_{\underline{\beta} \in \mathcal{P}\mathcal{M}} \frac{\left|\sum_{\forall i,\, C_i \subset \beta_i^i} h^{\varepsilon(\alpha,\underline{\beta}{:}\underline{C},\gamma)}_{\mathrm{ord}(\alpha,\underline{\beta}{:}\underline{C},\gamma)}\right|^2 1\!\!1_{[0,t]}(\underline{\beta})}{(a^{|\alpha|} b^{|\underline{\beta}|} c^{|\gamma|})}\, d\alpha\, d\gamma < \infty$$

for some $a \in (0,1)$, $b, c \in (0,+\infty)$. Then the operator

$$T_t = \lambda I + \sum_{i=1}^{\infty} \sum_{\varepsilon \in E^n} \int_{0<t_1<\ldots<t_n<t} h^{\varepsilon}_{t_1 \ldots t_n}\, dA^{\varepsilon_1}_{t_1} \ldots dA^{\varepsilon_n}_{t_n}$$

is well-defined as an operator on Φ with (dense) domain $\Phi(c+N^2b^2/(1-a))$, bounded from $\Phi(c+N^2b^2/(1-a))$ to Φ with norm at most $||T'||_{a,b,c}$. Furthermore, if $c < 1$ and $b = \frac{\sqrt{(1-c)(1-a)}}{N}$ the operator T_t is then a bounded operator on Φ, with the same bound for the norm.

Proof

Combine Lemma 2, Proposition 3 and Proposition 5. ■

We also have a uniqueness theorem for the representation of operators as non-commutative chaotic expansions.

Theorem 8 – *Let T be an operator on Φ having a representation of the form*

$$T = \lambda I + \sum_{i=1}^{\infty} \sum_{\varepsilon \in E^n} \int_{0<t_1<\ldots<t_n<t} h^{\varepsilon}_{t_1,\ldots,t_n}\, dA^{\varepsilon_1}_{t_1} \ldots dA^{\varepsilon_n}_{t_n}$$

for some $t \in \mathbb{R}^+ \cup \{+\infty\}$. Then T vanishes if and only if for all $n \in \mathbb{N}$, all $\varepsilon \in E^n$, almost all $(t_1,\ldots,t_n) \in \mathcal{P}_n$ one has $h^{\varepsilon}_{t_1\ldots t_n} = 0$.

Proof

We have proved in Proposition 3 that there is a one-to-one correspondance between non-commutative chaotic expansions and Maassen-Meyer kernels. In [At3], Theorem IV.6, the uniqueness of Maassen-Meyer kernels representation is proved, for any multiplicity of the Fock space. So one concludes. ∎

6 The case of iterated non-commutative stochastic integrals containing the time integrator

In all the results considered previously we have never considered the case where the iterated non-commutative stochastic integrals are containing the time process as an integrator. Indeed, we have only considered the integrators dA^i_j for $(i,j) \in \{0,1,\ldots,N\}^2 \setminus \{(0,0)\}$, avoiding the term $dA^0_0(t)$ which in fact corresponds to the time integrator dt. There are two reasons for that. The case of series of iterated non-commutative stochastic integrals without time integrator really corresponds to the notion of *non-commutative chaotic expansion*; that is, the representation of a given operator in terms of a series of iterated integrals of scalar operators with respect to the "quantum noises": the creation, annihilation and exchange processes. Another way to understand these series as non-commutative chaotic expansions is to see that, in some cases like the case of Hilbert-Schmidt operators on the Fock space (cf [At1]), these series can be obtained by iterating the integral representation of the operators; exactely like one can prove the chaotic representation property of the Brownian motion by iterating the predictable representation property. The second reason is that in the case where there is no time integrators, we have uniqueness of the series (Theorem 8); this uniqueness is lost when allowing the time integrator. But it is convenient in some problems (such as the application of the next section) to consider series with the time integrator. That is why we present here the results corresponding to this case. Let us first make precise some new notations.

Let $F = \{0,1,\ldots,N\}^2$. For $\varepsilon = (\varepsilon^1,\varepsilon^2) \in F$ the operator A^{ε}_t denotes, as previously, $A^{\varepsilon^1}_{\varepsilon^2}(t)$ for $\varepsilon \neq (0,0)$ and $A^0_0(t) = tI$ if $\varepsilon = (0,0)$. We now want to consider operators of the form

$$T_t = \lambda I + \sum_{n=1}^{\infty} \sum_{\varepsilon \in F^n} \int_{0<t_1<\ldots<t_n<t} h^{\varepsilon}_{t_1\ldots t_n}\, dA^{\varepsilon_1}_{t_1} \ldots dA^{\varepsilon_n}_{t_n}. \tag{6}$$

In the following we consider elements of $\mathcal{P}^{\mathcal{I}}$, $\mathcal{P}^{\mathcal{M}}$ and $\mathcal{P}$, so in order to distinguish them we will denote as previously elements of $\mathcal{P}^{\mathcal{I}}$ with small greek letters (ρ), elements of $\mathcal{P}^{\mathcal{M}}$ with underlined small greek letter $(\underline{\sigma})$ and elements of $\mathcal{P}$ with upperlined small greek letters $(\overline{\delta})$. A quadruple $(\rho,\underline{\sigma},\tau,\overline{\delta})$ in $\mathcal{P}^{\mathcal{I}} \otimes \mathcal{P}^{\mathcal{M}} \otimes \mathcal{P}^{\mathcal{I}} \otimes \mathcal{P}$ can be seen as a $(N+1) \otimes (N+1)$ matrix $(\alpha^i_j)_{i,j\in\{0,1,\ldots,N\}}$ of elements of $\mathcal{P}$ by putting $\alpha^0_0 = \delta$, $\alpha^i_0 = \rho_i$, $\alpha^i_j = \sigma^i_j$ and $\alpha^0_i = \tau_i$, for $i,j = 1,\ldots,N$. So, by

$\mathrm{ord}(\rho, \underline{\sigma}, \tau, \overline{\delta})$ we mean the ordered set $\cup_{i,j=0,\dots,N} \alpha_j^i$; by $\varepsilon(\rho, \underline{\sigma}, \tau, \overline{\delta})$ we mean the element of $\mathcal{P}^n$ (where $n = \#\mathrm{ord}(\rho, \underline{\sigma}, \tau, \overline{\delta})$) such that $\varepsilon(\rho, \underline{\sigma}, \tau, \overline{\delta})_k = (i,j)$ if the k-th smallest element of $\mathrm{ord}(\rho, \underline{\sigma}, \tau, \overline{\delta})$ is in α_j^i.

By using the same kind of proof as in Proposition 3, one can prove that an operator T_t given by a series of the form (6) admits a Maassen-Meyer kernel T_t which given by

$$T_t(\rho, \underline{\sigma}, \tau) = \int_{\mathcal{P}} h^{\varepsilon(\rho,\underline{\sigma},\tau,\overline{\delta})}_{\mathrm{ord}(\rho,\underline{\sigma},\tau,\overline{\delta})} \, d\overline{\delta} \, .$$

So we obtain the following easy extension of Theorem 7.

Theorem 9 – *Let $t \in I\!R^+ \cup \{+\infty\}$. For all $n \in I\!N$, all $\varepsilon \in \mathcal{E}^n$ let h^ε be a function on $\mathcal{P}_n$ satisfying the condition of Lemma 2. Suppose that the function h^ε is such that the quantity*

$$||T'||^2_{a,b,c} \stackrel{\mathrm{def}}{=}$$

$$= \int_{\mathcal{P}^{\mathcal{I}}} \int_{\mathcal{P}^{\mathcal{I}}} 1\!\!1_{[0,t]}(\alpha, \gamma) \sup_{\underline{\beta} \in \mathcal{P}^{\mathcal{M}}} \frac{\left| \sum_{\forall i,\, C_i \subset \beta_i^i} \int_{\mathcal{P}} h^{\varepsilon(\alpha, \underline{\beta}:C, \gamma, \overline{\delta})}_{\mathrm{ord}(\alpha, \underline{\beta}:C, \gamma, \overline{\delta})} 1\!\!1_{[0,t]}(\overline{\delta}) \, d\overline{\delta} \right|^2 1\!\!1_{[0,t]}(\beta)}{(a^{|\alpha|} b^{|\underline{\beta}|} c^{|\gamma|})} \, d\alpha \, d\gamma$$

is finite for some $a \in (0,1)$, $b, c \in (0, +\infty)$. Then the operator

$$T_t = \lambda I + \sum_{i=1}^{\infty} \sum_{\varepsilon \in F^n} \int_{0<t_1<\dots<t_n<t} h^{\varepsilon}_{t_1,\dots,t_n} \, dA^{\varepsilon_1}_{t_1} \dots dA^{\varepsilon_n}_{t_n}$$

is well-defined as an operator on Φ with (dense) domain $\Phi(c + N^2 b^2/(1-a))$, bounded from $\Phi(c + N^2 b^2/(1-a))$ to Φ with norm at most $||T'||_{a,b,c}$. Furthermore, if $c < 1$ and $b = \frac{\sqrt{(1-a)(1-c)}}{N}$ then the operator T_t is a bounded operator on Φ, with the same bound for the norm. ∎

7 An application to series of generalized integrators

In [CEH], Cohen, Eyre and Hudson are considering some generalized integrator processes. Let $\mathcal{K}$ be the space $\mathbb{C}^N$. For a $(N+1) \times (N+1)$ matrix H, that is an element of the space $\mathcal{J}$ of linear transformation of the space $\mathbb{C} \oplus \mathcal{K}$, they consider the process $(\Lambda_t(H))_{t \geq 0}$, defined on the exponential domain $\mathcal{E}$ by the identity

$$< \pi(g) \, , \, \Lambda_t(H) \pi(f) > = \int_0^t < \widetilde{g}(s) \, , \, H \widetilde{f}(s) > ds < \pi(g) \, , \, \pi(f) >,$$

where for $u \in \mathcal{K}$, $\widetilde{u} \stackrel{\mathrm{def}}{=} (1, u) \in \mathbb{C} \oplus \mathcal{K}$. This means that, with our notations, $\Lambda_t(H)$ is actually equal to $\sum_{i,j} H(i,j) \, A^i_j(t)$, where $H = (H(i,j))_{i,j}$. Now take $H^n \in \mathcal{J}^{\otimes n}$ of the form $H^n = H^n_1 \otimes \dots \otimes H^n_n$. Consider processes of the form

$$I_t(H^n) = \int_{0<t_1<\dots<t_n<t} d\Lambda_{t_1}(H^n_1) \, \dots \, d\Lambda_{t_n}(H^n_n)$$

$$= \sum_{\substack{l_1,\dots,l_n,k_1,\dots,k_n \\ \in \{0,1,\dots N\}}} \int_{0<t_1<\dots<t_n<t} H^n_1(k_1, l_1) \dots H^n_n(k_n, l_n) \times dA^{k_1}_{l_1}(t_1) \, \dots \, dA^{k_n}_{l_n}(t_n)$$

$$= \sum_{\varepsilon \in F^n} \int_{0<t_1<\dots<t_n<t} h^{\varepsilon}_{t_1 \dots t_n} \, dA^{\varepsilon^1}_{t_1} \, \dots \, dA^{\varepsilon^n}_{t_n}$$

where $h^{\varepsilon}_{t_1,\ldots,t_n} = H^n_1(\varepsilon^1_1,\varepsilon^2_1)\ldots H^n_n(\varepsilon^1_n,\varepsilon^2_n)$.

By extending it linearly the mapping I_t can be defined on the tensor space over $\mathcal{J}$: $J = \mathbb{C} \oplus \mathcal{J} \oplus \mathcal{J}^{\otimes 2} \oplus \ldots$.

In [CEH] such operators have been *formally* considered, and were proved to form an "Ito algebra" of which the product was described. This extended the formula given in [HP2] where only the purely conservation generalized processes case was considered. Indeed, the matrices H of $\mathcal{J}$ are composed of four elements: a complex number, a vector in $\mathcal{K}$, a linear form on $\mathcal{K}$ and a linear transformation of $\mathcal{K}$, each of them corresponding to the time, creation, annihilation and multidimensional conservation component of $\Lambda_t(H)$. So, when only the fourth component does not vanish, we are in the purely conservation case.

Our purpose here is, for elements $\mathcal{H}$ of the tensor algebra J, to give a meaning to $I_t(\mathcal{H})$ as a well-defined operator on the Fock space, a series of iterated non-commutative stochastic integrals.

Proposition 10 – *Let $\mathcal{H}$ be an element of the tensor algebra J over $\mathcal{J}$ of the form $\mathcal{H} = \oplus_n H^n$, with $H^n = \otimes_{i\leq n} H^n_i$ and each H^n_i beeing a $(N+1)\times(N+1)$ matrix. If $K = \sup\{|H^n_i(k,l)|\,;\, n \in \mathbb{N}\,;\, i \leq n\,;\, k,l = 0,1,\ldots,N\}$ is finite then, for all $t \in \mathbb{R}^+$, the operator*

$$I_t(\mathcal{H}) = \sum_{n=0}^{\infty} \int_{0<t_1<\ldots<t_n<t} d\Lambda_{t_1}(H^n_1)\,\ldots\, d\Lambda_{t_n}(H^n_n)$$

is well-defined as a bounded operator from $\Phi(r)$ to Φ, with any $r > N^2(K+1)^2$.

Proof

We have to apply Theorem 9 to the familly of scalar operators $h^{\varepsilon}_{t_1,\ldots,t_n} = H^n_1(\varepsilon^1_1,\varepsilon^1_2)\ldots H^n_n(\varepsilon^n_1,\varepsilon^n_2)$. The corresponding conditions of Lemma 2 are obviously satisfied.

Let $K = \sup\{H^n_i(k,l); n \in \mathbb{N}; i \leq n; k,l = 0,1,\ldots,N\} < \infty$. We then have $|h^{\varepsilon}_{t_1,\ldots,t_n}| \leq K^n$. Thus we get

$$\int_{\mathcal{P_I}}\int_{\mathcal{P_I}} \mathbb{1}_{[0,t]}(\alpha,\gamma) \sup_{\underline{\beta}\in\mathcal{P^M}} \frac{\left|\sum_{\forall i,\, C_i\subset\beta^i_i}\int_{\mathcal{P}} h^{\varepsilon(\alpha,\underline{\beta}:C,\gamma,\overline{\delta})}_{\mathrm{ord}(\alpha,\underline{\beta}:C,\gamma,\overline{\delta})}\, \mathbb{1}_{[0,t]}(\overline{\delta})\, d\overline{\delta}\right|^2 \mathbb{1}_{[0,t]}(\underline{\beta})}{(a^{|\alpha|} b^{|\underline{\beta}|} c^{|\gamma|})}\, d\alpha\, d\gamma$$

$$\leq \int_{\mathcal{P_I}}\int_{\mathcal{P_I}} \sup_{\underline{\beta}\in\mathcal{P^M}} \frac{\left(\sum_{\forall i,\, C_i\subset\beta^i_i}\int_{\mathcal{P}} K^{|\alpha|}\, K^{|\underline{\beta}:C|}\, K^{|\gamma|}\, K^{|\overline{\delta}|}\, \mathbb{1}_{[0,t]}(\overline{\delta})\, d\overline{\delta}\right)^2}{(a^{|\alpha|} b^{|\underline{\beta}|} c^{|\gamma|})} \times \mathbb{1}_{[0,t]}(\alpha,\gamma)\, d\alpha\, d\gamma\,.$$

$$= \int_{\mathcal{P_I}}\int_{\mathcal{P_I}} \mathbb{1}_{[0,t]}(\alpha,\gamma)\Big(\sup_{\underline{\beta}\in\mathcal{P^M}} \frac{\left(\sum_{\forall i,\, C_i\subset\beta^i_i} K^{|\underline{\beta}:C|}\right)^2}{b^{|\underline{\beta}|}}\Big)\, \frac{(K^{|\alpha|})^2}{a^{|\alpha|}}\, \frac{(K^{|\gamma|})^2}{c^{|\gamma|}} \times\Big(\sum_n \frac{t^n}{n!} K^n\Big)^2\, d\alpha\, d\gamma$$

$$\leq \Big(\int_{\mathcal{P_I}} \Big(\frac{K^2}{a}\Big)^{|\alpha|} \mathbb{1}_{[0,t]}(\alpha)\, d\alpha\Big)\Big(\int_{\mathcal{P_I}} \Big(\frac{K^2}{c}\Big)^{|\gamma|} \mathbb{1}_{[0,t]}(\gamma)\, d\gamma\Big)\, \exp(2tK)$$

$$\times \sup_{\beta \in \mathcal{PM}} \left(\frac{(K+1)^2}{b} \right)^{|\beta|}$$

$$= \exp\left(t(\frac{NK^2}{a} + \frac{NK^2}{c} + 2K)\right) \sup_{\beta \in \mathcal{PM}} \left(\frac{(K+1)^2}{b} \right)^{|\beta|}.$$

So taking any $a, c \in (0, +\infty)$, any $b > (K+1)^2$ one gets that the quantity $||T'||^2_{a,b,c}$ is finite. Taking a and c as close of 0 as possible, applying theorem 9, gives the result. ■

The authors are very grateful to Prof. P.-A. Meyer for pointing out an error in a first version of this article.

References

[At1] Attal S. : "Non-commutative chaotic expansion of Hilbert-Schmidt operators on Fock spâce", *Comm. Math. Phys.* **175**, (1996), p. 43-62.

[At2] Attal S. : "*Semimartingales non commutatives et applications aux endomorphismes browniens*", Thèse de Doctorat, Univ. L. Pasteur, Strasbourg (1993).

[At3] Attal S. : "Problèmes d'unicité dans les représentations d'opérateurs sur l'espace de Fock", *Séminaire de probabilités XXVI*, Springer Verlag L.N.M. 1526 (1992), p. 619-632.

[B-L] Belavkin V.P. & Lindsay J.M. : "The kernel of a Fock space operator II", *Quantum Prob. & Rel. Topics* IX, World Scientific p. 87-94.

[CEH] Cohen P.B., Eyre T.W.M. & Hudson R.L. : "Higher order Ito product formula, and generators of evolutions and flows", *Int. Journ. Theor. Phys.*, 34 (1995), p. 1481-1486.

[Der] Dermoune A. : "Formule de composition pour une classe d'opérateurs", *Séminaire de probabilités XXIV*, Springer Verlag L.N.M. 1426 (1990), p 397-401.

[HP1] Hudson R.L. & Parthasarathy K.R. : "Quantum Itô's formula and stochastic evolutions", *Comm. Math. Phys.* 93 (1984), p 301-323.

[HP2] Hudson R.L. & Parthasarathy K.R. : "The Casimir chaos map for $U(N)$", *Tatra Mountains Mathematicals Proceedings* 3 (1993), p 81-88.

[L-P] Lindsay J.M. & Parthasarathy K.R. : "Cohomology of power sets with applications in quantum probability", *Comm. in Math. Phys.* 124 (1989), p. 337-364.

[Maa] Maassen H. : "Quantum Markov processes on Fock space described by integral kernels", *Quantum Prob. & Appl.* II, Springer Verlag L.N.M. 1136 1985, p 361-374.

[Me1] Meyer P.A. : "*Quantum probability for probabilists*", second edition, Springer Verlag L.N.M. 1538, (1995).

[Me2] Meyer P.A. : "Eléments de probabilités quantiques", *Séminaire de Probabilités* XX, Springer Verlag L.N.M. 1204 (1986), p 186-312.

[Par] Parthasarathy K.R. : "*An introduction to quantum stochastic calculus*", Monographs in Mathematics, Birkhäuser, 1992.

p-variation for families of local times on lines

Haya Kaspi* and Jay Rosen†

1 Introduction

The local time process $(L_t^x)_{x\in S}$ for a Markov process with values in S measures, in a certain sense, the amount of time that the Markov process spends at each point up till time t. $(L_t^x)_{x\in S}$ is a family of continuous additive functionals which has been the subject of intensive investigation. Not all Markov processes have local times. In particular, Lévy processes can only have local times in one dimension since in higher dimensions they do not hit points. Nevertheless, one can study other families of continuous additive functionals and try to see which properties of the local time process $(L_t^x)_{x\in S}$ admit natural generalizations. The family of 'local times on lines' for the two-dimensional symmetric stable process X_t, which 'measures' the amount of time that X spends on each line up till time t, is in some ways the most natural extension of the family of local times at points. In this paper, the property of $(L_t^x)_{x\in R^1}$ which we plan to generalize is that of quadratic variation, or more generally p-variation, in the spatial variable. Aside from its intrinsic interest, we hope that this detailed study will pave the way for generalizations, both of other properties and to other families of continuous additive functionals.

The quadratic variation of the local time L_t^x of 1-dimensional Brownian motion W_s was studied in Bouleau and Yor [2] and Perkins [6]. They show that for any sequence of partitions π_n of $[a,b]$ with mesh size converging to zero,

$$\lim_{n\to\infty}\sum_{x_i\in\pi_n}(L_t^{x_i}-L_t^{x_{i-1}})^2=4\int_0^t 1_{[a,b]}(W_s)\,ds, \tag{1.1}$$

with convergence in probability. Similar results were obtained in [7] and [5] for the p-variation of the local times of 1-dimensional symmetric stable processes. The object of this paper is to generalize such results to the 'local times on lines' of 2-dimensional Brownian motion and symmetric stable processes. To better appreciate the results we shall obtain, we first reformulate (1.1). Let $c:[0,1]\mapsto R^1$ be a smooth curve, not necessarily monotone, and let $N(c\,|\,y)=\text{card}\{x\in[0,1]\,|\,c(x)=y\}$, the cardinality of the pre-image $c^{-1}(y)$. Using (1.1) on each interval of the complement of $\{s\,|\,c'(s)=0\}$ we see that for any sequence of partitions π_n of $[0,1]$ with mesh size converging to zero,

$$\lim_{n\to\infty}\sum_{s_i\in\pi_n}\left(L_t^{c(s_i)}-L_t^{c(s_{i-1})}\right)^2=4\int_0^t N(c\,|\,W_s)\,ds, \tag{1.2}$$

with convergence in probability.

*This research was supported, in part, by the Technion Promotion of Research Fund and VPR Fund—R. and M. Rochlin Research Fund.

†This research was supported, in part, by grants from the National Science Foundation and PSC-CUNY.

For any $0 \leq \theta < 2\pi$, let $e(\theta) = (\cos(\theta), \sin(\theta))$ denote the unit vector with angle θ, and $e(\theta^\perp) = (\sin(\theta), -\cos(\theta))$ denote the unit vector perpendicular to $e(\theta)$. We use $l_{a,\theta} = \{ae(\theta) + xe(\theta^\perp) \mid x \in R^1\}$ to denote the line such that $ae(\theta)$ is the foot of the perpendicular from the origin to $l_{a,\theta}$. If X_t denotes the symmetric stable process of index β in the plane, then $X_t^\theta = X_t \cdot e(\theta)$, the component of X_t in the direction of $e(\theta)$, is a real symmetric stable process of index β (just check the characteristic function). Let $L_t^{a,\theta}$ denote the local time of X_t^θ at $a \in R_+^1$.

Let $\gamma_s = (a_s, \theta_s)$ be a simple smooth curve, $\gamma : [0,1] \mapsto R_+^1 \times [0, 2\pi)$. For ease of notation we will sometimes write $\gamma(s)$ for γ_s.

Let $\Phi_\gamma : [0,1] \times R^1 \mapsto R^2$ be defined by $\Phi_\gamma(s,x) = a_s e(\theta_s) + xe(\theta_s^\perp))$ and let $N(\Phi_\gamma \mid y) = \text{card}\{(s,x) \in [0,1] \times R^1 \mid \Phi_\gamma(s,x) = y\}$, the cardinality of the pre-image $\Phi_\gamma^{-1}(y)$. Thus $N(\Phi_\gamma \mid y)$ is the number (possibly infinite) of parameter values s such that $y \in l_{\gamma_s}$.

Let $Q(0,1)$ denote the set of all partitions $\pi = \{0 = s_0 < s_1 \cdots < s_{k_\pi} = 1\}$ of $[0,1]$, and let $|\pi| = \sup_{1 \leq i \leq k_\pi}(s_i - s_{i-1})$ denote the length of the largest interval in π. ($|\pi|$ is called the mesh size of π).

Theorem 1 *If X_t is a planar Brownian motion, and $L_t^{a,\theta}$ denotes the local time of $X_t \cdot e(\theta)$, then*

$$\lim_{n\to\infty} \sum_{s_i \in \pi(n)} (L_t^{\gamma(s_i)} - L_t^{\gamma(s_{i-1})})^2 = 4 \int_0^t N(\Phi_\gamma \mid X_s)\, ds. \tag{1.3}$$

in L^2, uniformly both in $t \in [0,T]$ and $Q(0,1)$ as $|\pi(n)| \to 0$.

Note that for the special case of $\gamma(s) = (c(s), \theta_0)$ with θ_0 fixed, this reduces to (1.2) for the real Brownian motion $X_t \cdot e(\theta)$. When $\gamma(s) = (0, s)$, our theorem can be obtained from (1.1) using the skew-product representation for planar Brownian motion. However, already for the simple example $\gamma(s) = (1, s)$ our (1.3) is truly a two-dimensional result, and the formal similarity between Theorem 1 and (1.2) is thus rather striking.

Theorem 1 can essentially be proven using stochastic calculus and Tanaka's formula, and we will outline such a proof in section 4. However, we prefer to derive Theorem 1 as a special case of the Theorem 2 below for symmetric stable processes. We now introduce the notation needed for that Theorem. In what follows, X_t will denote the symmetric stable process in the plane of index β.

For $0 \leq \theta < 2\pi$, $a \in R_+^1$ let $\mu_{a,\theta}$ denote one-dimensional Lebesgue measure on $l_{a,\theta}$. Equivalently, if $h_{a,\theta} : R^1 \mapsto R^2$ is defined by $h_{a,\theta}(x) = ae(\theta) + xe(\theta^\perp)$ then $\mu_{a,\theta}$ is the measure induced by $h_{a,\theta}$ from Lebesgue measure λ^1 on the line: $\mu_{a,\theta}(A) = \lambda^1(h_{a,\theta}^{-1}(A))$. Thus

$$\int f(y)\, d\mu_{a,\theta}(y) = \int f(ae(\theta) + xe(\theta^\perp))\, dx.$$

We can easily check that $\mu_{a,\theta}$ is the Revuz measure of the CAF $L_t^{a,\theta}$ defined above.

Let $A_t^{\gamma,s}$ denote the CAF with Revuz measure $\widetilde{\mu}_{a,\theta,s}$ defined by

$$\int f(y)\, d\widetilde{\mu}_{a,\theta,s}(y) = \int f(a_s e(\theta_s) + xe(\theta_s^\perp))|a_s' + x\theta_s'|^{\beta-1}\, dx.$$

We note that

$$\int f(y)\, d\tilde{\mu}_{a,\theta,s}(y) = \int f(y)|a'_s + y \cdot e(\theta_s^{\perp})\theta'_s|^{\beta-1}\, d\mu_{a,\theta}(y) \tag{1.4}$$

so that

$$A_t^{\gamma,s} = \int_0^t |a'_s + X_r \cdot e(\theta_s^{\perp})\theta'_s|^{\beta-1}\, dL_r^{a_s,\theta_s}. \tag{1.5}$$

In Lemma 1 below we show that $s \mapsto A_t^{\gamma,s}$ is continuous in L^{2k}, and therefore $\int_0^1 (A_t^{\gamma,s})^k\, ds$ is well defined in L^2.

Let $\bar{c}(\beta) = (2k)!!(4c(\beta))^k$ where

$$c(\beta) = -\frac{\cos(\pi(\beta-1)/2)}{\pi(\beta-1)}\Gamma(2-\beta)$$

for $1 < \beta < 2$, and $c(2) = 1$.

Theorem 2 *If X_t is a symmetric stable process in the plane of index $\beta = 1 + 1/k$ with $k = 1, 2, \ldots$, and $L_t^{a,\theta}$ denotes the local time of $X_t \cdot e(\theta)$ then*

$$\lim_{n\to\infty} \sum_{s_i \in \pi(n)} (L_t^{\gamma(s_i)} - L_t^{\gamma(s_{i-1})})^{2k} = \bar{c}(\beta) \int_0^1 (A_t^{\gamma,s})^k\, ds \tag{1.6}$$

in L^2, uniformly both in $t \in [0,T]$ and $Q(0,1)$ as $|\pi(n)| \to 0$.

Remark 1. For the special case of $\gamma_s = (s,\theta)$, with θ fixed, so that $L_t^{\gamma(s_i)} = L_t^{s_i,\theta}$, this reduces to Theorem 1 of [7] for the p-variation of the local time of the 1-dimensional symmetric stable process X_t^{θ}.

For the case of planar Brownian motion, Theorem 2 says that

$$\lim_{n\to\infty} \sum_{s_i \in \pi(n)} (L_t^{\gamma(s_i)} - L_t^{\gamma(s_{i-1})})^2 = 4 \int_0^1 A_t^{\gamma,s}\, ds. \tag{1.7}$$

In [1] it is shown that for planar Brownian motion we can choose an a.s. continuous version of $\{L_t^{\theta}\,;\, (\theta,t) \in [0,2\pi) \times R_+^1\}$, so that by (1.5) we can choose an a.s. continuous version of $\{A_t^{\gamma,s}\,;\, (s,t) \in [0,1] \times R_+^1\}$. We note that $\int_0^1 A_t^{\gamma,s}\, ds$ is the CAF with Revuz measure $\nu = \int_0^1 \tilde{\mu}_{a,\theta,s}\, ds$. It is easily checked that $|a'_s + x\theta'_s|$ is $|J(\Phi_\gamma)(s,x)|$, the absolute value of the Jacobean of the map Φ_γ. We have that

$$\begin{aligned} \int f(y)\, d\nu(y) &= \int_0^1 \int f(y)\, d\tilde{\mu}_{a,\theta,s}(y)\, ds \\ &= \int_0^1 \int f(a_s e(\theta_s) + x e(\theta_s^{\perp}))|a'_s + x\theta'_s|\, dx\, ds \\ &= \int_0^1 \int f(a_s e(\theta_s) + x e(\theta_s^{\perp}))|J(\Phi_\gamma)(s,x)|\, dx\, ds \\ &= \int f(y) N(\Phi_\gamma \,|\, y)\, dy \end{aligned} \tag{1.8}$$

where the last step uses the Area Formula, see 3.2.3 of [3]. This shows that $d\nu(y) = N(\Phi_\gamma \,|\, y)\, dy$, which proves Theorem 1.

2 $A_t^{\gamma,s}$

Lemma 1 *If $\beta = 1+1/k$ with $k = 1,2,\ldots$, then for any s,t, we have $A_t^{\gamma,s} \in L^{2k}$ and $s \mapsto A_t^{\gamma,s}$ is continuous in L^{2k}.*

Proof of Lemma 1: We have

$$
\begin{aligned}
(2.1)\qquad & E\left(\{A_t^{\gamma,s}\}^{2k}\right) \\
&= (2k)! \int \int_{\sum_{i=1}^{2k} t_i \leq t} \cdots \int \prod_{i=1}^{2k} p_{t_i}(x_{i-1}, x_i) dt_i \, d\widetilde{\mu}_{a,\theta,s}(x_i) \\
&\leq C \int \prod_{i=1}^{2k} u^1(x_{i-1}, x_i) \, d\widetilde{\mu}_{a,\theta,s}(x_i) \\
&= C \int \prod_{i=1}^{2k} u^1\left(a_s e(\theta_s) + (y_i - y_{i-1}) e(\theta_s^{\perp})\right) |a_s' + y_i \theta_s'|^{1/k} \, dy_i \\
&\leq C \int \prod_{i=1}^{2k} u^1\left(a_s e(\theta_s) + (y_i - y_{i-1}) e(\theta_s^{\perp})\right) (1 + |y_i|^{1/k}) \, dy_i \\
&= C \int \prod_{i=1}^{2k} u^1\left(a_s e(\theta_s) + z_i e(\theta_s^{\perp})\right) (1 + |\sum_{j=1}^{i} z_j|^{1/k}) \, dz_i \\
&\leq C \int \prod_{i=1}^{2k} u^1\left(a_s e(\theta_s) + z_i e(\theta_s^{\perp})\right) (1 + |z_i|^2) \, dz_i \\
&= C \prod_{i=1}^{2k} \int u^1\left(a_s e(\theta_s) + z_i e(\theta_s^{\perp})\right) (1 + |z_i|^2) \, dz_i
\end{aligned}
$$

We claim that the one dimensional integral

$$
(2.2)\qquad \int_{R^1} u^1\left(ae(\theta) + ze(\theta^{\perp})\right) (1 + |z|^2) \, dz < \infty,
$$

which would show that (2.1) is finite, establishing the first part of our Lemma. To show (2.2) we first recall that each component of $X_1 = (X_1^{(1)}, X_1^{(2)})$ is a real-valued symmentic stable process of index β, and therefore has moments of order β', for any $\beta' < \beta$, [4], p.578. Consequently,

$$
(2.3)\qquad c_0 \stackrel{def}{=} E(|X_1|^{\beta'}) \leq E(|X_1^{(1)}|^{\beta'}) + E(|X_1^{(2)}|^{\beta'}) < \infty.
$$

Since, by scaling, $X_t = t^{1/\beta} X_1$ in law, we have $E(|X_t|^{\beta'}) = t^{\beta'/\beta} c_0$. Therefore

$$
(2.4)\qquad \int_{R^2} |x|^{\beta'} u^1(x) \, dx = \int_0^\infty e^{-t} E(|X_t|^{\beta'}) \, dt < \infty.
$$

$u^1(x)$ is spherically symmetric, and, by abuse of notation, we use $u^1(r)$ for the function on R^1_+ whose value is $u^1(re(\theta))$ for any θ. Using polar coordinates, we can now rewrite (2.4) as

$$
(2.5)\qquad \int_0^\infty |r|^{1+\beta'} u^1(r) \, dr < \infty
$$

which implies (2.2). In fact, (2.5) together with the monotonicity of $u^1(r)$ implies that for any $\delta > 0$

$$\sup_a \int_{R^1} u^1\left(ae(\theta) + ze(\theta^\perp)\right)\left(1 + |z|^{2+1/k-\delta}\right) dz < \infty. \tag{2.6}$$

The fact that $s \mapsto A_t^{\gamma,s}$ is continuous in L^{2k} now follows from these considerations, using the bound (3.9) below with $\gamma > 0$ small.

3 A Second Moment Proof

Proof of Theorem 2: We write, for $\tau \in Q(0,1)$

$$\begin{aligned}
& E\left(\left\{\bar{c}(\beta)\int_0^1 (A_t^{\gamma,r})^k\,dr - \sum_{r_i\in\tau}\left(L_t^{\gamma(r_i)} - L_t^{\gamma(r_{i-1})}\right)^{2k}\right\}^2\right) \\
&= \bar{c}^2(\beta)\int_0^1\int_0^1 E\left\{(A_t^{\gamma,r})^k\left(A_t^{\gamma,r'}\right)^k\right\} dr\,dr' \\
&- 2\bar{c}(\beta)\int_a^b \sum_i E\left\{\left(L_t^{\gamma(r_i)} - L_t^{\gamma(r_{i-1})}\right)^{2k}\left(A_t^{\gamma,r'}\right)^k\right\} dr' \\
&+ \sum_{i,j} E\left\{\left(L_t^{\gamma(r_i)} - L_t^{\gamma(r_{i-1})}\right)^{2k}\left(L_t^{\gamma(r_j)} - L_t^{\gamma(r_{j-1})}\right)^{2k}\right\} \\
&\doteq A - 2B_\epsilon + C_\epsilon, \quad \text{where } \epsilon \doteq |\tau|
\end{aligned} \tag{3.1}$$

We will show that as $\epsilon \to 0$, each of $A, B_\epsilon, C_\epsilon$ converge to

$$\begin{aligned}
&[(2k)!(2c(\beta))^k]^2 \sum_{\tilde{\pi}} \int_0^1 dr \int_0^1 dr' \\
&\int\cdots\int_{\sum_{i=1}^{2k} t_i \le t} \prod_{i=1}^{2k} p_{t_i}\left(x^{\tilde{\pi}(i-1)}_{c_{\tilde{\pi}}(i-1)}, x^{\tilde{\pi}(i)}_{c_{\tilde{\pi}}(i)}\right) dt_i \prod_{j=1}^{k} d\tilde{\mu}_{a,\theta,r}(x_j^1)\,d\tilde{\mu}_{a,\theta,r'}(x_j^2)
\end{aligned} \tag{3.2}$$

where the sum runs over all paths $\tilde{\pi}:\ \{1,\ldots,2k\} \to \{1,2\}$ which visit $1,2$ an equal number of times (i.e. k times each), and $c_{\tilde{\pi}}(i) = |\{j \le i \,|\, \tilde{\pi}(j) = \tilde{\pi}(i)\}|$.

The fact that A equals (3.2) is straightforward, so we turn to B_ϵ. We will write $\mu_{a,\theta,s}$ for μ_{a_s,θ_s}. We have

$$\begin{aligned}
&\bar{c}(\beta)E\left\{\left(L_t^{\gamma(r_l)} - L_t^{\gamma(r_{l-1})}\right)^{2k}(A_t^{\gamma,s})^k\right\} \\
&= ((2k)!)^2(2c(\beta))^k \sum_\pi \int\cdots\int_{\sum_{i=1}^{3k} t_i \le t} \prod_{i=1}^{3k} p_{t_i}\left(x^{\pi(i-1)}_{c_\pi(i-1)}, x^{\pi(i)}_{c_\pi(i)}\right) dt_i\, d\mu^{\pi(i)}_{a,\theta}\left(x^{\pi(i)}_{c_\pi(i)}\right)
\end{aligned}$$

where the sum runs over all paths $\pi:\ \{1,\ldots,3k\} \longrightarrow \{1,2\}$ which visit 2 exactly k times, and

$$\begin{aligned}
\mu^1_{a,\theta} &\doteq \mu_{a,\theta,r_l} - \mu_{a,\theta,r_{l-1}} \\
\mu^2_{a,\theta} &\doteq \tilde{\mu}_{a,\theta,r'}
\end{aligned} \tag{3.3}$$

Fix such a π. We intend to rewrite

$$(3.4) \qquad \int_{\sum_{i=1}^{3k} t_i \le t} \cdots \int \prod_{i=1}^{3k} p_{t_i}(x^{\pi(i-1)}_{c_\pi(i-1)}, x^{\pi(i)}_{c_\pi(i)}) dt_i \, d\mu^{\pi(i)}_{a,\theta}(x^{\pi(i)}_{c_\pi(i)})$$

as a sum of many terms, most of which will make 0 contribution to (3.1) in the $\epsilon \to 0$ limit. Eventually we will identify those terms that contribute to (3.1) in the $\epsilon \to 0$ limit, and show how they lead to (3.2). Our procedure involves three steps.

Step 1: Let $\phi : l(a_{r_l}, \theta_{r_l}) \mapsto l(a_{r_{l-1}}, \theta_{r_{l-1}})$ be defined by

$$\phi(a_{r_l} e(\theta_{r_l}) + x e(\theta^\perp_{r_l})) = a_{r_{l-1}} e(\theta_{r_{l-1}}) + x e(\theta^\perp_{r_{l-1}}).$$

With this notation we rewrite

$$\begin{aligned}
(3.5) \qquad & \int p_s(x,y) p_t(y,z) \left(d\mu_{a,\theta,r_l}(y) - d\mu_{a,\theta,r_{l-1}}(y) \right) \\
&= \int \{ p_s(x,y) p_t(y,z) - p_s(x,\phi(y)) p_t(\phi(y),z) \} \, d\mu_{a,\theta,r_l}(y) \\
&= \int \{ p_s(x,y) - p_s(x,\phi(y)) \} \, p_t(y,z) \, d\mu_{a,\theta,r_l}(y) \\
&\qquad + \int p_s(x,\phi(y)) \{ p_t(y,z) - p_t(\phi(y),z) \} \, d\mu_{a,\theta,r_l}(y) \\
&= \int \Delta_2 p_s(x,y) p_t(y,z) \, d\mu_{a,\theta,r_l}(y) \\
&\qquad + \int p_s(x,\phi(y)) \Delta_1 p_s(y,z) \, d\mu_{a,\theta,r_l}(y)
\end{aligned}$$

where

$$\begin{aligned}
\Delta_2 p_s(x,y) &= p_s(x,y) - p_s(x,\phi(y)), \\
\Delta_1 p_s(y,z) &= p_t(y,z) - p_t(\phi(y),z).
\end{aligned}$$

We proceed to rewrite the factors in (3.4), working with each factor in turn, in order of decreasing i. For the largest i with $\pi(i) = 1$ we use (3.5) to replace

$$(3.6) \qquad \int p_{t_i}(x^{\pi(i-1)}_{c_\pi(i-1)}, x^1_{c_\pi(i)}) p_{t_{i+1}}(x^1_{c_\pi(i)}, x^{\pi(i+1)}_{c_\pi(i+1)}) \, d\mu^1_{a,\theta}(x^1_{c_\pi(i)})$$

by a sum of two terms, which we then view as replacing (3.4) by two terms. In case the largest such i is $3k$, we simply rewrite

$$\begin{aligned}
(3.7) \qquad & \int p_{t_{3k}}(x^{\pi(3k-1)}_{c_\pi(3k-1)}, x^1_{c_\pi(3k)}) \, d\mu^1_{a,\theta}(x^1_{c_\pi(3k)}) \\
&= \int p_{t_{3k}}(x^{\pi(3k-1)}_{c_\pi(3k-1)}, x^1_{c_\pi(3k)}) \left(d\mu_{a,\theta,r_l}(x^1_{c_\pi(3k)}) - d\mu_{a,\theta,r_{l-1}}(x^1_{c_\pi(3k)}) \right) \\
&= \int \Delta_2 p_{t_{3k}}(x^{\pi(3k-1)}_{c_\pi(3k-1)}, x^1_{c_\pi(3k)}) \, d\mu_{a,\theta,r_l}(x^1_{c_\pi(3k)}).
\end{aligned}$$

We then proceed through decreasing i. At each stage that we find an i with $\pi(i) = 1$, we check to see if the previously handled factor, i.e. the factor involving $(x^1_{c_\pi(i)}, x^{\pi(i+1)}_{c_\pi(i+1)})$

is a p or a Δp. If it is a p, then we proceed as before using (3.5). If it is a Δp, we do not use (3.5), but rather simply write out $d\mu^1_{a,\theta}(x^1_{c_\pi(i)})$. In other words we use

$$\begin{aligned} (3.8) \qquad & \int p_{t_i}(x^{\pi(i-1)}_{c_\pi(i-1)}, x^1_{c_\pi(i)}) p_{t_{i+1}}(x^1_{c_\pi(i)}, x^{\pi(i+1)}_{c_\pi(i+1)})\, d\mu^1_{a,\theta}(x^1_{c_\pi(i)}) \\ &= \int p_{t_i}(x^{\pi(i-1)}_{c_\pi(i-1)}, x^1_{c_\pi(i)}) p_{t_{i+1}}(x^1_{c_\pi(i)}, x^{\pi(i+1)}_{c_\pi(i+1)})\, d\mu_{a,\theta,r_l}(x^1_{c_\pi(i)}) \\ &- \int p_{t_i}(x^{\pi(i-1)}_{c_\pi(i-1)}, \phi(x^1_{c_\pi(i)})) p_{t_{i+1}}(\phi(x^1_{c_\pi(i)}), x^{\pi(i+1)}_{c_\pi(i+1)})\, d\mu_{a,\theta,r_l}(x^1_{c_\pi(i)}) \end{aligned}$$

to generate two terms.

After we have proceeded in this manner for all i, we will have replaced (3.4) by many terms. Each term will have at least k factors of the form Δp. We claim that any term with more that k factors of the form Δp will be $o(\Delta r_l)$, hence such terms will make 0 contribution to (3.1) in the $\epsilon \to 0$ limit. To see this, let $u_\alpha(x)$ denote a generic symmetric, positive, monotone decreasing (in $|x|$) function on R^2 such that $u_\alpha(x) \leq Cu^1(x/2)$ for $|x| \geq 1$ and $u_\alpha(x) \leq C|x|^{-\alpha}$ for $|x| \leq 1$. We then have that $\int_0^T p_t(x)\,dt \leq cu_{1-1/k}(x)$ while for any $0 \leq \gamma \leq 1$

$$(3.9) \qquad \int_0^T |p_t(x+a) - p_t(x)|\,dt \leq c|a|^\gamma (u_{1-1/k+\gamma}(x+a) + u_{1-1/k+\gamma}(x)).$$

(Consider seperately $|x| \leq 4|a|$ and $|x| > 4|a|$.) We then use these estimates to bound the integrals in the term we are studying, once again proceeding in order of decreasing i. (Note: our integration is always over lines.) If our term has $j > k$ factors of the form Δp, we choose $\gamma < 1/k$, so that $u_{1-1/k+\gamma}(x)$ will be integrable on any line, but with $1 < j\gamma < 1 + 1/k$. Using the fact that $|y - \phi(y)| \leq c(1 + |x|)\Delta r_l$ for $y = a_{r_l} e(\theta_{r_l}) + x e(\theta^\perp_{r_l})$, we see that our term can be bounded by $c|x|^{j\gamma}|\Delta r_l|^{j\gamma}$ which establishes our claim. (The condition $j\gamma < 1 + 1/k$ guarantees that the factors of $|x|$ do not mount up sufficiently to destroy integrability; see the proof of Lemma 1).

We will say that a path π is even if its visits to 1 occur in even runs. A path will be called odd if it is not even. It is easily seen that when our proceedure is applied to any odd π, all the resulting terms will have $> k$ factors of the form Δp, hence such terms will make 0 contribution to (3.1) in the $\epsilon \to 0$ limit. Similarly, if π is even, the only resulting terms with k factors of the form Δp, will be those terms in which for each i such that $\pi(i)$ is an even-numbered visit to 1, we replace the factor p_{t_i} by a $\Delta_2 p_{t_i}$, while for each i such that $\pi(i)$ is an odd-numbered visit to 1, we retain the factor p_{t_i}, more precisely we use (3.8). We note that for such terms, the k factors of the form $\Delta_2 p_{t_i}$ are seperated from each other by p factors

Step 2: Let π be a fixed even path, and consider one of the terms generated in Step 1 which does make a contribution to (3.1) in the $\epsilon \to 0$ limit. As described above, such a term has precisely k factors of the form $\Delta_2 p$ and these k factors are seperated from each other by p factors. Let us rewrite every pair of the form $\Delta_2 p_s(x,y) p_t(y,z)$ using

$$\begin{aligned} (3.10) \qquad & \Delta_2 p_s(x,y) p_t(y,z) \\ &= \Delta_2 p_s(x,y) p_t(x,z) + \Delta_2 p_s(x,y)\,(p_t(y,z) - p_t(x,z))\,. \end{aligned}$$

Once more this allows us to rewrite our term as a sum of many terms. We now show that any term which contains a pair of the form $\Delta_2 p_s(x,y)\,(p_t(y,z) - p_t(x,z))$ will be

$o(\Delta r_l)$, hence such terms will make 0 contribution to (3.1) in the $\epsilon \to 0$ limit. To see this we proceed as above to bound our term by bounding the integrals, proceeding in order of decreasing i. We use (3.9) to bound the contribution of the z,t integrals of $|p_t(y,z) - p_t(x,z)|$ by $c|x-y|^\gamma$, with $\gamma = 3/(4k)$. We then use the bound

$$
\begin{aligned}
(3.11)\qquad & \int |y|^\gamma \left(\int_0^T |p_s(y+a) - p_s(y)|\, ds \right) d^1y \\
&= \int_{|y|\le 2|a|} |y|^\gamma \left(\int_0^T |p_s(y+a) - p_s(y)|\, ds \right) d^1y \\
&\quad + \int_{|y|>2|a|} |y|^\gamma \left(\int_0^T |p_s(y+a) - p_s(y)|\, ds \right) d^1y \\
&\le c|a|^\gamma \int_{|y|\le 2|a|} \left(u_{1-1/k}(y+a) + u_{1-1/k}(y) \right) d^1y \\
&\quad + c|a|^{\gamma'} \int_{|y|>2|a|} |y|^\gamma u_{1-1/k+\gamma'}(y)\, d^1y \\
&\le c|a|^{\gamma+1/k} + c|a|^{\gamma'} \int_{|y|>2|a|} u_{1-1/k+\gamma'-\gamma}(y)\, d^1y \\
&\le c(|a|^{7/(4k)} + |a|^{3/(2k)})
\end{aligned}
$$

by taking $\gamma' = 3/(2k)$. It is easily seen that this verifies our claim.

Step 3: We are now reduced to considering terms with precisely k factors of the form $\Delta_2 p_s(x, y_i)$, with no other factors containing the variable y_i. We now integrate each such factor with respect to $d\mu_{a,\theta,r_l}(y_i)$.

Note that

$$
\begin{aligned}
(3.12)\qquad & \int p_s(y,z)\, d\mu_{a,\theta,r}(z) \\
&= \int p_s(y - a_r e(\theta_r) - x e(\theta_r^\perp))\, dx \\
&= q_s(a_r - y \cdot e(\theta_r))
\end{aligned}
$$

where $q_s(x)$ denotes the density of the one dimensional symmetric stable process of index β. Setting

$$
(3.13)\qquad Q_s(y) = q_s\{a_{r_l} - y \cdot e(\theta_{r_l})\} - q_s\{a_{r_{l-1}} - y \cdot e(\theta_{r_{l-1}})\}
$$

we see from (3.12) that

$$
(3.14)\qquad \int \Delta_2 p_s(x_u^1, x_v^1) d\mu_{a,\theta,r_l}(x_v^1) = Q_s(x_u^1).
$$

In this manner we can see that the sum of all terms generated by (3.4) which contribute to (3.1) in the $\epsilon \to 0$ limit can be written as

$$
(3.15)\qquad \sum_{A\subseteq\{1\dots,k\}} (-1)^{|A|} \int\cdots\int_{0\le\sum_{i=1}^{2k} t_i + \sum_{j=1}^k \tau_j \le t} \prod_{i=1}^{2k} p_{t_i}(x_{c_{\bar\pi}(i),A}^{\bar\pi(i)}, x_{c_{\bar\pi}(i-1),A}^{\bar\pi(i-1)}) dt_i
$$

$$
\prod_{j=1}^k Q_{\tau_j}(x_{j,A}^1)\, d\mu_{a,\theta,r_l}(x_j^1)\, d\tilde\mu_{a,\theta,r'}(x_j^2)
$$

where $\bar{\pi}: \{1,\ldots,2k\} \longrightarrow \{1,2\}$ is the path visiting both 1 and 2 exactly k times induced by π as follows: since visits of π to 1 occur in pairs, we simply suppress one visit from each pair, and $x^2_{j,A} = x^2_j$ while $x^1_{j,A} = \phi(x^1_j)$ if $j \in A$ and $x^1_{j,A} = x^1_j$ if $j \notin A$. The methods of Step 1 show that we can replace this, up to terms which will contribute 0 to (3.1) in the $\epsilon \to 0$ limit by

(3.16)
$$\sum_{A\subseteq\{1,\ldots,k\}} (-1)^{|A|} \int_{0\leq\sum_{i=1}^{2k} t_i+\sum_{j=1}^{k}\tau_j\leq t} \cdots \int \prod_{i=1}^{2k} p_{t_i}(x^{\bar{\pi}(i)}_{c_{\bar{\pi}}(i)}, x^{\bar{\pi}(i-1)}_{c_{\bar{\pi}}(i-1)})dt_i$$
$$\prod_{j=1}^{k} Q_{\tau_j}(x^1_{j,A})\, d\mu_{a,\theta,r_l}(x^1_j)\, d\widetilde{\mu}_{a,\theta,r'}(x^2_j).$$

Note that

(3.17)
$$\begin{aligned}&Q_s(a_{r_l}e(\theta_{r_l}) + xe(\theta^\perp_{r_l}))\\ &= q_s\{0\} - q_s\{a_{r_{l-1}} - a_{r_l}e(\theta_{r_l})\cdot e(\theta_{r_{l-1}}) - xe(\theta^\perp_{r_l})\cdot e(\theta_{r_{l-1}})\}\\ &= q_s\{0\} - q_s\{a_{r_{l-1}} - a_{r_l}\cos(\theta_{r_l} - \theta_{r_{l-1}}) - x\sin(\theta_{r_l} - \theta_{r_{l-1}})\}\end{aligned}$$

while

(3.18)
$$\begin{aligned}&Q_s(a_{r_{l-1}}e(\theta_{r_{l-1}}) + xe(\theta^\perp_{r_{l-1}}))\\ &= q_s\{a_{r_l} - a_{r_{l-1}}e(\theta_{r_{l-1}})\cdot e(\theta_{r_l}) - xe(\theta^\perp_{r_{l-1}})\cdot e(\theta_{r_l})\} - q_s\{0\}\\ &= q_s\{a_{r_l} - a_{r_{l-1}}\cos(\theta_{r_{l-1}} - \theta_{r_l}) - x\sin(\theta_{r_{l-1}} - \ th_{r_l})\} - q_s\{0\}\\ &= q_s\{a_{r_{l-1}}\cos(\theta_{r_l} - \theta_{r_{l-1}}) - a_{r_l} - x\sin(\theta_{r_l} - \theta_{r_{l-1}})\} - q_s\{0\}\\ &= -Q_s(a_{r_l}e(\theta_{r_l}) + xe(\theta^\perp_{r_l})) + \Delta Q_s(a_{r_l}e(\theta_{r_l}) + xe(\theta^\perp_{r_l}))\end{aligned}$$

where

(3.19)
$$\begin{aligned}&\Delta Q_s(a_{r_l}e(\theta_{r_l}) + xe(\theta^\perp_{r_l}))\\ &= q_s\{a_{r_{l-1}}\cos(\theta_{r_l} - \theta_{r_{l-1}}) - a_{r_l} - x\sin(\theta_{r_l} - \theta_{r_{l-1}})\}\\ &-q_s\{a_{r_{l-1}} - a_{r_l}\cos(\theta_{r_l} - \theta_{r_{l-1}}) - x\sin(\theta_{r_l} - \theta_{r_{l-1}})\}.\end{aligned}$$

We record here Lemma 1 of [7].

Lemma 2

(3.20)
$$\int_0^t \mid q_t(x) - q_t(y) \mid\, dt \leq c\,\Big|\,|x|^{\beta-1} - |y|^{\beta-1}\Big| \leq c|x-y|^{\beta-1}$$

and

(3.21)
$$\int_0^T q_t(0) - q_t(x)dt = c(\beta)|x|^{\beta-1} + O\left(\frac{|x|^2}{T^{3/\beta-1}}\right)$$

where

(3.22)
$$c(\beta) = \int_0^\infty (q_t(0) - q_t(1))dt < \infty.$$

Actually, we need a slight refinement of (3.21). The error term $O\left(|x|^2/T^{3/\beta-1}\right)$ comes from a bound on $\int_T^\infty (q_t(0) - q_t(x))dt$. We can also bound this integral by

$2\int_T^\infty q_t(0)dt = C/T^{2/\beta-1}$. Interpolating these two bounds shows that for $\delta > 0$ sufficiently small we have

$$\int_0^T q_t(0) - q_t(x)dt = c(\beta)|x|^{\beta-1} + O\left(\frac{|x|^{\beta-1+\delta}}{T^{(3+\delta)/2\beta-1/2}}\right) \tag{3.23}$$

Using this we see that

$$\int_0^T |\Delta Q_\tau(a_{r_l}e(\theta_{r_l}) + xe(\theta_{r_l}^\perp))|\, d\tau = O\left(\frac{(1+|x|^{\beta-1+\delta})|\Delta r_l|^{\beta-1+\delta}}{T^{(3+\delta)/2\beta-1/2}}\right). \tag{3.24}$$

Thus, up to terms which will not contribute to (3.1) in the $\epsilon \to 0$ limit we can replace $Q_{\tau_j}(x^1_{j,A})$ by $(-1)^{1_A(j)}Q_{\tau_j}(x^1_j)$ in (3.16). Thus (3.16) can be replaced by

$$\begin{aligned}
&\sum_{A\subseteq\{1,\dots,k\}} \quad \int_{0\le\sum_{i=1}^{2k} t_i+\sum_{j=1}^{k}\tau_j\le t} \cdots \int \prod_{i=1}^{2k} p_{t_i}(x^{\bar\pi(i)}_{c_{\bar\pi}(i)}, x^{\bar\pi(i-1)}_{c_{\bar\pi}(i-1)})dt_i \\
&\qquad \prod_{j=1}^{k} Q_{\tau_j}(x^1_j)\, d\mu_{a,\theta,r_l}(x^1_j)\, d\tilde\mu_{a,\theta,r'}(x^2_j) \\
&= 2^k \int_{0\le\sum_{i=1}^{2k} t_i+\sum_{j=1}^{k}\tau_j\le t} \cdots \int \prod_{i=1}^{2k} p_{t_i}(x^{\bar\pi(i)}_{c_{\bar\pi}(i)}, x^{\bar\pi(i-1)}_{c_{\bar\pi}(i-1)})dt_i \\
&\qquad \prod_{j=1}^{k} Q_{\tau_j}(x^1_j)\, d\mu_{a,\theta,r_l}(x^1_j)\, d\tilde\mu_{a,\theta,r'}(x^2_j)
\end{aligned} \tag{3.25}$$

Furthermore, (3.23) tells us that

$$\begin{aligned}
&\int_0^T Q_\tau(a_{r_l}e(\theta_{r_l}) + xe(\theta_{r_l}^\perp))\, d\tau \\
&= \int_0^T q_\tau\{0\} - q_\tau\{a_{r_{l-1}} - a_{r_l}\cos(\theta_{r_l}-\theta_{r_{l-1}}) - x\sin(\theta_{r_l}-\theta_{r_{l-1}})\}\ , d\tau \\
&= c(\beta)|a_{r_{l-1}} - a_{r_l}\cos(\theta_{r_l}-\theta_{r_{l-1}}) - x\sin(\theta_{r_l}-\theta_{r_{l-1}})|^{\beta-1} \\
&\qquad\qquad + O\left(\frac{(1+|x|^{\beta-1+\delta})|\Delta r_l|^{\beta-1+\delta}}{T^{(3+\delta)/2\beta-1/2}}\right) \\
&= c(\beta)|a'_{r_l} + x\theta'_{r_l}|^{\beta-1}(\Delta r_l)^{\beta-1} + O\left(\frac{(1+|x|^{\beta-1+\delta})|\Delta r_l|^{\beta-1+\delta}}{T^{(3+\delta)/2\beta-1/2}}\right)
\end{aligned} \tag{3.26}$$

Using Lemma 3 of [7] now completes the proof that B_ϵ converges to (3.2).

Finally, the fact that C_ϵ converges to (3.2) follows using the methods employed above for B_ϵ together with the methods of [7]. This completes the proof of Theorem 1.

4 A Stochastic Calculus Proof

In this section we outline a proof of Theorem 1 using stochastic calculus. In fact, we only deal with convergence in probability, and for a fixed sequence $\pi(n)$ of partitions with $|\pi(n)| \to 0$. To emphasize that we are dealing with planar Brownian motion we use the notation $B_s = (B_s^1, B_s^2)$ in place of X_s, and write $B_s^\theta = B_s \cdot e(\theta)$, etc. By Tanaka's formula

$$\frac{1}{2}L_t^{\gamma(s)} = \left(B_t^{\theta(s)} - a(s)\right)^+ - \left(B_0^{\theta(s)} - a(s)\right)^+ - \int_0^t 1_{\{B_u^{\theta(s)} > a(s)\}} dB_u^{\theta(s)} . \tag{4.1}$$

As we will explain below, the only term which will contribute to the quadratic variation in the limit is the stochastic integeral term. Let

$$\hat{B}_t^{\gamma(s)} = \int_0^t 1_{\{B_u^{\theta(s)} > a(s)\}} dB_u^{\theta(s)}$$

and consider

$$V_n = \sum_{s_i \in \pi(n)} \left(\hat{B}_t^{\gamma(s_{i+1})} - \hat{B}_t^{\gamma(s_i)}\right)^2 .$$

By Ito's formula

$$\begin{aligned}
&\left(\hat{B}_t^{\gamma(s_{i+1})} - \hat{B}_t^{\gamma(s_i)}\right)^2 \\
&= 2\int_0^t \left(\hat{B}_u^{\gamma(s_{i+1})} - \hat{B}_u^{\gamma(s_i)}\right) [1_{\{B_u^{\theta(s_{i+1})} > a(s_{i+1})\}} dB_u^{\theta(s_i)} - 1_{\{B_u^{\theta(s_i)} > a(s_i)\}} dB_u^{\theta(s_i)}] \\
&+ \int_0^t du \left(1_{\{B_u^{\theta(s_{i+1})} > a(s_{i+1})\}} + 1_{\{B_u^{\theta(s_i)} > a(s_i)\}}\right) \\
&-2\int_0^t du \cdot 1_{\{B_u^{\theta(s_{i+1})} > a(s_{i+1})\}} 1_{\{B_u^{\theta(s_i)} > a(s_i)\}} [\cos\theta(s_{i+1})\cos\theta(s_i) + \sin\theta(s_{i+1})\sin\theta(s_i)] .
\end{aligned}$$

(4.2)

As we explain below, the only contribution to $\lim_{n\to\infty} V_n$ will come from the last two lines in (4.2) which we can rewrite as

$$\begin{aligned}
&\int_0^t \Big(1_{\{B_u^{\theta(s_{i+1})} > a(s_{i+1})\}} + 1_{\{B_u^{\theta(s_i)} > a(s_i)\}} \\
&\qquad -2\cos(\theta(s_{i+1}) - \theta(s_i)) 1_{\{B_u^{\theta(s_{i+1})} > a(s_{i+1}),\, B_u^{\theta(s_i)} > a(s_i)\}}\Big) du \\
&= \int_0^t \left(1_{\{B_u^{\theta(s_{i+1})} > a(s_{i+1}),\, B_u^{\theta(s_i)} < a(s_i)\}} + 1_{\{B_u^{\theta(s_i)} > a(s_i),\, B_u^{\theta(s_{i+1})} < a(s_{i+1})\}}\right) du \\
&\qquad + O(s_{i+1} - s_i)^2 .
\end{aligned}$$

Writing

$$\begin{aligned}
C_{s_i,s_{i+1}} &= \left\{x \in R^2 \,|\, x \cdot e(\theta(s_{i+1})) > a(s_{i+1}),\ x \cdot e(\theta(s_i)) < a(s_i)\right\} \\
&\qquad \bigcup \left\{x \in R^2 \,|\, x \cdot e(\theta(s_{i+1})) < a(s_{i+1}),\ x \cdot e(\theta(s_i)) > a(s_i)\right\}
\end{aligned}$$

we then see that

$$\lim_{n\to\infty} V_n = \lim_{n\to\infty} \int_0^t \sum_{s_i \in \pi(n)} 1_{C_{s_i,s_{i+1}}}(B_u)\, du. \tag{4.3}$$

$C_{s_i,s_{i+1}}$ is, in general, the cone contained between the lines $l_{\gamma(s_i)}$ and $l_{\gamma(s_{i+1})}$. (If the lines are parallel, $C_{s_i,s_{i+1}}$ is the strip between them). $C_{s_i,s_{i+1}}$ is not quite the same as $\widetilde{C}_{s_i,s_{i+1}} = \cup_{s_i \le s \le s_{i+1}} l_{\gamma(s)}$, but a detailed trigonometric calculation shows that in the $n \to \infty$ limit we can replace $C_{s_i,s_{i+1}}$ by $\widetilde{C}_{s_i,s_{i+1}}$ in (4.4) to obtain

$$\lim_{n\to\infty} V_n = \lim_{n\to\infty} \int_0^t \sum_{s_i\in\pi(n)} 1_{\widetilde{C}_{s_i,s_{i+1}}}(B_u)\,du \tag{4.4}$$

(It is mainly at this point that our proof is only an outline. We leave the trigonometric details to the dedicated reader). Now, if $N(\Phi_\gamma|x) = k < \infty$, it is easily seen that

$$\lim_{n\to\infty} \sum_{s_i\in\pi(n)} 1_{\widetilde{C}_{s_i,s_{i+1}}}(x) = N(\Phi_\gamma|x),$$

and since the arguments of section 2 and the end of section 1 show that $N(\Phi_\gamma|B_u) < \infty$ for a.e. u a.s., we see that

$$\lim_{n\to\infty} V_n = \int_0^t N(\Phi_\gamma|B_s)ds.$$

Noting the factor 1/2 in (4.1) will complete our proof, once we explain why the terms we have ignored do not contribute in the $n \to \infty$ limit.

To show that the stochastic integral term in (4.2) converges to 0 we write it as

$$\begin{aligned} &\int_0^t (\hat{B}_u^{\gamma(s_{i+1})} - \hat{B}_u^{\gamma(s_i)}) \left(1_{\{B_u^{\theta(s_{i+1})}>a(s_{i+1})\}} dB_u^{\theta(s_{i+1})} - 1_{\{B_u^{\theta(s_i)}>a(s_i)\}} dB_u^{\theta(s_i)}\right) \\ &= \int_0^t (\hat{B}_u^{\gamma(s_{i+1})} - \hat{B}_u^{\gamma(s_i)})\, 1_{\{B_u^{\theta(s_{i+1})}>a(s_{i+1}),\ B_u^{\theta(s_i)}>a(s_i)\}} \left(dB_u^{\theta(s_{i+1})} - dB_u^{\theta(s_i)}\right) \\ &+ \int_0^t (\hat{B}_u^{\gamma(s_{i+1})} - \hat{B}_u^{\gamma(s_i)})\, 1_{\{B_u^{\theta(s_{i+1})}>a(s_{i+1}),\ B_u^{\theta(s_i)}<a(s_i)\}} dB_u^{\theta(s_{i+1})} \\ &+ \int (\hat{B}_u^{\gamma(s_i)} - \hat{B}_u^{\gamma(s_{i+1})})\, 1_{\{B_u^{\theta(s_i)}>a(s_i),\ B_u^{\theta(s_{i+1})}<a(s_{i+1})\}} dB_u^{\theta(s_i)}\,. \end{aligned} \tag{4.5}$$

Summing over i in the first term, we get

$$\int_0^t \sum_{s_i\in\pi(n)} \left(\hat{B}_u^{\gamma(s_{i+1})} - \hat{B}_u^{\gamma(s_i)}\right) 1_{\{B_u^{\theta(s_{i+1})}>a(s_{i+1}), B_u^{\theta(s_i)}>a(s_i)\}}$$

$$(\cos\theta(s_{i+1}) - \cos\theta(s_i))dB_u^1 + (\sin\theta(s_{i+1}) - \sin\theta(s_i))dB_u^2\,.$$

But, since $s \to \hat{B}_u^{\gamma(s)}$ is continuous, it follows that

$$\begin{aligned} &\left|\sum_{s_i\in\pi(n)} \left(\hat{B}_u^{\gamma(s_{i+1})} - \hat{B}_u^{\gamma(s_i)}\right) 1_{\{B_u^{\theta(s_{i+1})}>a(s_{i+1}), B_u^{\theta(s_i)}>a(s_i)\}} (\cos\theta(s_{i+1}) - \cos(\theta(s_i))\right| \\ &\le \sum_{s_i\in\pi(n)} \left|\hat{B}_u^{\gamma(s_{i+1})} - \hat{B}_u^{\gamma(s_i)}\right| \, |\cos\theta(s_{i+1}) - \cos(\theta(s_i))| \\ &\le \sup_i \left|\hat{B}_u^{\gamma(s_{i+1})} - \hat{B}_u^{\gamma(s_i)}\right| \sum |\cos\theta(s_{i+1}) - \cos(\theta(s_i))| \end{aligned}$$

which is bounded by

$$\sup_i \left|\hat{B}_u^{\gamma(s_{i+1})} - \hat{B}_u^{\gamma(s_i)}\right| \quad \underset{a\le s\le b}{\text{Variation of}}\ \cos\theta(s)$$

which converges to 0 as $|\pi_n| \to 0$. The same is true with $(\sin\theta(s_{i+1}) - \sin\theta(s_i))$ replacing the cosine. Thus, by the dominated convergence theorem for stochastic integrals (we may assume that B^1, B^2 are bounded by localization) the above expression converges to 0.

As to the remaining expression in (4.5), summing over i and writing $B_u^{\theta(s_{i+1})}$ and $B_u^{\theta(s_i)}$ explicitly we obtain

$$\int_0^t \sum_{s_i\in\pi(n)} (\hat{B}_u^{\gamma(s_{i+1})} - \hat{B}_u^{\gamma(s_i)}) 1_{\{B_u^{\theta(s_{i+1})} > a(s_{i+1}),\ \hat{B}_u^{\theta(s_i)} < a(s_i)\}} (\cos\theta(s_{i+1})dB_u^1 + \sin\theta(s_{i+1})dB_u^2)$$

$$+\int_0^t \sum_{s_i\in\pi(n)} (\hat{B}_u^{\gamma(s_i)} - \hat{B}_u^{\gamma(s_{i+1})})\ 1_{\{B_u^{\theta(s_i)} > a(s_i),\ B_u^{\theta(s_{i+1})} < a(s_{i+1})\}} (\cos\theta(s_i)dB_u^1 + \sin\theta(s_i)dB_u^2)$$

which by the dominated convergence theorem for stochastic integrals (by localization if necessary to bound $\hat{B}_u^{\theta(s_i)}, \hat{B}_u^{\theta(s_{i+1})}$) converges, as $|\pi_n| \to 0$, to

$$\int_0^t \sum_{k=1}^{N(\Phi(\gamma)\,|\,B_u)} \left(\hat{B}_u^{\bar{\gamma}_k(u)} - \hat{B}_u^{\gamma_k(u)}\right)\ (\cos\theta(\gamma_k(u))dB_u^1 + \sin\theta(\gamma_k(u))dB_u^2)$$

where $\gamma_k(u)$ is the k-th of the $N(\Phi(\gamma)\,|\,B_u)$ lines on which B_u lies and $\theta(\gamma_k(u))$ is its angle. If $\gamma_k(u) = \gamma(s)$ for some s, then $\hat{B}_u^{\bar{\gamma}_k(u)}$ is the limit of $\hat{B}_u^{\gamma(t)}$ for a sequence $\{t_n\}$ that converges to s. But since $t \to \hat{B}_u^{\gamma(t)}$ is continuous, $\hat{B}_u^{\gamma_k(u)} = \hat{B}_u^{\bar{\gamma}_k(u)}$ and the whole expression is equal to 0.

We now return to (4.1) to show that the only term which will contribute to the quadratic variation in the limit is the stochastic integeral term. If

$$\psi_t(s) = \left(B_t^{\theta(s)} - a(s)\right)^+ - \left(B_0^{\theta(s)} - a(s)\right)^+$$

$$= (B_t^1\cos\theta(s) + B_t^2\sin\theta(s) - a(s))^+ - (B_0^1\cos\theta(s) + B_0^2\sin\theta(s) - a(s))^+,$$

we note that $s \to \psi_t(s)$ is Liptschitz 1 (by the smoothness of $s \to \gamma(s)$) and therefore

$$\lim_{n\to\infty} \sum_{s_i\in\pi(n)} (\psi_t(s_{i+1}) - \psi_t(s_i))^2 = 0\ .$$

Furthermore, since $s \to \hat{B}_t^{\gamma(s)}$ is continuous,

$$\lim_{n\to\infty} \sum_{s_i\in\pi(n)} (\phi_t(s_{i+1}) - \phi_t(s_i))\ (\hat{B}_t^{\gamma(s_{i+1})} - \hat{B}_t^{\gamma(s_i)}) = 0\ .$$

This completes our outline.

References

1. R. Bass, *Joint continuity and representations of additive functionals of d-dimensional Brownian motion*, Stochastic Process. Appl. **17** (1984), 211–227.

2. N. Bouleau and M.Yor, *Sur la variation quadratique de temps locaux de certaines semi-martingales*, C. R. Acad. Sc. Paris **292** (1981), 491–492.

3. H. Federer, *Geometric measure theory*, Springer-Verlag, New York, 1969.

4. W. Feller, *An introduction to probability theory and its applications, vol. ii*, John Wiley and Sons, New York, 1971.

5. M. Marcus and J. Rosen, *p-variation of the local times of symmetric stable processes and of Gaussian processes with stationary increments*, Ann. Probab. **20** (1992), 1685–1713.

6. E. Perkins, *Local time is a semimartingale*, Z. Wahrscheinlichkeitstheorie und Verw. Gebiete **60** (1982), 79–117.

7. J. Rosen, *p-variation of the local times of stable processes and intersection local time*, Seminar on Stochastic Processes, 1991 (Boston), Progress in Probability, vol. 33, Birkhauser, Boston, 1993, pp. 157–168.

Jay Rosen
Department of Mathematics
College of Staten Island, CUNY
Staten Island, NY 10314
jrosen3@mail.idt.net

Haya Kaspi
Department of Industrial Engineering and Management
Technion
Haifa, Israel 32000
iehaya3@techunix.technion.ac.il

LARGE DEVIATIONS FOR SOME POISSON RANDOM INTEGRALS

by

Zbigniew J. JUREK (Wroclaw) & Liming WU (Clermont-Ferrand)

In the theory of large deviations one of the main examples is Schilder's theorem. It gives the large deviation estimates for the convergence $\sqrt{\epsilon}W \Rightarrow \delta_0$ on $C([0,\infty), \mathbf{R}^d)$ as $\epsilon \to 0$, for the Brownian Motion W. In this paper we investigate analogous problems for $\epsilon N(f) := \epsilon \int f dN$ or $\epsilon \widetilde{N}(f) := \epsilon \int f d\widetilde{N}$, where N (resp. $\widetilde{N}$) is a (resp. compensated) Poisson point process and f is a deterministic function. We find that this large deviation estimation depends strongly on the tail behavior of f. This differs from the Brownian Motion case where only the norm of f in L^2 is involved. In particular,we get the large deviations principle for the Lévy class L distributions (called also self-decomposable measures). The question about large deviations for the multiple Poisson integrals is not discussed here. (The case of Brownian Motion is solved by Ledoux [L].)

1. Notation and basic terminology. Let S be a metric separable and complete space (or Polish space) with the Borel σ-field $\mathcal{S}$. A function $I : S \to [0, \infty]$ such that $\{I \le L\}$ are compact subsets of S for all $L > 0$, is called *good rate function.* We say that a family $\{\mu_\epsilon, \epsilon > 0\}$ of Borel probability measures on S satisfies *large deviations principle* with the good rate I and the *speed* $\lambda(\epsilon)$ provided

$$\overline{\lim_{\epsilon\to 0}} \frac{1}{\lambda(\epsilon)} \log \mu_\epsilon(F) \le -\inf_{s\in F} I(s) \tag{1.1}$$

for all closed subsets F in S; and

$$-\inf_{s\in G} I(s) \le \underline{\lim_{\epsilon\to 0}} \frac{1}{\lambda(\epsilon)} \log \mu_\epsilon(G) \tag{1.2}$$

for all open sets G in S.
[Here: $\lambda(\epsilon) > 0$ and $\lambda(\epsilon) \to +\infty$ as $\epsilon \to 0$. Also we adopt convention $\inf \phi = +\infty$ throughout this paper]. In short, we write that (μ_ϵ) satisfies *LDP*. Note that (1.1) and (1.2) roughly mean that $\mu_\epsilon(A) \simeq \exp[-\lambda(\epsilon) \inf_{s\in A} I(s)]$.

We require the following variant of comparison technique in large deviation theory (see e.g. [DS, Exercice 2.1.20, p.47-49] for other versions), whose proof is left to the reader.

Comparison Lemma : *Let $(X^n_\epsilon, X_\epsilon, n \in \mathbf{N}, \epsilon > 0)$ be a family of random variables valued in a Polish space S with metric $d(\cdot,\cdot)$, defined on a probability space $(\Omega, \mathcal{F}, \mathbf{P})$. Assume*

(i) for each $n \in \mathbf{N}$, $\mathbf{P}(X^n_\epsilon \in \cdot)$ satisfies as $\epsilon \to 0$, the LDP on S with speed $\lambda(\epsilon)$ and good rate function $I_n(x)$;

(ii) there is a good rate function I on S such that $\forall L \geq 0$,

(1.3)
$$\sup_{x\in[I\leq L]} |I_n(x) - I(x)| \longrightarrow 0, \quad \textit{as } n \to \infty \,;$$

(iii) for every $\delta > 0$,

(1.4)
$$\lim_{n\to\infty} \overline{\lim_{\epsilon\to 0}} \frac{1}{\lambda(\epsilon)} \log \mathbf{P}\left(d(X_\epsilon^n, X_\epsilon) > \delta\right) = -\infty \,.$$

Then $\mathbf{P}(X_\epsilon \in \cdot)$, as $\epsilon \to 0$, satisfies the LDP on S with speed $\lambda(\epsilon)$ and good rate function $I(x)$ given in (ii) above.

Let $(E, \mathcal{E}, \rho)$ be a σ-finite measure space and $(\Omega, \mathcal{F}, \mathbf{P})$ be a probability space. A mapping

(1.5)
$$N : \Omega \to \left\{ \sum_i \delta_{x_i} \text{ (at most countable)} : x_i \in E \right\} ,$$

where δ_x denotes Dirac measure, is called *Poisson point process with intensity measure ρ* if

(i)
$$\mathbf{P}[N(A) = k] = e^{-\rho(A)} \frac{(\rho(A))^k}{k!} , \quad k = 0, 1, 2, \dots ,$$

for all $A \in \mathcal{E}$ such that $0 < \rho(A) < \infty$;

(ii) for $k \geq 1$ and $A_j \in \mathcal{E}$, $j = 1, 2, ..., k$, pair-wise disjoint, and $0 < \rho(A_j) < \infty$, random variables $N(A_1)$, $N(A_2)$, ..., $N(A_k)$ are independent.

We shall often denote the integral $\int_E f dN$ also by $N(f)$, where f is a ρ-integrable function. If $\widetilde{N}$ is the compensated Poisson point process, i.e., $\widetilde{N} := N - \rho$, then the random integral $\widetilde{N}(f) := \int_E f d\widetilde{N}$ exists for $f \in L^2(\rho)$; cf, [JS].

2. Large deviations for integrals $N(f)$ on $\mathbf{R}^d$ for bounded f. The main results will be preceeded by two auxilliary steps.

Step 1. *If $0 < \rho(A) < \infty$ then the probability measures $\mu_\epsilon(\cdot) := \mathbf{P}[\epsilon N(A) \in \cdot]$ on $\mathbf{R}$, $\epsilon > 0$, satisfy LDP with the speed $\lambda(\epsilon) = \epsilon^{-1}|\log \epsilon|$ and the rate function*

$$I(x) = \begin{cases} +\infty , & \text{for } x < 0 \,; \\ x , & \text{for } x \geq 0 \,. \end{cases}$$

P r o o f. Since for $x > 0$ we have

(2.1)
$$\lim_{\epsilon\to 0} \frac{\epsilon}{|\log \epsilon|} [\frac{x}{\epsilon}] \log[\frac{x}{\epsilon}] = x \,,$$

therefore the Stirling formula implies that

(2.2)
$$\lim_{\epsilon\to 0} \frac{\epsilon}{|\log \epsilon|} \log([\frac{x}{\epsilon}]!) = x \,,$$

where $[\cdot]$ denotes the integral part. From the inequalities

$$e^{-\rho(A)}\frac{(\rho(A))^{[\frac{x}{\epsilon}]+1}}{([\frac{x}{\epsilon}]+1)!} \le \mathbf{P}[\epsilon N(A) > x] = \sum_{j\ge[\frac{x}{\epsilon}]+1} e^{-\rho(A)}\frac{(\rho(A))^j}{j!} \le \frac{\rho(A)^{[\frac{x}{\epsilon}]+1}}{([\frac{x}{\epsilon}]+1)!}$$

and (2.1) with (2.2) we conclude that for all $x > 0$

$$\lim_{\epsilon\to 0}\frac{\epsilon}{|\log\epsilon|}\log \mathbf{P}[\epsilon N(A) > x] = -x \ . \tag{2.3}$$

For a closed F in $\mathbf{R}$, with $\inf_{s\in F} I(s) = x > 0$ and all $0 < \delta < x$, one has

$$\mathbf{P}[\epsilon N(A) \in F] \le \mathbf{P}[\epsilon N(A) \ge x] \le P[\epsilon N(A) > x-\delta] \ .$$

Hence by (2.3) we conclude

$$\overline{\lim_{\epsilon\to 0}}\frac{\epsilon}{|\log\epsilon|}\log \mathbf{P}[\epsilon N(A) \in F] \le -\inf_{s\in F} I(s) + \delta \ ,$$

for all $0 < \delta < x$, which proves the upper bound (1.1). If $\inf_{s\in F} I(s) = 0$ then (1.1) holds automatically.

For an open set $G \ni s$, let us choose $\delta > 0$ such that $(s-\delta, s+\delta) \subseteq G$. Then

$$\mathbf{P}[\epsilon N(A) \in G] \ge \mathbf{P}[N(A) \in \epsilon^{-1}(s-\delta, s+\delta)] \ge e^{-\rho(A)}\frac{\rho(A)^{[\frac{s}{\epsilon}]+1}}{([\frac{s}{\epsilon}]+1)!} \ ,$$

$$\text{whenever} \quad \epsilon([\frac{s}{\epsilon}]+1) \in (s-\delta, s+\delta) \ .$$

Of course, the last claim is true for all sufficiently small ϵ, and by (2.2) we get

$$\underline{\lim_{\epsilon\to 0}}\frac{\epsilon}{|\log\epsilon|}\log \mathbf{P}[\epsilon N(A) \in G] \ge -s$$

for any $s \in G$. Hence follows the lower bound (1.2) and the proof of Step 1 is completed.
Also note that the rate function does not depend on set A.

Step 2. *If $0 < \rho(A_l) < \infty$, and A_l's are pair-wise disjoint $l = 1, 2, ..., k$, then the probability measures $\mathbf{P}[\epsilon(N(A_1), ..., N(A_k)) \in \cdot]$ on $\mathbf{R}^k$, satisfy LDP with the speed $|\log\epsilon|/\epsilon$ and the rate function*

$$I((x_1, ..., x_k)) = \begin{cases} x_1 + x_2 + ... + x_k \ , & \text{if} \quad x_l \ge 0 \ , \ l = 1, 2, , ..., k \ ; \\ +\infty \ , & \textit{otherwise} \ . \end{cases}$$

P r o o f. It follows from Step 1 and Lemma 2.8 from [LS].

Theorem 1. *Let $(E, \mathcal{E}, \rho)$ be a finite measure space and $f : E \to \mathbf{R}^d$ be measurable and bounded function. Let $K_f :=$ conv(supp $\rho\circ f^{-1}$) be the convex hull spanned by the support of the measure $\rho\circ f^{-1}$ on $\mathbf{R}^d$, and*

$$q_{K_f}(x) := \inf\{c > 0 : \ c^{-1}x \in K_f\} \ , \quad \forall x \in \mathbf{R}^d \setminus \{0\} \ , \quad q_{K_f}(0) := 0 \ ,$$

be the Minkowski functional of the set K_f. *Then the measures* $\mathbf{P}[\epsilon N(f) \in \cdot]$ *satisfy* *LDP on* $\mathbf{R}^d$ *with the speed* $\lambda(\epsilon) = |\log \epsilon|/\epsilon$ *and the rate function* $I(x) = q_{K_f}(x)$.

P r o o f. Suppose first that f is a simple $\mathbf{R}^d$-valued function, i.e., for $e \in E$

$$f(e) = \sum_{j=1}^{m} x^j 1_{A_j}(e) \ , \quad x^j \in \mathbf{R}^d \ , \quad \rho(A_j) > 0 \ ,$$

and $A_j \in \mathcal{E}$ are pair-wise disjoint. From Step 2 and the contraction principle we infer that $\mathbf{P}[\epsilon N(f) \in \cdot]$ satisfy *LDP* on $\mathbf{R}^d$ with $\lambda(\epsilon)$ as before and the rate function

$$\begin{aligned} I_f(u) &:= \inf \left\{ y_1 + \ldots + y_m : \ y_i \geq 0 \, , \sum_{j=1}^{m} x^j y_j = u \right\} \\ &= \inf \left\{ \nu(E) : \ \nu \in \mathcal{M}^+(E) \ \text{ and } \ \int_E f d\nu = u \right\} \end{aligned}$$

which is equal to zero for $u = 0$, and for $u \neq 0$,

$$\begin{aligned} &= \inf \left\{ c > 0 : \ \text{ there is } \ \nu \in \mathcal{M}_1^+(E) \ \text{ and } \ \int_E f d\nu = \frac{u}{c} \right\} \\ &= \inf \left\{ c > 0 : \ c^{-1} u \in K_f \right\} \ , \end{aligned}$$

because for simple f we have

$$K_f = \left\{ \sum_{j=1}^{m} x^j \lambda_j : \ \sum_{j=1}^{m} \lambda_j = 1 \text{ and } \lambda_j \geq 0 \right\} = \left\{ \int_E f d\nu : \ \nu \in \mathcal{M}_1^+(E) \right\} \tag{2.4}$$

where $\mathcal{M}^+$ and $\mathcal{M}_1^+$ denote the sets of non-negative and probability measures, respectively.

For a general bounded and measurable function $f : \ E \to \mathbf{R}^d$, let us choose $f_n : \ E \to \mathbf{R}^d$, $n \geq 1$, simple, measurable such that $\|f - f_n\|_\infty := \sup_{s \in E} \|f(s) - f_n(s)\| \to 0$. Since for each $\delta > 0$

$$\begin{aligned} \overline{\lim_{\epsilon \to 0}} \frac{1}{\lambda(\epsilon)} \log \mathbf{P}[\epsilon \|N(f_n) - N(f)\| \geq \delta] &\leq \overline{\lim_{\epsilon \to 0}} \frac{1}{\lambda(\epsilon)} \log \mathbf{P}[\epsilon N(E) \|f_n - f\|_\infty \geq \delta] \\ &= - \frac{\delta}{\|f_n - f\|_\infty} \ , \end{aligned}$$

where the last equality follows from Step 1, we conclude that

$$\lim_{n \to \infty} \overline{\lim_{\epsilon \to 0}} \frac{1}{\lambda(\epsilon)} \log \mathbf{P}[\epsilon \|N(f_n) - N(f)\| \geq \delta] = -\infty \ .$$

And it is easy to see that $I_n := q_{\mathbf{K}_{f_n}}$ converges to $I := q_{\mathbf{K}_f}$ in the sense of (1.3). Now the comparison lemma completes the proof of Theorem 1.

Remark 1. The assumption in Theorem 1, that ρ is finite can be replaced by $\rho(f \neq 0) < \infty$.

Corollary 1. *Let E be locally compact metric, separable space and ρ a Radon measure such that $\rho(U) > 0$ for all open sets $U \neq \emptyset$. Let $M_R^+(E)$ be the space of nonnegative Radon measures on E, equipped with the vague convergence topology. Then $\mathbf{P}[\epsilon N \in \cdot]$ satisfy LDP on $\mathcal{M}_R^+(E)$ with the speed $\lambda(\epsilon) = |\log \epsilon|/\epsilon$, $\epsilon > 0$, and the rate function*

$$I(\nu) := \nu(E)\ , \quad \nu \in \mathcal{M}_R^+(E)\ .$$

P r o o f. Let $K_n \uparrow E$, $K_n \subset K_{n+1}$ are compact such that $0 < \rho(K_n) < \infty$. We know that the space $\mathcal{M}_R^+(E)$ with *vague* topology is the projective limit space of sequences $\mathcal{M}_R^+(K_n)$ with the *weak* topology. By Dawson-Gärtner (1987) we can and do assume that E is *compact*. For every $\nu \in \mathcal{M}_R^+(E)$ fixed, the sets

$$U(\nu, f, \delta) := \left\{\nu' \in \mathcal{M}_R^+(E) : \ |\int_E f d\nu - \int_E f d\nu'| < \delta\right\}\ ,$$

where $\delta > 0$, $f:\ E \to \mathbf{R}^d$ is continuous, $d \geq 1$, form a basis of neighbourhoods of ν. Since

$$\begin{aligned} q_{\mathbf{K}_f}(u) &= \inf\left\{\nu(E) :\ \nu \in \mathcal{M}^+(E) \ \text{ and } \ \int_E f d\nu = u\right\} \\ &= \inf\left\{I(\nu) :\ \nu \in \mathcal{M}^+(E) \ \text{ and } \ \int_E f d\nu = u\right\}\ , \end{aligned}$$

$\forall u \in \mathbf{R}^d$, by the proof of Theorem 1, we get by Theorem 1

$$\begin{aligned} \lim_{\epsilon \to 0} \begin{pmatrix}\inf \\ \sup\end{pmatrix} \frac{1}{\lambda(\epsilon)} \log \mathbf{P}[\epsilon N \in U(\nu, f, \delta)] &= \lim_{\epsilon \to 0} \begin{pmatrix}\inf \\ \sup\end{pmatrix} \frac{1}{\lambda(\epsilon)} \log \mathbf{P}[|\epsilon N(f) - \nu(f)| < \delta] \\ &\in [-\inf_{u:\ |u-\nu(f)|<\delta} q_{\mathbf{K}_f}(u), -\inf_{u:\ |u-\nu(f)|\leq\delta} q_{\mathbf{K}_f}(u)] \\ &= [-\inf_{\nu' \in U(\nu,f,\delta)} I(\nu')\ ,\ -\inf_{\nu' \in \overline{U(\nu,f,\delta)}} I(\nu')]\ . \end{aligned}$$

This implies the *weak LDP* by [DS, p. 46, (v)]. Furthermore, for any $L > 0$, $[I \leq L] =: K_L$ is compact in $\mathcal{M}_R^+(E)$ and by Step 1,

$$\limsup_{\epsilon \to 0} \frac{1}{\lambda(\epsilon)} \log \mathbf{P}(\epsilon N \notin K_L) = \limsup_{\epsilon \to 0} \frac{1}{\lambda(\epsilon)} \log \mathbf{P}(\epsilon N(E) > L) \leq -L\ .$$

This exponential tightness with the *weak LDP* shown before gives the desired *LDP*.

Remark 2. This corollary is one counterpart of the classical Schilder theorem about the Brownian Motion, and it complements a result in [GW], recalled below. In the setting of the Corollary 1, let N^ϵ be the Poisson point process with intensity measure $\epsilon\rho$. Then the laws of N^ϵ satisfy the LDP with the same rate function $I(\nu) = \nu(E)$, but with a different speed $\lambda(\epsilon) = |log\epsilon|$.

Theorem 2. *Assume that $(E, \mathcal{E}, \rho)$ is an infinite measure space, i.e., $\rho(E) = \infty$, and $f : E \to \mathbf{R}^d$ be measurable bounded and square integrable, i.e., $f \in L^2(\rho) \cap L^\infty$. Then for the compensated Poisson point process $\widetilde{N} = N - \rho(dx)$, we have that $\mathbf{P}[\epsilon\widetilde{N}(f) \in \cdot]$, $\epsilon > 0$, satisfy LDP on $\mathbf{R}^d$ with the speed $\lambda(\epsilon) = |\log \epsilon|/\epsilon$ and the rate $I(x) = q_{K_f}(x)$.*

P r o o f. Note that the integral $\widetilde{N}(f)$ is well defined for $f \in L^2(\rho) \cap L^\infty$, cf. [JS]. Taking $f_n := 1_{[\|f\| > n^{-1}]} f$, $n \geq 1$, we have $\rho(f_n \neq 0) < \infty$. Since $\epsilon\widetilde{N}(f_n) := \epsilon N(f_n) - \epsilon \int_E f_n d\rho$, $\epsilon > 0$, and the last term goes to zero as $\epsilon \to 0$, therefore by Theorem 1 and Remark 1, we conclude that $\mathbf{P}[\epsilon\widetilde{N}(f_n) \in \cdot]$ satisfy LDP on $\mathbf{R}^d$ with the rate $q_{K_{f_n}}(\cdot)$. But $q_{K_{f_n}}(x) \to q_{K_f}(x)$ uniformly over compact sets.

By the comparison Lemma, in order to complete the proof it is enough to show that

$$\lim_{n\to\infty} \overline{\lim_{\epsilon\to 0}} \frac{1}{\lambda(\epsilon)} \log \mathbf{P}[\epsilon\|\widetilde{N}(f_n) - \widetilde{N}(f)\| \geq \delta] = -\infty \tag{2.5}$$

for each $\delta > 0$. Since (for Euclidean norm)

$$\overline{\lim_{\epsilon\to 0}} \frac{1}{\lambda(\epsilon)} \log \mathbf{P}[\epsilon\|\widetilde{N}(f - f_n)\| \geq \delta] \leq \max_{1\leq i\leq d} \overline{\lim_{\epsilon\to 0}} \frac{1}{\lambda(\epsilon)} \log \mathbf{P}[\epsilon|\widetilde{N}(f^i - f^i_n)| \geq \delta]$$

we can and do assume below that f is a real-valued function.

Let $F := f - f_n = f1_{[|f| \leq n^{-1}]}$, so $\|F\|_\infty \leq n^{-1}$ and let

$$\wedge(t) := \overline{\lim_{\epsilon\to 0}} \frac{1}{\lambda(\epsilon)} \log \mathbf{E}[e^{\lambda(\epsilon)\epsilon t\widetilde{N}(F)}] , \quad t \in \mathbf{R} . \tag{2.6}$$

Then we have

$$\begin{aligned}
\wedge(t) &= \overline{\lim_{\epsilon\to 0}} \frac{1}{\lambda(\epsilon)} \int_{[|f| \leq n^{-1}]} [e^{\epsilon\lambda(\epsilon)tf} - 1 - \epsilon\lambda(\epsilon)tf]\, d\rho \\
&\leq \overline{\lim_{\epsilon\to 0}} \frac{1}{\lambda(\epsilon)} \int_{[|f| \leq n^{-1}]} e^{\epsilon\lambda(\epsilon)|tf|} \frac{1}{2} (\epsilon\lambda(\epsilon))^2 t^2 f^2 d\rho \\
&\leq \overline{\lim_{\epsilon\to 0}} [\frac{1}{2} \epsilon |\log\epsilon|\, t^2 \exp(|\log\epsilon|\, |t| n^{-1}) \int_E f^2 d\rho] \\
&= \frac{1}{2} t^2 \int_E f^2 d\rho\, \overline{\lim_{\epsilon\to 0}} |\log\epsilon| \epsilon^{1-|t|n^{-1}} = 0 , \quad \text{for} \quad |t| < n .
\end{aligned}$$

Hence for $t = n - 1$ we obtain

$$\begin{aligned}
&\overline{\lim_{\epsilon\to 0}} \frac{1}{\lambda(\epsilon)} \log \mathbf{P}[\epsilon\widetilde{N}(f - f_n) \geq \delta] \\
&= \overline{\lim_{\epsilon\to 0}} \frac{1}{\lambda(\epsilon)} \log \mathbf{P}[\exp(\lambda(\epsilon)t\epsilon\widetilde{N}(f - f_n) \geq e^{t\lambda(\epsilon)\delta}] \\
&\leq \overline{\lim_{\epsilon\to 0}} \frac{1}{\lambda(\epsilon)} \log(e^{-t\lambda(\epsilon)\delta} \mathbf{E}[e^{\lambda(\epsilon)t\epsilon\widetilde{N}(F)}]) \\
&= -\delta t + \wedge(t) = -\delta(n-1) .
\end{aligned}$$

Similarly taking $t = -(n-1)$, we obtain

$$\overline{\lim_{\epsilon\to 0}} \frac{1}{\lambda(\epsilon)} \log \mathbf{P}(\epsilon\widetilde{N}(f-f_n) \le -\delta) \le -\delta(n-1) .$$

These two estimations lead to

$$\lim_{n\to\infty} \overline{\lim_{\epsilon\to 0}} \frac{1}{\lambda(\epsilon)} \log \mathbf{P}[|\epsilon\widetilde{N}(f-f_n)| \ge \delta] = -\infty.$$

Thus we conclude (2.5) and the proof of Theorem 2 is complete.

Corollary 2. *Suppose $f:\ E \to \mathbf{R}^d$ belongs to $L^2(\rho)$ and for $\lambda(\epsilon) = |\log\epsilon|/\epsilon$, $\epsilon > 0$, one has*

$$\overline{\lim_{\epsilon\to 0}} \frac{1}{\lambda(\epsilon)} \log \mathbf{P}[\epsilon\|\widetilde{N}(f)\| \ge a] < 0 , \tag{2.7}$$

for some $a > 0$. Then $f \in L^\infty$.

P r o o f. Without loss of generality we assume that f is a real-valued function. Furthermore note that

$$-I(a) := \overline{\lim_{\epsilon\to 0}} \frac{1}{\lambda(\epsilon)} \log \mathbf{P}[\epsilon|\widetilde{N}(f)| \ge a] = -aI(1) ,$$

and therefore $I(a) > 0$ for some $a > 0$ is equivalent to $I(a) > 0$ for all $a > 0$. Let us assume that $f \notin L^\infty$ and choose $r > 2/I(1)$ such that $\rho(f_r := f1_{[r\le|f|\le r+1]} \neq 0) > 0$. Observe

$$\mathbf{P}[\epsilon|\widetilde{N}(f-f_r)| \le 1] > \frac{1}{2} , \quad \text{for all sufficiently small } \epsilon > 0 . \tag{2.8}$$

By the independence of $\widetilde{N}(f_r)$ and $\widetilde{N}(f-f_r)$, (2.8) implies that

$$\mathbf{P}[\epsilon|\widetilde{N}(f)| \ge 1] \ge \mathbf{P}[\epsilon|\widetilde{N}(f_r)| \ge 2] \cdot \mathbf{P}[\epsilon|\widetilde{N}(f-f_r)| \le 1] \ge 2^{-1}\mathbf{P}[\epsilon|\widetilde{N}(f_r)| \ge 2] .$$

Hence with Theorem 2 we get

$$-I(1) \ge \lim_{\epsilon\to 0} \frac{1}{\lambda(\epsilon)} \log \mathbf{P}[\epsilon|\widetilde{N}(f_r)| \ge 2] = -\inf_{|x|\ge 2} q_{K_{f_r}}(x) \ge -\frac{2}{r} ,$$

which contradicts the selection of r, and the proof is complete.

Corollary 3. *Let (E,ξ,ρ) be σ-finite measure space, a function $f \in L^2(E,\mathcal{E},\rho;\mathbf{R}^d)$ and $\widetilde{N}$ be the compensated Poisson point process. For $\lambda(\epsilon) := |\log\epsilon|/\epsilon$, $\epsilon > 0$, define*

$$a := -\overline{\lim_{\epsilon\to 0}} \frac{1}{\lambda(\epsilon)} \log \mathbf{P}[\epsilon\|\widetilde{N}(f)\| > 1]$$

$$b := \sup\left\{\alpha > 0 : \mathbf{E}[\exp(\alpha\|\widetilde{N}(f)\| \log(1+\|\widetilde{N}(f)\|))] < \infty\right\}.$$

Then $a = b = \|f\|_\infty^{-1}$.

Proof. For each $0 < \eta < a$ we have

$$\frac{1}{\lambda(\epsilon)} \log \mathbf{P}[\|\widetilde{N}(f)\| > \epsilon^{-1}] \le -\eta ,$$

for all sufficiently small ϵ, i.e., $\mathbf{P}[\|\widetilde{N}(f)\| > s] \le \exp(-\eta s \log s)$, for all sufficiently large s. Hence $\mathbf{E}[\exp(\eta - \delta)\widetilde{\mathbf{N}}(f)\log(1 + \|\widetilde{N}(f)\|)] < \infty$ for all $\delta > 0$ such that $\eta - \delta > 0$. Thus $\eta \le b$ and hence $a \le b$. One gets the converse inequality using Tschebyshev's inequality.

If $\|f\|_\infty < \infty$, then Theorem 2 gives $a = \|f\|_\infty^{-1}$. In fact, we have

$$a = \inf_{\|x\|>1} q_{K_f}(x) = \inf_{\|x\|\ge 1} q_{K_f}(x) = \|f\|_\infty^{-1} .$$

For $f \notin L_\infty$, Corollary 2 justifies $a = 0$. Thus the proof is completed.

3. Large deviations on R under exponential integrability. In this section we consider the case of $f \notin L^\infty(E, \mathcal{E}, \rho; \mathbf{R})$. Let us introduce parameters

$$\gamma^+ := \sup\left\{\alpha \ge 0 : \int_{[f\ge 1]} e^{\alpha f} d\rho < \infty\right\} \text{and } \gamma^- := \sup\left\{\alpha \ge 0 : \int_{[f\le -1]} e^{-\alpha f} d\rho < \infty\right\} .$$

Arguing as in the proof of Corollary 3 we infer that

$$\gamma^+ = -\overline{\lim}_{k\to\infty} \frac{1}{k} \log \rho(f \ge k) , \quad \text{and} \tag{3.1}$$

$$\gamma^- = -\overline{\lim}_{k\to\infty} \frac{1}{k} \log \rho(f \le -k) . \tag{3.2}$$

Theorem 3. *Assume that $f \in L^2(\rho)$, and $\gamma^+, \gamma^- > 0$. Then $\mathbf{P}[\epsilon\widetilde{N}(f) \in \cdot]$, $\epsilon > 0$, satisfy LDP on $\mathbf{R}$ with the speed $\lambda(\epsilon) = \epsilon^{-1}$ and the rate function*

$$I(x) = \begin{cases} \gamma^+ x , & \text{for} \quad x \ge 0 ; \\ -\gamma^- x , & \text{for} \quad x < 0 . \end{cases}$$

Proof. Let us write $f = f \cdot 1_{[|f|<1]} + f \cdot 1_{[f\ge 1]} + f \cdot 1_{[f\le -1]} =: f_0 + f_1 + f_{-1}$. From Theorem 1, for each $\delta > 0$

$$\lim_{\epsilon\to 0} \frac{\epsilon}{|\log \epsilon|} \log \mathbf{P}[\epsilon|\widetilde{N}(f_0)| > \delta] = -\frac{1}{\|f_0\|_\infty} ,$$

and hence

$$\lim_{\epsilon\to 0} \epsilon \log \mathbf{P}[\epsilon|\widetilde{N}(f_0)| > \delta] = -\infty . \tag{3.3}$$

Case 1. Now let us assume that $f \geq 1$ on E, i.e., $f = f_1$. For the upper bound (in *LDP*) note

$$\begin{aligned}\Lambda_1(t) &:= \lim_{\epsilon\to 0} \epsilon \log \mathbf{E}[\exp t\widetilde{N}(\epsilon f_1)/\epsilon] \\ &= \lim_{\epsilon\to 0} \epsilon \int_{[f\geq 1]} (e^{tf_1} - 1 - tf_1)d\rho \\ &= \lim_{\epsilon\to 0} \epsilon \int_{[f\geq 1]} e^{tf} d\rho = \begin{cases} 0\,, & \text{for} \quad t < \gamma^+ \\ +\infty, & \text{for} \quad t > \gamma^+ \,. \end{cases}\end{aligned}$$

Hence its Fenchel-Legendre transformation is given by

$$I_1(x) := \Lambda_1^*(x) = \sup_{t\in\mathbf{R}}[xt - \Lambda_1(t)] = \sup_{t<\gamma^+} xt = \begin{cases} \gamma^+ x\,, & \text{for} \quad x \geq 0\,, \\ +\infty\,, & \text{for} \quad x < 0\,. \end{cases} \tag{3.4}$$

By Ellis-Gärtner Theorem ([DS], Thm.2.2.4), the upper bound of large deviations holds for $\mathbf{P}[\epsilon\widetilde{N}(f_1) \in \cdot]$, $\epsilon > 0$, on $\mathbf{R}$ with the speed $\lambda(\epsilon) = \epsilon^{-1}$ and with the rate function $I_1(x)$.

For the lower bound observe that $f_1 \in L^1(\rho)$ and instead of $\widetilde{N}(f_1)$ we can consider $N(f_1)$. Furthermore, using inequality $N(f_1) \geq tN(f_1 > t)$ we get, for $a > 0$ and $t \geq 1$,

$$\begin{aligned}\mathbf{P}[\epsilon N(f_1) > a] &\geq \mathbf{P}[t\epsilon N(f_1 > t) > a] \\ &\geq \sum_{k>a/\epsilon t} e^{-\rho(f_1>t)}(\rho(f_1 > t))^k/k! \\ &\geq e^{-\rho(f_1>t)}(\rho(f_1 > t))^{[\frac{a}{\epsilon t}]+1}/([\frac{a}{\epsilon t}] + 1)!\end{aligned}$$

Choose now $t = t(\epsilon)$, a positive function of ϵ verifying $\lim_{\epsilon\to 0} \epsilon t(\epsilon) = +\infty$ and

$$\gamma^+ = -\lim_{\epsilon\to 0} \frac{1}{t(\epsilon)} \log \rho[f > t(\epsilon)]$$

(possible by (3.1)!). Hence using the above inequality and (2.2), we get

$$\underline{\lim}_{\epsilon\to 0}\, \epsilon \log \mathbf{P}[\epsilon N(f_1) > a] \geq \lim_{\epsilon\to 0} \epsilon([\frac{a}{\epsilon t(\epsilon)}] + 1) \log \rho[f_1 > t(\epsilon)] = -\gamma^+ a\,.$$

Since we used only one term estimate we infer the lower bound for $\mathbf{P}[\epsilon N(f_1) \in (a, b)]$ as well. In other words, the measures $\mathbf{P}[\epsilon\widetilde{N}(f_1) \in \cdot]$, $\epsilon > 0$, satisfy *LDP* with the speed $\lambda(\epsilon) = \epsilon^{-1}$ and the rate function $I_1(x)$.

Case 2. Applying Case 1 for $-f$ and observing that $(-f)_1 = -f_{-1}$, $\widetilde{N}(f_{-1}) = -\widetilde{N}((-f)_1)$ we conclude *LDP* for $\mathbf{P}[\epsilon N(f_{-1}) \in \cdot]$, $\epsilon > 0$, with the speed $\lambda(\epsilon) = \epsilon^{-1}$, $\epsilon > 0$, and the rate function

$$I_{-1}(x) = \begin{cases} +\infty\,, & \text{for} \quad x > 0\,; \\ \gamma^-|x|\,, & \text{for} \quad x \leq 0\,. \end{cases} \tag{3.5}$$

Finally since $\widetilde{N}(f) = \widetilde{N}(f_0) + \widetilde{N}(f_1) + \widetilde{N}(f_{-1})$ is a sum of independent variables, (3.3),(3.4) and (3.5) imply the LDP in Theorem 3 by [LS, Lemma 2.8].

4. Applications to the class L distributions. For the basic information on the class L (or selfdecomposable) distributions cf. [JM] p. 177-182. For the purpose of this application, let us recall that

$$(4.1) \qquad \mu \in L \ \text{ iff } \ \mu \stackrel{d}{=} Z(0) := \int_{(0,\infty)} e^{-s} dY(s) \ , \quad \mathbf{E} \log(1 + \|Y(1)\|) < \infty \ ,$$

and Y is a Lévy process. Of course,

$$(4.2) \qquad \mathbf{P}[Z(t) = \int_t^\infty e^{-s} dY(s) \in \cdot] \to \delta_0(\cdot) \ , \text{as} \ \ t \to \infty \ ,$$

and it is "natural" to ask for LDP for probability distributions in (4.2).

Let Y be without Gaussian component and shift, i.e., $Y(1) \stackrel{d}{=} [0, 0, M]$ (these are the parameters in the Lévy-Khintchine formula of $Y(1)$; M is the spectral Lévy measure). Then

$$(4.3) \qquad Y(t) = \int_0^t \int_{\mathbf{R}^d \setminus \{0\}} x \widetilde{N}(dx, ds) \ , \quad t \geq 0 \ ,$$

where $\widetilde{N}$ is a Poisson point process with the compensator $\rho(dt, dx) = dt \times dM$ on $E := [0, \infty) \times \mathbf{R}^d \setminus \{0\}$ (dt= Lebesque measure); see [JS]. Thus

$$(4.4) \qquad Z(0) = \int_E e^{-t} x \widetilde{N}(dx, dt) = \widetilde{N}(f_0) \ , \quad \text{for} \quad f_0(t, x) := e^{-t} x \ .$$

¿From (4.1) we also have that

$$e^{-t} Z(0) = \int_0^\infty e^{-(t+s)} dY(s) = \int_t^\infty e^{-u} dY(u - t) \stackrel{d}{=} \int_t^\infty e^{-u} dY(u) = Z(t) \ .$$

Hence (4.2) with (4.4) is equivalent to

$$\mathbf{P}[e^{-t} \widetilde{N}(f_0) \in \cdot] \to \delta_0 \ , \quad \text{as} \quad t \to \infty \ .$$

All the above we can summarize in the following

Theorem 4. *Let* $Y(\cdot)$ *be a real Lévy process with* $Y(1) \stackrel{d}{=} [0, 0, M]$. *Assume that* $suppM$ *is compact. Then* $\mathbf{P}[\int_{(t,\infty)} e^{-s} dY(s) \in \cdot]$ *satisfy* LDP *on* $\mathbf{R}$, *as* $t \to \infty$ *with the speed* $\lambda(t) := te^t$ *and the rate function*

$$(4.5) \qquad I_M(x) := \begin{cases} x/b \ , & for \quad 0 < b := sup(supp\, M) \ , \quad x > 0 \ ; \\ x/a \ , & for \quad inf(supp\, M) =: a < 0 \ , x < 0 \ ; \\ +\infty \ , & otherwise \ . \end{cases}$$

P r o o f. Note that f_0 defined in (4.4) belongs to $L^\infty(\rho)$ iff $\operatorname{supp} M$ is bounded in $\mathbf{R}$. In this case $f_0 \in L^2(\rho)$ as well. Since $\operatorname{supp}(\rho \circ f_0^{-1}) = f_0(\operatorname{supp} \rho) = [a, b]$ where $a = \inf(\operatorname{supp} M)$, $b = \sup(\operatorname{supp} M)$, Theorem 2 gives the conclusion of Theorem 4.

Remark 3. If $Y(1) \stackrel{d}{=} [0, \sigma^2, 0]$, i.e., $Y(t)$ is a Brownian Motion, then $Z(t) \stackrel{d}{=} \frac{1}{\sqrt{2}} e^{-t} Y(1)$. In other words, $\sqrt{2e^{2t}}Z \stackrel{d}{=} [0, \sigma^2, 0] = N(0, \sigma^2)$. By an easy calculation, $\mathbf{P}[\int_{(t,\infty)} e^{-s} dY(s) \in \cdot]$ satisfy LDP on $\mathbf{R}$, as $t \to \infty$, with the speed $\lambda(t) = 2e^{2t}$ and the rate $I(x) = x^2/2\sigma^2$, $x \in \mathbf{R}$ (well known!).

Corollary 4. *If* $Y(1) \stackrel{d}{=} [0, \sigma^2, M]$ *and* $\operatorname{supp} M$ *is compact then*

$$\mathbf{P}[Z(t) = \int_{(t,\infty)} e^{-s} dY(s) \in \cdot]$$

satisfy LDP *(on* $\mathbf{R}$*) with speed* $\lambda(t) := te^t$ *and the rate function* $I_M(x)$ *(i.e., Gaussian part does not contribute to the rate function).*

P r o o f. Let us write $Y(t) = Y^1(t) + Y^2(t)$, where Y^1 and Y^2 are independent Lévy processes such that $Y^1(1) \stackrel{d}{=} [0, \sigma^2, 0]$ and $Y^2(1) \stackrel{d}{=} [0, 0, M]$. Defining

$$Z^i(t) := \int_{(t,\infty)} e^{-s} dY^i(s) \ , \quad i = 1, 2$$

one has $Z(t) = Z^1(t) + Z^2(t)$ (with two independent summands). By Remark 3, $\mathbf{P}(Z^1(t) \in \cdot]$ satisfy LDP with the speed $2e^{2t} >> te^t$ (as $t \to \infty$), it is then negligible for the large deviations with speed te^t. Consequently, $\mathbf{P}[Z(t) \in \cdot]$ satisfy the same LDP as $\mathbf{P}[Z^2(t) \in \cdot]$. Then the Corollary follows from Theorem 4.

Since in the theory of large deviations often one needs the existence of exponential moments we complete this section with the following facts about class L distributions (on Banach spaces).

Lemma. *Let* $Y(1) \stackrel{d}{=} [a, R, M]$, Y *be a Banach space valued Lévy process. Then for any* $\lambda > 0$

$$\text{(4.6)} \qquad \mathbf{E}[\exp \lambda \| \int_0^\infty e^{-s} dY(s) \|] < \infty \quad \textit{iff} \quad \int_{[\|x\| > a]} \|u\|^{-1} e^{\lambda \|u\|} dM(u) < \infty$$

for all $a > 0$. *In particular, it is so whenever one has* $\mathbf{E}[\exp \lambda \|Y(1)\|] < \infty$.

P r o o f. For an infinitely divisible measure $\nu = [b, S, K]$ (on a Banach space B) and submultiplicative (or subadditive) functions $\Phi : \ E \to [0, \infty)$

$$\int_B \Phi(\|x\|) \nu(dx) < \infty \quad \text{iff} \quad \int_{\|x\| > a} \Phi(\|x\|) K(dx) < \infty \ , \quad \text{for all} \quad a > 0 \ .$$

(see: [JM], p. 36). If M is the Lévy spectral measure of $Y(1)$, the integral $\int_{(0,\infty)} e^{-s}dY(s)$ (class L distribution) has the Lévy spectral measure $\overline{M}$ given by

$$\overline{M}(A) := \int_0^\infty M(e^t A)dt\ , \quad A \text{ is a Borel subset (in } B) \ .$$

(cf. [JM] p.120). Hence

$$\begin{aligned}\int_{[\|x\|\geq a]} e^{\lambda\|x\|}d\overline{M}(x) &= \int_B\int_0^\infty 1_{[\|x\|\geq a]}(e^{-t}x)\exp(\lambda e^{-t}\|x\|)dtM(dx)\\ &= \int_{[\|x\|\geq a]}\Big(\int_0^{ln(\|u\|/a)} \exp(\lambda e^{-t}\|u\|)dtM(du)\\ &= \int_{[\|x\|\geq a]}\Big(\int_a^{\|u\|} (e^{\lambda s}/s)ds\Big)M(du)\ .\end{aligned}$$

Since $\int_a^s e^{\lambda y}/ydy \sim e^{\lambda s}/s$, as $s \to \infty$, and M is finite on $[\|x\| > a]$ we conclude the proof of Lemma.

Corollary 5. *For $Y(1) \stackrel{d}{=} [0, \sigma^2, M]$ on $\mathbf{R}$ let us assume that the limits*

$$\gamma^+ := -\overline{\lim}_{a\to+\infty}a^{-1}\log M(x > a)\ , \quad \gamma^- := -\overline{\lim}_{a\to+\infty}a^{-1}\log M(x < -a)$$

are finite and strictly positive. Then $\mathbf{P}[Z(t) = \int_{(t,\infty)} e^{-s}dY(s) \in \cdot]$ satisfy LDP (on $\mathbf{R}$) with the speed $\lambda(t) := e^t$ and the rate function $I(x) := \gamma^+x$, for $x \geq 0$ and $I(x) := -\gamma^-x$, for $x < 0$.

P r o o f. As in the proof of Corollary 4 (or Theorem 4) we can assume that $\sigma^2 = 0$ and

$$Z(t) \stackrel{d}{=} e^{-t}Z(0) = e^{-t}\int_0^\infty\int_{\mathbf{R}^*} e^{-s}x\widetilde{N}(ds, dx)\ , \quad t \geq 0$$

where $\rho(ds, dx) := ds \times M(dx)$ is the compensator. Observe for $f(s, x) := e^{-s}x$,

$$\begin{aligned}\overline{\lim}_{a\to+\infty}\ a^{-1}\log\rho((s,x):\ e^{-s}x > a) &= \sup\left\{\lambda > 0 :\ \int_{[f\geq 1]} e^{\lambda f}d\rho < +\infty\right\}\\ &= \sup\left\{\lambda > 0 :\ \int_{[x>1]} e^{\lambda x}/xM(dx) < \infty\right\}\\ &= \sup\left\{\lambda > 0 :\ \int_{[x>1]} e^{\lambda x}M(dx) < \infty\right\}\\ &= \overline{\lim}_{a\to+\infty}\ a^{-1}\log M(x > a) = -\gamma^+\ .\end{aligned}$$

Consequently we proved that

$$\text{(4.7)} \qquad \overline{\lim}_{a\to\infty}a^{-1}\log\rho((s,x):\ e^{-s}x > a) = -\gamma^+\ ,$$

and by similar arguments we also have

(4.8) $$\overline{\lim}_{a\to+\infty} a^{-1} \log \rho((s,x) : e^{-s}x < -a) = -\gamma^- .$$

Now applying Theorem 3 we conclude the *LDP* described in Corollary 5.

Remark 4. The two main results of this paper, Theorem 2 and 3, show that the behavior of the tail probability of $\widetilde{N}(f)$ (an element in the first chaos of the Poisson point process N), depends strongly on that of f. This is essentially different from the Brownian Motion case. A further interesting question is to investigate the large deviations of multiple random integrals (or element in the chaos of order ≥ 2), similarly to the work of Ledoux [L] on the Wiener space.

Acknowledgement: We are grateful to a referee for his careful reading and suggested improvements in the first version. The first author was supported, in part, by KBN grant, 1995-1997, Warsaw. The second author was partially supported by the NSF of China and Y.D.Fok's Foundation. And we both benefited from the cooperation program between Université Blaise Pascal and the University of Wroclaw.

References

[DG] D.W. Dawson and J. Gärtner (1987), Long time fluctuation of weakly interacting diffusions. *Stochastics* 20, pp. 247-308.

[DS] J.D. Deuschel and D.W. Stroock (1989), *Large deviations*, Academic Press, New York.

[GW] M.Z. Guo and L. Wu (1995), Several large deviation estimations for the Poisson point processes, Advances in Math., Beijing, Vol. 24, 4, 313-319.

[JS] J. Jacod and A.N. Shiryaev (1987), *Limit theorems for stochastic processes*, Springer-Verlag, Berlin & New York.

[JM] Z.J. Jurek and J.D. Mason (1993), *Operator-limit distributions in probability theory*, J. Wiley, New York.

[L] M. Ledoux (1990) A note on large deviations for Wiener chaos, *Séminaire de Proba. XXIV, LNM 1426, p1-14.*

[LS] J. Lynch and J. Sethuraman (1987), Large deviations for processes with independent increments, *Ann. Probab.* 15, pp. 610-627.

Institute of Mathematics
University of Wroclaw
Pl. Grunwaldzki 2/4
50-384 WROCLAW, Poland
[zjjurek@math.uni.wroc.pl]

Laboratoire de Mathématiques Appliquées
CNRS-UMR 6620
Université Blaise Pascal
63177 AUBIERE Cedex, France
[wuliming@ucfma.univ-bpclermont.fr]
and (in quittance of)
Department of Mathematics
Wuhan University, 430072-HUBEI, China

Formes de Dirichlet sur un Espace de Wiener-Poisson. Application au grossissement de filtration

Laurent DENIS,[1] Axel GRORUD,[2] Monique PONTIER[3]

résumé

Il existe une construction classique d'une structure de Dirichlet sur un espace de Wiener. On construit ici une structure analogue sur un espace de Poisson et sur l'espace produit Wiener-Poisson. Cette construction permet de donner une condition simple sur une variable terminale qui rend possible le grossissement initial de la filtration naturelle par cette variable. On analyse la stratégie financière optimale d'un agent qui a une information anticipant le marché en grossissant la filtration engendrée par les prix.

abstract

A Dirichlet structure is built on a Wiener-Poisson space. A simple condition is given on a terminal random variable such that the initial enlargement of the natural filtration can be done. We study the optimal financial strategy of an insider trader enlarging the filtration generated by the assets prices with his anticipating information.

1 Introduction

La motivation initiale de ce travail a été la modélisation du délit d'initié dans un marché dont les prix comportent des sauts, en prolongement d'un travail précédent [10]. Sur un espace de probabilité filtré $(\Omega, (\mathcal{F}_t, t \in [0,T]), \mathbb{P})$, la dynamique des prix est régie par un mouvement brownien W de dimension m et un processus ponctuel N de dimension n ayant la propriété de représentation prévisible :

$$S_t^i = S_0^i + \int_0^t S_s^i(b_s^i ds + \sigma_s^i dW_s) + \int_0^t S_{s-}^i \int_O \phi^i(x,s) N(dx,ds), 0 \leq t \leq T, i = 1, \cdots, d. \tag{1}$$

On se place selon le point de vue d'un investisseur "initié" : il connait des informations sur le futur, représentées par une variable aléatoire $L \in L^1(\Omega, \mathcal{F}_T, \mathbb{R}^\kappa)$, (par exemple, des échanges auront lieu et il sait à quelle date). On note $\mathcal{Y}$ la filtration "naturelle" de l'initié régularisée à droite : $\mathcal{Y}_t = \cap_{s>t}(\mathcal{F}_s \vee \sigma(L)), t \in [0,T]$.

[1]L. D. : U.M.R. CNRS 6633, Département de Mathématiques Université du Maine, Avenue Olivier Messiaen,
BP 535 , 72017 LE MANS cedex. e-mail ldenis@univ-lemans.fr

[2]A. G. : L.A.T.P.,Université de Provence, Projet OMEGA INRIA, 39 rue Joliot-Curie, 13453 MARSEILLE cedex 13 ; e-mail : axel@gyptis.univ-mrs.fr.

[3]M.P. : L.S.P, U.M.R. CNRS C 5583 Université Paul Sabatier, 118 route de Narbonne, 31 062 TOULOUSE cedex 04 ; e-mail : pontier@cict.fr.

Le problème qui se pose alors est que W et N ne sont plus nécessairement des semi-martingales pour la nouvelle filtration. La méthode de grossissement initial d'une filtration permet de trouver les conditions sur L pour que $W_t = B_t + A_t$ où B est un $\mathcal{Y}-$brownien et A un $\mathcal{Y}-$processus croissant et pour que N_t admette un compensateur $\mathcal{Y}$-prévisible. Dans le cas continu, Marc Yor [23], Mireille Chaleyat-Maurel et Thierry Jeulin [6] ont traité le cas où L est une variable gaussienne. Toujours sur un espace de Wiener, Thierry Jeulin [17] et Marc Yor [24] donnent explicitement le mouvement brownien de la filtration grossie dans le cas où la variable aléatoire L est un temps d'atteinte ou le temps local du brownien initial. Plus généralement, Jean Jacod [14] et Shiqi Song [22] ont résolu le problème si la famille des lois conditionnelles $Q_t(\omega, .)$ de L sachant $\mathcal{F}_t$ est dominée presque sûrement par une mesure non-aléatoire.

Ce travail donne la construction d'une structure de Dirichlet sur l'espace de Wiener-Poisson. On utilise les résultats de Bouleau-Hirsch [3] pour obtenir une structure de Dirichlet conditionnelle ce qui permet alors de donner une condition simple sur L (l'hypothèse $\boldsymbol{H_C}$, section 4.2) pour obtenir la continuité absolue de ces lois conditionnelles, y compris dans le cas de variables aléatoires L vectorielles, ce qui prolonge au cas vectoriel les travaux de [3] et [4] dans une structure plus simple. Nous appliquons cette construction aux marchés financiers, ce qui prolonge aux espaces de Wiener-Poisson les résultats obtenus dans [10].

Après quelques rappels, notations (section 2), dans la section 3 on définit une structure de Dirichlet sur l'espace $(\Omega, \mathcal{F}, \mathbb{P})$ d'abord par un rappel rapide sur la partie brownienne puis, sur l'espace de Poisson on construit une structure de Dirichlet par une succession de produits indépendants (cf. [4]) ; enfin on effectue le produit de ces deux structures. Puis, dans la section 4, on utilise cette construction et la propriété d'indépendance des accroissements de W et N pour obtenir une structure de Dirichlet conditionnelle et étudier la loi conditionnelle de L sachant $\mathcal{F}_t$. L'hypothèse $\boldsymbol{H_C}$ permet d'appliquer les résultats de J. Jacod sur le grossissement initial de filtration et de construire un $(\mathcal{Y}, \mathbb{P})$-mouvement brownien et la $(\mathcal{Y}, \mathbb{P})$-intensité du processus N. Enfin, la section 5 donne un théorème de représentation des $(\mathcal{Y}, Q)$-martingales pour une probabilité Q équivalente à $\mathbb{P}$ telle que W est un $(\mathcal{Y}, Q)$-mouvement brownien et la $(\mathcal{Y}, Q)$-intensité du processus N est la $(\mathcal{F}, \mathbb{P})$-intensité du processus N, ainsi qu'un théorème d'existence de probabilités neutres au risque pour le marché défini en (1). On en déduit une caractérisation des probabilités neutres au risque équivalentes à la probabilité Q, ce qui étend aussi des résultats connus pour ce modèle de marché financier même sans investisseur informé (cf. [1]).

2 Notations et définitions

On considère W un mouvement brownien standard de dimension m défini sur son espace canonique $(\Omega^W, \mathcal{F}^W, (\mathcal{F}_t^W, t \in [0;T]), \mathbb{P}^W)$, où $\Omega^W = C([0,T]; \mathbb{R}^m)$ et la filtration est relative à W.

On note $(\Omega^N, \mathcal{F}^N, \mathbb{P}^N)$ un espace de probabilité qui sera construit dans la section 3 en même temps qu'un processus ponctuel marqué $(Z_n, T_n), n \in \mathbb{N}$, (cf [5]) noté N, d'intensité $\nu(x)dxds$, de compensateur $\tilde{N}(dx, ds) = N(dx, ds) - \nu(x)dxds$, où $x \in O$ ouvert de $\mathbb{R}^n$. On suppose que ν est strictement positive sur O et de classe $\mathcal{C}^1$, que $\nu(O) = +\infty$ et qu'il existe une suite d'ouverts disjoints $O_i, \cup_i O_i = O$, tels que pour tout $i \in \mathbb{N}$, $\nu(O_i) < \infty$.

On se place désormais sur l'espace de probabilité filtré produit :

$$(\Omega, \mathcal{F}, (\mathcal{F}_t, t \in [0;T]), \mathbb{P}) = (\Omega^W \times \Omega^N, \mathcal{F}^W \otimes \mathcal{F}^N, \mathcal{F}_t^W \otimes \mathcal{F}_t^N, t \in [0;T]), \mathbb{P}^W \otimes \mathbb{P}^N).$$

3 Structures de Dirichlet

3.1 Structure de Dirichlet sur l'espace de Wiener

Sur l'espace de Wiener Ω^W, on définit la structure de Dirichlet $(\Omega^W, \mathcal{F}^W, \mathbb{P}^W, \mathbb{D}^W, \mathcal{E}^W)$ de la façon suivante :

Pour $h \in \Omega^W$ tel que $\dot{h}$ est dans $L^2([0;T], \mathbb{R}^m)$ on note $w(h) = \int_0^T \dot{h}(s)dW_s$.
$\mathcal{S}$ désigne l'ensemble des fonctionnelles de Wiener à valeurs réelles simples (cf. P.Malliavin [19] ou D.Nualart [20]) :

$$\begin{aligned} \mathcal{S} = \{F &\in L^2(\Omega)/\exists n \in \mathbb{N}, f \in C_b^\infty(\mathbb{R}^n), \text{tels que :} \\ F &= f(w(h_1), \cdots, w(h_n)), \text{ avec } \dot{h}_1, \cdots, \dot{h}_n \in L^2([0;T]; \mathbb{R}^m)\} \end{aligned}$$

Pour $F \in \mathcal{S}$ on définit $D^W F \in L^2(\Omega \times [0;T]; \mathbb{R}^m)$ par

$$D_t^W F = \sum_{i=1}^{i=n} \frac{\partial f}{\partial x_i}(w(h_1), \cdots, w(h_n))\dot{h}_i(t).$$

D^W est le gradient stochastique usuel associé à W. On note $\mathbb{D}_T^W$ l'espace de Sobolev construit à l'aide de D^W des fonctionnelles sur Ω^W. Alors l'opérateur carré du champ et la forme de Dirichlet sont définis sur $\mathbb{D}_T^W$ respectivement par :

$$\Gamma_T^W(F) = \|D_.^W F\|^2 \ ; \ \mathcal{E}_T^W(F) = \frac{1}{2}E[\|D_.^W F\|^2],$$

où $\|.\|$ désigne la norme dans $L^2([0;T]; \mathbb{R}^m)$.
On dispose d'un critère classique d'absolue continuité ([3]) :

Proposition 3.1 *Soit $F \in (\mathbb{D}_T^W)^\kappa$. Alors $F * (\det(\Gamma_T^W(F)) \cdot P^W)$, la mesure image de $det(\Gamma_T^W(F)) \cdot P^W$ par F, est absolument continue par rapport à la mesure de Lebesgue.*

Preuve: Le corollaire II 5.2.3. de [3] donne le résultat pour toute forme de Dirichlet $(\mathbb{D}, \mathcal{E})$ dont l'opérateur carré du champ est construit à partir de dérivées directionnelles. □

3.2 Structure de Dirichlet sur l'espace de Poisson

3.2.1 Construction de l'espace

On désigne par dx la mesure de Lebesgue sur $\mathbb{R}^n$ et par λ la mesure $\nu(x)dx$ sur O. Dans un premier temps, on supposera que $\lambda(O) < +\infty$.
On construit une mesure de Poisson à valeurs dans $\mathbb{R}^+ \times O$ et d'intensité $\nu(x)dtdx$:

1. Soit $(\Omega', \mathcal{F}', P')$ un espace de probabilité sur lequel est défini un processus de Poisson de paramètre $\lambda(O)$ que l'on note M. La suite des instants de sauts de ce processus sera notée par $(T_i)_{i \in \mathbb{N}}$, on a ainsi :

$$\forall t \geq 0, \ M_t = \sum_{i=1}^{+\infty} 1_{\{T_i \leq t\}}.$$

2. Soit $(O, \mathcal{B}, \lambda/\lambda(O))$ l'espace de probabilité où $\mathcal{B}$ désigne la tribu borélienne sur O et on note
$$(U, \mathcal{G}, Q) = (O, \mathcal{B}, \lambda/\lambda(O))^{\otimes \mathbb{N}}.$$
On notera $(Z_k)_{k\in\mathbb{N}}$ la suite des applications coordonnées sur U, il est clair que les variables aléatoires $(Z_k)_{k\in\mathbb{N}}$, à valeurs dans O, sont indépendantes et de loi commune $\lambda/\lambda(O)$.

3. On pose $(\Omega^N, \mathcal{F}^N, P^N) = (\Omega' \times U, \mathcal{F}' \otimes \mathcal{G}, P' \otimes Q)$, et on définit sur cet espace produit la mesure aléatoire à valeurs dans $\mathbb{R}^+ \times O$ par :
$$\forall w = (w', x) \in \Omega^N,\ N(w) = \sum_{i=1}^{+\infty} \delta_{(T_i(w'), Z_i(x))}.$$

On reconnait alors que N est une mesure de Poisson à valeurs dans $\mathbb{R}^+ \times O$ et d'intensité $\nu(x)dxdt$ (Cf. [12]). Enfin, $(\mathcal{F}_t^N)_{t\geq 0}$ désignera la filtration naturelle de N. On retrouve ainsi le processus ponctuel marqué introduit en 2 dans le cas $\lambda(O) < +\infty$.

3.2.2 Structure de Dirichlet associée

On définit sur $C_b^\infty(O)$ la forme bilinéaire symétrique par :
$$\forall f, g \in C_b^\infty(O),\ e(f,g) = \frac{1}{\lambda(O)} \int_O (\sum_{i=1}^n \frac{\partial f}{\partial x_i}(x) \cdot \frac{\partial g}{\partial x_i}(x))\, \nu(x)dx.$$

Il est facile (Cf. [9]) de vérifier que la forme (C_b^∞, e) est fermable dans $L^2(O, \mathcal{B}, \lambda/\lambda(O))$. On note $(H(O), e)$ sa fermeture, c'est une structure de Dirichlet sur $L^2(O, \mathcal{B}, \lambda/\lambda(O))$ et elle admet pour opérateur carré du champ la forme bilinéaire :
$$\forall f, g \in H(O)^2,\ \gamma(f,g) = \sum_{i=1}^n \frac{\partial f}{\partial x_i} \cdot \frac{\partial g}{\partial x_i}.$$

Remarquons enfin que :
$$H(O) = \{f \in L^2(O, \mathcal{B}, \lambda/\lambda(O));\ \forall i \in \{1, \cdots, n\} \frac{\partial f}{\partial x_i} \in L^2(O, \mathcal{B}, \lambda/\lambda(O))\}.$$

En suivant Bouleau-Hirsch ([3]) on définit une structure de Dirichlet sur $L^2(\Omega^N, \mathcal{F}^N, P^N)$ comme étant la structure produit de $L^2(\Omega', \mathcal{F}', P')$ (structure élémentaire) et de la structure $(H(O), e)^{\otimes \mathbb{N}}$ (structure produit infini). Il s'agit donc d'une structure de Dirichlet sur $L^2(\Omega', \mathcal{F}', P') \otimes L^2(U, \mathcal{G}, Q)$ que l'on assimile de façon naturelle à $L^2(\Omega^N, \mathcal{F}^N, P^N)$. On notera $(\mathbb{D}^N, \mathcal{E}^N)$ cette structure. Comme simple conséquence des résultats dus à Bouleau-Hirsch, on a :

Proposition 3.2 $(\mathbb{D}^N, \mathcal{E}^N)$ *est une forme de Dirichlet locale telle que* $1 \in \mathbb{D}$ *et* $\mathcal{E}(1) = 0$.

De plus, elle admet un opérateur carré du champ que l'on note Γ^N. □

De fait le problème est situé dans un modèle d'horizon fini : l'espace $\mathbb{D}^N$ est alors "trop gros", c'est pour cela qu'on préfère le restreindre aux fonctionnelles des trajectoires en temps fini.

Définition 3.3 : *Pour tout $t > 0$, on note $\mathbb{D}_t^N$ l'ensemble des éléments de $\mathbb{D}^N$ qui sont $\mathcal{F}_t$ mesurables.*

On a alors de façon triviale :

Proposition 3.4 *$(\mathbb{D}_t^N, \mathcal{E}^N)$ est un forme de Dirichlet sur $L^2(\Omega^N, \mathcal{F}_t^N, P^N)$, locale et qui admet pour carré du champ la restriction de Γ^N à $\mathbb{D}_t^N$, notée Γ_t^N.* □

Mais comme on sait expliciter (voir [3]) les éléments d'une structure produit, on sait ici aussi expliciter les éléments de $\mathbb{D}_t^N$, ce qui donne :

Proposition 3.5 *Soit $F \in L^2(\Omega^N, \mathcal{F}_t^N, P^N)$, alors $F \in \mathbb{D}_t^N$ si et seulement si F peut s'écrire :*

$$F = a \cdot \mathbf{1}_{\{M_t=0\}} + \sum_{i=1}^{+\infty} f_i(T_1, \cdots, T_i, Z_1, \cdots, Z_i)\mathbf{1}_{\{M_t=i\}},$$

où $a \in \mathbb{R}$ et

1. *pour tout $i \in \mathbb{N}$, f_i est telle que pour P'-presque tout $w \in \Omega'$, pour tout $j \in \{1, \cdots, i\}$ et pour presque tout $(x_1, \cdots, x_{j-1}, x_{j+1}, \cdots, x_i)$ dans O^{i-1}, l'application*
$$x \in O \longrightarrow f_i(T_1, \cdots, T_i, x_1, \cdots, x_{j-1}, x, x_{j+1}, \cdots x_i)$$
appartient à $H(O)$,

2. $\sum_{i=1}^{+\infty} E(\sum_{j=1}^{i} \| \nabla_{i+j} f_i(T_1, \cdots, T_i, Z_1, \cdots, Z_i) \|^2)\mathbf{1}_{\{M_t=i\}}) < +\infty.$

De plus, on a : $\Gamma_t^N(F) = \sum_{i=1}^{+\infty}(\sum_{j=1}^{i} \| \nabla_{i+j} f_i(T_1, \cdots, T_i, Z_1, \cdots, Z_i) \|^2)\mathbf{1}_{\{M_t=i\}}$.

Remarque 3.6 : *1- Ici, $\nabla_{i+j} f_i$ désigne le gradient de f_i pris par rapport à la variable n-dimensionnelle x_j ; la norme $\|.\|$ désigne la norme vectorielle.*

2- Les gradients se font toujours par rapport aux amplitudes des sauts (et non par rapport aux instants de saut).

3- Lorsque que ce n'est pas précisé, "presque tout" se rapporte à la mesure de Lebesgue sur l'espace considéré.

4- Enfin, on retrouve ici le même opérateur que celui introduit dans le chapitre 9 de [2] pour un processus de Poisson d'intensité la mesure de Lebesgue sur un ouvert.

Preuve : D'une part, si $F = a \cdot \mathbf{1}_{\{M_t=0\}} + \sum_{i=1}^{+\infty} f_i(T_1, \cdots, T_i, Z_1, \cdots, Z_i)\mathbf{1}_{\{M_t=i\}}$, cette fonctionnelle est clairement $\mathcal{F}_t$-mesurable, puisque pour tout i, sur l'événement $\{M_t = i\}$ le temps T_i est inférieur ou égal à t et donc F ne dépend que du passé avant t. Ensuite, les propriétés 1 et 2 montrent que F est élément de $\mathbb{D}^N$.

Réciproquement, par définition un élément de $\mathbb{D}_t^N$ est une fonctionnelle F sur $\Omega' \times U$ telle que pour tout $u \in U, F(.,u) \in H'$ et pour tout $w \in \Omega', F(w,.) \in H(O)^{\otimes \mathbb{N}}$. Par ailleurs, la restriction $F(w,u)\mathbf{1}_{\{M_t=i\}}$ ne doit dépendre que de $T_1, \cdots, T_i, Z_1, \cdots, Z_i$ pour être $\mathcal{F}_t$-mesurable et doit donc être de la forme annoncée dans les propriétés 1 et 2 pour que $F(w,.) \in H(O)^{\otimes \mathbb{N}}$. □

On va maintenant s'attacher à donner un critère d'absolue continuité ; pour cela on introduit quelques notations :
Soit $\kappa \in \mathbb{N}$ et $F = (F_1, \cdots, F_\kappa) \in (\mathbb{D}_t^N)^\kappa$, on note par $\Gamma_t^N(F)$ la matrice $\kappa \times \kappa$:

$$\Gamma_t^N(F) = (\Gamma_t^N(F_i, F_j))_{i,j \in \{1, \cdots, \kappa\}},$$

et det désignera le déterminant.

Théorème 3.7 *Soit $F \in (\mathbb{D}_t^N)^\kappa$. Alors $F * (\det(\Gamma_t^N(F)) \cdot P^N)$, la mesure image de $\det(\Gamma_t^N(F)) \cdot P^N$ par F, est absolument continue par rapport à la mesure de Lebesgue.*

Preuve : On remarque d'abord que $\Gamma_t^N(F) = 0$ sur $\{M_t = 0\}$. En fait, il suffit de montrer que pour tout $i \in \mathbb{N}$, $F * (\det(\Gamma_t^N(F))\mathbf{1}_{\{M_t=i\}} \cdot P^N)$ est absolument continue.

Soit donc $i \in \mathbb{N}$ fixé. D'après la proposition précédente, on peut écrire :

$$F.\mathbf{1}_{\{M_t=i\}} = f(T_1, \cdots, T_i, Z_1, \cdots, Z_i)\mathbf{1}_{\{M_t=i\}},$$

où f appartient au domaine de la structure de Dirichlet produit $(L^2(\Omega', \mathcal{F}', P') \otimes (H(O), e)^{\otimes i})^\kappa$.

Soit alors B, un ensemble borélien de $\mathbb{R}^\kappa$ et de mesure de Lebesgue nulle. On a :

$$\int 1_B(F(w))(\det(\Gamma_t^N(F(w)))^{1/2})\mathbf{1}_{\{M_t=i\}}\, dP^N(w) =$$

$$\frac{1}{\lambda(O)^i}\int_{\Omega'}\int_{O^i} \quad 1_B(f(T_1(w'), \cdots, T_i(w'), x_1, \cdots, x_i)) \cdot$$

$$Jf(w', x_1, \cdots, x_i)\nu(x_1)\cdots\nu(x_i)\, dx_1 \cdots dx_i dP'(w'),$$

Jf désigne la quantité $\det[(\sum_{j=1}^i \nabla_j f_k \cdot \nabla_j f_l)_{k,l \in \{1,\cdots,\kappa\}}]^{1/2}$.

La formule de la co-aire pour les fonctions de $\mathbb{R}^{i \times n}$ à valeurs dans $\mathbb{R}^\kappa$ (Cf. [3] II.5 ou [7]) assure alors que :

$$\int_O \cdots \int_O 1_B(f(T_1(w'), \cdots, T_i(w'), x_1, \cdots, x_i)) \cdot Jf(w', x_1, \cdots, x_i)\, dx_1 \cdots dx_i = 0,$$

pour P'-presque tout w' et, comme ν est strictement positive sur O :

$$\int_{O^i} 1_B(f(T_1(w'), \cdots, T_i(w'), x_1, \cdots, x_i)) \cdot Jf(w', x_1, \cdots, x_i)\nu(x_1)\cdots\nu(x_i)\, dx_1 \cdots dx_i = 0,$$

pour P'-presque tout w' et donc

$$\int 1_B(F(w))(\det(\Gamma^N(F(w))))^{1/2}\mathbf{1}_{\{M_t=i\}}\, dP^N(w) = 0.$$

Ainsi la mesure image de $[det(\Gamma_t^N(F))]^{\frac{1}{2}} \cdot P^N$ par F, est absolument continue par rapport à la mesure de Lebesgue, donc aussi son produit par la fonction positive $[det(\Gamma_t^N(F))]^{\frac{1}{2}}$, ce qui achève la preuve. □

3.2.3 Cas d'un nombre infini de sauts

Dans ce qui précède, on a supposé que l'on avait un nombre fini de sauts sur chaque intervalle fini de temps (hypothèse $\lambda(O) < \infty$), en fait les hypothèses 2 sont que O est un ouvert de $\mathbb{R}^n$ tel que $\lambda(O) = +\infty$ et tel qu'il existe un suite $(O_i)_{i \in \mathbb{N}}$ d'ouverts disjoints tels que :

$$O = \bigcup_{i=1}^{+\infty} O_i, \text{ avec } \forall i \in \mathbb{N},\ \lambda(O_i) < +\infty$$

Alors, pour tout $i \in \mathbb{N}$, on peut se donner un espace de probabilité $(\Omega_i, \mathcal{F}_i, P_i)$ sur lequel sont définies, comme en 3.2.2, des mesures de Poisson N^i, à valeurs dans $[0, T] \times O_i$ d'intensité $\nu(x)dxds$, ainsi que la structure de Dirichlet $(\mathbb{D}_t^i, \mathcal{E}_t^i)$ admettant pour opérateur carré du champ Γ_t^i.

On pose alors $(\Omega^N, \mathcal{F}^N, P^N) = \otimes_{i=1}^{+\infty}(\Omega_i, \mathcal{F}_i, P_i)$ et on définit la mesure aléatoire N par :

$$\forall w = (w_1, \cdots, w_i, \cdots),\ N(w) = \sum_{i=1}^{+\infty} N^i(w_i),$$

c'est à dire la mesure de Poisson N à valeurs dans $[0,T] \times O$ et d'intensité $\nu(x)dtdx$ définie en 2 (Cf. [12], pages 42-43).

Introduisons à présent quelques notations :
Pour tout $i \in \mathbb{N}$, on pose $\Omega^i = \prod_{j\in\mathbb{N}, j\neq i} \Omega_j$ et $P^i = \prod_{j\in\mathbb{N}, j\neq i} P_i$. De façon naturelle, on identifie Ω^N et $\Omega_i \times \Omega^i$.
La structure de Dirichlet considérée est :

$$(\mathbb{D}_t^N, \mathcal{E}_t^N) = \otimes_{i=1}^{+\infty}(\mathbb{D}_t^i, \mathcal{E}_t^i).$$

On a, en suivant toujours Bouleau-Hirsch ([3]) :

Proposition 3.8 *1. $(\mathbb{D}_t^N, \mathcal{E}_t^N)$ est une structure de Dirichlet sur $L^2(\Omega^N, \mathcal{F}^N, P^N)$ locale et telle que $\mathcal{E}_t^N(1) = 0$.*

2. Soit $F \in L^2(\Omega^N, \mathcal{F}^N, P^N)$, F appartient à $\mathbb{D}_t^N$ si et seulement si pour tout $i \in \mathbb{N}$, pour presque tout $w^i \in \Omega^i$, $F(\cdot, w^i) \in \mathbb{D}_t^i$ et :

$$\mathcal{E}_t^N(F) = \sum_{i=1}^{+\infty} \int_{\Omega^i} \mathcal{E}_t^i(F(\cdot, w^i))\, dP^i(w^i) < +\infty.$$

3. $(\mathbb{D}_t^N, \mathcal{E}_t^N)$ admet pour carré du champ, Γ_t^N : $\forall F \in \mathbb{D}_t^N$, $\Gamma_t^N(F) = \sum_{i=1}^{+\infty} \Gamma_t^i(F)$, où $\Gamma_t^i(F)$ désigne l'application : $(w_i, w^i) \in \Omega_i \times \Omega^i \longrightarrow \Gamma_t^i(F(\cdot, w^i))(w_i)$. □

On a de plus comme dans le théorème 3.7 un critère d'absolue continuité :

Théorème 3.9 *Soit $\kappa \in \mathbb{N}$ et $F \in (\mathbb{D}_t^N)^\kappa$. Alors $F * (\det(\Gamma_t^N(F)) \cdot P^N)$, la mesure image de $\det(\Gamma_t^N(F)) \cdot P^N$ par F, est absolument continue par rapport à la mesure de Lebesgue.*

Preuve : Remarquons d'abord que les matrices $\Gamma_t^N(F)$ et $\Gamma_t^i(F)$ sont des matrices symétriques positives. On pose :

$$A = \{\omega \in \Omega^N,\ \det(\Gamma_t^N(F))(\omega) > 0\},$$

et pour tout $i \in \mathbb{N}$ et tout $\omega^i \in \Omega^i$:

$$A_i(\omega^i) = \{\omega_i \in \Omega_i,\ \det(\Gamma_t^i(F(\cdot, \omega^i))(\omega_i)) > 0\}.$$

Comme $\Gamma_t^N(F)$ est symétrique et positive :

$$A = \{\omega \in \Omega^N,\ \forall u \in \mathbb{R}^\kappa - \{0\}, u^* \cdot \Gamma_t^N(F)(\omega) \cdot u > 0\},$$

et

$$A_i(\omega^i) = \{\omega_i \in \Omega_i,\ \forall u \in \mathbb{R}^\kappa - \{0\}, u^* \cdot \Gamma_t^i(F(\cdot, \omega^i))(\omega_i) \cdot u > 0\}.$$

Comme $\Gamma_t^N(F) = \sum_{i=1}^{+\infty} \Gamma_t^i(F)$, on a :

$$A = \bigcup_{i=1}^{+\infty} \{\omega = (\omega_i, \omega^i) \in \Omega^N,\ \omega_i \in A_i(\omega^i)\}.$$

Soit alors B un ensemble borélien de $\mathbb{R}^\kappa$, de mesure de Lebesgue nulle :

$$\int_A \mathbf{1}_B(F(\omega))dP^N(\omega) \leq \sum_{i=1}^{+\infty} \int_{\Omega^i} \int_{A_i(\omega^i)} \mathbf{1}_B(F(\omega_i, \omega^i))dP_i(\omega_i)\, dP^i(\omega^i),$$

mais grâce au théorème 3.7, on sait que pour tout i :

$$\int_{A_i(\omega^i)} \mathbf{1}_B(F(\omega_i, \omega^i)) det(\Gamma_t^N(F(\omega_i, \omega^i)))\, dP_i(\omega_i) = 0.$$

Comme $det(\Gamma_t^N(F(\omega_i, \omega^i)))$ est strictement positif sur l'ensemble d'intégration par définition, on a également pour tout i :

$$\int_{A_i(\omega^i)} \mathbf{1}_B(F(\omega_i, \omega^i))dP_i(\omega_i) = 0,$$

d'où la somme en i est nulle, et :

$$\int_A \mathbf{1}_B(F(\omega))dP^N(\omega) = 0.$$

Toujours parce que sur A, $det(\Gamma_t^N(F(\omega))) > 0$, il vient

$$\int_A \mathbf{1}_B(F(\omega)) det(\Gamma_t^N(F(\omega)))\, dP^N(\omega) = 0,$$

ce qui achève la preuve. □

3.3 Espace de Dirichlet produit

On a obtenu en 3.1 et 3.2 deux structures de Dirichlet pour tout $t > 0$ respectivement sur les espaces de Wiener et de Poisson :

$$(\Omega^W, \mathcal{F}^W, \mathbb{P}^W, \mathbb{D}_t^W, \mathcal{E}^W) \text{ et } (\Omega^N, \mathcal{F}^N, \mathbb{P}^N, \mathbb{D}_t^N, \mathcal{E}^N).$$

Suivant encore [3], chapitre V, section 2, on définit la structure produit :

$$\begin{aligned} & (\Omega = \Omega^W \times \Omega^N, \mathcal{F} = \mathcal{F}^W \otimes \mathcal{F}^N, \mathbb{P} = \mathbb{P}^W \otimes \mathbb{P}^N, \mathbb{D}_t, \mathcal{E}_t), \\ \mathbb{D}_t \;=\; & \{f \in L^2(\Omega, \mathbb{P}) / \mathbb{P}^N \text{ presque sûrement } f(., n) \in \mathbb{D}_t^W, \\ & \mathbb{P}^W \text{ presque sûrement } f(\omega, .) \in \mathbb{D}_t^N \text{ et } E(\Gamma_t^W(f) + \Gamma_t^N(f)) < \infty\} \qquad (2) \end{aligned}$$

et l'opérateur $\mathcal{E}_t(f) = E(\Gamma_t^W(f) + \Gamma_t^N(f))$.

Les deux structures étudiées précédemment ayant les "bonnes" propriétés (de Markov, locales, avec opérateur carré du champ...), la structure produit admet les mêmes propriétés avec l'opérateur carré du champ défini par :

$$\Gamma_t(f)(\omega, n) = \Gamma_t^W(f)(\omega) + \Gamma_t^N(f)(n).$$

Théorème 3.10 *Soit $F \in \mathbb{D}_t^\kappa$. Alors $F * (\det(\Gamma_t(F)) \cdot P)$, la mesure image de $det(\Gamma_t(F)) \cdot P$ par F, est absolument continue par rapport à la mesure de Lebesgue.*

Preuve: On reproduit la preuve du théorème 3.9 sur l'espace de Dirichlet produit $(\mathbb{D}_t, \mathcal{E})$ en remplaçant le produit infini par le produit des deux espaces $(\mathbb{D}_t^W, \mathcal{E}^W)$ et $(\mathbb{D}_t^N, \mathcal{E}^N)$. □

4 Structure de Dirichlet conditionnelle et grossissement de filtration

La construction qui suit est vraie en toute généralité dès que l'on a une structure produit (en temps). Elle s'applique en particulier très naturellement à l'espace de Wiener, en effet l'espace de Wiener $C_0([0,T],\mathbb{R})$ peut être vu comme le produit de $C_0([0,t],\mathbb{R})$ et de $C_0([0,T-t],\mathbb{R})$ de façon usuelle grâce à l'indépendance des accroissements du mouvement brownien et il est aisé de voir que la structure de Dirichlet sur $C_0([0,T],\mathbb{R})$ est le produit des structures sur $C_0([0,t],\mathbb{R})$ et $C_0([0,T-t],\mathbb{R})$.
Cette démonstration s'applique aussi au cas où l'on aura un mouvement brownien et une mesure de Poisson indépendants car la structure de Dirichlet sera le produit de deux structures pouvant être elles-mêmes considérées comme produit de deux structures : une sur l'intervalle de temps $[0,t]$ et l'autre sur $[t,T]$.

4.1 Application aux lois conditionnelles

On fixe $0 < t < T$. On peut alors considérer deux mesures de Poisson et leur structures de Dirichlet associées :

1. La première notée N^1, mesure de Poisson sur $[0,t] \times O$ d'intensité $\nu(x)dxds$ et définie sur l'espace de probabilité $(\Omega^1, \mathcal{F}_t^1, P^1)$. On notera T_i^1 et Z_i^1 les instants et les amplitudes des sauts et M^1 le processus de Poisson associé. On définit sur $L^2(\Omega^1, \mathcal{F}_t^1, P^1)$ la structure de Dirichlet $(\mathbb{D}_t^N, \mathcal{E}_t^N)$ admettant pour carré du champ Γ_t^N.

2. La deuxième notée N^2, mesure de Poisson sur $[0,T-t] \times O$ d'intensité $\nu(x)dxds$ et définie sur l'espace de probabilité $(\Omega^2, \mathcal{F}_{T-t}^2, P^2)$. De même T_i^2 et Z_i^2 désignent les instants et les amplitudes des sauts et M^2 le processus de Poisson associé. On définit sur $L^2(\Omega^2, \mathcal{F}_{T-t}^2, P^2)$ la structure de Dirichlet $(\mathbb{D}_{T-t}^N, \mathcal{E}_{T-t}^N)$ admettant pour carré du champ Γ_{T-t}^N.

On remarque que $(\Omega^N, \mathcal{F}_T^N, P^N) = (\Omega^1, \mathcal{F}_t^1, P^1) \otimes (\Omega^2, \mathcal{F}_{T-t}^2, P^2)$ et on considère la mesure aléatoire N à valeurs dans $[0,T] \times O$ définie par :

$$\forall w = (w_1, w_2) \in \Omega^1 \times \Omega_2, \; N(w) = \sum_{i=1}^{M^1(w_1)} \delta_{(T_i^1(w_1), Z_i^1(w_1))} + \sum_{i=1}^{M^2(w_2)} \delta_{(t+T_i^2(w_2), Z_i^2(w_2))}.$$

Il est clair que N est une mesure de Poisson sur $[0,T] \times O$ d'intensité $\nu(x)dsdx$, on notera T_i et Z_i les instants et amplitudes des sauts.
On peut alors comme précédemment considérer la structure de Dirichlet associée $(\mathbb{D}_T^N, \mathcal{E}_T^N)$ dont le carré du champ est Γ_T^N.
Mais sur $L^2(\Omega^N, \mathcal{F}_T^N, P^N)$, on peut aussi considérer la structure produit $(\mathbb{D}_t^N, \mathcal{E}_t^N) \otimes (\mathbb{D}_{T-t}^N, \mathcal{E}_{T-t}^N)$ et on a le résultat naturel suivant :

Proposition 4.1

$$(\mathbb{D}_T^N, \mathcal{E}_T^N) = (\mathbb{D}_t^N, \mathcal{E}_t^N) \otimes (\mathbb{D}_{T-t}^N, \mathcal{E}_{T-t}^N).$$

Preuve : Supposons d'abord que $\lambda(O) < +\infty$:
Soit $F \in \mathbb{D}_t$ et $G \in \mathbb{D}_{T-t}$ tels que :

$$F = \mathbf{1}_{\{M_t^1=i\}} \cdot f(T_1^1, \cdots, T_i^1, Z_1^1, \cdots, Z_i^1) \text{ et } G = \mathbf{1}_{\{M_{T-t}^2=j\}} \cdot g(T_1^2, \cdots, T_j^2, Z_1^2, \cdots, Z_j^2),$$

où f et g vérifient les hypothèses de la proposition 3.5 ; on a alors :

$$F \cdot G = \mathbf{1}_{\{M_T=i+j\}} h(T_1, \cdots, T_{i+j}, Z_1, \cdots, Z_{i+j}),$$

où $h : [0,T]^{i+j} \times O^{i+j} \longrightarrow \mathbb{R}$ est définie par :
$\forall (t_1, \cdots, t_{i+j}) \in [0,T]^{i+j}$, $\forall (x_1, \cdots, x_{i+j}) \in O^{i+j} = O^i \times O^j$,

$$h(t_1, \cdots, t_{i+j}, x_1, \cdots, x_{i+j}) = \begin{cases} f(t_1, \cdots, t_i, x_1, \cdots, x_i) \cdot g(t_{i+1}, \cdots, t_{i+j}, x_{i+1}, \cdots, x_{i+j}) \\ \text{si } t_i \leq t \leq t_{i+1} \\ \\ 0 \text{ sinon.} \end{cases}$$

Il est alors aisé de vérifier que $F.G$ appartient à $\mathbb{D}_T^N$ et que de plus :

$$\Gamma_T^N(F \cdot G) = \Gamma_t^N(F) \cdot G^2 + F^2 \cdot \Gamma_{T-t}^N(G).$$

Par densité, on conclut alors que : $(\mathbb{D}_t^N, \mathcal{E}_t^N) \otimes (\mathbb{D}_{T-t}^N, \mathcal{E}_{T-t}^N) \subset (\mathbb{D}_T^N, \mathcal{E}_T^N)$.

Montrons alors l'inclusion inverse :
Soit $F \in \mathbb{D}_T^N$ tel qu'il existe $i \in \mathbb{N}$ tel que : $F = f(T_1, \cdots, T_i, Z_1, \cdots, Z_i)\mathbf{1}_{\{M_t=i\}}$, où f vérifie les bonnes hypothèses. On a alors :

$$F = \sum_{j=0}^{i} \mathbf{1}_{\{M_t^1=j\}} \cdot \mathbf{1}_{\{M_{T-t}^2=i-j\}} \cdot f(T_1^1, \cdots, T_j^1, t+T_1^2, \cdots, t+T_{i-j}^2, Z_1^1, \cdots, Z_i^1, Z_1^2, \cdots, Z_{i-j}^2).$$

A partir de là, il est aisé de vérifier l'inclusion inverse.

On suppose maintenant que $\lambda(O) = +\infty$ et on va étendre les résultats obtenus, sachant qu'il existe un suite $(O_i)_{i \in \mathbb{N}}$ d'ouverts deux à deux disjoints tels que (cf 2) :

$$O = \bigcup_{i=1}^{+\infty} O_i, \ \forall i \in \mathbb{N}, \ \lambda(O_i) < +\infty.$$

On a alors :

$$(\mathbb{D}_t^N, \mathcal{E}_t^N) = \otimes_{i=1}^{+\infty} (\mathbb{D}_t^i, \mathcal{E}_t^i), \text{ et } (\mathbb{D}_{T-t}^N, \mathcal{E}_{T-t}^N) = \otimes_{i=1}^{+\infty} (\mathbb{D}_{T-t}^i, \mathcal{E}_{T-t}^i)$$

ce qui donne naturellement :

$$\begin{aligned} (\mathbb{D}_T^N, \mathcal{E}_T^N) &= (\otimes_{i=1}^{+\infty} (\mathbb{D}_t^i, \mathcal{E}_t^i)) \otimes (\otimes_{i=1}^{+\infty} (\mathbb{D}_{T-t}^i, \mathcal{E}_{T-t}^i)) \\ &= \otimes_{i=1}^{+\infty} ((\mathbb{D}_t^i, \mathcal{E}_t^i) \otimes (\mathbb{D}_{T-t}^i, \mathcal{E}_{T-t}^i)) = (\mathbb{D}_t^N, \mathcal{E}_t^N) \otimes (\mathbb{D}_{T-t}^N, \mathcal{E}_{T-t}^N). \end{aligned}$$

□

Remarque 4.2 : *Comme il est naturel pour une structure produit, si $F \in \mathbb{D}_T^N$ on note $\Gamma_t^N(F)$ (resp. $\Gamma_{T-t}^N(F)$) l'application :*

$$(w_1, w_2) \in \Omega^1 \times \Omega^2 \longrightarrow \Gamma_t^N(F(\cdot, w_2))(w_1) \, (\textit{resp. } \Gamma_{T-t}^N(F(w_1, \cdot))(w_2)).$$

On a alors l'égalité : $\forall F \in \mathbb{D}_T^N,\ \Gamma_T^N(F) = \Gamma_t^N(F) + \Gamma_{T-t}^N(F)$

Pour l'espace afférent au mouvement brownien, on a de même :

$$(\mathbb{D}_T^W, \mathcal{E}_T^W) = (\mathbb{D}_t^W, \mathcal{E}_t^W) \otimes (\mathbb{D}_{T-t}^W, \mathcal{E}_{T-t}^W) \text{ et } \forall F \in \mathbb{D}_T^W,\ \Gamma_T^W(F) = \Gamma_t^W(F) + \Gamma_{T-t}^W(F)\ P^W\text{-p.s.}$$

Enfin, on définit $\mathbb{D}_T$ *comme produit de* $\mathbb{D}_T^W$ *et* $\mathbb{D}_T^N$ *de façon naturelle, c'est à dire :*

$$(\mathbb{D}_T, \mathcal{E}_T) = (\mathbb{D}_T^W, \mathcal{E}_T^W) \otimes (\mathbb{D}_T^N, \mathcal{E}_T^N),$$

et le mouvement brownien étant comme le processus de Poisson à accroissements indépendants :

$$(\mathbb{D}_T, \mathcal{E}_T) = (\mathbb{D}_t, \mathcal{E}_t) \otimes (\mathbb{D}_{T-t}, \mathcal{E}_{T-t}),\ \forall F \in \mathbb{D}_T,\ \Gamma_T(F) = \Gamma_t(F) + \Gamma_{T-t}(F)\ P\text{-p.s.}$$

Théorème 4.3 *Soit* $\kappa \in \mathbb{N}^*$, *et* $L \in \mathbb{D}_T^\kappa$. *Alors sous la mesure* $det(\Gamma_{T-t}(L)) \cdot P$, *la loi conditionnelle de* L *sachant* $\mathcal{F}_t$ *est absolument continue par rapport à la mesure de Lebesgue.*

Preuve : Soit A un ensemble borélien de $\mathbb{R}$, de mesure de Lebesgue nulle. On a :

$$\begin{aligned} E(\mathbf{1}_A(L) \cdot det(\Gamma_{T-t}(L))|\mathcal{F}_t) &= \int_{\Omega^2} \mathbf{1}_A(L(w_1, w_2)) \cdot det(\Gamma_{T-t}(L(w_1, \cdot)(w_2)))dP^2(w_2) \\ &= 0, \end{aligned}$$

car pour P^1-presque tout w_1, $L(w_1, \cdot)$ appartient à $\mathbb{D}_{T-t}^\kappa$ et on a que l'image par $L(w_1, \cdot)$ de la mesure $(\Gamma_{T-t}(L(w_1, \cdot)).P^2$ est absolument continue par rapport à la mesure de Lebesgue en utilisant le théorème 3.10. □

On en déduit le corollaire :

Corollaire 4.4 *Soit* $L \in \mathbb{D}_T$ *tel que* $\Gamma_{T-t}(L) > 0$ P*-p.s. alors la loi conditionnelle de* L *sachant* $\mathcal{F}_t$ *est absolument continue par rapport à la mesure de Lebesgue.* □

Exemples : Supposons L totalement discontinu :

$$L = \sum_{T_n \le T} \phi(Z_n, T_n) = \int_0^T \int_O \phi(y, s) N(dy, ds) = N(\phi)$$

avec $\phi \in H(O)$. Alors la condition du corollaire est :

$$\int_t^T \int_O \|\nabla\phi(x, s)\|^2 N(dx, ds) = \sum_{t < T_n \le T} \|\nabla\phi(Z_n, T_n)\|^2 > 0.$$

Elle est vérifiée s'il existe des sauts dans l'intervalle de temps $[t, T]$ (ce qui est presque sûrement vrai sous l'hypothèse choisie que $\nu(O) = \infty$) et dès que $\nabla\phi(x, s)$ est non identiquement nul sur $O \times [0, T]$, c'est à dire que ϕ varie en x.

On n'est plus obligé de supposer que la mesure $\nu(O) = +\infty$ lorsque, par exemple $L = \int_0^T \sigma_s dW_s + N_T(\phi)$, il suffit alors d'avoir
$\Gamma_{T-t}(L) = \int_t^T |\sigma_s + \int_t^T D_s\sigma_u dW_u|^2 ds + N_{T-t}(\|\nabla\phi\|^2) > 0$ pour pouvoir opérer le grossissement de filtration.

4.2 Grossissement de filtration

Construisons maintenant sur la filtration grossie $\mathcal{Y}$ un $(\mathcal{Y},\mathbb{P})$-mouvement brownien et un $(\mathcal{Y},\mathbb{P})$-processus ponctuel.

On note $\boldsymbol{H_C}$ l'hypothèse

$$L \in \mathbb{D}_T^\kappa \text{ telle que } det(\Gamma_{T-t}(L)) > 0,\ \mathbb{P} \text{ presque sûrement pour tout } t \in [0,T[. \quad (3)$$

Proposition 4.5 *Sous l'hypothèse* $\boldsymbol{H_C}$

i/ *sur* $\Omega \times [0,T[\times \mathbb{R}^\kappa$*, il existe une version mesurable de la densité conditionnelle* $(\omega,t,x) \mapsto p(\omega,t,x)$ *qui est une* $(\mathcal{F},\mathbb{P})$*-martingale et se représente par*

$$p(\omega,t,x) = p_L(x) + \int_0^t \alpha(s,x)dW_s + \int_0^t \int_O \beta(y,s,x)\tilde{N}(dy,ds),$$

où pour tout x $s \mapsto \alpha(s,x)$ *et* $s \mapsto \beta(.,s,x)$ *sont des processus* $\mathcal{F}$*-prévisibles.*

ii/ *si* M *est une* $(\mathcal{F},\mathbb{P})$*-martingale locale continue égale à*

$$M_0 + \int_0^t u_s dW_s + \int_0^t \int_O v(y,s)\tilde{N}(dy,ds),$$

alors le crochet $d\langle M,p\rangle_t$ *est égal à* $\langle \alpha,u\rangle_t dt + \int_O \beta(y,t,x)v(y,t)\nu(y)dydt$ *et le processus*

$$\tilde{M}_t = M_t - \int_0^t \frac{(\langle \alpha(.,.,x),u\rangle_s + \int_O \beta(y,s,x)v(y,s)\nu(y)dy)|_{x=L}}{p(s,L)} ds, 0 \le t < T$$

est une $(\mathcal{Y},\mathbb{P})$*-martingale locale.*

Preuve : **i/** On utilise l'article de Jacod [14] dont les résultats ne dépendent que de l'hypothèse suivante : la famille $(Q_t(\omega,.), t > 0)$, loi de L sachant $\mathcal{F}_t$ est dominée par une mesure $q_t(dx)$, ici la mesure de Lebesgue sur $\mathbb{R}^\kappa$; on obtient donc (lemme 1.8 [14]) l'existence de l'application

$$(\omega,t,x) \mapsto p(\omega,t,x)$$

mesurable, càdlàg en t telle que $\forall t$ $p(\omega,t,x)dx$ est une version de la loi de L sachant $\mathcal{F}_t$ et pour tout x, $p(.,.,x)$ est une $(\mathcal{F},\mathbb{P})$-martingale. Soit le $\mathcal{F}$-temps d'arrêt $T_a^x = \inf\{t > 0 : p(.,t,x) \le a\}$; sur $[0,T_0^x[$ on a $p(.,t,x) > 0$. Le corollaire 1.11 de [14] montre que $T_0^L = +\infty$ p.s. et par conséquent pour tout t, $p(.,t,L) > 0$.
Par le théorème de représentation prévisible (les processus W et $\tilde{N}$ ont tous les deux la propriété de représentation prévisible) relatif au processus à accroissements indépendants $(W,\tilde{N})$, on obtient l'existence d'un couple (α,β) qui pour tout $(x \in \mathbb{R}, y \in O)$ est un processus prévisible vectoriel tel que :

$$p(.,t,x) = p(0,x) + \int_0^t \alpha(s,x)dW_s + \int_0^t \int_O \beta(y,s,x)\tilde{N}(dy,ds),$$

ii/ Le Théorème 5.1 de [14] permet de conclure : pour tout M, $\mathcal{F}$-martingale locale de la forme $M_t = M_0 + \int_0^t u_s dW_s + \int_0^t \int_O v(y,s)\tilde{N}(dy,ds)$, alors

$$\tilde{M}_t = M_t - \int_0^t \frac{(\langle \alpha(.,.,x),u\rangle_s + \int_O \beta(y,s,x)v(y,s)\nu(y)dy)|_{x=L}}{p(s,L)} ds, 0 \le t < T$$

est une $\mathcal{Y}$-martingale locale. □

On obtient en corollaire d'une part que le processus vectoriel

$$B_t = W_t - \int_0^t \frac{\alpha(u, L)}{p(u, L)} du, \ 0 \leq t < T$$

est un mouvement brownien sur l'espace de probabilité filtré $(\Omega, \mathcal{Y}, \mathbb{P})$, d'autre part, puisque

$$\overline{N}_t = \tilde{N}_t - \int_0^t \frac{\int_O \beta(y, u, L)\nu(y)dy}{p(u, L)} du, 0 \leq t < T$$

est une $\mathcal{Y}$-martingale locale, cela signifie que sur l'espace de probabilité $(\Omega, \mathcal{Y}, \mathbb{P})$, l'intensité prévisible du processus ponctuel N est $\nu(y)(1 + \frac{\beta(y,u,L)}{p(u^-,L)})$.

Remarque 4.6 *L'hypothèse $\boldsymbol{H_C}$ est suffisante pour assurer que la $(\mathcal{F}, \mathbb{P})$-martingale est une $(\mathcal{Y}, \mathbb{P})$-semi-martingale ; elle n'est pas nécessaire. En effet, si par exemple $L = \inf\{0 < t \leq T / \log S_t^1 = a\}$ avec a un réel fixé et S_t^1 le prix au temps t d'un actif dans un marché à coefficients constants (cf section 5), la loi de L est équivalente à la mesure de Lebesgue sur $\mathbb{R}_*^+$, de densité la dérivée en t de $t \mapsto \int_0^1 \frac{1}{\sqrt{2\pi t}} e^{-\frac{y^2}{2t}} dy$ et la loi de L conditionnellement en $\mathcal{F}_t$, elle, admet une densité strictement positive seulement sur $[t, +\infty[$ puisqu'avant les choses sont connues : la loi conditionnelle est absolument continue par rapport à la loi de L (sans être équivalente) ; de plus, les calculs montrent que $\boldsymbol{H_\xi}$ n'est pas vérifiée. Néanmoins, dans [17] ou [24] est donné explicitement un $(\mathcal{Y}, \mathbb{P})$-mouvement brownien.*

Ou encore, si $L = N_T$ est le nombre de sauts dans $[0; T]$ alors L ne vérifie pas l'hypothèse $\boldsymbol{H_C}$ mais la loi conditionnelle de L sachant $\mathcal{F}_t$ est quand même absolument continue par rapport à la loi de de L. Dans ce cas, on peut donc utiliser le lemme 1.8 de [14] pour effectuer le grossissement de filtration.

5 Application au marché financier

Le modèle est celui d'un marché financier de $d = m + 1$ actions, dont la dynamique est guidée par le mouvement brownien W de dimension m et le processus ponctuel N. Les prix $(S_t^i, i = 0, \cdots, d)$ suivent la dynamique :

$$\begin{aligned} dS_t^0 &= S_t^0 r_t dt, \\ dS_t^i &= S_t^i \left[b_t^i dt + \sigma_t^i dW_t + \int_O \phi^i(x, t) N(dx, dt) \right], S_0^i = x^i, i = 1, \cdots, d. \end{aligned} \quad (4)$$

On note $(\mathcal{F}_t)_{(0 \leq t \leq T)}$ la filtration engendrée par (W, N).

Soit u_t une sélection prévisible d'un vecteur unitaire de $Im(\sigma_t)^\perp$, l'espace de dimension 1 orthogonal de $Im(\sigma_t)$ et A_t l' ensemble $\{y \in O / \frac{\langle u_t, b_t - r_t \mathbf{1} \rangle}{\langle \phi(y,t), u_t \rangle} < 0\}$.

On note **H1** l'hypothèse :

(i) b, σ, ϕ sont prévisibles ,

(ii) $[\sigma_t \ \phi(y, t)]^* [\sigma_t \ \phi(y, t)]$ est inversible,

(iii) $\langle u_t, b_t - r_t \mathbf{1} \rangle \neq 0$ et $\nu(A_t) \neq 0$, $d\mathbb{P} \otimes dt$ presque sûrement.

L'investisseur initié a des informations sur le futur, représentées par une variable aléatoire $L \in L^1(\Omega, \mathcal{F}_T, \mathbb{R}^\kappa)$, on note $\mathcal{Y}$ la filtration "naturelle" de l'initié régularisée à droite : $\mathcal{Y}_t = \cap_{s > t}(\mathcal{F}_s \vee \sigma(L)), t \in [0, T[$.

Sous l'hypothèse $\boldsymbol{H_C}$ on peut appliquer la proposition 4.5 et on récrit la dynamique des prix relativement à la filtration $(\mathcal{Y}_t)_{t\in[0;A]}$ pour $A < T$. Ayant noté :

$$l_s = (\frac{\alpha^i(s,L)}{p(s^-,L)}, i = 1, m) \; ; \; \delta(y,s) = \frac{\beta(y,s,L)}{p(s^-,L)}, \; y \in O, \; s \in [0,T[, \tag{5}$$

on obtient pour $0 \leq t \leq A < T$ et $i = 1, \cdots, d$ une expression des prix comme $(\mathcal{Y}, \mathbb{P})$-semi-martingales :

$$dS_t^i = S_t^i \left((b_t^i + \sigma_t^i l_t) dt + \sigma_t^i dB_t + \int_O \phi^i(y,t) N(dy,dt) \right).$$

5.1 Représentation des $(\mathcal{Y}, Q)$-martingales

On montre l'existence d'une probabilité Q équivalente à $\mathbb{P}$ pour laquelle une propriété de représentation des $(\mathcal{Y}, Q)$-martingales est vérifiée. C'est un cas non classique puisque $\mathcal{Y}$ est plus grande que la filtration naturelle de (W, N).

Proposition 5.1 *Soit $A < T$. On suppose que l_t vérifie l'hypothèse $\boldsymbol{H_W}$:*

$$\exists C, \exists k > 0, \forall s \in [0,A], E[\exp k \|l_s\|^2] < C \tag{6}$$

et $\delta(y,t)$ vérifie l'hypothèse $\boldsymbol{H_N}$:

$$E[\exp \int_0^A \int_O (1 + \delta(y,s)) \log(1 + \delta(y,s)) - \delta(y,s)) \nu(y) dy ds] < +\infty$$

$$ou \quad E[\exp \int_0^A \int_O (\log(1 + \delta(y,s)) - \frac{\delta(y,s)}{1 + \delta(y,s)}) N(dy,ds)] < +\infty. \tag{7}$$

Soit (M^1, M^2) solution de l'équation

$$dM_t^1 = -M_1 l_t dB_t, \; M_0^1 = 1 \; ; \; dM_t^2 = M_{t-}^2 \int_O \delta(y,t) \tilde{N}(dy,dt), M_0^2 = 1 \; ; \; t \in [0,A].$$

On pose $M = M^1 M^2$.
Alors, M est une $(\mathcal{Y}, \mathbb{P})$-martingale uniformément intégrable égale à

$$M_t = \exp[-\int_0^t l_s dW_s - \frac{1}{2}\int_0^t \|l_s\|^2 ds + \sum_n \log(1 + \delta(Z_n, T_n)) - \delta(Z_n, T_n)].$$

On pose $Q = M\mathbb{P}$. Le processus $W_t = B_t + \int_0^t l_s ds$ est un $(\mathcal{Y}, Q)$-mouvement brownien. Le processus N est de $(\mathcal{Y}, Q)$-intensité $\nu(y)$.

Preuve : La preuve est classique, voir par exemple [18]. □

Du fait qu'alors le $(\mathcal{Y}, Q)$-processus (W, N) est à accroissements indépendants, on a le corollaire immédiat :

Corollaire 5.2 *Sous la probabilité Q, les tribus $\mathcal{F}_0$ et $\mathcal{F}_t$ sont indépendantes pour tout $t \leq A$.*

Le théorème 4.34 page 176 de [15] permet enfin, du fait de cette indépendance, d'obtenir le théorème de représentation de martingale :

Proposition 5.3 *Sous les hypothèses $\boldsymbol{H_C}$, $\boldsymbol{H_W}$ et $\boldsymbol{H_N}$, alors, pour tout $A < T$, et pour toute martingale locale Z relative à $(\mathcal{Y}_t, t \leq A; Q)$, il existe un unique couple prévisible (χ, ζ) tel que*

$$Z_t = E_Q(Z/\mathcal{Y}_0) + \int_0^t \chi_s dW_s + \int_0^t \int_O \zeta(y,s)(N(dy,ds) - \nu(y) dy ds), t \leq A.$$

5.2 Probabilités neutres au risque

On montre dans ce paragraphe que, si l'on se restreint à l'intervalle $[0, A], A < T$, il existe une famille de probabilités neutres au risque ; en effet, les processus (l, δ) n'existent que sur $[0, A], A < T$, car en T, la loi de L sachant $\mathcal{F}_T$ est une mesure de Dirac et $\lim_{t \to T} p(t, L)$ n'existe pas. On peut définir une stratégie optimale pour l'agent informé entre 0 et $A < T$; entre 0 et T il y a en général une stratégie d'arbitrage. Plus précisément l'existence de probabilités neutres au risque implique l'absence d'opportunité d'arbitrage strictement avant T.

On écrit l'équation des prix actualisés sous Q en remplaçant le processus W par $\tilde{W}_. = W_. - \int_0^. \xi_s ds$ et en changeant l'intensité de N en $\eta(y, t)\nu(y)$:

$$\begin{aligned} d\tilde{S}_t^i &= \tilde{S}_t^i(b_t^i - r_t + \sigma_t^i \xi_t)dt + \int_O \tilde{S}_t^i \phi^i(y,t)\eta(y,t)\nu(y)dydt \\ &+ \tilde{S}_t^i \sigma_t^i d\tilde{W}_t + \int_O \tilde{S}_t^i \phi^i(y,t)(N(dy,dt) - \eta(y,t)\nu(y)dydt). \end{aligned} \tag{8}$$

Pour que ces prix soient des martingales il suffit de trouver ξ et η qui permettent d'annuler $d\mathbb{P} \otimes dt$ presque sûrement les termes :

$$b_t^i - r_t + \sigma_t^i \xi_t + \int_O \phi^i(y,t)\eta(y,t)\nu(y)dy, i = 1, \cdots, d, .$$

soit :

$$\sigma_t \xi_t + \int_O \phi(y,t)\eta(y,t)\nu(y)dy = -(b_t - r_t \mathbf{1}). \tag{9}$$

$d\mathbb{P} \otimes dt$ presque sûrement. Si de plus $\eta(y, t) > 0$ $d\mathbb{P} \otimes dy \otimes dt$ presque sûrement, on pourra alors utiliser η pour un changement de probabilité.

Proposition 5.4 *On suppose les hypothèses* **H1** *et* $\boldsymbol{H_C}$. *Alors les solutions prévisibles de l'équation (9) sont de la forme*

$$\begin{aligned} \xi_t &= (\sigma_t^* \sigma_t)^{-1} \sigma_t^* \left[\langle u_t, b_t - r_t \mathbf{1} \rangle \int_O \frac{\phi(y,t)}{\langle \phi(y,t), u_t \rangle} \psi(y,t)\nu(y)dy - (b_t - r_t \mathbf{1}) \right] \\ \eta(y,t) &= -\frac{\langle u_t, b_t - r_t \mathbf{1} \rangle}{\langle \phi(y,t), u_t \rangle} \psi(y,t), \end{aligned} \tag{10}$$

où ψ est un processus prévisible sur $\mathbb{R}^d \times O$, tel que $\int_O \psi(s,y)\nu(y)dy = 1$.

Preuve: En premier lieu, l'hypothèse $\boldsymbol{H1}$ (ii) assure que ces processus sont bien définis, car σ_t est de rang plein et $\langle \phi(y,t), u_t \rangle \neq 0$ $d\mathbb{P} \otimes dt$ presque sûrement par définition de ϕ.

On montre ensuite que ce couple $(\xi_t, \eta(., t))$ donné dans la proposition est solution de l'équation (9) : on fait la somme de

$$\sigma_t \xi_t = \sigma_t (\sigma_t^* \sigma_t)^{-1} \sigma_t^* \left[\langle u_t, b_t - r_t \mathbf{1} \rangle \int_O \frac{\phi(y,t)}{\langle \phi(y,t), u_t \rangle} \psi(y,t)\nu(y)dy - (b_t - r_t \mathbf{1}) \right]$$

et de

$$\int_O \phi(y,t)\eta(y,t)\nu(y))dy = -\langle u_t, b_t - r_t \mathbf{1} \rangle \int_O \frac{\phi(y,t)\psi(y,t)}{\langle u_t, \phi(y,t) \rangle} \nu(y)dy.$$

Puis, en utilisant que ψ vérifie $\int_O \psi(t,y)\nu(y)dy = 1$ et en remarquant que la matrice $I_d = \sigma_t(\sigma_t^*\sigma_t)^{-1}\sigma_t^* + u_t u_t^*$, on vérifie que le couple (ξ, η) proposé est une solution de l'équation (9).

Réciproquement, si $(\xi_t, \eta(.,t))$ est une solution prévisible de l'équation (9), on remarque que le vecteur

$$\int_O \phi(y,s)\eta(y,s)\nu(y)dy + (b_s - r_s\mathbf{1})$$

appartient presque sûrement à l'image de la matrice σ_s, donc est orthogonal à u_s, d'où il vient :

$$\int_O \langle u_s, \phi(y,s)\rangle \eta(y,s)\nu(y)dy = -\langle u_s, (b_s - r_s\mathbf{1})\rangle.$$

Il existe alors un processus prévisible ψ sur $\mathbb{R}^d \times O$, tel que $\int_O \psi(s,y)\nu(y)dy = 1$ et presque sûrement :

$$\langle u_s, \phi(y,s)\rangle \eta(y,s) = -\langle u_s, (b_s - r_s\mathbf{1})\rangle \psi(y,s).$$

Le processus η est alors de la forme annoncée en (10) qui est bien définie, on l'a vu plus haut, grâce à l'uniforme ellipticité de la matrice introduite dans l'hypothèse **H1**. Reportant ensuite cette expression de η dans (9), il vient :

$$\sigma_s\xi_s = -\int_O \frac{\phi(y,t)}{\langle \phi(y,t), u_t\rangle}\psi(y,t)\nu(y)dy - (b_s - r_s\mathbf{1}).$$

On multiplie à gauche cette expression par la matrice $(\sigma_t^*\sigma_t)^{-1}\sigma_t^*$ (dont l'existence là aussi est assurée par l'hypothèse **H1**), et il vient l'expression annoncée en (10). Ainsi toute solution prévisible de l'équation (9) est de la forme annoncée dans la proposition. □

On montre alors à l'aide de cette proposition l'existence d'une famille de probabilités neutres au risque, paramétrée par cette fonction ψ.

Proposition 5.5 *Soit $A < T$.*

1. Il existe une famille de processus prévisibles ψ sur $\mathbb{R}^d \times O$ tel que :
$\int_O \psi(s,y)\nu(y)dy = 1$ *et* $\frac{\langle u_t, b_t - r_t\mathbf{1}\rangle}{\langle \phi(y,t), u_t\rangle}\psi(y,t) < 0$, $d\mathbb{P}\otimes dy \otimes dt$ *presque sûrement.*

2. Soient ξ et η définis par (10) à partir de ce processus ψ. On suppose que :
ξ_t vérifie l'hypothèse $\boldsymbol{H_\xi}$:

$$E_Q[\exp\frac{1}{2}\int_0^A \|\xi_s\|^2 ds] < +\infty \ ou\ \exists C, \exists k > 0, \forall s \in [0,A], E_Q[\exp k\|\xi_s\|^2] < C \quad (11)$$

et $\eta(y,t)$ vérifie l'hypothèse $\boldsymbol{H_\eta}$:

$$E_Q[\exp \int_0^A \int_O (\eta(y,s)\log\eta(y,s) - \eta(y,s) + 1)\nu(y)dyds] < +\infty$$
$$ou\quad E_Q[\exp \int_0^A \int_O (\log\eta(y,s) - \frac{\eta(y,s)-1}{\eta(y,s)})N(dy,ds)] < +\infty. \quad (12)$$

Soit (M^1, M^2) solution de l'équation, pour $t \in [0,A]$,

$$dM_t^1 = M_t^1\xi_t dW_t,\ M_0^1 = 1\ ;\ dM_t^2 = M_{t-}^2\int_O (\eta(y,t)-1)(N(dy,dt) - \nu(y)dydt),\ M_0^2 = 1.$$

On pose $M^\psi = M^1 M^2$.
Alors, M^ψ *est une* $(\mathcal{Y}, Q)$*-martingale uniformément intégrable égale à*

$$M_t^\psi = \exp[\int_0^t \xi_s dW_s - \frac{1}{2}\int_0^t \|\xi_s\|^2 ds + \sum_n \log \eta(Z_n, T_n) + 1 - \eta(Z_n, T_n)].$$

On pose $R^\psi = M^\psi.Q$. *Le processus* $\tilde{W}_t = W_t - \int_0^t \xi_s ds$ *est un* $(\mathcal{Y}, R^\psi)$*-mouvement brownien et le processus* N *est de* $(\mathcal{Y}, R^\psi)$*-intensité* $\eta(y,u)\nu(y)$.

3. Les prix actualisés sont des $(\mathcal{Y}, R^\psi)$*-martingales locales, c'est à dire que* R^ψ *est une probabilité neutre au risque.*

Preuve :

1. Le processus $\eta(y,t)$ obtenu dans la proposition précédente n'est pas nécessairement positif ; or, pour faire un changement de probabilité, on en a besoin afin que $\eta\nu$ soit effectivement une intensité. ***H1*** (iii) dit que $\eta(.,t) \neq 0$ $d\mathbb{P} \otimes dt$ presque sûrement. Rappelons que $A_t = \{y \in O / \frac{\langle u_t, b_t - r_t \mathbf{1}\rangle}{\langle \phi(y,t), u_t\rangle} < 0\})$ et $\nu(A_t) > 0$ (hypothèse **H1** (iii)) ; soit ψ_1 un processus prévisible et borné sur $\Omega \times \mathbb{R}_+ \times O$ tel que $\psi_1(t,.) \in L^1(\nu)$ $d\mathbb{P} \otimes dt$ presque sûrement et $\int_{A_t} |\psi_1(t,y)|\nu(y)dy \neq 0$ $d\mathbb{P} \otimes dt$ presque sûrement.

On choisit une constante C_t telle que

$$C_t \int_{A_t} |\psi_1(t,y)|\nu(y)dy > \int_{A_t^c} |\psi_1(t,y)|\nu(y)dy$$

(A_t^c est le complémentaire de A_t dans O), et on pose $\psi(t,y) = g_t^{-1}(C_t|\psi_1(t,y)|\mathbf{1}_{A_t} - |\psi_1(t,y)|\mathbf{1}_{A_t^c})$, où g_t est le facteur de normalisation ($g_t > 0$ par choix de C_t), et ψ vérifie les conditions demandées.

2. Le fait que ξ et η soient solutions de l'équation (9) et les hypothèses $\boldsymbol{H_\xi}$ et $\boldsymbol{H_\eta}$ permettent de conclure : ce qui concerne le mouvement brownien est tout à fait classique ; ce qui concerne le processus ponctuel de Poisson découle des théorèmes III.1 page 185 ou III.7 page 190 de [18] et du fait que $\eta > 0$, ce qui assure que le saut de la martingale dont M_2 est l'exponentielle de Doléans-Dade est strictement supérieur à -1.

3. C'est une conséquence de la proposition 5.4. □

Remarque 5.6 *1. En l'absence d'initiation, sous les hypothèses* **H1**, $\boldsymbol{H_\xi}$ *et* $\boldsymbol{H_\eta}$, *on obtient une probabilité neutre au risque équivalente à* $\mathbb{P}$ *de la forme* $M^\psi.\mathbb{P}$.
2. Dans [16], la condition $\eta > 0$ *demandait l'hypothèse* $\frac{\langle u_t, b_t - r_t \mathbf{1}\rangle}{\langle \phi(y,t), u_t\rangle} < 0$ *que l'on retrouve ici sous la forme plus générale donnée en* **H1** *(iii).*

5.3 Caractérisation des probabilités neutres au risque

On note $\mathcal{R}$ l'ensemble des probabilités neutres au risque pour l'agent initié équivalentes à la probabilité Q, c'est à dire l'ensemble des probabilités R équivalentes à la probabilité Q, telles que les prix actualisés sont des $((\mathcal{Y}_t, R), t \leq A)$-martingales.

Soit donc $R \in \mathcal{R}, R = Z.Q$ et $R^\psi = M^\psi.Q$ autre élément de $\mathcal{R}$ donné par la section précédente. Les prix étant à la fois des $(\mathcal{Y}, R)$-martingales et des $(\mathcal{Y}, R^\psi)$-martingales, on obtient que pour tout, i, (cf [21] page 109) :

$$[S^i, Z.(M^\psi)^{-1}] \text{ est une } R^\psi\text{-martingale.}$$

C'est à dire que l'on a pour la partie continue $\langle S^i, Z.(M^\psi)^{-1}\rangle = 0$ et la somme des sauts $\sum \Delta S^i \Delta(Z(M^\psi)^{-1})$ est une R^ψ-martingale, donc le compensateur est nul. On obtient donc :

Proposition 5.7 *Toute probabilité R dans $\mathcal{R}$ ensemble des probabilités neutres au risque pour l'agent initié équivalentes à Q, est de la forme $Z_A.Q$ où la $(\mathcal{Y}, Q)$-martingale Z vérifie :*

$$dZ_s = Z_s\chi_s dW_s + \int_O Z_{s-}(\zeta(y,s) - 1)(N(dy,ds) - \nu(y)dyds, s \leq A, Z_0 \in L^1(\mathcal{Y}_0, Q),$$

avec $\chi_t = (\sigma_t^*\sigma_t)^{-1}\sigma_t^*\left[\langle u_t, b_t - r_t\mathbf{1}\rangle \int_O \frac{\phi(y,t)}{\langle\phi(y,t),u_t\rangle}\psi(y,t)\nu(y)dy - (b_t - r_t\mathbf{1})\right]$ *et*

$$\int_O \phi(y,t)\zeta(y,t)\nu(y)dy = -\langle u_t, b_t - r_t\mathbf{1}\rangle \int_O \frac{\phi(y,t)\psi(y,t)}{\langle u_t, \phi(y,t)\rangle}\nu(y)dy$$

pour tout ψ, processus prévisible sur $\mathbb{R}^d \times O$, tel que $\int_O \psi(s,y)\nu(y)dy = 1$ et $\frac{\langle u_t, b_t - r_t\mathbf{1}\rangle}{\langle\phi(y,t),u_t\rangle}\psi(y,t) < 0$, $d\mathbb{P}\otimes dy \otimes dt$ presque sûrement.

Preuve: D'après la proposition 5.3, notant $Z_t = E_Q(Z/\mathcal{Y}_t)$, et puisque $Z > 0$, il existe un unique couple prévisible (χ, ζ) tel que

$$Z = E_Q(Z/\mathcal{Y}_0) + \int_0^A Z_s\chi_s dW_s + \int_0^A \int_O Z_{s-}(\zeta(y,s) - 1)(N(dy,ds) - \nu(y)dyds).$$

Puis, on utilise la proposition 5.1 :

$$dM_t^\psi = M_t^\psi \xi_t^\psi dW_t + M_{t-}^\psi \int_O (\eta^\psi(y,t) - 1)(N(dy,dt) - \nu(y)dydt) \;;\; t \in [0,A]$$

et l'écriture des semi-martingales S^i sous Q :

$$S_t^i = S_0^i + \int_0^t S_s^i(b_s^i ds + \sigma_s^i dW_s) + \int_0^t S_{s-}^i \int_O \phi^i(x,s)N(dx,ds), 0 \leq t \leq T, i = 1,\cdots,d.$$

Pour la partie continue du crochet, il vient $ds \otimes d\mathbb{P}$-presque sûrement, pour tout $i = 1,\cdots,d$:

$$S_s^i Z_s (M_s^\psi)^{-1}\sigma_s^i(\chi_s - \xi_s^\psi) = 0,$$

soit $\chi = \xi^\psi$ puisque par l'hypothèse **H1**(ii) la matrice σ est de rang plein.

Pour les sauts, notons que les sauts de S^i sont $\Delta S^i = S_{t-}^i\phi^i(y,t)$. Puisque $Z_t = Z_{t-}\zeta(y,t)$ et $M_t^\psi = M_{t-}^\psi\eta^\psi(y,t)$, $Z_t(M_t^\psi)^{-1} = Z_{t-}(M_t^\psi)^{-1}\frac{\zeta(y,s)}{\eta^\psi(y,t)}$ et les sauts de $Z_t(M_t^\psi)^{-1} = \Delta(Z_t(M_t^\psi)^{-1}) = Z_{t-}(M_t^\psi)^{-1}(\frac{\zeta(y,t)}{\eta^\psi(y,t)} - 1)$. Il vient donc $dt \otimes d\mathbb{P}$-presque sûrement, pour tout $i = 1,\cdots,d$:

$$\int_O S_{t-}^i Z_{t-}(M_t^\psi)^{-1}\phi^i(y,t).(\frac{\zeta(y,t)}{\eta^\psi(y,t)} - 1)\eta^\psi(y,t)\nu(y)dy = 0,$$

soit $\int_O \phi^i(y,t).(\zeta(y,s) - \eta^\psi(y,t))\nu(y)dy = 0$ pour tout $i = 1,\cdots,d$. □

Remarquons que si l'on applique la matrice σ_t^* à ce vecteur, on retrouve l'unicité de χ puisque pour tout ψ,
$\sigma_t^* \int_O \phi(y,t).\zeta(y,t)\nu(y)dy = -\langle u_t, b_t - r_t\mathbf{1}\rangle \int_O \frac{\psi(y,t)}{\langle u_t,\phi(y,t)\rangle}\nu(y)dy.$

6 Conclusion

La construction des structures de Dirichlet présentée ici a d'autres applications, par exemple donner des conditions d'existence de la densité de la loi d'une diffusion guidée par (W, N) comme cela est fait dans [2].

Pour les applications au modèle financier en présence d'agent informé, on aurait pu procéder comme dans un précédent travail [11]. En effet les hypothèses $\boldsymbol{H_C}$, $\boldsymbol{H_W}$ et $\boldsymbol{H_N}$ impliquent l'existence d'une probabilité Q_0 équivalente à $\mathbb{P}$ sous laquelle (W, N) est indépendant de $\sigma(L)$ (Cf. [8]). Ceci est suffisant pour écrire l'équation des prix sur $(\Omega, \mathcal{Y}, \mathbb{P})$ et obtenir la représentation prévisible comme ci-dessus.

Avec seulement l'hypothèse que la loi conditionnelle de L sachant $\mathcal{F}_t$ est absolument continue par rapport à le loi de L on ne sait pas démontrer un théorème de représentation de martingale.

Remerciements : Ce travail est redevable à plusieurs collègues, spécialement Francis Hirsch, Thierry Jeulin, Jean Jacod, sans compter les auditeurs de nos différents exposés.

References

[1] I.BARDHAN, X. CHAO , "On martingale measures when asset returns have unpredictable jumps", Stoch. Proc. and their Applic. 63, 1996, 35-54.

[2] K. BICHTELER, J-B. GRAVEREAUX et J. JACOD, "Malliavin's Calculus for Processes with Jumps", Gordon and Breach Sc. Pub., New York 1987.

[3] N. BOULEAU and F. HIRSCH, "Dirichlet Forms and Analysis on Wiener Space", Walter de Gruyter, Berlin, 1991.

[4] N. BOULEAU, "Constructions of Dirichlet structures", dans *Potential Theory-ICPT 94*, Král/Lukěs/Netuka/Veselý eds., Walter de Gruyter, 1996.

[5] P. BREMAUD, "Point Processes and Queues", Springer-Verlag, 1981.

[6] M. CHALEYAT-MAUREL et T. JEULIN, "Grossissement gaussien de la filtration brownienne", Séminaire de Calcul Stochastique 1982-83, Paris, Lecture Notes in Mathematics 1118, 59-109, Springer-Verlag, 1985.

[7] H. FEDERER, "Geometric Measure Theory", Springer-Verlag, Berlin-Heidelberg-New-York, 1969.

[8] H. FOLLMER and P. IMKELLER, "Anticipation cancelled by a Girsanov transformation : a paradox on Wiener space", Ann. Inst. Henri Poincaré, 29(4), 1993, 569-586.

[9] M. FUKUSHIMA, Y. OSHIMA, M. TAKEDA "Dirichlet Forms and Symmetric Markov Processes", de Gruyter studies in Math., 1994.

[10] A. GRORUD et M. PONTIER, "Comment détecter le délit d'initié ? " ,CRAS, t.324, Serie 1, p.1137-1142, 1997.

[11] A. GRORUD, M. PONTIER, "Insider trading in a Continuous Time Market Model",I.J.T.A.F., vol. 1 (3), p. 315-330, July 1998.

[12] N. IKEDA, S. WATANABE "Stochastic Differential Equations and Diffusions Processes", Second Edition, North-Holland, Amsterdam-Oxford-New-York, 1989.

[13] J. JACOD, "Calcul Stochastique et Problèmes de Martingales", Lecture Notes in Mathematics 714, Springer-Verlag, 1979.

[14] J. JACOD, "Grossissement initial, Hypothèse H' et Théorème de Girsanov", Séminaire de Calcul Stochastique 1982-83, Paris, Lecture Notes in Mathematics 1118, Springer-Verlag 1985, 15-35.

[15] J. JACOD, A.N. SHIRYAEV, "Limit Theorems for Stochastic Processes", Springer-Verlag, 1987.

[16] M. JEANBLANC-PIQUE et M. PONTIER, "Optimal Portfolio for a Small Investor in a Market Model with Discontinuous Prices" Economica 22, Paris, 1994, 287-310.

[17] T. JEULIN, "Semi-martingales et grossissement de filtration", Lecture Notes in Mathematics 833, Springer-Verlag 1980.

[18] D. LEPINGLE et J. MEMIN, "Sur l'intégrabilité uniforme des martingales exponentielles", Z. Wahrs. verw. Geb. 42, 1978, 175-203.

[19] P. MALLIAVIN, "Stochastic calculus of variations and hypoelliptic operators", Proceedings of the International Symposium on Stochastic Differential Equations, Kyoto, 1976, Kinokuniya-Wiley,1978,195-263.

[20] D. NUALART, "Analysis on Wiener space and anticipating stochastic calculus", Ecole de Probabilités de Saint-Flour, 1995.

[21] P. PROTTER, "Stochastic Integration and Diffential Equations", Springer-Verlag, 1990.

[22] S. SONG, "Grossissement de filtrations et problèmes connexes", thèse de doctorat de l'université de Paris VII, 29 Octobre 1987.

[23] M. YOR, "Grossissement de filtrations et absolue continuité de noyaux", Séminaire de Calcul Stochastique 1982-83, Paris, Lecture Notes in Mathematics 1118, 6-14, Springer-Verlag 1985.

[24] M. YOR, "Some Aspects of Brownian Motion", vol. II, Birkhaüser, 1997.

Saturations of Gambling Houses

A. Maitra and W. Sudderth [1]
School of Statistics, University of Minnesota
Minneapolis,Minnesota 55455

Abstract

Suppose that X is a Borel subset of a Polish space. Let $\mathbb{P}(X)$ be the set of probability measures on the Borel σ-field of X. We equip $\mathbb{P}(X)$ with the weak topology. A *gambling house* Γ on X is a subset of $X \times \mathbb{P}(X)$ such for each $x \in X$, the section $\Gamma(x)$ of Γ at x is nonempty. Assume moreover that Γ is an analytic subset of $X \times \mathbb{P}(X)$. Then we can associate with Γ optimal reward operators G_Γ, R_Γ, and M_Γ as follows:

$$(G_\Gamma u)(x) = \sup\{\textstyle\int u\, d\gamma : \gamma \in \Gamma(x)\}, \qquad x \in X,$$

$$(R_\Gamma u)(x) = \sup \int u(x_t)\, dP_\sigma, \qquad x \in X,$$

where u is a bounded, Borel measurable function on X, the sup in the definition of R_Γ is over all measurable strategies σ available in Γ at x and Borel measurable stop rules t (including $t \equiv 0$), x_t is the terminal state and P_σ the probability measure on H, the space of infinite histories, induced by σ;

$$(M_\Gamma g)(x) = \sup \int g\, dP_\sigma, \qquad x \in X,$$

where g is a bounded, Borel measurable function on H and the sup is over all measurable strategies σ available in Γ at x. The aim of this article is to describe the "largest" houses or "saturations" for which the associated operators are the same as the corresponding operators for the original house. Our methods are constructive and will show that the saturations are again analytic gambling houses.

1 INTRODUCTION

The point of departure of this article is a beautiful result of Dellacherie and Meyer [5, 38] in gambling theory. We will describe this result in the framework of the Dubins-Savage ([6]) theory.

Let X be a Borel subset of a Polish space, and let $\mathbb{P}(X)$ be the set of probability measures on the Borel σ-field of X. Give $\mathbb{P}(X)$ the topology of weak convergence, so $\mathbb{P}(X)$ is again a Borel subset of a Polish space (see ([10], 17E) for details). A *gambling house* on X is a subset Γ of $X \times \mathbb{P}(X)$ such that each section $\Gamma(x)$ of Γ at x is nonempty. A strategy σ *available in* Γ *at* x is a sequence $\sigma_o, \sigma_1, \ldots$ such that $\sigma_o \in \Gamma(x)$ and, for $n \geq 1, \sigma_n$ is a universally measurable function on X^n into $\mathbb{P}(X)$ such that $\sigma_n(x_1, x_2, \ldots, x_n) \in \Gamma(x_n)$ for every $x_1, x_2, \ldots, x_n \in X$. Such a σ defines a unique probability measure on the Borel subsets of the history space $H = X^N$, where N is the set of positive integers and H is given the product topology. We will use the same symbol σ for this probability measure. (See ([3], 7.45) for the existence of

[1]Research supported by National Science Foundation Grant DMS-9703285.

this measure.) If σ is a strategy available in Γ at x and $x' \in X$, then the *conditional strategy* $\sigma[x']$ is the strategy defined as follows:

$$(\sigma[x'])_0 = \sigma_1(x')$$

and, for $n \geq 1$,

$$(\sigma[x'])_n(x_1, x_2, \ldots, x_n) = \sigma_{n+1}(x', x_1, x_2, \ldots, x_n)$$

for $x_1, x_2, \ldots, x_n \in X$. Note that $\sigma[x']$ is available in Γ at x'. The set of strategies (and also the measures induced on H by these strategies) available in Γ at x will be denoted by $\Sigma_\Gamma(x)$.

A *stop rule* is a universally measurable function t on H into $\omega = N \cup \{0\}$ such that $t(h) = k$ and $h \equiv_k h'$ imply $t(h') = k$, where $h \equiv_k h'$ means that h and h' agree through the first k coordinates. In particular, if $t(h) = 0$ for some h, then t is identically zero. If t is a stop rule such that $t \geq 1$ and $x \in X$, then the *conditional stop rule* $t[x]$ is defined by

$$t[x](h) = t(xh) - 1, \; h \in H,$$

where xh is the history obtained by catenating x and h. Note that $t[x]$ is again a stop rule. A pair $\pi = (\sigma, t)$ where $\sigma \in \Sigma_\Gamma(x)$ and t is a stop rule is said to be a *policy* available at x.

In the sequel, none of the results would be affected if we had restricted ourselves to Borel measurable stop rules.

A *measurable leavable gambling problem* is a triple (X, Γ, u), where X is a Borel subset of a Polish space, Γ is a gambling house which is an analytic subset of $X \times \mathbb{P}(X)$, and u is a bounded, upper analytic function on X, that is, $[u > a]$ is an analytic subset of X for every real a. Such structures with Γ and u both Borel measurable were introduced by Strauch ([19]); the extension to analytic gambling house and upper analytic utility functions is due to Meyer and Traki ([17]).

If Γ is an analytic gambling house on X, then $\Sigma_\Gamma(x) \neq \phi$ for each x, courtesy of the von-Neumann selection theorem ([10], 29.9). Furthermore the set $\Sigma_\Gamma = \bigcup_{x \in X}\{x\} \times \Sigma_\Gamma(x)$ is analytic in $X \times \mathbb{P}(H)$, as was established by Dellacherie ([4], Theorem 3).

The *optimal reward operator* for a measurable leavable gambling problem (X, Γ, u) is defined by

$$(R_\Gamma u)(x) = \sup_\pi \int u(h_t)\, d\sigma, \qquad x \in X, \tag{1.1}$$

where h_t abbreviates $h_{t(h)}$ and the sup is taken over all policies $\pi = (\sigma, t)$ available in Γ at x.

The Fundamental Theorem of Gambling (see([17]) or ([14], Theorem 4.8) provides another description of R_Γ as follows. First we need a definition. We say that a bounded function g on X is Γ – *excessive* if it is upper analytic and $\int g\, d\gamma \leq g(x)$ for every $\gamma \in \Gamma(x)$ and $x \in X$.

Theorem 1.1. *(Fundamental Theorem of Gambling) If Γ is an analytic gambling house on X and u is a bounded, upper analytic function on X, then $R_\Gamma u$ is the least Γ-excessive function g such that $g \geq u$.*

Note that the function $R_\Gamma u$ can be defined for *every* house Γ for which $\Sigma_\Gamma(x) \neq \phi$ for each x and any bounded, universally measurable function u on X by using (1.1).

We associate with each analytic house Γ on X a house Γ^c as follows:

$$\Gamma^c(x) = \{\sigma\psi_t^{-1} : \sigma \in \Sigma_\Gamma(x) \text{ and } t \text{ is a bounded stop rule}\}, \qquad x \in X,$$

where $\psi_t(h) = h_t$, $h \in H$. (For $t = 0$, $\sigma\psi_t^{-1}$ is defined to be $\delta(x)$ if $\sigma \in \Sigma_\Gamma(x)$.) In other words, $\Gamma^c(x)$ is the set of distributions of the terminal state induced by policies $\pi = (\sigma, t)$ available in Γ at x for which t is bounded. As we will prove in section 3, Γ^c is an analytic subset of $X \times \mathbb{P}(X)$.

If L_1 and L_2 are operators that map bounded functions on X (respectively H) to bounded functions on X, then we will write $L_1 \sim L_2$ if $L_1 = L_2$ on bounded, Borel measurable functions on X (respectively H); and we will write $L_1 \approx L_2$ if $L_1 = L_2$ on bounded, upper analytic functions on X (respectively H).

We are now ready to state the result of Dellacherie and Meyer which was mentioned in the first paragraph.

Theorem 1.2. *Suppose that Γ is an analytic gambling house on X. Then the largest gambling house Γ' such that $R_{\Gamma'} \sim R_\Gamma$ is defined by*

$$\Gamma^s(x) = \overline{\text{sco}}\,\Gamma^c(x), \qquad x \in X, \tag{1.2}$$

where $\overline{\text{sco}}\,\Gamma^c(x)$ denotes the (total variation) norm closure of the strong convex hull of $\Gamma^c(x)$. In particular, Γ^s is an analytic gambling house.

Recall that if $M \subseteq \mathbb{P}(X)$, then the strong convex hull of M,written sco M, is the set of all $\nu \in \mathbb{P}(X)$ such that there is $\mu \in \mathbb{P}(\mathbb{P}(X))$ with $\mu^*(M) = 1$ and

$$\nu(B) = \int_{\mathbb{P}(X)} \eta(B)\,\mu(d\eta)$$

for every Borel subset B of X,where μ^* is the outer measure induced by μ. For $M \subseteq \mathbb{P}(X)$, we say that M is *strongly convex* if $M = \text{sco}\,M$.

The gambling house Γ^s is called the saturation of the gambling house Γ. Dellacherie and Meyer [5] define the saturation of Γ to be the largest house having the same excessive functions as Γ. It is easy to see that their definition is equivalent to the one given above. The statement of Theorem 1.2 differs slightly from the formulation of Dellacherie and Meyer. In place of the gambling house Γ^c they have a "house" consisting of sub-probability measures and they remark ([5], p.192) that their proof depends critically on allowing sub-probability measures in their construction. They pay a price for this: they need to perform the operation of hereditary closure on the strong convex hull before taking the norm-closure and then intersect the result with $\mathbb{P}(X)$. In fairness, we must point out that they are aware, as they remark ([5], p.183), that they could have worked with Γ^c but chose not to do so as the proof that Γ^c is analytic is laborious. It turns out that proving the analyticity of Γ^c is not so hard after all, as we shall see presently.

Given a gambling house Γ, there are other optimal reward operators of interest. The aim of this article is to construct "largest" houses or "saturations" keeping those operators invariant in the spirit of Theorem 1.2. We will now define two such operators.

Dubins et al. ([8]) define a *measurable non-leavable gambling problem* to be a triple (X, Γ, u^*), where X is a Borel subset of a Polish space, Γ an analytic gambling house on X, u a bounded, upper analytic function on X and

$$u^*(h) = \limsup_n u(h_n), \qquad h \in H.$$

The *optimal reward operator* for the nonleavable gambling problem is defined by

$$(V_\Gamma u)(x) = \sup_{\sigma \in \Sigma_\Gamma(x)} \int u^* \, d\sigma, \qquad x \in X.$$

Note that V_Γ is defined even when Γ is not analytic just so long as $\Sigma_\Gamma(x)$ is nonempty for each $x \in X$.

For any set $\Sigma \subseteq X \times \mathbb{P}(H)$ such that $\Sigma(x)$ is nonempty for each x and any bounded, upper analytic function g on H, we define

$$(M_\Sigma g)(x) = \sup_{\sigma \in \Sigma(x)} \int g \, d\sigma, \qquad x \in X.$$

We will also write M_Γ for M_Σ in case $\Sigma = \Sigma_\Gamma$ for a gambling house Γ on X. In this case, we will say that Σ is a *global gambling house* on X.

Here are the main results of the paper.

Theorem 1.3. *Suppose that Γ is an analytic gambling house on X. Then the largest gambling house Γ' such that $M_\Gamma \sim M_{\Gamma'}$ is*

$$\Gamma^{s_1}(x) = \overline{\text{sco}}\, \Gamma(x), \qquad x \in X.$$

In consequence, Γ^{s_1} is analytic.

In the sequel, we will write $\overline{\text{sco}}\, \Gamma$ for Γ^{s_1}.

Theorem 1.4. *Let Γ be an analytic gambling house on X. Then the largest global gambling house Σ such that $M_\Sigma \sim M_{\Sigma_\Gamma}$ is $\Sigma_{\overline{\text{sco}}\,\Gamma}$.*

Theorem 1.5. *Suppose that Γ is an analytic gambling house on X. Then the largest set $\Sigma \subseteq X \times \mathbb{P}(H)$ such that $\Sigma(x)$ is nonempty for each x and $M_\Sigma \sim M_{\Sigma_\Gamma}$ is $\Sigma_{\overline{\text{sco}}\,\Gamma}$.*

Theorem 1.6. *For $X = \{0, 1\}$, there is a Borel gambling house Γ on X such that there is no largest gambling house Γ' such that $V_\Gamma \sim V_{\Gamma'}$.*

A word about notation. Throughout the paper, the operations of forming the strong convex hull and the (variation) norm closure will be performed (vertical) sectionwise on subsets of $X \times \mathbb{P}(X)$ or $X \times \mathbb{P}(H)$. Thus if Γ is a gambling house on X, then $\text{sco}\, \Gamma$ is the gambling house whose x-section is the strong convex hull of $\Gamma(x)$; or if Σ is a subset of $X \times \mathbb{P}(H)$ then $\overline{\text{sco}}\, \Sigma$ is the subset of $X \times \mathbb{P}(H)$ whose x-section is the norm-closure of the strong convex hull of $\Sigma(x)$ and $\overline{\Sigma}$ is the subset of $X \times \mathbb{P}(H)$ whose x-section is the norm closure of $\Sigma(x)$.

For X countable, there are versions of Theorems 1.2 and 1.4 in Maitra and Sudderth ([13], 6.8.16 and 6.8.21). A finitely additive version of Theorem 1.2 is in Armstrong ([1]).

The article is organized as follows. Section 2 contains a summary of the properties of the Mokobodzki capacity and related results. In section 3 we prove that Γ^c is an analytic gambling house. Section 4 is devoted to results in gambling theory. The proofs of the theorems stated in this section are in section 5.

2 THE MOKOBODZKI CAPACITY

As in the Dellacherie-Meyer proof of Theorem 1.2, the ad hoc capacity of Mokobodzki will play a crucial role in our proofs. We will also need the effective analogue of the Mokobodzki capacity as defined by Louveau ([12]). In this section we summarize the properties of the Mokobodzki capacity and prove some consequences of these properties and other related results which will be used in the sequel.

Let X be a compact metric space and let λ be a probability measure on the Borel σ-field of X. For $A \subseteq \mathbb{P}(X)$, define

$$I(A;\lambda) = I(A) = \inf\{\sup_{\eta \in A} \eta(f) + \lambda(1-f) : f \in \Phi\},$$

where Φ is the set of Borel measurable functions on X into $[\,0,\,1]$ and we write $\mu(f)$ for $\int f\,d\mu$ when $f \in \Phi$ and $\mu \in \mathbb{P}(X)$.

Theorem 2.1. *([5], 35) For fixed $\lambda \in \mathbb{P}(X)$,*
(a) $I(\cdot;\lambda) = I$ is a capacity on $\mathbb{P}(X)$.
(b) If A is an analytic, strongly convex subset of $\mathbb{P}(X)$, then

$$I(A) = \sup_{\eta \in A} (\eta \wedge \lambda)(1),$$

where 1 is the function that is identically equal to one on X and $\wedge$ is the minimum operation in the lattice of bounded, signed measures.

Corollary 2.2. *([5], 34). If A is a strongly convex, analytic subset of $\mathbb{P}(X)$, then*

$$\lambda \in norm - cl(A) \iff (\forall f \in \Phi)(\lambda(f) \leq \sup_{\eta \in A} \eta(f)),$$

where norm stands for the total variation norm.

Proof. The 'only if' part is easy. For the 'if' part, the hypothesis is equivalent to the statement that $I(A) \geq 1$. Hence, for each n, there is $\eta_n \in A$ such that $\eta_n \wedge \lambda(1) > 1 - \frac{1}{n}$ by virtue of Theorem 2.1. Now

$$\eta_n \wedge \lambda = \lambda - (\eta_n - \lambda)^-,$$

hence

$$\lambda(1) - (\eta_n - \lambda)^-(1) > 1 - \frac{1}{n}$$

so that, since $\lambda(1) = 1$,

$$(\eta_n - \lambda)^-(1) < \frac{1}{n}.$$

Also

$$\eta_n \wedge \lambda = \eta_n - (\lambda - \eta_n)^-,$$

from which it follows that, since $(\eta_n - \lambda)^+ = (\lambda - \eta_n)^-$,

$$(\eta_n - \lambda)^+(1) < \frac{1}{n}.$$

Hence,

$$\| \eta_n - \lambda \| = (\eta_n - \lambda)^+(1) + (\eta_n - \lambda)^-(1) < \frac{2}{n},$$

so $\| \eta_n - \lambda \| \to 0$ as $n \to \infty$. Consequently, $\lambda \in norm - cl(A)$. □

Lemma 2.3. *If* Γ *is an analytic gambling house on* X*, then* $\operatorname{sco}\Gamma$ *and* $\overline{\operatorname{sco}}\,\Gamma$ *are both analytic.*

Proof. First note that the set

$$E = \{(x,\mu) \in X \times \mathbb{P}(\mathbb{P}(X)) : \mu(\Gamma(x)) = 1\}$$

is analytic ([10], 29.26). Consequently, the set $\operatorname{sco}\Gamma$ is analytic, since it is the projection to the first two coordinates of the analytic set

$$\{(x,\nu,\mu) \in X \times \mathbb{P}(X) \times \mathbb{P}(\mathbb{P}(X)) : (x,\mu) \in E \\ \text{and } \nu(\cdot) = \int \eta(\cdot)\mu\,(d\eta)\}.$$

The set $\overline{\operatorname{sco}}\,\Gamma$ is analytic, since it is the projection to the first two coordinates of the analytic set

$$\{(x,\nu,(\mu_n)) \in X \times \mathbb{P}(X) \times \mathbb{P}(X)^N : \mu_n \in \operatorname{sco}\Gamma(x),\ n = 1,2,\ldots \\ \text{and } \| \mu_n - \nu \| \to 0 \text{ as } n \to \infty\}.$$

To see that the above set is analytic, use the fact from ([7]) that the map $\mu \to \| \mu \|$ is Borel measurable on the space of bounded signed measures (on X) equipped with the weak topology. □

Lemma 2.4. *Let* $\mu_n, \mu \in \mathbb{P}(X)$ *be such that* $\mu_n \to \mu$ *in norm. Then, for any bounded, upper analytic function* g *on* X*,* $\int g\,d\mu_n \to \int g\,d\mu$*.*

Proof. Choose $\nu \in \mathbb{P}(X)$ such that μ_n, μ are absolutely continuous with respect to ν. Then there is a bounded, Borel measurable function f on X such that $f = g\,a.s(\nu)$. Hence

$$\int g\,d\mu_n = \int f\,d\mu_n$$

and

$$\int g\,d\mu = \int f\,d\mu.$$

The conclusion now follows from the fact that $\int f\,d\mu_n \to \int f\,d\mu$, since $\mu_n \to \mu$ in norm. □

We now turn to the effective analogue of the capacity I. Effective descriptive set theory takes place in recursively presented Polish spaces (see [18] for details). We will take X to be the recursively presented compact metric space 2^ω and λ to be a Δ^1_1 probability measure. For $A \subseteq \mathbb{P}(X)$, let

$$J(A;\lambda) = J(A) = \inf\{\sup_{\eta\in A} \eta(f) + \lambda(1-f) : f \text{ is } \Delta^1_1 - \text{ recursive on } X \text{ into } [0,1]\}.$$

Theorem 2.5. *([12], 2.4, 2.12(a), 2.14(a)) If* A *is a* Σ^1_1 *subset of* $\mathbb{P}(X)$*, then*

$$I(A) = \inf\{J(C) : C \text{ is } \Delta^1_1 - \text{ recursive and } A \subseteq C\}.$$

An immediate consequence is

Corollary 2.6. *If A is a Σ_1^1 subset of $\mathbb{P}(X)$, $\lambda \in \mathbb{P}(X)$ is a Δ_1^1 measure and $I(A) < 1$, then there is a Δ_1^1-recursive function f on X into [0, 1] such that*

$$\sup_{\eta \in A} \eta(f) + \lambda(1-f) < 1.$$

For the next result, we need a coding of $\Delta_1^1(\alpha)$-recursive functions on $X = 2^\omega$ into [0, 1], that is, a set W and a function U with the following properties:
(i) W is a Π_1^1 subset of ω^ω,
(ii) U is a Π_1^1-recursive partial function on $\omega^\omega \times \omega \times X$ into [0, 1],
(iii) if $(\alpha, n) \in W$ and $(\forall x)(U(\alpha, n, x)$ is defined), then $U(\alpha, n, .)$ is a $\Delta_1^1(\alpha)$-recursive function on X into [0, 1] , and
(iv) if g is a $\Delta_1^1(\alpha)$-recursive function on X into [0, 1], then there is n such that $(\alpha, n) \in W$ and $(\forall x)(g(x) = U(\alpha, n, x))$.

Such a coding is easy to construct from the coding of $\Delta_1^1(\alpha)$ subsets of $X\times$ [0, 1] (see ([11], p.13)). For the next result, regard 2^ω as a Π_1^0 subset of ω^ω.

Theorem 2.7. *Let Γ be a Σ_1^1 gambling house on $X = 2^\omega$. Suppose that $x \to \mu_x$ is a Δ_1^1-recursive function on a Δ_1^1 set $E \subseteq X$ into $\mathbb{P}(X)$. Assume that $I(\Gamma(x); \mu_x) = I_x(\Gamma(x)) < 1$ for all $x \in E$. Then there is a Δ_1^1-recursive function $f : E \times X \to [0,1]$ such that*

$$\sup_{\eta \in \Gamma(x)} \int f(x,y)\, \eta(dy) + \int (1 - f(x,y))\, \mu_x(dy) < 1$$

for each $x \in E$.

Proof. Let

$$P(x,n) \leftrightarrow x \in E \,\&(x,n) \in W \,\&(\forall y)(U(x,n,y) \text{ is defined}) \\ \&\left(\sup_{\eta \in \Gamma(x)} \int U(x,n,\cdot\,)\, d\eta + \int (1 - U(x,n,\cdot\,))\, d\mu_x < 1 \right).$$

It is easy to check that P is Π_1^1. It follows by relativizing Corollary 2.6 that $(\forall x \in E)(\exists n)\, P(x,n)$. So, by Kreisel's selection theorem ([18], 4B.5), there is a Δ_1^1-recursive function $\varphi : E \to \omega$ such that $(\forall x \in E)\, P(x, \varphi(x))$. Set

$$f(x,y) = U(x, \varphi(x), y), \qquad x \in E,\, y \in X.$$

It is now easily verified that f satisfies the assertions of the theorem. □

The bold-face version of this theorem is obtained by replacing Σ_1^1 by analytic and Δ_1^1-recursive by Borel measurable.

3 THE GAMBLING HOUSE Γ^c

The present section contains the proof that Γ^c is analytic whenever Γ is an analytic gambling house on X. We start with a technical result.

Lemma 3.1. *([14], Lemma 2.2) Suppose that X and Y are Borel subsets of Polish spaces. Then there is a Borel measurable mapping $(x, \mu) \to \mu[x]$ from $X \times \mathbb{P}(X \times Y)$ to $\mathbb{P}(Y)$ such that $\mu[x]$ is a version of the μ-regular conditional distribution on Y given x.*

For $\mu \in \mathbb{P}(X \times X)$ or $\mu \in \mathbb{P}(H)$, $\mu\pi_i^{-1}$ will denote the μ-distribution of the i-th coordinate, $i \geq 1$; $\mu[x]$ is a version of the μ-regular conditional distribution of the remaining coordinates given that the first coordinate is x such that $\mu[x]$ is jointly Borel measurable in μ and x, as is guaranteed by Lemma 3.1.

Suppose now that Γ is an analytic gambling house on X. If $\Delta \subseteq X \times \mathbb{P}(X)$, denote by Δ^* the subset of $X \times \mathbb{P}(X)$ whose x-section $\Delta^*(x)$ is $\Delta(x) \bigcup \{\delta(x)\}$. Define next an operator χ that takes subsets of $X \times \mathbb{P}(X)$ to subsets of $X \times \mathbb{P}(X)$ as follows:for $\Delta \subseteq X \times \mathbb{P}(X)$and $x \in X$, the x-section of $\chi(\Delta)$,namely, $\chi(\Delta)(x)$, is defined as the set of all $\gamma \in \mathbb{P}(X)$ such that $(\exists \mu \in \mathbb{P}(X \times X))$ satisfying these three conditions

(i) $\mu\pi_1^{-1} \in \Gamma(x)$,

(ii) $\mu\pi_2^{-1} = \gamma$, and

(iii) $(\mu\pi_1^{-1})^*(\{ x' \in X : \mu[x'] \in \Delta^*(x') \}) = 1$.

Here $(\mu\pi_1^{-1})^*$ is the outer measure induced by $\mu\pi_1^{-1}$.

We also define an operator ψ that takes subsets of $X \times \mathbb{P}(X)$ to subsets of $X \times \mathbb{P}(X) \times \mathbb{P}(X \times X)$ by letting the x-section of $\psi(\Delta)$, namely, $\psi(\Delta)(x)$, be the set of all pairs (γ, μ) in $\mathbb{P}(X) \times \mathbb{P}(X \times X)$ satisfying conditions (i), (ii), and (iii) above.

Lemma 3.2. *If Δ is an analytic subset of $X \times \mathbb{P}(X)$, then $\chi(\Delta)$ is an analytic gambling house on X.*

Proof. First observe that $\psi(\Delta)$ is the intersection of three sets, the first of which is clearly analytic and the second Borel. The third is analytic by virtue of the fact that Δ^* is analytic and ([10], 29.26). So $\psi(\Delta)$ is analytic. Since $\chi(\Delta)$ is the projection to the first two coordinates of $\psi(\Delta)$, it follows that $\chi(\Delta)$ is analytic. To see that $\chi(\Delta)$ is a gambling house, note that for each $x \in X$, $\chi(\Delta)(x) \supseteq \chi(\emptyset) = \Gamma(x)$ and so $\chi(\Delta)(x)$ is nonempty. This completes the proof. □

Define by induction on n subsets Γ_n of $X \times \mathbb{P}(X)$ as follows:

$$\Gamma_0 = \emptyset, \qquad \text{and} \qquad \Gamma_{n+1} = \chi(\Gamma_n), \qquad n \geq 0.$$

It is easy to see that $\Gamma_n \subseteq \Gamma_{n+1}$ since χ is monotone. Also, by Lemma 3.2, the gambling houses Γ_n are analytic. Finally, set

$$\Gamma_\infty = \bigcup_{n \geq 0} \Gamma_n.$$

Then Γ_∞ is an analytic house on X.

Here is the main result of this section.

Theorem 3.3. *If Γ is an analytic gambling house on X, then*

$$\Gamma^c(x) = \Gamma_\infty(x) \cup \{\delta(x)\}$$

for each $x \in X$. Consequently, Γ^c is analytic.

Proof. For a policy $\pi = (\sigma, t)$ available in Γ at x, denote by $\bar{\gamma}(\pi)$ the distribution of the terminal state h_t under σ.

To start with, recall that if $\pi = (\sigma, t)$ is available in Γ at x and $t \equiv 0$ then $\bar{\gamma}(\pi) = \delta(x)$. So to prove the inclusion $\subseteq$, it will suffice to show that if $\pi = (\sigma, t)$ is available in Γ at x and $1 \leq t \leq n$, then $\bar{\gamma}(\pi) \in \Gamma_n(x)$. The proof is by induction on n. For $n = 1$ the assertion is clear. So suppose the assertion is true for $n = m$. Let $\pi = (\sigma, t)$ be available in Γ at x_0 and suppose that $1 \leq t \leq m + 1$. It is easy to verify that $x \to \bar{\gamma}(\sigma[x], t[x])$ is σ_0 - measurable. Furthermore, $\bar{\gamma}(\sigma[x], t[x]) = \delta(x)$ if $t[x] \equiv 0$ while if $t[x] \neq 0$, then $1 \leq t[x] \leq m$, so that $\bar{\gamma}(\sigma[x], t[x]) \in \Gamma_m(x)$ by virtue of the inductive hypothesis. It follows that $\bar{\gamma}(\pi) \in \Gamma_{m+1}(x_0)$, since $\sigma_0 \in \Gamma(x_0)$ and

$$\bar{\gamma}(\pi)(B) = \int \bar{\gamma}(\sigma[x], t[x])(B)\, \sigma_0(dx)$$

for every Borel subset B of X.

For the reverse inclusion $\supseteq$, we will prove, again by induction, that there is a universally measurable function $\phi_n : \Gamma_n \to \mathbb{P}(X)$ such that $\phi_n(x, \gamma) \in \Sigma_\Gamma(x)$ for all $(x, \gamma) \in \Gamma_n$, and a universally measurable function $t_n : \Gamma_n \times H \to N$ with $t_n(x, \gamma, \cdot)$ a stop rule on H and $t_n \leq n$ for all $(x, \gamma) \in \Gamma_n$ such that whenever $(x, \gamma) \in \Gamma_n$, $\gamma = \bar{\gamma}(\phi_n(x, \gamma), t_n(x, \gamma, \cdot)$.

For $n \geq 1$, fix a universally measurable function f_n from $\chi(\Gamma_n)$ to $\mathbb{P}(X \times X)$ such that $(x, \gamma, f_n(x, \gamma)) \in \psi(\Gamma_n)$ for every $(x, \gamma) \in \chi(\Gamma_n)$. The existence of f_n is guaranteed by the von Neumann selection theorem ([10], 29.9). Use the theorem one more time to fix a universally measurable selector f^* for Γ. For each $x \in X$, let $\sigma^*(x)$ be the strategy that uses as initial gamble $f^*(x)$ and thereafter uses $f^*(y)$ when the current state is y. Note that $\sigma^*(x) \in \Sigma_\Gamma(x)$.

To start the induction, set

$$(\phi_1(x, \gamma))_0 = \gamma$$

and the conditional strategy

$$(\phi_1(x, \gamma))[x_1] = \sigma^*(x_1)$$

for $(x, \gamma) \in \Gamma_1 = \Gamma$. (Note that a strategy σ is completely determined by the specification of σ_0 and the collection of conditional strategies $\sigma[x], x \in X$.) Let

$$t_1(x, \gamma, h) = 1 \qquad \text{for } (x, \gamma) \in \Gamma_1 \text{ and } h \in H.$$

Suppose now that ϕ_n, t_n have been defined. Set $\phi_{n+1} = \phi_n$ on Γ_n and $t_{n+1} = t_n$ on $\Gamma_n \times H$. We will next define $\phi_{n+1}(x, \gamma)$ and $t_{n+1}(x, \gamma, \cdot)$ for $(x, \gamma) \in \Gamma_{n+1} - \Gamma_n$. First let

$$(\phi_{n+1}(x, \gamma))_0 = f_n(x, \gamma)\pi_1^{-1}$$

and define the conditional strategies by

$$(\phi_{n+1}(x, \gamma))[x_1] = \begin{cases} \phi_n(x_1, (f_n(x, \gamma))[x_1]), & \text{if } (f_n(x, \gamma))[x_1] \in \Gamma_n(x_1), \\ \sigma^*(x_1), \text{ otherwise.} & \end{cases}$$

Note that $\phi_{n+1}(x, \gamma) \in \Sigma_\Gamma(x)$ for all $(x, \gamma) \in \Gamma_{n+1}$. Define $t_{n+1}(x, \gamma, \cdot)$ to be the stop rule whose conditional stop rules are given by

$$(t_{n+1}(x, \gamma, \cdot))[x_1] = \begin{cases} t_n(x_1, (f_n(x, \gamma))[x_1]), \cdot), & \text{if } (f_n(x, \gamma))[x_1] \in \Gamma_n(x_1), \\ 0, & \text{otherwise.} \end{cases}$$

It is easy to verify that ϕ_{n+1} and t_{n+1} are universally measurable and that $t_{n+1}(x,\gamma,\cdot)$ is a stop rule with $t_{n+1} \le n+1$ for every $(x,\gamma) \in \Gamma_{n+1}$.

Finally, let $\gamma \in \Gamma_{n+1}(x)$. If $\gamma \in \Gamma_n(x)$, then

$$\bar{\gamma}(\phi_{n+1}(x,\gamma), t_{n+1}(x,\gamma,\cdot)) = \bar{\gamma}(\phi_n(x,\gamma), t_n(x,\gamma,\cdot)) = \gamma,$$

by the inductive hypothesis. So suppose that $\gamma \in \Gamma_{n+1}(x) - \Gamma_n(x)$. Let B be a Borel subset of X and define $A = \{x_1 : (f_n(x,\gamma))[x_1] \in \Gamma_n(x_1)\}$. Then

$$\begin{aligned}
&\bar{\gamma}(\phi_{n+1}(x,\gamma), t_{n+1}(x,\gamma,\cdot))(B) \\
&\quad = \int \bar{\gamma}((\phi_{n+1}(x,\gamma))[x_1], t_{n+1}(x,\gamma,\cdot))[x_1])(B)\,(\phi_{n+1}(x,\gamma))_0(dx_1) \\
&\quad = \int \bar{\gamma}((\phi_{n+1}(x,\gamma))[x_1], t_{n+1}(x,\gamma,\cdot))[x_1])(B)\,(f_n(x,\gamma)\pi_1^{-1})(dx_1) \\
&\quad = \int_A \bar{\gamma}((\phi_{n+1}(x,\gamma))[x_1], t_{n+1}(x,\gamma,\cdot))[x_1])(B)\,(f_n(x,\gamma)\pi_1^{-1})(dx_1) \\
&\qquad + \int_{A^c} \bar{\gamma}(\sigma^*(x_1), t^*)(B)\,(f_n(x,\gamma)\pi_1^{-1})(dx_1) \\
&\quad = \int_A ((f_n(x,\gamma))[x_1])(B)\,(f_n(x,\gamma)\pi_1^{-1})(dx_1) \\
&\qquad + \int_{A^c} \delta(x_1)(B)\,(f_n(x,\gamma)\pi_1^{-1})(dx_1) \\
&\quad = \int ((f_n(x,\gamma))[x_1])(B)(f_n(x,\gamma)\pi_1^{-1})(dx_1) \\
&\quad = (f_n(x,\gamma)\pi_2^{-1})(B) \\
&\quad = \gamma(B),
\end{aligned}$$

where t^* is the identically zero stop rule and the fourth equality is by virtue of the fact that $t^* \equiv 0$. Consequently, $\bar{\gamma}(\phi_{n+1}(x,\gamma), t_{n+1}(x,\gamma,\cdot)) = \gamma$. This completes the proof. □

4 Gambling

We turn to gambling theory in this section. Let Γ be a gambling house on X. We associate with Γ an operator G_Γ as follows: for any bounded, upper analytic function f on X

$$(G_\Gamma f)(x) = \sup_{\gamma \in \Gamma(x)} \int f\,d\gamma, \qquad x \in X.$$

It is easy to verify that if Γ is analytic and f is upper analytic, then $G_\Gamma f$ is upper analytic.

Lemma 4.1. *If Γ is an analytic gambling house on X, then $R_\Gamma \approx G_{\Gamma^c}$. Consequently, $R_\Gamma f$ is upper analytic if f is a bounded, upper analytic function on X.(Recall that $R_\Gamma \approx G_{\Gamma^c}$ means that R_Γ and G_{Γ^c} agree on bounded, upper analytic functions on X.)*

Proof. Let f be a bounded, upper analytic function on X. Then, by the change of variable theorem, for any $x \in X$,

$$(G_{\Gamma^c} f)(x) = \sup \left\{ \int f(h_t)\, d\sigma : \sigma \in \Sigma(x) \,\&\, t \text{ a bounded stop rule} \right\} \leq (R_\Gamma f)(x).$$

For the reverse inequality, let $\pi = (\sigma, t)$ be a policy available in Γ at x. Then, by the Dominated Convergence Theorem,

$$\lim_n \int f(h_{t \wedge n})\, d\sigma = \int f(h_t)\, d\sigma$$

where $(t \wedge n)(h) = \min\{t(h), n\}$, $h \in H$. It follows that

$$\int f(h_t)\, d\sigma \leq (G_{\Gamma^c} f)(x).$$

Now take the sup over all policies π available at x to get

$$(R_\Gamma f)(x) \leq (G_{\Gamma^c} f)(x).$$

The second assertion is now a consequence of Theorem 3.3. □

Lemma 4.2. *Let Γ be an analytic gambling house on X. If g is a bounded Γ-excessive function on X, then g is Γ^s-excessive.*

Proof. We first prove that g is Γ^c-excesssive. Fix $x_0 \in X$ and $\gamma \in \Gamma^c(x_0)$. Choose a policy $\pi = (\sigma, t)$ available in Γ at x with t bounded such that $\gamma = \sigma\psi_t^{-1}$. Since g is Γ-excessive, it follows that, under σ, $g(x_0)$, $g(h_1)$, $g(h_2)$, ... is a supermartingale. So, by the optional sampling theorem ([6], 2.12.2),

$$\int g(h_t)\, d\sigma \ \leq\ g(x_0)$$

Hence, by the change of variable theorem,

$$\int g\, d\gamma \leq g(x_0)$$

This shows that g is Γ^c-excessive. It is now easy to verify first that g is $sco\,\Gamma^c$-excessive and then that g is $\overline{sco}\,\Gamma^c$-excessive by using Lemma 2.4. So, by (1.2), g is Γ^s-excessive. □

Lemma 4.3. *If Γ is an analytic gambling house on X, then $R_\Gamma \approx R_{\Gamma^s}$.*

Proof. Let f be a bounded, upper analytic function on X. Note that Γ^s is analytic by Theorem 3.3 and Lemma 2.3. Moreover, $\Gamma \subseteq \Gamma^c \subseteq \Gamma^s$, so $R_\Gamma f \leq R_{\Gamma^s} f$. For the reverse inequality, recall that $R_\Gamma f$ is Γ-excessive by virtue of the Fundamental Theorem applied to Γ. Hence, by Lemma 4.2, $R_\Gamma f$ is Γ^s-excessive. Also $R_\Gamma f \geq f$. So, by the Fundamental Theorem applied to Γ^s, $R_\Gamma f \geq R_{\Gamma^s} f$. □

The next theorem states that the operator M_Γ is determined by G_Γ for analytic gambling houses. The proof is based on a number of results proved elsewhere. We will now summarize these results.

Lemma 4.4. *If Γ and Γ' are analytic gambling houses on X and $G_\Gamma \approx G_{\Gamma'}$, then $R_\Gamma \approx R_{\Gamma'}$*

Proof. Since $G_\Gamma \approx G_{\Gamma'}$, it follows that the class of Γ-excessive functions is exactly the same as the class of Γ'-excessive functions. Now use the Fundamental Theorem to see that $R_\Gamma \approx R_{\Gamma'}$. □

With each analytic gambling house Γ on X, we associate the operator T_Γ as follows:

$$(T_\Gamma u)(x) = \sup_\pi \int u(h_t)\, d\sigma, \qquad x \in X,$$

where u is a bounded, upper analytic function on X and the sup is taken over all policies $\pi = (\sigma, t)$ available in Γ at x such that $t \geq 1$.

The next lemma shows that T_Γ is closely related to R_Γ.

Lemma 4.5. *If Γ is an analytic gambling house on X, then $T_\Gamma \approx G_\Gamma \circ R_\Gamma$.*

Proof. Let g be a bounded, upper analytic function on X. Fix $x_0 \in X$. Let $\pi = (\sigma, t)$ be a policy available in Γ at x_0 with $t \geq 1$. Then

$$\int g(h_t)\, d\sigma = \int [\int g(h'_{t[x]})\sigma[x](dh')]\sigma_0(dx) \leq \int (R_\Gamma g)(x)\, \sigma_0(dx) \leq (G_\Gamma(R_\Gamma g))(x_0),$$

where the last inequality is by virtue of the fact that $\sigma_0 \in \Gamma(x_0)$. Now take the sup of the left side over all $\pi = (\sigma, t)$ available in Γ at x_0 with $t \geq 1$ to get

$$(T_\Gamma g)(x_0) \leq (G_\Gamma(R_\Gamma g))(x_0).$$

For the reverse inequality, fix $\epsilon > 0$. Choose $\gamma_0 \in \Gamma(x_0)$ such that

$$\int (R_\Gamma g)\, d\gamma_0 \geq (G_\Gamma(R_\Gamma g))(x_0) - \epsilon/2.$$

Next use a selection theorem (see, for example, [15], Lemma 2.1) to choose $\gamma_x \in \Gamma^c(x)$ such that $x \to \gamma_x$ is universally measurable and

$$\int g\, d\gamma_x > (R_\Gamma g)(x) - \epsilon/2$$

for each $x \in X$.

Now define a policy $\pi = (\sigma, t)$ available in Γ atx_0 as follows:

$$\sigma_0 = \gamma_0$$

and

$$\sigma[x] = \begin{cases} \phi_n(x, \gamma_x), & \text{if } (x, \gamma_x) \in \Gamma_n - \Gamma_{n-1}, n \geq 1, \\ \sigma^*(x), & \text{otherwise,} \end{cases}$$

where $\phi_n, \sigma^*(x)$ are defined in the proof of Theorem 3.3 and Γ_n is defined just before the statement of Theorem 3.3; t is defined so that

$$t[x] = \begin{cases} t_n(x, \gamma_x, \cdot), & \text{if } (x, \gamma_x) \in \Gamma_n - \Gamma_{n-1}, n \geq 1 \\ 0, & \text{otherwise,} \end{cases}$$

where t_n is defined in the proof of Theorem 3.3. Note that $t \geq 1$. Consequently,

$$\begin{aligned}
(T_\Gamma g)(x_0) &\geq \int g(h_t)\, d\sigma \\
&= \int [\int g(h'_{t[x]})\, \sigma[x](dh')]\, \sigma_0(dx) \\
&= \int [\int g(y)\, \gamma_x(dy)]\, \gamma_0(dx) \\
&\geq \int (R_\Gamma g)(x)\, \gamma_0(dx) - \epsilon/2 \\
&\geq (G_\Gamma(R_\Gamma))(x_0) - \epsilon,
\end{aligned}$$

where the second equality holds because $\overline{\gamma}(\phi_n(x, \gamma_x), t_n(x, \gamma_x, \cdot)) = \gamma_x$ if $\gamma_x \in \Gamma_n - \Gamma_{n-1}$ for some $n \geq 1$ and $t[x] = 0$ otherwise. Since ϵ is arbitrary, we have: $(T_\Gamma g)(x_0) \geq (G_\Gamma(R_\Gamma g))(x_0)$. This completes the proof.

□

The next result is a characterization of $V_\Gamma u$ for bounded, upper analytic u.

Lemma 4.6. *([8], Theorem 7.1) If Γ is an analytic gambling house on X and u is a bounded, upper analytic function on X, then $V_\Gamma u$ is the largest bounded, upper analytic function v on X such that $T_\Gamma(u \wedge v) = v$, where $u \wedge v$ is the pointwise minimum of u and v.*

An immediate consequence of Lemmas 4.4-4.6 is

Lemma 4.7. *If Γ and Γ' are analytic gambling houses on X such that $G_\Gamma \approx G'_\Gamma$, then $V_\Gamma \approx V_{\Gamma'}$.*

We now define a class $\mathcal{F}$ of relatively simple functions on H, which will be used to approximate bounded, upper analytic functions. Let $\mathcal{F}$ be the set of all $f : H \to [0,1]$ such that f takes on finitely many values and $[f \geq c]$ is a countable intersection of Borel, open sets for each $c \in [0,1]$, where "open" refers to the product topology on H when X is given the discrete topology.

Let $X^* = \bigcup_{n \geq 0} X^n$ and $H^* = (X^*)^N$. For $h \in H$, we will write $p_n(h)$ for $(h_1, h_2, \cdots, h_n)$. Let $\phi : H \to H^*$ be defined by setting

$$\phi(h) = (p_1(h), p_2(h), \cdots).$$

Set $H' = \phi(H)$. The next lemma gives a representation for elements of $\mathcal{F}$.

Lemma 4.8. *If $f \in \mathcal{F}$, then there is a Borel measurable function $u : X^* \to [0,1]$ such that*

$$f(h) = \limsup_n u(p_n(h))$$

for every $h \in H$; that is, $f = u^* \circ \phi$.

Proof. Suppose that f assumes the values $a_1, a_2, \cdots, a_m$ with $a_1 < a_2 < \cdots < a_m$. By Lemma 6.6 in ([15]), we can choose, for each $i = 1, 2, \cdots, m$, a Borel subset S_i of X^* such that

$$[f \geq a_i] = \{h \in H : p_n(h) \in S_i \text{ for infinitely many } n\}, \tag{4.1}$$

and $X^* = S_1 \supseteq S_2 \supseteq \cdots \supseteq S_m$.

Write $[S_i\, i.o.]$ for the set on the right side of (4.1) and e for the empty sequence in X^*. Define u on X^* as follows:

$$u(e) = 0$$

and, for $p \in X^* - \{e\}$,

$$u(p) = a_i \qquad \text{if } p \in S_i - S_{i+1}\,, i = 1, 2, \cdots, m,$$

where $S_{m+1} = \emptyset$.

To complete the proof, suppose that $f(h) = a_i$. Then $h \in [S_i\, i.o.]$ and $h \notin [S_{i+1}\, i.o.]$. It follows that $h \in [S_i - S_{i+1}\, i.o.]$, so that $u^*(\phi(h)) \geq a_i$. But $u^*(\phi(h)) \leq a_i$ because $h \notin [S_{i+1}\, i.o.]$. Hence $u^*(\phi(h)) = a_i$. □

The next result implies that the operator M_Γ is determined by its values on the class $\mathcal{F}$.

Lemma 4.9. *([16],Theorem 10.1) If Γ is an analytic house on X and $g : H \to [0, 1]$ is upper analytic, then*

$$(Mg)(x) = \inf\{(Mf)(x) \,:\, f \in \mathcal{F} \; and \; f \geq g\}$$

for every $x \in X$.

Theorem 4.10. *If Γ and Γ' are analytic gambling houses on X such that $G_\Gamma \approx G_{\Gamma'}$, then $M_\Gamma \approx M_{\Gamma'}$.*

Proof. By virtue of Lemma 4.9, it suffices to prove that $M_\Gamma = M_{\Gamma'}$ on the class $\mathcal{F}$. So fix $x_0 \in X$ and $f \in \mathcal{F}$. By Lemma 4.8, choose a Borel measurable function $u : X^* \to [0, 1]$ such that $f = u^* \circ \varphi$.

Now consider the nonleavable gambling problems (X^*, Γ^*, u^*) and (X^*, Γ'^*, u^*), where

$$\Gamma^*(p) = \{\gamma\varphi_p^{-1} : \gamma \in \Gamma(l(p))\},\, p \in X^*,$$

$$\varphi_p(x) = px,\, x \in X,\, p \in X^*,$$

and $l(p)$ is the last coordinate of p if $p \neq e$, while $l(e) = x_0$.
Similarly, define Γ'^* from Γ'. Observe that, if σ is a strategy available in Γ^* or Γ'^* at e, then $\sigma(H') = 1$. It is then easy to verify that

$$(M_\Gamma f)(x_0) = (V_{\Gamma^*} u)(e)$$

and

$$(M_{\Gamma'} f)(x_0) = (V_{\Gamma'^*} u)(e).$$

Consequently, the proof will be complete as soon as we establish that $V_{\Gamma^*} \approx V_{\Gamma'^*}$. This in turn will be proved, courtesy of Lemma 4.7, if we can show that $G_{\Gamma^*} \approx G_{\Gamma'^*}$. So let g be a bounded, upper analytic function on X^*. Then, for any $p \in X^*$,

$$(G_{\Gamma^*} g)(p) = (G_\Gamma g_p)(l(p))$$

and

$$(G_{\Gamma'^*} g)(p) = (G_{\Gamma'} g_p)(l(p))$$

where $g_p(x) = g(px)$, $x \in X$. It follows that $G_{\Gamma^*} g = G_{\Gamma'^*} g$. This completes the proof. □

The converse of Theorem 4.10 holds in a very strong sense.

Lemma 4.11. *Suppose that* Γ, Γ' *are gambling houses on* X *such that* $\Sigma_{\Gamma(x)}$ *and* $\Sigma_{\Gamma'}(x)$ *are nonempty for each* $x \in X$ *(so* M_Γ *and* $M_{\Gamma'}$ *are defined on bounded, universally measurable functions on* H*). If* $M_\Gamma \sim M_{\Gamma'}$ *, then* $G_\Gamma \sim G_{\Gamma'}$*.*

Proof. Suppose that g is a bounded, Borel measurable function on X. Define $\tilde{g}$ on H by

$$\tilde{g}(h) = g(h_1).$$

Then, for any $x \in X$,

$$(G_\Gamma g)(x) = (M_\Gamma \tilde{g})(x) = (M_{\Gamma'} \tilde{g})(x) = (G_{\Gamma'} g)(x).$$

□

For the next theorem, recall that Φ is the set of Borel measurable functions from X into $[0,1]$.

Theorem 4.12. *Suppose that* Γ *is an analytic gambling house on* X*. If* f *is an upper analytic function on* X *into [0, 1], then*

$$(G_\Gamma f)(x) = \inf\{(G_\Gamma g)(x) : g \in \Phi \; and \, g \geq f\}$$

for each $x \in X$*.*

Proof. Fix $x_0 \in X$ and $\epsilon > 0$. Let

$$E = \{(x, a) \in X \times [0,1] : f(x) \geq a\}.$$

Then E is analytic. By Corollary 4.4 in ([15]), there is a Borel subset B of $X \times$ [0, 1] such that $B \supseteq E$ and

$$\sup_{\gamma \in \Gamma(x_0)} (\gamma \times \lambda)(B) \leq \sup_{\gamma \in \Gamma(x_0)} (\gamma \times \lambda)(E) + \epsilon, \tag{4.2}$$

where λ is Lebesgue measure on [0, 1]. Define g on X into [0, 1] by

$$g(x) = \lambda(B_x).$$

Then g is Borel measurable and $g \geq f$. It follows from (4.2) that

$$(Gg)(x_0) \leq (Gf)(x_0) + \epsilon.$$

This completes the proof. □

An immediate consequence of the previous result is:

Corollary 4.13. *Suppose that* Γ, Γ' *are analytic gambling houses on* X *such that* $G_\Gamma \sim G_{\Gamma'}$*. Then* $G_\Gamma \approx G_{\Gamma'}$*.*

We conclude this section with a result on the randomization of strategies. First we introduce some notation.

For $\mu \in \mathbb{P}(H)$ and $n \geq 0$, $(\mu)^n$ will denote the μ-probability distribution of the first $(n+1)$ coordinates and $\mu(x_1, x_2, \dots, x_n)$ will denote a version of the μ-conditional distribution of x_{n+1} given $x_1, x_2, \dots, x_n$ which is jointly Borel measurable in $\mu, x_1, x_2, \dots, x_n$, as guaranteed by Lemma 3.1.

Theorem 4.14. *Let Γ be an analytic gambling house on X. Then*

$$\operatorname{sco}\Sigma_\Gamma \subseteq \Sigma_{\operatorname{sco}\Gamma}.$$

Proof. Fix $x_0 \in X$ and $\nu \in \operatorname{sco}\Sigma_\Gamma(x_0)$. So there is a probability measure m on the Borel subsets of $\Sigma_\Gamma(x_0)$ such that

$$\nu(E) = \int_{\Sigma_\Gamma(x_0)} \mu(E)\, m(d\mu)$$

for Borel subsets E of H. We have to define a strategy σ available in $\operatorname{sco}\Gamma$ at x_0 such that $\nu = \sigma$.

Let

$$\sigma_0(B) = \int_{\Sigma_\Gamma(x_0)} (\mu)^0(B)\, m(d\mu)$$

for Borel subsets B of X. Plainly, $\sigma_0 \in \operatorname{sco}\Gamma(x_0)$ and $\sigma_0 = (\nu)^0$.

Suppose that σ_0, σ_1, ..., σ_{n-1} have been defined so that $\sigma_i(x_1, x_2, \dots, x_i) \in \operatorname{sco}\Gamma(x_i)$ for all x_1, x_2, ..., $x_i \in X$ and $(\sigma)^i$ = the σ-distribution of the first $(i+1)$ coordinates $= (\nu)^i$, $i = 0, 1, 2, \dots, n-1$. We will now define σ_n. By von Neumann's selection theorem ([10], 29.9), fix an analytically measurable selector ψ for Γ. Let

$$\varphi(\mu, x_1, x_2, \dots, x_n) = \begin{cases} \mu(x_1, x_2, \dots, x_n) & \text{if } \mu \in \Sigma_\Gamma(x_0)\ \&\ \mu(x_1, x_2, \dots, x_n) \in \Gamma(x_n) \\ \psi(x_n) & \text{if } \mu \in \Sigma_\Gamma(x_0)\ \&\ \mu(x_1, x_2, \dots, x_n) \notin \Gamma(x_n). \end{cases}$$

Then φ is universally measurable. Next define a probability measure P on the Borel subsets of $\Sigma(x_0) \times X^n$ such that

$$P(S \times B) = \int_S (\mu)^{n-1}(B)\, m(d\mu),$$

for Borel subsets S of $\Sigma(x_0)$ and B of X^n. Fix a version $P(\cdot \mid x_1, x_2, \dots, x_n)$ of the P-regular conditional probability on the Borel subsets of $\Sigma(x_0)$ given $x_1, x_2, \dots, x_n$. Finally set

$$\sigma_n(x_1, x_2, \dots, x_n)(B) = \int_{\Sigma(x_0)} \varphi(\mu, x_1, x_2, \dots, x_n)(B)\, P(d\mu \mid x_1, x_2, \dots, x_n)$$

for Borel subsets B of X and $x_1, x_2, \dots, x_n \in X$. Then σ_n is universally measurable and $\sigma_n(x_1, x_2, \dots, x_n) \in sco\Gamma(x_n)$.

Now let A be a Borel subset of X^n and B be a Borel subset of X. Abbreviate $(x_1, x_2, \dots, x_n)$ by $\vec{x}$ in the following calculation:

$$\begin{aligned}
(\nu)^n(A \times B) &= \int_{\Sigma(x_0)} (\mu)^n(A \times B)\, m(d\mu) \\
&= \int_{\Sigma(x_0)} \left[\int_A \varphi(\mu, \vec{x})(B)\, (\mu)^{n-1}(d\vec{x})\right] m(d\mu) \\
&= \int_{\Sigma(x_0) \times A} \varphi(\mu, \vec{x})(B)\, dP \\
&= \int_A \left[\int_{\Sigma(x_0)} \varphi(\mu, \vec{x})(B)\, P(d\mu \mid \vec{x})\right] (\nu)^{n-1}(d\vec{x}) \\
&= \int_A \sigma_n(\vec{x})(B)\, (\sigma)^{n-1}(d\vec{x}) \\
&= (\sigma)^n(A \times B),
\end{aligned}$$

where the fourth equality is by virtue of the fact that the marginal of P on X^n is $(\nu)^{n-1}$ and the penultimate equality is by the inductive hypothesis. It follows that $(\nu)^n = (\sigma)^n$ and the induction is complete. So $\nu = \sigma$, hence $\nu \in \Sigma_{\text{sco}\,\Gamma}(x_0)$. □

The reverse inclusion $\Sigma_{sco\Gamma} \subseteq sco\Sigma_\Gamma$ is also true. See Aumann ([2]) and especially, Feinberg ([9], Theorem 5.2), who proved the inclusion in the context of dynamic programming. But we will have no use for the result in this article.

5 Proofs of Theorems in Section 1

Lemma 5.1. *Suppose that Γ is an analytic gambling house on X. Then the largest gambling house Γ' such that $G_\Gamma \sim G_{\Gamma'}$ is $\overline{sco}\Gamma$.*

Proof. It is easy to verify that $G_\Gamma f = G_{sco\Gamma} f$ for $f \in \Phi$. Now $sco\Gamma$ is analytic by Lemma 2.3 and, obviously, $sco\Gamma(x)$ is strongly convex for each $x \in X$. So it follows from the 'only if' part of Corollary 2.2 that $G_{sco\Gamma} f = G_{\overline{\text{sco}}\,\Gamma} f$ for $f \in \Phi$. Consequently, $G_\Gamma \sim G_{\overline{\text{sco}}\,\Gamma}$. Suppose next that Γ' is a gambling house on X such that $G_{\Gamma'} \sim G_\Gamma$. Then $G_{\Gamma'} = G_{\overline{\text{sco}}\,\Gamma}$ on Φ. It nows follows from the 'if' part of Corollary 2.2 that $\Gamma' \subseteq \overline{sco}\Gamma$. □

Proof of Theorem 1.2. The final assertion of Theorem 1.2 follows from Theorem 3.3 and Lemma 2.3. By Lemma 4.3, $R_\Gamma \sim R_{\Gamma^s}$. Suppose then that Γ' is a gambling house on X such that $R_{\Gamma'} \sim R_\Gamma$. Let $x_0 \in X$ and $\gamma \in \Gamma'(x_0)$. Then, for any $g \in \Phi$,

$$\int g\, d\gamma \leq (R_{\Gamma'} g)(x_0) = (R_\Gamma g)(x_0) = (G_{\Gamma^c} g)(x_0)$$

where the last equality is by virtue of Lemma 4.1. Consequently,

$$(\forall g \in \Phi)(\gamma(g) \leq \sup_{\eta \in \Gamma^c(x_0)} \eta(g) = \sup_{\eta \in sco\Gamma^c(x_0)} \eta(g)),$$

from which it follows by Corollary 2.2 that $\gamma \in \overline{\text{sco}}\,\Gamma^c(x_0) = \Gamma^s(x_0)$. Since $\gamma \in \Gamma'(x_0)$ and x_0 were arbitrary, we have : $\Gamma' \subseteq \Gamma^s$. This completes the proof. □

Proof of Theorem 1.3. It follows from Lemma 5.1 and Corollary 4.13 that $G_\Gamma \approx G_{\overline{\text{sco}}\,\Gamma}$ since both Γ and $\overline{\text{sco}}\,\Gamma$ are analytic gambling houses. Hence, by Theorem 4.10, $M_\Gamma \sim M_{\overline{\text{sco}}\,\Gamma}$. Suppose next that Γ' is a gambling house on X such that $M_{\Gamma'} \sim M_\Gamma$. Then, by Lemma 4.11, $G_{\Gamma'} \sim G_\Gamma$, , so $\Gamma' \subseteq \overline{sco}\Gamma$ by Lemma 5.1. □

Proof of Theorem 1.4. By Theorem 1.3, $M_{\Sigma_\Gamma} \sim M_{\Sigma_{\overline{\text{sco}}\,\Gamma}}$. Suppose then that Σ is a global gambling house on X such that $M_\Sigma \sim M_{\Sigma_\Gamma}$ Choose a gambling house Γ' on X such that $\Sigma = \Sigma_{\Gamma'}$. Then $M_\Gamma \sim M_{\Gamma'}$, so $\Gamma' \subseteq \overline{\text{sco}}\,\Gamma$ by Theorem 1.3. Hence, $\Sigma = \Sigma_{\Gamma'} \subseteq \Sigma_{\overline{\text{sco}}\,\Gamma}$. □

Proof of Theorem 1.5. As observed above, $M_{\Sigma_\Gamma} \sim M_{\Sigma_{\overline{\text{sco}}\,\Gamma}}$ Suppose now that $\Sigma \subseteq X \times \mathbb{P}(H)$ such that $M_\Sigma \sim M_{\Sigma_\Gamma}$. By arguing as in the proof of Lemma 5.1, it is easy to show that $\Sigma \subseteq \overline{sco}\Sigma_\Gamma$. To complete the proof, it will suffice to show that $\overline{\text{sco}}\,\Sigma_\Gamma \subseteq \Sigma_{\overline{\text{sco}}\,\Gamma}$. By Theorem 4.14, $sco\Sigma_\Gamma \subseteq \Sigma_{sco\Gamma} \subseteq \Sigma_{\overline{\text{sco}}\,\Gamma}$. So we will be done as soon as we show that $\Sigma_{\overline{\text{sco}}\,\Gamma}(x)$ is norm-closed for each $x \in X$.

Fix $x_0 \in X$ and let $\mu_k \in \Sigma_{\overline{\text{sco}}\,\Gamma}(x_0), k \geq 1$, and assume that $\mu_k \to \mu$ in norm. In order to show that $\mu \in \Sigma_{\overline{\text{sco}}\,\Gamma}(x_0)$, we must prove that $(\mu)^0 \in \overline{sco}\Gamma(x_0)$ and for each $n \geq 1, \mu(x_1, x_2, \ldots, x_n) \in \overline{sco}\Gamma(x_n)$ almost surely $((\mu)^{n-1})$.

To see that $(\mu)^0 \in \overline{sco}\Gamma(x_0)$, note that $(\mu_k)^0 \in \overline{sco}\Gamma(x_0)$ and $(\mu^k)^0 \to (\mu)^0$ in norm. Suppose next that for some $n \geq 1$ there is a Borel set $E \subseteq X^n$ such that $(\mu)^{n-1}(E) > 0$ and $\mu(x_1, x_2, \ldots, x_n) \notin \overline{sco}\Gamma(x_n)$ for all $(x_1, x_2, \ldots, x_n) \in E$. Write $\vec{x}$ for $(x_1, x_2, \ldots, x_n)$ and, using the notation of sections 2 and 4, define

$$I_{\vec{x}}(A) = I(A; \mu(\vec{x}))$$

for $A \subseteq \mathbb{P}(X)$. Since $\mu(\vec{x}) \notin \overline{sco}\Gamma(x_n)$ for $\vec{x} \in E$, it follows that $I_{\vec{x}}(sco\Gamma(x_n)) < 1$ for all $\vec{x} \in E$ by virtue of Corollary 2.2. Hence, by the bold-face version of Theorem 2.7, there is a Borel measurable function $g : E \times X \to [0, 1]$ such that

$$\sup_{\gamma \in \text{sco}\,\Gamma(x_n)} \int g(\vec{x}, y)\gamma(dy) < \int g(\vec{x}, y)\mu(\vec{x})(dy),$$

for $\vec{x} \in E$. Hence

$$\int_{E\times X} gd(\mu)^n = \int_E \left[\int g(\vec{x}, y)\mu(\vec{x})(dy)\right] d(\mu)^{n-1} > \int_E (G_{\text{sco}\,\Gamma}g_{\vec{x}})(x_n)d(\mu)^{n-1}, \quad (5.1)$$

where $g_{\vec{x}}$ is the $\vec{x}$-section of g.

On the other hand, for $k \geq 1$,

$$\int_{E\times X} g\, d(\mu_k)^n = \int_E \left[\int g(\vec{x}, y)\mu_k(\vec{x})(dy)\right] d(\mu_k)^{n-1} \leq \int_E (G_{\text{sco}\,\Gamma}g_{\vec{x}})(x_n)d(\mu_k)^{n-1}, \quad (5.2)$$

since $\mu_k(\vec{x}) \in \overline{\text{sco}}\,\Gamma(x_n)$ almost surely $((\mu_k)^{n-1})$ and $G_{sco\Gamma} \sim G_{\overline{\text{sco}}\,\Gamma}$. Now, as is easily verified, the function $\vec{x} \to (G_{sco\Gamma}g_{\vec{x}})(x_n)$ is bounded and upper analytic. Since $(\mu_k)^i \to (\mu)^i$ in norm for each $i \geq 0$, by letting $k \to \infty$ in (5.2), we get, by virtue of Lemma 2.4,

$$\int_{E\times X} g\, d(\mu)^n \leq \int_E (G_{sco\Gamma}g_{\vec{x}})(x_n)\, d(\mu)^{n-1},$$

which contradicts (5.1). It follows that $\mu(\vec{x}) \in \overline{sco}\Gamma(x_n)$ almost surely $((\mu)^{n-1})$ This completes the proof. □

Corollary 5.2. *If Γ is an analytic gambling house on X, then*

$$\overline{\text{sco}}\,\Sigma_\Gamma = \Sigma_{\overline{\text{sco}}\,\Gamma} = \overline{\Sigma}_{sco\Gamma}.$$

In particular, $\overline{\text{sco}}\,\Sigma_\Gamma$ and $\overline{\Sigma}_{sco\Gamma}$ are global gambling houses on X.

Proof. The first equality is implicit in the proof of Theorem 1.5. The other equality is proved by observing that $\text{sco}\,\Sigma_\Gamma \subseteq \Sigma_{\text{sco}\,\Gamma}$ (Theorem 4.14), so $\overline{\text{sco}}\,\Sigma_\Gamma \subseteq \overline{\Sigma}_{\text{sco}\,\Gamma}$. On the other hand, $\Sigma_{\text{sco}\,\Gamma} \subseteq \Sigma_{\overline{\text{sco}}\,\Gamma}$ and hence $\overline{\Sigma}_{\text{sco}\,\Gamma} \subseteq \Sigma_{\overline{\text{sco}}\,\Gamma}$, since $\Sigma_{\overline{\text{sco}}\,\Gamma}(x)$ is norm-closed for each $x \in X$, as was observed in the course of proving Theorem 1.5. □

Proof of Theorem 1.6. Let $X = \{0, 1\}$ and define a gambling house Γ on X as follows:

$$\Gamma(0) = \{\delta(0)\} = \Gamma(1)$$

Then, for any real-valued function u on X,

$$(V_\Gamma u)(0) = u(0) = (V_\Gamma u)(1).$$

For each $a \in (0, 1)$, define a gambling house Γ^a thus:

$$\Gamma^a(0) = \{\delta(0)\}; \Gamma^a(1) = \{(1-b)\delta(0) + b\delta(1) : 0 \leq b \leq a\}.$$

It is easy to check that $V_{\Gamma^a} = V_\Gamma$ for every $a \in (0, 1)$. Towards a contradiction, assume that there is a largest house Γ^*. Then $\Gamma^* \supseteq \Gamma^a$ for each $a \in (0, 1)$. In particular,

$$\Gamma^*(1) \supseteq \{(1-b)\delta(0) + b\delta(1) : 0 \leq b < 1\}.$$

Now consider the following strategy σ available in Γ^* at 1 :

$$\sigma_0 = \frac{1}{2^2}\delta(0) + (1 - \frac{1}{2^2})\delta(1)$$

and, for $n \geq 1$,

$$\sigma_n(x_1, x_2, \ldots, x_n) = \begin{cases} \delta(0) & \text{if } x_n = 0 \\ \frac{1}{(n+2)^2}\delta(0) + \left(1 - \frac{1}{(n+2)^2}\right)\delta(1) & \text{if } x_n = 1. \end{cases}$$

Then

$$\sigma(\{h \in H : h_n = 1 \text{ for all } n \geq 1\}) = \prod_{n=0}^{\infty}(1 - \frac{1}{(n+2)^2}) = p(say) > 0.$$

Hence, for any $u : X \to \mathcal{R}$,

$$\int u^* d\sigma = (1-p)u(0) + pu(1).$$

Consequently,

$$(V_{\Gamma^*} u)(1) \geq (1-p)u(0) + pu(1),$$

so $V_{\Gamma^*} \not\sim V_\Gamma$, as can be seen by a suitable choice of u. This yields the desired contradiction. □

The following result is a close analogue of Theorem 1.2.

Theorem 5.3. *If Γ is an analytic gambling house on X, then the largest gambling house Γ' on X such that $T_{\Gamma'} \sim T_\Gamma$ is $\overline{\text{sco}}\,\Gamma_\infty$.*

We omit the proof.

References

[1] T.E. Armstrong, Full houses and cones of excessive functions, *Indiana Univ. Math. J.* 29(1980), 737-746.

[2] R.J. Aumann, Mixed and behavior strategies in infinite extensive games, *Ann. Math. Studies* 53(1964), 627–650.

[3] D.P. Bertsekas and S.E. Shreve, *Stochastic Optimal Control : The Discrete Time Case*, Academic Press, New York, 1978.

[4] C. Dellacherie, Quelques résultats sur les maisons de jeux analytiques, *Séminaire de Probabilités XIX, Strasbourg 1983-84*, Lecture Notes in Math., vol. 1123, Springer-Verlag, Berlin and New York, 1985, pp. 222–229.

[5] C. Dellacherie and P. A. Meyer, *Probabilitiés et Potentiel*, Hermann, Paris, Chapter XI, 1983.

[6] L.E. Dubins and L.J. Savage, *Inequalities for Stochastic Processes*, Dover, New York, 1976.

[7] L.E. Dubins and D.A. Freedman, Measurable sets of measures, *Pacific J. Math.* 14(1964), 1211–1222.

[8] L. Dubins, A.Maitra, R. Purves and W. Sudderth, Measurable, nonleavable gambling problems, *Israel J. Math.* 67(1989), 257–271.

[9] E. A. Feinberg, On measurability and representation of strategic measures in Markov decision processes, in : T. S. Ferguson, L. S. Shapley and J. B. MacQueen (editors), *Statistics, Probability and Game Theory,* Lecture Notes-Monograph Series, vol. 30, Institute of Math. Statist., Hayward, CA, 1996, pp. 29–43.

[10] A. S. Kechris, *Classical Descriptive Set Theory,* Springer-Verlag, Berlin and New York, 1995.

[11] A. Louveau, Ensembles analytiques et boreliens dans les espaces produits, *Astérisque* 78(1980), 1–84.

[12] A. Louveau, Recursivity and capacity theory, *Proc. Sympos. Pure Math.*, vol. 42, Amer. Math. Soc., Providence, R.I., 1985, pp. 285–301.

[13] A. P. Maitra and W.D. Sudderth, *Discrete Gambling and Stochastic Games,* Springer, New York, 1996.

[14] A. Maitra, R. Purves and W. Sudderth, Leavable gambling problems with unbounded utilities, *Trans. Amer. Math. Soc.* 320(1990), 543–567.

[15] A. Maitra, R. Purves and W. Sudderth, A Borel measurable version of Konig's Lemma for random paths, *Ann. Probab.* 19(1991), 423–451.

[16] A. Maitra, R. Purves and W. Sudderth, A capacitability theorem in measurable gambling theory, *Trans. Amer. Math. Soc.* 333(1992), 221–249.

[17] P.A. Meyer and M. Traki, Reduites et jeux de hasard, *Séminaire de Probabilitiés VII Strasbourg 1971-72*, Lecture Notes in Math., vol. 321, Springer-Verlag, Berlin and New York, 1973, pp. 155–171.

[18] Y.N. Moschovakis, *Descriptive Set Theory,* North-Holland, Amsterdam, 1980.

[19] R.E. Strauch, Measurable gambling houses, *Trans. Amer. Math. Soc.* 126(1967), 64–72. (correction, 130(1968), 184.)

School of Statistics, University of Minnesota, Minneapolis, Minnesota 55455.

CONVERGENCE OF A 'GIBBS-BOLTZMANN' RANDOM MEASURE FOR A TYPED BRANCHING DIFFUSION

by

Simon C. Harris

Department of Mathematical Sciences,
University of Bath, Bath, BA2 7AY, United Kingdom.

1. Introduction

We consider certain 'Gibbs-Boltzmann' random measures which are derived from the positions of particles in the typed branching diffusion introduced in Harris and Williams[**6**]. We prove that, as time progresses, these random measures almost surely converge to deterministic normal distributions (corresponding to the type distributions of the 'dominant' particles contributing to the measure at large times). The random measures considered are closely linked to some martingales of fundamental importance in the study of the long-term behaviour of the branching diffusion. The method of proof relies on a martingale expansion and the study of the behaviour of various families of martingales.

(1.1) The Branching Model

The typed branching diffusion we consider has particles which independently move in space according to a Brownian motion with variance controlled by the particle's type process. The type of each particle evolves as an Ornstein-Uhlenbeck process and also controls the rate at which births occur. This model was introduced in Harris and Williams[**6**], a paper which forms the foundations for this work. Although the paper deals entirely with one family of such branching diffusions, analogous results and similar martingale methods may well be applicable in a variety of other typed branching diffusions where the spatial Brownian motion and the breeding rate are controlled by a type process moving as a finite state Markov chain or sufficiently ergodic Markov process.

Consider the typed branching diffusion where, for time $t \geq 0$,

$N(t)$ is the number of particles alive,

$X_k(t)$ in $\mathbb{R}$ is the spatial position of the kth-born particle,

$Y_k(t)$ in $\mathbb{R}$ is the 'type' of the kth-born particle,

$$\big(N(t); X_1(t), \ldots, X_{N(t)}; Y_1(t), \ldots, Y_{N(t)}\big) \text{ is the current state of the system.}$$

The *type* moves on the real line as an Ornstein-Uhlenbeck process associated with the differential operator (generator)

$$\mathcal{Q}_\theta := \frac{\theta}{2}\left(\frac{\partial^2}{\partial y^2} - y\frac{\partial}{\partial y}\right)$$

where θ is a positive real parameter considered as the *temperature* of the system. The *spatial* motion of a particle of type y is a driftless Brownian motion with variance

$$A(y) := ay^2, \qquad \text{where } a \geq 0.$$

The *breeding* of a type y particle occurs at a rate

$$R(y) := ry^2 + \rho, \qquad \text{where } r, \rho \geq 0,$$

and we have one child born at these times (binary splitting). A child inherits its parent's current type and (spatial) position then moves off *independently* of all others. Particles live forever (once born!).

The model has a very different behaviour for low temperature parameter values and *throughout* this paper we consider only values above the critical temperature, that is $\theta > 8r$. All the above parameters of the model are considered as fixed for the rest of this paper, unless otherwise stated. We use $\mathbb{P}^{x,y}$ and $\mathbb{E}^{x,y}$ with $x, y \in \mathbb{R}$ to represent probability and expectation when the Markov process starts with an initial state $(N; \mathbf{X}, \mathbf{Y}) = (1; x; y)$.

(1.2) **Convergence of a 'Gibbs-Boltzmann' random measure**

Let $\alpha, \lambda \in \mathbb{R}$. For $t \geq 0$ and $1 \leq j \leq N(t)$ we define

$$J_{\alpha,\lambda}(t,j) := \frac{\exp\left(\alpha Y_j(t)^2 + \lambda X_j(t)\right)}{\sum_{k=1}^{N(t)} \exp\left(\alpha Y_k(t)^2 + \lambda X_k(t)\right)}$$

so that we have

$$J_{\alpha,\lambda}(\cdot,\cdot) \geq 0 \qquad \text{and} \qquad \sum_{k=1}^{N(t)} J_{\alpha,\lambda}(t,k) = 1.$$

We consider $J_{\alpha,\lambda}(t)$ as a *random probability measure on* $\mathbb{R}$ with a mass of size $J_{\alpha,\lambda}(t,j)$ at type position $Y_j(t)$ for each $j = 1, \ldots, N(t)$.

Under certain parameter constraints, this random probability measure almost surely converges to a certain (deterministic) normal distribution. The very crude large-deviation heuristics in Harris and Williams[**6**] go some way to explaining why this convergence may be anticipated, as well as providing some motivation for looking at these random measures. These heuristics lead us to suspect that the distribution of the types of the 'majority' of particles which are to be found with spatial positions in the 'vicinity of $\gamma_\lambda t$' is normal with variance $(2\psi_\lambda^+)^{-1}$. It is precisely these particles that the $J_{\cdot,\lambda}$ random measures end up concentrating on.

Before stating the result, we need to introduce a couple of key definitions (the significance of which will become clearer in later sections):

$$\mu_\lambda := \frac{\sqrt{\theta(\theta - 8r - 4a\lambda^2)}}{2}, \qquad \psi_\lambda^\pm := \frac{1}{4} \pm \frac{\mu_\lambda}{2\theta}$$

$$\tilde{\lambda}(\theta) := -\sqrt{\frac{2(\theta - 8r)(\theta\rho + 2\rho^2 + r\theta)}{a(\theta + 4\rho)^2}}$$

(1.3) **Theorem.**

Suppose $\alpha < 1/4$ and $|\lambda| < \tilde{\lambda}(\theta)$, then for each starting law $\mathbb{P}^{x,y}$, the random probability measure $J_{\alpha,\lambda}(t)$ almost surely weakly converges to a (deterministic) normal distribution with mean 0 and variance $\sigma^2_{\alpha,\lambda} := \{2(\psi_\lambda^+ - \alpha)\}^{-1}$, denoted by

$$J_{\alpha,\lambda}(t) \overset{a.s.}{\Rightarrow} N(0, \sigma^2_{\alpha,\lambda}) \qquad as \qquad t \to \infty.$$

Equivalently, for every continuous bounded function $f : \mathbb{R} \mapsto \mathbb{R}$

$$\int_{\mathbb{R}} f(y) J_{\alpha,\lambda}(t, dy) := \sum_{k=1}^{N(t)} f\left(Y_k(t)\right) J_{\alpha,\lambda}(t,k) \to \int_{\mathbb{R}} f(y) \frac{e^{-y^2/2\sigma^2_{\alpha,\lambda}}}{\sqrt{2\pi\sigma^2_{\alpha,\lambda}}} \, dy$$

almost surely as $t \to \infty$.

We shall actually prove some stronger limit theorems which will combine to yield this theorem. The methods we shall employ will require the study of the long-term behaviour of various martingales for the branching diffusion. In fact, study of these martingales will essentially yield the asymptotic behaviour of the normalization constants in the above Gibbs-Boltzmann random measures, as well as identifying the normal distribution limit of the measures themselves. The reader should also see Chauvin and Rouault [**4**] concerning Gibbs-Boltzmann random measures in the branching random walk.

2. Martingales and The Main Convergence Theorem

Define

$$\lambda_{\min} := -\sqrt{\frac{\theta - 8r}{4a}}. \tag{2.1}$$

Let $\lambda \in \mathbb{R}$, with the following convention which we always use for λ:

$$\lambda_{\min} < \lambda < 0. \tag{2.2}$$

(Note that $\lambda_{\min}$ is the point beyond which ψ_λ^- is no longer a real number.)

In Harris and Williams[**6**], we proved the almost sure speed of the spatially left-most particle by making use of the following martingales:

(2.3) **Lemma. The 'ground-state' martingales.** *For $t \geq 0$, let*

$$Z_\lambda^-(t) := \sum_{k=1}^{N(t)} \exp\left(\psi_\lambda^- Y_k(t)^2 + \lambda\left[X_k(t) + c_\lambda^- t\right]\right), \tag{2.4}$$

where

$$c_\lambda^- := -\left(\rho + \theta\psi_\lambda^-\right)/\lambda, \tag{2.5}$$

This defines a martingale Z_λ^- (under each $\mathbb{P}^{x,y}$ measure).

Since the martingale is non-negative it must converge. It is easy to check that the function c^- is convex on $(\lambda_{\min}, 0)$, and achieves its minimum at the unique point $\tilde{\lambda}(\theta)$. We used this simple geometric fact and an idea of Neveu[**10**] in proving the following:

(2.6) **Theorem. Convergence of the 'ground-state' martingales.** *The martingale Z_λ^- is uniformly integrable and has an almost sure strictly positive limit if $\lambda \in \left(\tilde{\lambda}(\theta), 0\right)$.*

Similar martingales have been studied for standard branching Brownian motion and they are also strongly linked to travelling waves of the related FKPP reaction-diffusion equation (see McKean[8],[9] and Neveu[**10**] for example). The two-type branching Brownian motion model of Champneys et al. [**3**] is also closely related to our current continuous-type model and indeed most of the ideas of this paper should translate to models where the type of each particle evolves as a finite state irreducible Markov chain.

(2.7) **The 'one-particle picture'**

We now remind the reader how we can go about calculating certain expectations for branching diffusion by making use of a *'one-particle picture'* as follows:

Let (ξ, η) be a process behaving like a single particle's space and type motions in the branching model described above. Thus, ξ is a Brownian motion controlled by an Ornstein-Uhlenbeck process η, and (ξ, η) has formal generator $\mathcal{H}$, where

$$(\mathcal{H}F)(x,y) = \frac{1}{2}A(y)\frac{\partial^2 F}{\partial x^2} + (\mathcal{Q}_\theta F)(x,y) = \frac{1}{2}A(y)\frac{\partial^2 F}{\partial x^2} + \frac{\theta}{2}\left(\frac{\partial^2 F}{\partial y^2} - y\frac{\partial F}{\partial y}\right).$$

Of course, η is an autonomous Markov process with generator $\mathcal{Q}_\theta$ and with (standard normal) invariant density

$$\phi(y) := (2\pi)^{-\frac{1}{2}}\exp\left(-\tfrac{1}{2}y^2\right).$$

For functions h_1, h_2 on $\mathbb{R}$, we define the $L^2(\phi)$ inner product:

$$\langle h_1, h_2\rangle_\phi := \int_{\mathbb{R}} h_1(y)h_2(y)\phi(y)dy.$$

Also recall from Harris and Williams[**6**] that we made important use of the following lemma:

(2.8) **Lemma: 'From One to Many'.** *For any non-negative Borel function f on $\mathbb{R}\times\mathbb{R}$, we have*

$$\mathbb{E}^{x,y}\left(\sum_{k=1}^{N(t)} f(X_k(t),Y_k(t))\right) = \mathbb{E}^{x,y}\left(\exp\left(\int_0^t R(\eta_s)\,ds\right)\,f(\xi_t,\eta_t)\right).$$

Now, we try to find functions f and real constants E that will give us a *martingale* of the form

$$\sum_{k=1}^{N(t)} f(Y_k(t))e^{\lambda X_k(t)-Et}$$

Exploiting lemma 2.8 tells us that

$$\mathbb{E}^{x,y}\sum_{k=1}^{N(t)} f(Y_k(t))e^{\lambda X_k(t)-Et} = \mathbb{E}^{x,y} f(\eta_t)e^{\lambda\xi_t-Et+\int_0^t R(\eta_s)\,ds}$$

and utilising the standard exponential martingale for a Brownian motion we have

$$\mathbb{E}^{x,y}\left(e^{\lambda\xi_t}\middle|\,\sigma(\eta_s:s\le t)\right) = e^{\lambda x+\frac{1}{2}\lambda^2\int_0^t A(\eta_s)\,ds}$$

Then combining these observations and looking for a martingale requires that

$$f(y) = \mathbb{E}^y f(\eta_t)e^{\int_0^t\left\{R(\eta_s)+\frac{1}{2}\lambda^2A(\eta_s)-E\right\}ds}$$

and now the Feynman-Kac formula suggests

$$\left\{\frac{\theta}{2}\left(\frac{\partial^2}{\partial y^2}-y\frac{\partial}{\partial y}\right)+R(y)+\tfrac{1}{2}\lambda^2A(y)-E\right\}f(y)=0.$$

(2.9) **Eigenfunctions for a linear differential operator**

Define the differential operator

$$\mathcal{L}_\lambda := \frac{\theta}{2}\left(\frac{d^2}{dy^2}-y\frac{d}{dy}\right)+(r+\tfrac{1}{2}a\lambda^2)y^2+\rho$$

which is essentially self-adjoint with respect to the $L^2(\phi)$ inner-product $<\cdot,\cdot>_\phi$. This should remind you of the harmonic oscillator equation, a point which now enables us to perform further explicit calculations.

Consider $\lambda\in(\lambda_{\min},0)$ fixed. There is a set of ortho-normal eigenfunctions for the self-adjoint operator $\mathcal{L}_\lambda$ represented by

$$\mathcal{L}_\lambda\Psi_{n,\lambda} = E_{n,\lambda}\Psi_{n,\lambda} \qquad \forall n\in\{0,1,\dots\},$$
$$\langle\Psi_{n,\lambda},\Psi_{m,\lambda}\rangle_\phi = \delta_{n,m} \qquad \forall m,n\in\{0,1,\dots\},$$

with

$$\Psi_{n,\lambda}(y) := h_{n,\lambda}(y)\exp\left\{\psi_\lambda^- y^2\right\},$$
$$E_{n,\lambda} := E_\lambda - n\mu_\lambda,$$

and

$$h_{n,\lambda}(y) := \sqrt{\frac{\mu_\lambda^{\frac{1}{2}}}{\theta^{\frac{1}{2}} n! 2^n}} H_n\left(\sqrt{\frac{\mu_\lambda}{\theta}} y\right),$$
$$E_\lambda := \rho + \theta\psi_\lambda^- = -\lambda c_\lambda^-,$$

where H_n is the nth Hermite polynomial so that

$$H_n''(z) - 2zH_n'(z) + 2nH_n(z) = 0,$$
$$H_n(z) = (-1)^n e^{z^2} \frac{d^n}{dz^n}\left(e^{-z^2}\right),$$

so in particular, $H_0(z) \equiv 1$, $H_1(z) = 2z$, $H_2(z) = 4z^2 - 2$, etc.

The eigenfunctions are complete; they form an ortho-normal basis for $L^2(\phi)$. Given any $f \in L^2(\phi)$ we have the $L^2(\phi)$ convergent expansion

$$f(y) = \sum_{i=0}^{\infty} f_i \Psi_{i,\lambda}(y), \qquad f_i := \langle f, \Psi_{i,\lambda}\rangle_\phi .$$

(In fact, later on we will need to make use of certain 'smooth' functions that have *uniformly* convergent eigenfunction expansions.)

There is another strictly positive 'eigenfunction' of $\mathcal{L}_\lambda$ satisfying

$$\mathcal{L}_\lambda \Psi_{\lambda,+} = E_\lambda^+ \Psi_{\lambda,+}$$

given by

$$\Psi_{\lambda,+}(y) := e^{\psi_\lambda^+ y^2}, \qquad E_\lambda^+ := \rho + \theta\psi_\lambda^+,$$

but we note that it is *not* normalisable, that is $\Psi_{\lambda,+} \notin L^2(\phi)$. However, this 'eigenfunction' will still give rise to a martingale which proves to be of important use later on.

(2.10) **Other martingales.**

Combining the above ideas with the branching-property yields a further family of martingales that will be very helpful in understanding the type-space behaviour of the particles.

(2.11) **Lemma.** *Let* $\lambda \in (\tilde{\lambda}(\theta), 0)$.
(a) For each $n \in \{0, 1, \ldots\}$ *and* $t \in [0, \infty)$,

$$Z_{n,\lambda}(t) := \sum_{k=1}^{N(t)} e^{n\mu_\lambda t} h_{n,\lambda}(Y_k(t)) e^{\psi_\lambda^- Y_k(t)^2 + \lambda X_k(t) - E_\lambda t}$$

defines a martingale $Z_{n,\lambda}$ *for each* $\mathbb{P}^{x,y}$ *starting law.*

(b) For $t \in [0, \infty)$,

$$Z_\lambda^+(t) := \sum_{k=1}^{N(t)} e^{\psi_\lambda^+ Y_k(t)^2 + \lambda X_k(t) - E_\lambda^+ t}$$

defines a martingale Z_λ^+ for each $\mathbb{P}^{x,y}$ starting law.

We now suggest the motivation for studying the long-term behaviour of these martingales in our present context.

(2.12) Main Convergence Theorem.

We are interested in studying processes of the form

$$\sum_{k=1}^{N(t)} f(Y_k(t)) e^{\lambda X_k(t) - E_\lambda t} \qquad (t \geq 0).$$

Now, for 'nice' functions f that are square integrable with respect to the standard normal distribution, at least formally, we can write f as its eigenfunction expansion so *suggesting* that

$$\begin{aligned}\sum_{k=1}^{N(t)} f(Y_k(t)) e^{\lambda X_k(t) - E_\lambda t} &= \sum_{k=1}^{N(t)} \left\{ \sum_{n=0}^{\infty} f_n \Psi_{n,\lambda}(Y_k(t)) \right\} e^{\lambda X_k(t) - E_\lambda t} \\ &= \sum_{n=0}^{\infty} f_n e^{-n\mu_\lambda t} Z_{n,\lambda}(t)\end{aligned}$$

If we further restrict our attention to functions of the form $f(y) = p_n(y)\Psi_{0,\lambda}(y)$ where p_n is a polynomial of degree n, then the previous eigenfunction expansion becomes *exact* with

$$\sum_{k=1}^{N(t)} f(Y_k(t)) e^{\lambda X_k(t) - E_\lambda t} = f_0 Z_\lambda^-(t) + f_1 e^{-\mu_\lambda t} Z_{1,\lambda}(t) + \ldots f_n e^{-n\mu_\lambda t} Z_{n,\lambda}(t).$$

Later on we prove that $e^{-n\mu_\lambda t} Z_{n,\lambda}(t) \to 0$ almost surely for all $n \geq 1$ (see corollary 3.5) and whence

$$\sum_{k=1}^{N(t)} f(Y_k(t)) e^{\lambda X_k(t) - E_\lambda t} \to f_0 Z_\lambda^-(\infty)$$

where we have

$$f_0 = \langle f, \Psi_{0,\lambda} \rangle_\phi = \int_{\mathbb{R}} p_n(y) \frac{e^{-\frac{y^2}{2\sigma^2}}}{\sqrt{2\pi\sigma^2}} \, dy$$

and $\sigma^2 = \theta/(2\mu_\lambda)$. In particular, recalling that for $\lambda \in (\tilde{\lambda}, 0)$ we have $Z_\lambda^-(\infty) > 0$ almost surely (see theorem 2.6), we find that the moments of the corresponding random measure converge almost surely to the moments of a (deterministic) normal distribution. Yet, it is well known that the convergence of moments to the moments of a normal distribution implies weak convergence to the normal distribution (see Breiman[**2**], for example). Thus, the polynomial p_n can be replaced by any bounded continuous function p and the convergence will still hold.

It should now seem at least plausible that we can further extend this convergence to cover all continuous functions f that are (safely) square integrable with respect to the standard normal distribution to give the following theorem:

(2.13) **Theorem.** *Let* $\lambda \in \left(\tilde{\lambda}(\theta), 0\right)$ *and* $\alpha < 1/4$. *For each* $\mathbb{P}^{x,y}$ *starting law and every continuous bounded function* $f : \mathbb{R} \mapsto \mathbb{R}$, *we have*

$$\sum_{k=1}^{N(t)} f\left(Y_k(t)\right) e^{\alpha Y_k(t)^2 + \lambda\left(X_k(t) + c_\lambda^- t\right)} \overset{a.s.}{\to} f_0\, Z_\lambda^-(\infty).$$

where

$$f_0 := \int_{\mathbb{R}} f(y) e^{\alpha y^2} \Psi_{0,\lambda}(y) \phi(y)\, dy$$

Simply combining this result with the known convergence of the 'ground-state' martingales from Theorem 2.6 will yield the 'Gibbs-Boltzmann' random measure convergence to the (deterministic) normal distribution in Theorem 1.3.

3. Martingale Convergence Results

We first present a theorem which gives sufficient criteria for the convergence of the Hermite polynomial based martingales.

(3.1)**Theorem.** *Let* $n \in \mathbb{N}$ *and* $\lambda \in \left(\tilde{\lambda}(\theta), 0\right)$. *For each starting law,* $\mathbb{P}^{x,y}$, *the* n^{th} *Hermite 'additive' martingale,* $Z_{n,\lambda}^-$, *converges almost surely and in* $\mathcal{L}^\alpha$ *for* $\alpha \in (1, 2]$ *if the following inequalities hold simultaneously:*

$$\lambda(c_\lambda^- - c_{\alpha\lambda}^-) + n\mu_\lambda < 0,$$
$$\alpha\psi_\lambda^- < \psi_{\alpha\lambda}^+.$$

(3.2)**Definitions.** *Let* α_λ^* *to be the* α *value which minimises* $c_{\alpha\lambda}^-$ *subject to the constraints* $\alpha\psi_\lambda^- \leq \psi_{\alpha\lambda}^+$ *and* $\alpha \in [1, 2]$. *Further, let* n_λ^* *to be the largest integer* n *satisfying* $n < -\lambda(c_\lambda^- - c_{\alpha_\lambda^*\lambda}^-)/\mu_\lambda$.

These values now get the 'best' from the theorem as follows.

(3.3) **Corollary.** *Let* $\lambda \in (\tilde{\lambda}(\theta), 0)$. *For each starting law,* $\mathbb{P}^{x,y}$,

$$\{Z_{0,\lambda}, Z_{1,\lambda}, \ldots, Z_{n^*_\lambda,\lambda}\}$$

is a set of uniformly integrable martingales, where, for all $\alpha < \alpha^*_\lambda$,

$$Z_{n,\lambda}(t) \to Z_{n,\lambda}(\infty) \quad a.s. \text{ and in } \mathcal{L}^\alpha \text{ for all } n = 0, \ldots, n^*_\lambda.$$

[**Remarks.** The result for the ground-state martingale, Z^-_λ, was also given in Harris and Williams[**6**] but this proof would not cover the other signed martingales. The reader can check that the integer n^*_λ does indeed take non-zero values for some choices of parameters in the model. Some large-deviation heuristics suggest that this result is the best possible, see Harris[**7**] and further papers. We conjecture that the conditions given in Theorem 3.1 are necessary as well as sufficient for the convergence and, in particular, for $n > n^*_\lambda$ the martingales $Z_{n,\lambda}$ fail to converge.]

We can also give bounds on the growth of all the martingales as follows:

(3.4) **Theorem.** *Let* $n \in \mathbb{N}$ *and* $\lambda \in (\tilde{\lambda}(\theta), 0)$. *If* $\alpha \in (1, 2]$ *with*

$$\beta := \lambda(c^-_\lambda - c^-_{\alpha\lambda}) + n\mu_\lambda > 0,$$
$$\alpha\psi^-_\lambda < \psi^+_{\alpha\lambda},$$

then for all $\epsilon > 0$ *and for every starting law,* $\mathbb{P}^{x,y}$,

$$e^{-(\epsilon+\beta)t} Z_{n,\lambda}(t) \to 0 \qquad a.s.$$

[**Remarks.** This theorem is only useful when $n > n^*_\lambda$, otherwise Theorem 3.1 can be applied and the martingale actually converges. The 'best' control on the rate of growth of the martingales in this theorem is again found with α^*_λ.]

The next corollary was used in the previous section's discussion of a restricted version of the convergence Theorem 2.13. The actual proof of Theorem 2.13 will require elements from the proof of Theorems 3.1 and 3.4 as well as further work to enlarge the space of functions for which the convergence holds.

(3.5) **Corollary.** *Let* $n \in \mathbb{N}$ *and* $\lambda \in (\tilde{\lambda}(\theta), 0)$. *For every starting law,* $\mathbb{P}^{x,y}$,

$$e^{-n\mu_\lambda t} Z_{n,\lambda}(t) \to 0 \qquad a.s.$$

In Git and Harris [**5**], we will show that the ground-state martingales with parameters $\lambda \in [\lambda_{\min}, \tilde{\lambda}(\theta)]$ tend to zero almost surely (so cannot be uniformly integrable). The other positive martingales Z^+_λ for $\lambda \in [\lambda_{\min}, 0]$ also tend to zero almost surely and study of the rate of this convergence in [**5**] will give almost sure outer bounds on the asymptotic shape in the space-type plane of the branching particle system, whilst some large-deviation results will prove this bound is actually attained.

4. Proofs of Martingale Convergence Results

For details of the standard martingale results relied upon throughout this section, see Rogers and Williams **[12]**&**[13]** or Revuz and Yor **[11]**.

Proof of Theorem 3.1 and Theorem 3.4.

We have the Hermite martingales

$$Z_{n,\lambda}(t) = \sum_{k=1}^{N(t)} e^{n\mu_\lambda t}\Psi_n\left(Y_k(t)\right) e^{\lambda\left(X_k(t)+c_\lambda^- t\right)}.$$

Clearly,

$$Z_{n,\lambda}(t+s) = \sum_{k=1}^{N(s)} e^{\lambda\left(X_k(s)+c_\lambda^- s\right)+n\mu_\lambda s} W_k^{0,y_k}(t)$$

where the $W_k^{0,y_k}(t)$ are independent conditional on $\mathcal{F}_s$ and each look like $Z_{n,\lambda}(t)$ when the branching process is started with one particle at $(x,y)=(0,y_k)$, with $y_k = Y_k(s)$, run for time t. Then,

$$Z_{n,\lambda}(s+t) - Z_{n,\lambda}(s) = \sum_{k=1}^{N(s)} e^{\lambda\left(X_k(s)+c_\lambda^- s\right)+n\mu_\lambda s}\{W_k^{0,y_k}(t) - W_k^{0,y_k}(0)\}$$

where $W_k^{0,y_k}(0) = \Psi_n\big(Y_k(s)\big)$.

Conditional on $\mathcal{F}_s$, $\{W_k^{0,y_k}(t) - W_k^{0,y_k}(0)\}$ are independent and the martingale property gives

$$\mathbb{E}\{W_k^{0,y_k}(t) - W_k^{0,y_k}(0)\} = 0.$$

We now make use of the following important lemma, which was drawn to our attention by a paper of Biggins [1] which studies related (complex valued) martingales for the branching random walk.

(4.1) **Lemma.** *If X_i are independent and $\mathbb{E}X_i = 0$, or they are martingale differences, then for $\alpha \in [1,2]$,*

$$\mathbb{E}\left|\sum_i X_i\right|^\alpha \leq 2^\alpha \sum_i \mathbb{E}|X_i|^\alpha .$$

$Z_{n,\lambda}(s+t) - Z_{n,\lambda}(t)$ is a martingale null at $s=0$, then $|Z_{n,\lambda}(s+t) - Z_{n,\lambda}(t)|^\alpha$ is a submartingale for $\alpha \in [1,2]$, hence $\mathbb{E}|Z_{n,\lambda}(s+t) - Z_{n,\lambda}(t)|^\alpha$ is non-decreasing in s. We are interested in finding out when the martingales are $\mathcal{L}^\alpha$ bounded. Now,

$$\begin{aligned}\mathbb{E}|Z_{n,\lambda}(Ns+t) - Z_{n,\lambda}(t)|^\alpha &= \mathbb{E}\left|\sum_{j=0}^{N-1}\{Z_{n,\lambda}((j+1)s+t) - Z_{n,\lambda}(js+t)\}\right|^\alpha \\ &\leq 2^\alpha \sum_{j=0}^{N-1} \mathbb{E}\left|Z_{n,\lambda}((j+1)s+t) - Z_{n,\lambda}(js+t)\right|^\alpha\end{aligned}$$

as we have martingale differences of the $Z_{n,\lambda}$ martingale so can apply lemma 4.1. Also,

$$|Z_{n,\lambda}(s+t) - Z_{n,\lambda}(s)|^\alpha = \left|\sum_{k=1}^{N(s)} e^{\lambda\left(X_k(s)+c_\lambda^- s\right)+n\mu_\lambda s}\{W_k^{0,y_k}(t) - W_k^{0,y_k}(0)\}\right|^\alpha ,$$

where the entries on the last summation are mean-zero and independent conditional on $\mathcal{F}_s$; hence, applying Lemma 4.1 conditional on $\mathcal{F}_s$, we get

$$\mathbb{E}\Big\{ |Z_{n,\lambda}(s+t) - Z_{n,\lambda}(t)|^\alpha \Big|\mathcal{F}_s\Big\}$$
$$\leq 2^\alpha \sum_{k=1}^{N(s)} e^{\alpha\lambda\left(X_k(s)+c_\lambda^- s\right)+n\alpha\mu_\lambda s}\mathbb{E}|W_k(t) - W_k(0)|^\alpha ,$$

where W_k looks like $Z_{n,\lambda}$ started from one particle at $(0, y_k)$ where $y_k = Y_k(s)$.

Now we want an estimate (small times will do) to bound the $\mathcal{L}^\alpha$ norm. Currently, we are interested in having n *fixed* to try and get the best bounds for a single Hermite martingale (in a later result, at this point we shall employ a bound that holds uniformly over all $n \in \mathbb{N}$). The following lemma (proved in section 5) works effectively.

(4.2) **Lemma.** *Let $n \in \mathbb{N}$ be fixed. Given $\epsilon > 0$, there exists $K \in \mathbb{R}$ and $T > 0$ such that for all $\alpha \in [1,2]$,*

$$\mathbb{E}^{0,y}\left(|Z_{n,\lambda}(t) - Z_{n,\lambda}(0)|^\alpha\right) \leq K\, e^{\alpha(\psi_\lambda^- +\epsilon)y^2} \qquad \forall t \in [0,T], \forall y.$$

Returning to the previous inequality,

$$\mathbb{E}\left\{ |Z_{n,\lambda}(s+t) - Z_{n,\lambda}(s)|^\alpha \middle|\mathcal{F}_s\right\}$$
$$\leq 2^\alpha \sum_{k=1}^{N(s)} e^{\alpha\lambda\left(X_k(s)+c_\lambda^- s\right)+n\alpha\mu_\lambda s}\mathbb{E}^{0,y_k}|Z_{n,\lambda}(t) - Z_{n,\lambda}(0)|^\alpha$$
$$\leq \tilde{K} \sum_{k=1}^{N(s)} e^{\alpha\left(\psi_\lambda^- +\epsilon\right)Y_k(s)^2+\alpha\lambda\left(X_k(s)+c_\lambda^- s\right)+n\alpha\mu_\lambda s}$$

where this holds $\forall\alpha \in [1,2], \forall t \in [0,T], \forall s \geq 0$.

Hence,

$$\mathbb{E}|Z_{n,\lambda}(s+t) - Z_{n,\lambda}(s)|^\alpha \leq \tilde{K}\,\mathbb{E}\left(\sum_{k=1}^{N(s)} e^{\alpha\left(\psi_\lambda^- +\epsilon\right)Y_k(s)^2+\alpha\lambda\left(X_k(s)+c_\lambda^- s\right)+n\alpha\mu_\lambda s}\right)$$
$$= \tilde{K}\exp\left(\alpha s\left\{\lambda(c_\lambda^- - c_{\alpha\lambda}^-) + n\mu_\lambda\right\}\right)\mathbb{E}\left(\sum_{k=1}^{N(s)} e^{\alpha\left(\psi_\lambda^- +\epsilon\right)Y_k(s)^2+\alpha\lambda\left(X_k(s)+c_{\alpha\lambda}^- s\right)}\right)$$

We can now calculate the last expectation above explicitly, using Lemma 2.8 and a change of measure between OU processes (see Harris and Williams [**6**] pp 137–138). In particular, the value is bounded by a constant for all times s if

$$\alpha(\psi_\lambda^- + \epsilon) < \psi_{\alpha\lambda}^+$$

(otherwise there is an explosion at some finite time and the bound is useless for our purposes). In this case then

$$\mathbb{E}\left|Z_{n,\lambda}(s+t) - Z_{n,\lambda}(s)\right|^\alpha \le K' e^{\alpha\left(\lambda c_\lambda^- + n\mu_\lambda - \lambda c_{\alpha\lambda}^-\right)s}, \qquad \forall s.$$

Defining

$$\beta := \lambda(c_\lambda^- - c_{\alpha\lambda}^-) + n\mu_\lambda$$

then for all $t > 0$, $N \in \mathbb{N}$ and $s \in [0,T]$,

$$\begin{aligned}
\mathbb{E}\left|Z_{n,\lambda}(Ns+t) - Z_{n,\lambda}(t)\right|^\alpha &\le 2^\alpha \sum_{j=0}^{N-1} \mathbb{E}\left|Z_{n,\lambda}((j+1)s+t) - Z_{n,\lambda}(js+t)\right|^\alpha \\
&\le 2^\alpha K' \sum_{j=0}^{N-1} e^{\alpha\beta(js+t)} \\
&= 2^\alpha K' e^{\alpha\beta t}\left(\frac{1 - e^{N\alpha\beta s}}{1 - e^{\alpha\beta s}}\right) \\
&\le \begin{cases} 2^\alpha K' e^{\alpha\beta\left(t+Ns\right)} & \text{if } \beta > 0, \\ 2^\alpha K'(1 - e^{\alpha\beta s})^{-1} e^{\alpha\beta t} & \text{if } \beta < 0. \end{cases}
\end{aligned}$$

If we have a case where $\alpha \in (1,2]$ satisfies both $\alpha\psi_\lambda^- < \psi_{\alpha\lambda}^+$ and $\beta = \lambda\left(c_\lambda^- - c_{\alpha\lambda}^-\right) + n\mu_\lambda < 0$ then it follows that we have $\mathcal{L}^\alpha$ boundedness for $Z_{n,\lambda}$, hence Doob's $\mathcal{L}^p$ inequality reveals that the martingale $Z_{n,\lambda}$ converges almost surely and in $\mathcal{L}^\alpha$ (so it is a uniformly integrable martingale). This completes the proof of Theorem 3.1.

Otherwise, suppose we have a case where $\alpha \in (1,2]$ satisfies $\alpha\psi_\lambda^- < \psi_{\alpha\lambda}^+$ and $\beta = \lambda\left(c_\lambda^- - c_{\alpha\lambda}^-\right) + n\mu_\lambda > 0$. Then there exists a K' such that for all $t > 0$, $N \in \mathbb{N}$ and $s \in [0,T]$,

$$\mathbb{E}\left|Z_{n,\lambda}(Ns) - Z_{n,\lambda}(0)\right|^\alpha \le 2^\alpha K' \frac{e^{\alpha\beta N s}}{e^{\alpha\beta s} - 1}.$$

Doob's submartingale inequality tells us that for any $\epsilon > 0$

$$\mathbb{P}\left(\sup_{u\in[t,s+t]} \left|Z_{n,\lambda}(u) - Z_{n,\lambda}(0)\right| > \epsilon\right) \le \frac{\mathbb{E}\left|Z_{n,\lambda}(s+t) - Z_{n,\lambda}(0)\right|^\alpha}{\epsilon^\alpha}$$

so, for a fixed $s \in [0,T]$ and all $N \in \mathbb{N}$,

$$\begin{aligned}
\mathbb{P}\left(\sup_{u\in[(N-1)s,Ns]} e^{-\delta u}|Z_{n,\lambda}(u)-Z_{n,\lambda}(0)|>\epsilon\right) \\
&\leq \mathbb{P}\left(\sup_{u\in[(N-1)s,Ns]} |Z_{n,\lambda}(u)-Z_{n,\lambda}(0)|>e^{\delta(N-1)s}\epsilon\right) \\
&\leq \frac{\mathbb{E}|Z_{n,\lambda}(Ns)-Z_{n,\lambda}(0)|^{\alpha}}{\epsilon^{\alpha}e^{\alpha\delta(N-1)s}} \\
&\leq \left\{\frac{2^{\alpha}K'e^{\alpha\delta s}}{\epsilon^{\alpha}(e^{\alpha\beta s}-1)}\right\} e^{-\alpha(\delta-\beta)Ns}
\end{aligned}$$

When $\delta > \beta$, we can sum over the N and apply a Borel-Cantelli argument to conclude that $e^{-\delta u}|Z_{n,\lambda}(u) - Z_{n,\lambda}(0)| > \epsilon$ only finitely many times, yet since $\epsilon > 0$ was arbitrary this yields

$$e^{-\delta u} Z_{n,\lambda}(u) \to 0 \qquad a.s.$$

as required. □

Proof of Theorem 2.13.

Suppose $f \in L^2(\phi)$ with the eigenfunction expansion coefficients

$$f_n := \int_{\mathbb{R}} f(y)\Psi_n(y)\phi(y)\,dy = \int_{\mathbb{R}} \left\{e^{-\frac{y^2}{4}} f(y)\right\} \phi_n(y)\,dy$$

where $\phi_n(y) := e^{-\frac{y^2}{4}}\Psi_n(y)$ and $n \in \{0,1,\ldots\}$. Suppose also that the eigenfunction expansion

$$e^{-\frac{y^2}{4}} f(y) = \sum_{n=0}^{\infty} f_n\phi_n(y)$$

is uniformly convergent so that for all $\epsilon > 0$ there exists $M_\epsilon \in \mathbb{N}$ such that

$$\left|e^{-\frac{y^2}{4}} f(y) - \sum_{n=0}^{m} f_n\phi_n(y)\right| < \epsilon \qquad \forall y \in \mathbb{R}, \forall m \geq M_\epsilon.$$

Then for all $m \geq M_\epsilon$ and all $t > 0$,

$$\begin{aligned}
&\left|\sum_{k=1}^{N(t)} f\left(Y_k(t)\right) e^{\lambda\left(X_k(t)+c_\lambda^- t\right)} - \sum_{n=0}^{m} f_n e^{-n\mu_\lambda t} Z_{n,\lambda}(t)\right| \\
&\qquad = \left|\sum_{k=1}^{N(t)} \left\{f\left(Y_k(t)\right) - \sum_{n=0}^{m} f_n\Psi_n\left(Y_k(t)\right)\right\} e^{\lambda\left(X_k(t)+c_\lambda^- t\right)}\right| \\
&\qquad \leq \sum_{k=1}^{N(t)} \left|f\left(Y_k(t)\right) - \sum_{n=0}^{m} f_n\Psi_n\left(Y_k(t)\right)\right| e^{\lambda\left(X_k(t)+c_\lambda^- t\right)} \\
&\qquad \leq \epsilon \sum_{k=1}^{N(t)} e^{\frac{1}{4}Y_k(t)^2+\lambda\left(X_k(t)+c_\lambda^- t\right)}.
\end{aligned}$$

We now let ϵ decrease with time sufficiently fast that

$$\epsilon_t \sum_{k=1}^{N(t)} e^{\frac{1}{4}Y_k(t)^2+\lambda\left(X_k(t)+c_\lambda^- t\right)} \to 0 \qquad a.s.$$

This choice of ϵ_t is possible by a simple comparison with the Z_λ^+ martingale which is positive and hence must converge. It then only remains to show that whenever $M_t \to +\infty$ as $t \to \infty$, we also have

$$\sum_{n=0}^{M_t} f_n e^{-n\mu_\lambda t} Z_{n,\lambda}(t) \to f_0 Z_{0,\lambda}(\infty) \qquad a.s.$$

Now we proceed along very similar lines to those found in the proof of Theorems 3.1 and 3.4. There we found that for $\alpha \in [1,2]$

$$\mathbb{E}\left\{ |Z_{n,\lambda}(s+t) - Z_{n,\lambda}(s)|^\alpha \middle| \mathcal{F}_s \right\}$$
$$\leq 2^\alpha \sum_{k=1}^{N(s)} e^{\alpha\lambda\left(X_k(s)+c_\lambda^- s\right)+n\alpha\mu_\lambda s} \mathbb{E}^{0,y_k} |Z_{n,\lambda}(t) - Z_{n,\lambda}(0)|^\alpha$$

but this time we proceed onwards utilising the following bound (proved in section 5) that is uniform over all the Hermite martingales.

(4.3) **Lemma.** *There exists $K \in \mathbb{R}$ and $T > 0$ such that for all $\alpha \in [1,2]$,*

$$\mathbb{E}^{0,y}\left(|Z_{n,\lambda}(t) - Z_{n,\lambda}(0)|^\alpha \right) \leq K\, n^{\frac{\alpha}{2}}\, e^{\frac{\alpha}{4}y^2} \qquad \forall t \in [0,T], \forall y \in \mathbb{R}, \forall n \in \mathbb{N}.$$

Hence we get

$$\mathbb{E}\,|Z_{n,\lambda}(s+t) - Z_{n,\lambda}(s)|^\alpha$$
$$\leq \tilde{K}\, n^{\frac{\alpha}{2}}\, e^{\alpha\{\lambda(c_\lambda^- - c_{\alpha\lambda}^-)+n\mu_\lambda\}s}\, \mathbb{E}\left(\sum_{k=1}^{N(s)} e^{\frac{\alpha}{4}Y_k(s)^2+\alpha\lambda\left(X_k(s)+c_{\alpha\lambda}^- s\right)} \right)$$

where now to keep the last expectation bounded over all s we require that $\alpha/4 < \psi_{\alpha\lambda}^+$. When this is the case, the submartingale inequality yields

$$\mathbb{P}\left(\sup_{u\in[(l-1)s,ls]} e^{-n\mu_\lambda u}\, |Z_{n,\lambda}(u) - Z_{n,\lambda}(0)| > \epsilon \right) \leq C\, n^{\frac{\alpha}{2}}\, \epsilon^{-\alpha}\, e^{\alpha\lambda(c_\lambda^- - c_{\alpha\lambda}^-)ls}$$

for some constant $C \in \mathbb{R}$. Then

$$\sum_{l=1}^{\infty}\sum_{n=1}^{\infty} \mathbb{P}\left(\sup_{u\in[(l-1)s,ls]} e^{-n\mu_\lambda u}\, |f_n|\, |Z_{n,\lambda}(u) - Z_{n,\lambda}(0)| > \frac{\epsilon}{n^{3/2}} \right)$$
$$\leq \sum_{l=1}^{\infty}\sum_{n=1}^{\infty} C\, \epsilon^{-\alpha}\, |f_n|^\alpha\, n^{2\alpha}\, e^{\alpha\lambda\left(c_\lambda^- - c_{\alpha\lambda}^-\right)ls}$$
$$= C\, \epsilon^{-\alpha} \left(\sum_{n=1}^{\infty} |f_n|^\alpha\, n^{2\alpha} \right) \left(\sum_{l=1}^{\infty} e^{\alpha\lambda\left(c_\lambda^- - c_{\alpha\lambda}^-\right)ls} \right)$$

which is finite if we can choose an α such that

$$\alpha/4 < \psi^+_{\alpha\lambda}, \quad \alpha\lambda(c^-_\lambda - c^-_{\alpha\lambda}) < 0,$$

$$\sum_{n=1}^{\infty} |f_n|^\alpha n^{2\alpha} < \infty,$$

the first two of which are certainly satisfied for α near to 1. The Borel-Cantelli Lemma then says (almost surely) that for only *finitely many* pairs of (l, n) is

$$\sup_{u\in[(l-1)s,ls]} e^{-n\mu_\lambda u} |f_n| \, |Z_{n,\lambda}(u) - Z_{n,\lambda}(0)| > \frac{\epsilon}{n^{3/2}}$$

so there exists a (random) $T \in \mathbb{R}$ s.t.

$$e^{-n\mu_\lambda t} |f_n| \, |Z_{n,\lambda}(t) - Z_{n,\lambda}(0)| \le \frac{\epsilon}{n^{3/2}} \qquad \forall n \ge 1, t > T$$

hence

$$\left| \sum_{n\ge 1} f_n e^{-n\mu_\lambda t} Z_{n,\lambda}(t) \right| \le \sum_{n\ge 1} e^{-n\mu_\lambda t} |f_n| \, |Z_{n,\lambda}(t)| \le \sum_{n\ge 1} \frac{\epsilon}{n^{3/2}} < \infty \qquad \forall t > T.$$

Since this is true for all $\epsilon > 0$, we have

$$\left| \sum_{n\ge 1} f_n e^{-n\mu_\lambda t} Z_{n,\lambda}(t) \right| \to 0 \qquad \text{a.s.}$$

Finally, we give some explicit functions that enable us to satisfy the above conditions and complete the proof of the theorem. Consider the function

$$f_\tau(y) = (1+\tau^2)^{-1} e^{\left(\psi^-_\lambda + \left\{\frac{\tau^2}{1+\tau^2}\right\}\frac{\mu_\lambda}{\theta}\right) y^2}$$

where $|\tau| < 1$ so that $f \in L^2(\phi)$. Using, for example, Mehler's formula we find that

$$e^{-\frac{1}{4}y^2} f(y) = \sum_{n=0}^{\infty} f_{2n} \phi_{2n}(y)$$

where the coefficients are given by

$$f_{2n} = \left(\frac{\theta}{\mu_\lambda}\right)^{\frac{1}{4}} \frac{\tau^{2n} (2n)!^{\frac{1}{2}}}{2^n n!}.$$

This eigenfunction expansion is uniformly convergent for each $|\tau| < 1$ which can readily be checked using the uniform bound (see, for example, Szegö **[14]**)

$$|H_n(x)| \le K 2^{\frac{n}{2}} (n!)^{\frac{1}{2}} e^{\frac{x^2}{2}} \qquad \forall n, x \tag{4.4}$$

so that we have a constant C such that $|\phi_n(y)| \le C$ for all n and y, the Weierstrass test and the ratio test. From Stirling's formula, $n! \sim \left(\sqrt{2\pi n}\right) n^n e^{-n}$, we also find that

$$f_{2n} \sim \frac{\tau^{2n}}{(n\pi)^{\frac{1}{4}}}$$

so this geometric decay of the coefficients ensures that all the requirements of the previous arguments hold for f_τ when $|\tau| < 1$. This proves Theorem 2.13 for every $\alpha < 1/4$ in the special case when the bounded continuous function is constant.

Next, it is easy to check that for any $q \in \mathbb{N}$, $y^q f_\tau(y)$ will also have an eigenfunction expansion that satisfies all our requirements. We therefore have that the moments of the $J_{\alpha,\lambda}(t)$ probability measure all converge to the moments of the required normal distribution, which implies weak convergence of the measure. □

5. Proof of Lemmas 4.2 and 4.3.

We go through the proof of lemma 4.3, which is a modification of the proof of the corresponding lemma used in Harris and Williams [**6**] combined with the use of raising and lowering operators and uniform bounds for the relevant eigenfunctions.

The branching process has state space

$$\mathcal{I} := \bigcup_{n\geq 1}(\{n\} \times \mathbb{R}^n \times \mathbb{R}^n).$$

Its formal generator $\mathcal{G}$ is given by

$$\mathcal{G} = \mathcal{G}_A + \mathcal{G}_\theta + \mathcal{G}_R, \tag{5.1}$$

where for $n \geq 1, \mathbf{x}, \mathbf{y} \in \mathbb{R}^n$, we have

$$\begin{aligned}
(\mathcal{G}_A F)(n;\mathbf{x};\mathbf{y}) &= \sum_{k=1}^{n} \tfrac{1}{2}A(y_k)\frac{\partial^2 F}{\partial x_k^2},\\
(\mathcal{G}_\theta F)(n;\mathbf{x};\mathbf{y}) &= \sum_{k=1}^{n} \frac{\theta}{2}\left\{\frac{\partial^2 F}{\partial y_k^2} - y_k\frac{\partial F}{\partial y_k}\right\}, \qquad (5.2)\\
(\mathcal{G}_R F)(n;\mathbf{x};\mathbf{y}) &= \sum_{k=1}^{n} R(y_k)\Big\{F\big(n+1;(\mathbf{x},x_k);(\mathbf{y},y_k)\big) - F(n;\mathbf{x};\mathbf{y})\Big\},
\end{aligned}$$

where $(\mathbf{x}, x_k) := (x_1, \ldots, x_n, x_k) \in \mathbb{R}^{n+1}$.

(5.3)**Proposition. Local-martingale condition.** *If $F : [0,\infty) \times \mathcal{I} \to \mathbb{R}$ and*

$$\left\{\left(\frac{\partial}{\partial t} + \mathcal{G}\right)F\right\}(t;n;\mathbf{x};\mathbf{y}) = 0 \qquad \text{for } t \geq 0, n \geq 1, \mathbf{x}, \mathbf{y} \in \mathbb{R}^n,$$

then $F\big(t; N(t); \mathbf{X}(t); \mathbf{Y}(t)\big)$ is a local martingale.

We know that

$$h_{m,\lambda}(t;n;\mathbf{x};\mathbf{y}) := \sum_{k=1}^{n} \Psi_{m,\lambda}(y_k)e^{\lambda x_k - E_{m,\lambda}t} \tag{5.4}$$

leads to the martingale $Z_{m,\lambda}(t) = h_{m,\lambda}\big(t; N(t); \mathbf{X}(t); \mathbf{Y}(t)\big)$.

Now, $Z_{m,\lambda}$ jumps only when a new particle is born; but any jump of $Z_{m,\lambda}$ is of magnitude no greater than the largest magnitude of the individual particles contributions, therefore, introducing the stopping times

$$V_n := \inf\left\{ t : \sum_{k=1}^{N(t)} \left|\Psi_{m,\lambda}(Y_k(t))\right| e^{\lambda X_k(t) - E_{m,\lambda} t} \geq n \right\},$$

then $Z_{m,\lambda}$ *stopped at* V_n never exceeds $2n$. Thus, $Z_{m,\lambda}$ *is locally in* $\mathcal{L}^2$ (relative to any $\mathbb{P}^{x,y}$), so can now conclude that

$$Z_{m,\lambda}(t)^2 - A_m(t) \quad \text{is a local martingale}$$

where

$$A_m(t) := \int_0^t \left\{ \left(\mathcal{G} + \frac{\partial}{\partial t}\right) \left((h_{m,\lambda})^2\right) \right\} (s; N(s); \mathbf{X}(s); \mathbf{Y}(s))\, ds.$$

It is easy to calculate that

$$\text{(5.5)} \qquad \frac{dA_m}{dt}(t) = \sum_{k=1}^{N(t)} \Bigg(\left\{ (a\lambda^2 + r) Y_k(t)^2 + \rho \right\} \Psi_{m,\lambda}(Y_k(t))^2 + \theta \left\{ \frac{d\Psi_{m,\lambda}}{dy}(Y_k(t)) \right\}^2 \Bigg) e^{2\lambda X_k(t) - 2E_{m,\lambda} t}$$

Now, utilising the raising and lowering operators

$$\mathcal{H}_\lambda := 2\psi_\lambda^+ y - \frac{d}{dy} \qquad \mathcal{H}_\lambda^\dagger := \frac{d}{dy} - 2\psi_\lambda^- y$$

where

$$\mathcal{H}_\lambda \Psi_{m,\lambda} = \sqrt{\frac{2\mu_\lambda}{\theta}}(m+1)\Psi_{m+1,\lambda} \qquad \mathcal{H}_\lambda^\dagger \Psi_{m,\lambda} = \sqrt{\frac{2\mu_\lambda}{\theta}}\, m \Psi_{m-1,\lambda}$$

and the uniform bound for the eigenfunctions (see (4.4))

$$\text{(5.6)} \qquad \Psi_{m,\lambda}(y) \leq K e^{\frac{y^2}{4}} \qquad \forall n \in \mathbb{N}, y \in \mathbb{R}$$

it is relatively straight forward to show that

$$\frac{dA_m}{dt}(t) \leq C\, m \sum_{k=1}^{n} e^{\frac{y_k^2}{2} + 2\lambda x_k - 2E_{m,\lambda} t} \qquad \forall m \in \mathbb{N}.$$

The one-particle picture and a change of measure methods (see Harris and Williams[**6**]) can now give bounds on $\mathbb{E}^{0,y} A_m(t)$ and in particular show that it is finite for small t.

We can use Fatou's Lemma to deduce from the fact that $(Z_{m,\lambda})^2 - A_m$ is a local martingale that

$$\text{(5.7)} \qquad \mathbb{E}^{0,y}\left[Z_{m,\lambda}(t)^2\right] \leq \mathbb{E}^{0,y} A_m(t) + \Psi_{m,\lambda}(y)^2.$$

Combining these ideas with the monotonicity of $\mathcal{L}^p$ norms now leads to Lemma 4.3. Sacrificing the uniformity over m for better control of the exponential growth bound in (5.6) will give Lemma 4.2 instead.

Acknowledgement. We thank the referee for their comments. We especially thank David Williams for originally suggesting this model and, moreover, for all the encouragement and inspiration he has given over the years. We wish him all the best for his 'retirement' years!

References

[1] Biggins, J. (1992) Uniform convergence in the branching random walk, *Ann. Probab.*, **20**, 137-151

[2] Breiman, L. (1968) *Probability.* Addison-Wesley, London.

[3] Champneys, A, Harris, S. C., Toland, J. F., Warren, J. & Williams, D (1995) Analysis, algebra and probability for a coupled system of reaction-diffusion equations, *Phil. Trans. Roy. Soc. London (A)*, **350**, 69–112.

[4] Chauvin, B. & Rouault, A. (1997) Boltzmann-Gibbs weights in the branching random walk. *Classical and Modern Branching Processes* (ed. Athreya, Krishna, et al.), IMA Vol. Math. Appl., **84**, pp 41–50. Springer, New York.

[5] Git, Y. & Harris, S.C. (2000) Large-deviations and martingales for a typed branching diffusion: II, (In preparation).

[6] Harris, S.C. & Williams, D. (1996) Large-deviations and martingales for a typed branching diffusion : I, *Astérisque*, **236**, 133–154.

[7] Harris, S.C. (2000) A typed branching diffusion, a reaction-diffusion equation and travelling-waves. (In preparation).

[8] McKean, H. P. (1975) Application of Brownian motion to the equation of Kolmogorov-Petrovskii-Piskunov. *Comm. Pure Appl. Math.* **28**, 323–331.

[9] McKean, H. P. (1976) Correction to the above. *Comm. Pure Appl. Math.* **29**, 553–554.

[10] Neveu, J. (1987) Multiplicative martingales for spatial branching processes. *Seminar on Stochastic Processes* (ed. E.Çinlar, K.Chung and R.Getoor), Progress in Probability & Statistics. **15**. pp. 223–241. Birkhäuser, Boston.

[11] Revuz, D. & Yor, M. (1991) *Continuous martingales and Brownian motion.* Springer, Berlin.

[12] Rogers, L.C.G. & Williams, D. (1994) *Diffusions, Markov processes and martingales. Volume 1: Foundations.* (Second Edition). Wiley, Chichester and New York.

[13] Rogers, L.C.G. & Williams, D. (1987) *Diffusions, Markov processes and martingales. Volume 2: Itô Calculus.* Wiley, Chichester and New York.

[14] Szegö, G. (1967) *Orthogonal Polynomials* (Third Edition). American Mathematical Society Colloquium Publications, Volume XXIII.

TIME DEPENDENT SUBORDINATION AND MARKOV PROCESSES WITH JUMPS

Masao NAGASAWA[1] and Hiroshi TANAKA[2]

Abstract

Bochner's subordination is extended to time-inhomogeneous Markov processes and the Feynman-Kac formula is generalized to the time-dependent subordination. As an application it is shown that stochastic differential equations with jumps can be directly solved with the help of the time-dependent subordination and consequently that the equation of motion for relativistic quantum particles is solved.

1. Introduction

For a prescribed drift coefficient $\boldsymbol{b}(s, x)$ and a potential function $c(s, x)$, we prepare a pair of operators

$$A_s^c = \frac{1}{2}(\sigma\nabla + \boldsymbol{b}(s, x))^2 + c(s, x)\,\mathrm{I},$$

and

$$\widehat{A}_t^c = \frac{1}{2}(\sigma\nabla - \boldsymbol{b}(t, x))^2 + c(t, x)\,\mathrm{I},$$

which are formal adjoint of each other, and set

$$M_s^c = -\sqrt{-A_s^c + \kappa^2\mathrm{I}} + \kappa\mathrm{I}, \quad \widehat{M}_t^c = -\sqrt{-\widehat{A}_t^c + \kappa^2\mathrm{I}} + \kappa\mathrm{I}.$$

We then consider

$$\frac{\partial\varphi}{\partial s} + M_s^c\varphi = 0, \tag{1.1}$$

$$-\frac{\partial\widehat{\varphi}}{\partial t} + \widehat{M}_t^c\widehat{\varphi} = 0, \tag{1.2}$$

---

[1] Institut für Mathematik der Universität Zürich Irchel, Winterthurerstr. 190, CH-8057 Zürich Switzerland

[2] Department of Mathematics, Japan Women's University, 2-8-1 Mejirodai Bunkyo-ku Tokyo Japan

which is the equation of motion of Nagasawa (1996, 1997) for a relativistic (spinless) quantum particle(s). The movement of a relativistic quantum particle is described by Markov processes of pure-jumps $\{Y(t), t \in [a, b], \mathrm{Q}\}$ such that the distribution of $Y(t)$ is given by

$$\mathrm{Q}[Y(t) \in dx] = \widehat{\varphi}(t, x)\varphi(t, x)dx,$$

where $\varphi(t, x)$ and $\widehat{\varphi}(t, x)$ are solutions of equations (1.1) and (1.2), respectively. In the Schrödinger representation we have

$$\mathrm{Q}[f(Y(t))] = \int \widehat{\varphi}_a(x)dxq(a, x; t, y)dyf(y)q(t, y; b, z)\varphi_b(z)dz,$$

where $q(s, x; t, y)$ is the fundamental solution of the pair of equations (1.1) and (1.2) which are in duality with respect to $dtdx$, and $\{\widehat{\varphi}_a, \varphi_b\}$ is a prescribed entrance-exit law, for details cf. Nagasawa (1996, 1997). Nagasawa-Tanaka (1998, 1999) discussed the existence and uniqueness of solutions of equation (1.1) in terms of stochastic differential equations of pure-jumps. The objective of the present article is to solve equation (1.1) more directly, through extending Bochner's subordination to temporally inhomogeneous diffusion processes and generalizing the Feynman-Kac formula to the time-dependent subordination.

2. Time-Dependent Subordination

2.1 *Bochner's Subordination*

We begin with a remark that it is immediate to construct a pure-jump Markov process $\{Y_t, t \in [a, b], \mathrm{P}\}$ with the fractional power generator

$$M = -\sqrt{-A + \kappa^2 \mathrm{I}} + \kappa \mathrm{I},$$

where

$$A = \frac{1}{2} \sum_{i,j=1}^{d} (\sigma\sigma^T)^{ij}(x) \frac{\partial^2}{\partial x_i \partial x_j} + \sum_{i=1}^{d} b^i(x) \frac{\partial}{\partial x_i}, \tag{2.1}$$

which does not depend on time. In fact, we apply the subordination of Bochner (1949) to the semi-group P_t of the temporally homogeneous diffusion process $\{X(t), t \geq 0, \mathrm{P}\}$ with the generator A in (2.1), i.e., we set

$$Y_t = X(Z(t)), \quad t \in [0, \infty), \tag{2.2}$$

where $\{Z(t), t \in [0, \infty), \mathrm{P}\}$ is the subordinator of Sato (1990), which is independent of the diffusion process $X(t)$ (cf. also Vershik-Yor (1995)). Then the subordinate process

$\{Y_t = X(Z(t)), t \geq 0, \mathrm{P}_x\}$ is a temporally homogeneous Markov process of pure-jumps with the transition probability

$$\mathrm{Q}_t f(x) = \mathrm{P}_x[f(X(Z(t)))] = \int_0^\infty \mathrm{P}_s f(x) \mathrm{P}[Z(t) \in ds],$$

and the generator M of the semi-group Q_t has the expression

$$Mf(x) = \int_0^\infty \{\mathrm{P}_r f(x) - f(x)\} \nu(dr),$$

where $\nu(dr)$ is the Lévy measure of $Z(t)$. However, if the coefficients of the operator in (2.1) depend on time, Bochner's subordination in (2.2) is no longer applicable.

2.2. *Time Dependent Subordination*

A typical example of time-dependent coefficients appears in the equation of motion in (1.1). We consider a stochastic process governed by

$$M_s = -\sqrt{-A_s + \kappa^2 \mathrm{I}} + \kappa \mathrm{I},$$

with

$$A_s = \frac{1}{2} \sum_{i,j=1}^{d} (\sigma\sigma^T)^{ij}(s, x) \frac{\partial^2}{\partial x_i \partial x_j} + \sum_{i=1}^{d} b^i(s, x) \frac{\partial}{\partial x_i}. \tag{2.3}$$

Let $B(t)$ be a d-dimensional Brownian motion and $Z(t)$ be a subordinator which is independent of the Brownian motion, and define the inverse function of $Z(t)$ by

$$Z^{-1}(t) = \inf\{s : Z(s) > t\}, \tag{2.4}$$

which is right-continuous in t. We denote by $X_{t_0,x}(t)$ the unique solution of a stochastic differential equation

$$X(t) = x + \int_{t_0}^{t} \sigma(t_0 + Z^{-1}(s - t_0), X(s)) dB(s) + \int_{t_0}^{t} \boldsymbol{b}(t_0 + Z^{-1}(s - t_0), X(s)) ds. \tag{2.5}$$

The key point in equation (2.5) is the inverse function $Z^{-1}(s - t_0)$ in the time parameter of the coefficients $\sigma(s, x)$ and $\boldsymbol{b}(s, x)$. We assume that the entries of the matrix $\sigma(s, x)$ and vector $\boldsymbol{b}(s, x)$ are bounded and continuous in (s, x), Lipschitz continuous in x for each fixed s, and the Lipschitz constants are bounded in s, so that equation (2.5) has a unique solution. We then set

$$Y_{t_0,x}(t) = X_{t_0,x}(t_0 + Z(t - t_0)),$$

which will be called *time-dependent subordination of the solution* $X_{t_0,x}(t)$ of equation (2.5). It is clear that $Y_{t_0,x}(t)$ satisfies

$$Y_{t_0,x}(t) = x + \int_{t_0}^{t_0 + Z(t-t_0)} \sigma(t_0 + Z^{-1}(s - t_0), X(s))dB(s)$$

$$+ \int_{t_0}^{t_0 + Z(t-t_0)} b(t_0 + Z^{-1}(s - t_0), X(s))ds.$$

To avoid notational complexity, let us set $t_0 = 0$, and denote $Y(t) = X(Z(t))$, where $X(t)$ is a solution of equation (2.5) with $t_0 = 0$, that is,

$$X(t) = x + \int_0^t \sigma(Z^{-1}(s), X(s))dB(s) + \int_0^t b(Z^{-1}(s), X(s))ds.$$

Then $Y(t) = X(Z(t))$ satisfies

$$Y(t) = x + \int_0^{Z(t)} \sigma(Z^{-1}(s), X(s))dB(s) + \int_0^{Z(t)} b(Z^{-1}(s), X(s))ds.$$

Putting $Z^{-1}(s) = u$ formally, we obtain a stochastic differential equation for $Y(t)$

$$Y(t) = x + \int_0^t \sigma(u, Y(u))dB(Z(u)) + \int_0^t b(u, Y(u))dZ(u),$$

which, however, does not give the right expression, but a more careful treatment of jumps of the subordinator $Z(t)$ will prove that $X(Z(t))$ satisfies

$$X(Z(t)) = x + \sum_{0<s\leq t} \int_{Z(s-)}^{Z(s)} \sigma(s, X(u))dB(u) + \sum_{0<s\leq t} \int_{Z(s-)}^{Z(s)} b(s, X(u))du, \tag{2.6}$$

where we assume $Z(t)$ is a pure-jump process.

On the other hand, let $W(dw)$ be the Wiener measure on the space Ω_c of all continuous paths, and $\xi_t(s,x,w)$ be the unique solution of

$$\xi(t) = x + \int_0^t \sigma(s, \xi(u))dw(u) + \int_0^t b(s, \xi(u))du, \tag{2.7}$$

where s is fixed. Then, as will be shown in Section 4, equation (2.6) is equivalent to the stochastic differential equation of pure-jumps

$$Y(t) = x + \int_{(0,t]\times(0,\infty)\times\Omega_c} \{\xi_\theta(s, Y(s-), w) - Y(s-)\}N(dsd\theta dw),$$

that was discussed in Nagasawa-Tanaka (1998), where $N(dsd\theta dw)$ is a Poisson random measure with the mean measure $ds\nu(d\theta)W(dw)$, and $\nu(d\theta)$ is the Lévy measure of the subordinator $Z(t)$.

3. Lemmas

Let $Z(t)$ be a right continuous non-decreasing function on $[0, \infty)$ such that

$$Z(t) = \beta t + \sum_{0<u\leq t} \{Z(u) - Z(u-)\}, \tag{3.1}$$

where $\beta \geq 0$ is a constant.

Lemma 3.1. *Let* $Z(t)$ *be given in* (3.1), *and define* $Z^{-1}(t)$ *by* (2.4). *Then*

$$\int_0^{Z(t)} f(Z^{-1}(s), g(s))ds$$

$$= \beta \int_0^t f(u, g(Z(u)))du + \sum_{0<s\leq t} \int_{Z(s-)}^{Z(s)} f(s, g(u))du, \tag{3.2}$$

for any $\mathbf{R}^d$*-valued continuous function* $g(s)$ *on* $[0, \infty)$ *and real-valued continuous function* $f(s, x)$ *on* $[0, \infty)\times\mathbf{R}^d$. *In applications, equation* (3.2) *is often expressed in another form as*

$$\int_0^{Z(t)} f(Z^{-1}(s), g(s))ds = \beta \int_0^t f(u, g(Z(u)))du$$

$$+ \sum_{0<s\leq t} \int_0^{\Delta Z(s)} f(s, g(Z(s-) + u))du,$$

where $\Delta Z(s) = Z(s) - Z(s-)$.

Proof. (i) At the first step, we assume that the set of jump times of $Z(t)$ has no finite accumulation point, and denote them by

$$0 < s_1 < s_2 < s_3 < \dots ,$$

in natural order. Let us assume $\beta > 0$. The case $\beta = 0$ is simpler, and can be handled in the same way. If $0 \leq t < s_1$, then $Z(t) = \beta t$ and $Z^{-1}(s) = s/\beta$ for $0 \leq s < Z(t)$. Therefore, we have

$$\int_0^{Z(t)} f(Z^{-1}(s), g(s))ds = \beta \int_0^t f(u, g(Z(u)))du. \tag{3.3}$$

When $s_1 \leq t < s_2$, we have

$$\int_0^{Z(t)} f(Z^{-1}(s), g(s))ds = \int_0^{Z(s_1-)} f(Z^{-1}(s), g(s))ds + \int_{Z(s_1-)}^{Z(s_1)} f(s_1, g(s))ds$$

$$+ \int_{Z(s_1)}^{Z(s_1)+\beta(t-s_1)} f(s_1 + \frac{s - Z(s_1)}{\beta}, g(s))ds,$$

on the right-hand side of which the first integral is equal to

$$\beta \int_0^{s_1} f(u, g(Z(u)))du,$$

in view of equation (3.3), in the second integral we have applied the property that $Z^{-1}(s)$ remains constant in the interval $(Z(s_1-), Z(s_1)]$ and hence $Z^{-1}(s) = Z^{-1}(Z(s_1)) = s_1$, and the third integral is equal to

$$\beta\int_{s_1}^{t} f(u, g(Z(u)))du,$$

for which we put $u = s_1 + (s - Z(s_1))/\beta$ and $s = Z(s_1) + \beta(u - s_1) = Z(u)$. Therefore,

$$\int_0^{Z(t)} f(Z^{-1}(s), g(s))ds = \beta\int_0^{t} f(u, g(Z(u)))du + \int_{Z(s_1-)}^{Z(s_1)} f(s_1, g(u))du.$$

Repeating the same argument, we have, for $s_n \leq t < s_{n+1}$,

$$\int_0^{Z(t)} f(Z^{-1}(s), g(s))ds = \beta\int_0^{t} f(u, g(Z(u)))du + \sum_{k=1}^{n}\int_{Z(s_k-)}^{Z(s_k)} f(s_k, g(u))du, \tag{3.4}$$

which yields equation (3.2) in the special case of no accumulation point, and also

$$\int_0^{Z(t)} f(Z^{-1}(s), g(s))ds$$

$$= \beta\int_0^{t} f(u, g(Z(u)))du + \sum_{k=1}^{n}\int_0^{\Delta Z(s_k)} f(s_k, g(Z(s_k-) + u))du.$$

(ii) In the general case, we set

$$Z_\varepsilon(t) = \beta t + \sum_{0<s\leq t} \{Z(s) - Z(s-)\} 1_{(\varepsilon,\infty)}(Z(s) - Z(s-)),$$

for $\varepsilon > 0$. Then $Z_\varepsilon(t)$ satisfies the condition of the first case, and hence

$$\int_0^{Z_\varepsilon(t)} f(Z_\varepsilon^{-1}(s), g(s))ds$$

$$= \beta\int_0^{t} f(u, g(Z_\varepsilon(u)))du + \sum_{0<s\leq t}\int_{Z_\varepsilon(s-)}^{Z_\varepsilon(s)} f(s_k, g(u))du,$$

in view of equation (3.4). Since $Z_\varepsilon(t) \uparrow Z(t)$ and $Z_\varepsilon^{-1}(t) \downarrow Z^{-1}(t)$ as $\varepsilon \downarrow 0$, we have equation (3.2). This completes the proof.

Let $B(t)$ be a d-dimensional Brownian motion, and $Z(t)$ be a subordinator which is independent of the Brownian motion. By definition the subordinator $Z(t)$ is expressed as in equation (3.1) and its Lévy measure $\nu(d\theta)$ satisfies

$$\int_{(0,\infty)} (1 \wedge \theta)\nu(d\theta) < \infty.$$

We define

$$\mathcal{F}_t(B) = \sigma\{B(s)\,;\, 0 \le s \le t\},\, \mathcal{F}(Z) = \sigma\{Z(t)\,;\, t \ge 0\},\, \textit{and}\;\; \mathcal{F}_t = \mathcal{F}_t(B) \vee \mathcal{F}(Z).$$

Then $B(t)$ is an $\{\mathcal{F}_t\}$-Brownian motion.

Let $g(s)$ be an $\mathbf{R}^d$-valued continuous $\{\mathcal{F}_t\}$-adapted process, and $f(s, x)$ be a real-valued continuous function on $[0, \infty) \times \mathbf{R}^d$. Then $f(Z^{-1}(t), g(t))$ is a right-continuous $\{\mathcal{F}_t\}$-adapted process. Therefore, the Itô integral

$$\int_0^t f(Z^{-1}(s), g(s))dB(s)$$

is well-defined.

Lemma 3.2. *Let $g(s)$ be a $\mathbf{R}^d$-valued continuous $\{\mathcal{F}_t\}$-adapted process, and $f(s, x)$ be a real-valued continuous function on $[0, \infty) \times \mathbf{R}^d$. Let $Z(t)$ be a subordinator of the form in equation* (3.1) *with the Lévy measure $\nu(dr)$. Then*

$$\int_0^{Z(t)} f(Z^{-1}(s), g(s))dB(s) = \int_0^t f(u, g(Z(u)))d\widetilde{B}(u)$$

$$+ \sum_{0<s\le t} \int_{Z(s-)}^{Z(s)} f(s, g(u))dB(u), \qquad (3.5)$$

where $\widetilde{B}(t)$ is the continuous part of the Lévy process $B(Z(t))$, which is equal to $\sqrt{\beta}B(t)$ in law.

Proof. We can and will proceed as in the proof of Lemma 3.1. (i) We first assume $\nu((0, \infty)) < \infty$, and denote the jump times of $Z(t)$ as $0 < s_1 < s_2 < s_3 < \dots$. The equation (3.5) then turns out to be

$$\int_0^{Z(t)} f(Z^{-1}(s), g(s))dB(s) = \int_0^t f(u, g(Z(u)))d\widetilde{B}(u)$$

$$+ \sum_{0<s_k\le t} \int_{Z(s_k-)}^{Z(s_k)} f(s_k, g(u))dB(u). \tag{3.6}$$

Let us prove equation (3.6).

If $0 \le t < s_1$, then $Z(t) = \beta t$ and $Z^{-1}(s) = s/\beta$ for $0 \le s < \beta s_1$. Therefore, we have

$$\int_0^{Z(t)} f(Z^{-1}(s), g(s))dB(s) = \int_0^{\beta t} f(\frac{s}{\beta}, g(s))dB(s) = \int_0^t f(u, g(\beta u))dB(\beta u).$$

Hence, defining

$$\widetilde{B}_0(u) = B(\beta u), \quad for \quad u \le s_1,$$

we have, for $0 \le t < s_1$,

$$\int_0^{Z(t)} f(Z^{-1}(s), g(s))dB(s) = \int_0^t f(u, g(Z(u)))d\widetilde{B}_0(u). \tag{3.7}$$

Let $t \ge 0$ and $s_1 \le s_1 + t < s_2$. Then $Z(s_1 + t) = Z(s_1) + \beta t$, and

$$\int_0^{Z(s_1+t)} f(Z^{-1}(s), g(s))dB(s) = \int_0^{Z(s_1-)} f(Z^{-1}(s), g(s))dB(s)$$

$$+ \int_{Z(s_1-)}^{Z(s_1)} f(s_1, g(s))dB(s) + \int_{Z(s_1)}^{Z(s_1+t)} f(Z^{-1}(s), g(s))dB(s), \tag{3.8}$$

where, in view of equation (3.7), the first integral is equal to

$$\int_0^{s_1} f(u, g(Z(u)))d\widetilde{B}_0(u),$$

and to the second integral we have applied that $Z^{-1}(s) = Z^{-1}(Z(s_1)) = s_1$ for $s \in (Z(s_1-), Z(s_1)]$. In the third integral on the right-hand side of (3.8), we have $Z^{-1}(s) = s_1 + (s - Z(s_1))/\beta$, and hence it is equal to

$$\int_{Z(s_1)}^{Z(s_1)+\beta t} f(s_1 + \frac{s - Z(s_1)}{\beta}, g(s))dB(s) = \int_{s_1}^{s_1+t} f(u, g(Z(u)))d\widetilde{B}_1(u), \tag{3.9}$$

where we set, for $s_1 \leq u \leq s_2$,

$$\widetilde{B}_1(u) = B(Z(s_1) + \beta(u - s_1)) - B(Z(s_1)). \tag{3.10}$$

We can verify equation (3.9), going back to the definition of the stochastic integral. In fact, let

$$Z(s_1) = t_0 < t_1 < \ldots < t_n = Z(s_1) + \beta t,$$

and set

$$u_k = s_1 + \frac{t_k - Z(s_1)}{\beta}.$$

Then

$$t_k = Z(s_1) + \beta(u_k - s_1),$$

and

$$s_1 = u_0 < u_1 < \ldots < u_n = s_1 + t.$$

By definition, the left-hand side of equation (3.9) is the limit of

$$\sum_k f(s_1 + \frac{t_{k-1} - Z(s_1)}{\beta}, g(t_{k-1}))\{B(t_k) - B(t_{k-1})\}$$

$$- \sum_k f(u_{k-1}, g(Z(s_1) + \beta(u_{k-1} - s_1)))$$

$$\times \{B(Z(s_1) + \beta(u_k - s_1)) - B(Z(s_1) + \beta(u_{k-1} - s_1))\}$$

where $Z(s_1) + \beta(u_k - s_1) = Z(u_k)$, if $s_1 + t < s_2$, and hence,

$$= \sum_k f(u_{k-1}, g(Z(u_{k-1})))\{B(Z(u_k)) - B(Z(u_{k-1}))\},$$

from which we get the right-hand side of equation (3.9) in view of (3.10). Thus we have, for $s_1 + t < s_2$,

$$\int_0^{Z(s_1+t)} f(Z^{-1}(s), g(s))dB(s) = \int_0^{s_1} f(u, g(Z(u)))d\widetilde{B}_0(u)$$

$$+ \int_{Z(s_1-)}^{Z(s_1)} f(s_1, g(s))dB(s) + \int_{s_1}^{s_1+t} f(u, g(Z(u)))d\widetilde{B}_1(u). \tag{3.11}$$

Moreover, defining $\widetilde{B}(u)$ by

$$\widetilde{B}(u) = \widetilde{B}_0(u), \qquad \textit{for } 0 \leq u \leq s_1,$$

$$= \widetilde{B}(s_1) + \widetilde{B}_1(u), \quad \textit{for } s_1 \leq u \leq s_2,$$

we have, for $s_1 + t < s_2$,

$$\int_0^{Z(s_1+t)} f(Z^{-1}(s), g(s))dB(s)$$

$$= \int_0^{s_1+t} f(u, g(Z(u)))d\widetilde{B}(u) + \int_{Z(s_1-)}^{Z(s_1)} f(s_1, g(s))dB(s).$$

Applying the same argument, we have, for $s_2 \leq t < s_3$,

$$\int_0^{Z(t)} f(Z^{-1}(s), g(s))dB(s)$$

$$= \int_0^{t} f(u, g(Z(u)))d\widetilde{B}(u) + \sum_{k=1}^{2} \int_{Z(s_k-)}^{Z(s_k)} f(s_k, g(u))dB(u), \tag{3.12}$$

where we set

$$\widetilde{B}_2(u) = B(Z(s_2) + \beta(u - s_2)) - B(Z(s_2)), \ \textit{for } s_2 \leq u \leq s_3,$$

and then define $\widetilde{B}(u)$ by

$$\widetilde{B}(u) = \widetilde{B}_0(u), \qquad for\ 0 \le u \le s_1,$$

$$= \widetilde{B}(s_1) + \widetilde{B}_1(u), \quad for\ s_1 \le u \le s_2,$$

$$= \widetilde{B}(s_2) + \widetilde{B}_2(u), \quad for\ s_2 \le u \le s_3. \tag{3.13}$$

It is clear that $\widetilde{B}(u)$ is the continuous part of $B(Z(u))$ and equal to $\sqrt{\beta}B(u)$ in law. Repeating this procedure we obtain equation (3.6).

(ii) For the case of $\nu((0, \infty)) = \infty$. Let $N(dtd\theta)$ be a Poison random measure which is independent of $B(t)$, with the mean measure $dt\nu(d\theta)$, and set

$$Z(t) = \beta t + \int_{(0,t]\times(0,\infty)} \theta N(dsd\theta),$$

$$Z_\varepsilon(t) = \beta t + \int_{(0,t]\times(\varepsilon,\infty)} \theta N(dsd\theta). \tag{3.14}$$

Then we have

$$Z_\varepsilon(t) \Rightarrow Z(t), \quad as\ \varepsilon \downarrow 0, \tag{3.15}$$

$$Z_\varepsilon^{-1}(t) \Rightarrow Z^{-1}(t), \quad as\ \varepsilon \downarrow 0, \tag{3.16}$$

and

$$B(Z_\varepsilon(t)) \Rightarrow B(Z(t)), \quad as\ \varepsilon \downarrow 0, \tag{3.17}$$

where we may take $\varepsilon = 1/n$ and "$\Rightarrow$" denotes the uniform convergence on each finite time-interval almost surely. Let us denote by ν_ε the Lévy measure of $Z_\varepsilon(t)$. Then $\nu_\varepsilon((0, \infty)) < \infty$. Therefore, in view of equation (3.6), we have

$$\int_0^{Z_\varepsilon(t)} f(Z_\varepsilon^{-1}(s), g(s))dB(s) = \int_0^t f(u, g(Z_\varepsilon(u)))d\widetilde{B}_\varepsilon(u)$$

$$+ \sum_{0<s\le t} \int_{Z_\varepsilon(s-)}^{Z_\varepsilon(s)} f(s, g(u))dB(u), \tag{3.18}$$

where $\widetilde{B}_\varepsilon(u)$ is defined by (3.13) with $Z_\varepsilon(t)$ in place of $Z(t)$. We notice that $\widetilde{B}_\varepsilon(t)$ and $\widetilde{B}(t)$ are the continuous parts of $B(Z_\varepsilon(t))$ and $B(Z(t))$, respectively. Let us define

$$J(t) = \lim_{n \to \infty} \sum_{\substack{0 < s \le t \\ \Delta Z(s) > 1/n}} \{B(Z(s)) - B(Z(s-))\},$$

where $\Delta Z(s) = Z(s) - Z(s-)$, and set

$$J_\varepsilon(t) = \sum_{0 < s \le t} \{B(Z_\varepsilon(s)) - B(Z_\varepsilon(s-))\}.$$

Then

$$B(Z(t)) = \widetilde{B}(t) + J(t), \quad \textit{and} \quad B(Z_\varepsilon(t)) = \widetilde{B}_\varepsilon(t) + J_\varepsilon(t).$$

Moreover,

$$\widetilde{B}_\varepsilon(t) \Rightarrow \widetilde{B}(t), \quad \textit{as} \quad \varepsilon \downarrow 0,$$
$$J_\varepsilon(t) \Rightarrow J(t), \quad \textit{as} \quad \varepsilon \downarrow 0. \tag{3.19}$$

Therefore, we have equation (3.5), making $\varepsilon \downarrow 0$ in equation (3.18), because of (3.15), (3.16), (3.17) and (3.19). This completes the proof.

4. Stochastic Differential Equations with Jumps

Let $X_{t_0,x}(t)$ be the solution of a stochastic differential equation

$$X(t) = x + \int_{t_0}^{t} \sigma(t_0 + Z^{-1}(u - t_0), X(u))dB(u)$$

$$+ \int_{t_0}^{t} b(t_0 + Z^{-1}(u - t_0), X(u))du, \tag{4.1}$$

where $B(t)$ is a d-dimensional Brownian motion and $Z(t)$ is a subordinator which is independent of the Brownian motion. Generalizing the subordination in (2.2), we set

$$Y_{t_0,x}(t) = X_{t_0,x}(t_0 + Z(t - t_0)), \tag{4.2}$$

which will be called *time-dependent* (or *time-inhomogeneous*) *subordination of the solution* $X_{t_0,x}(t)$ of equation (4.1). Then $X_{t_0,x}(t_0 + Z(t - t_0))$ satisfies

$$X_{t_0,x}(t_0 + Z(t - t_0)) = x + \int_{t_0}^{t_0+Z(t-t_0)} \sigma(t_0 + Z^{-1}(u - t_0), X_{t_0,x}(u))dB(u)$$

$$+ \int_{t_0}^{t_0+Z(t-t_0)} b(t_0 + Z^{-1}(u - t_0), X_{t_0,x}(u))du. \qquad (4.3)$$

We first consider the case $t_0 = 0$, to avoid notational complexity. Let $X(t)$ be the unique solution of equation

$$X(t) = x + \int_0^t \sigma(Z^{-1}(u), X(u))dB(u) + \int_0^t b(Z^{-1}(u), X(u))du. \qquad (4.4)$$

Then $X(Z(t))$ satisfies

$$X(Z(t)) = x + \int_0^{Z(t)} \sigma(Z^{-1}(u), X(u))dB(u) + \int_0^{Z(t)} b(Z^{-1}(u), X(u))du,$$

and hence by Lemmas 3.1 and 3.2

$$X(Z(t)) = x + \int_0^t \sigma(u, X(Z(u)))d\widetilde{B}(u) + \sum_{0<s\le t} \int_{Z(s-)}^{Z(s)} \sigma(s, X(u))dB(u)$$

$$+ \beta\int_0^t b(u, X(Z(u)))du + \sum_{0<s\le t} \int_0^{\Delta Z(s)} b(s, X(Z(s-) + u))du. \qquad (4.5)$$

We first treat the case that $\nu((0, \infty)) < \infty$, and denote the jump times of $Z(t)$ by $D(\omega) = \{0 < \tau_1 < \tau_2 < \tau_3 < \dots,\ where\ \tau_i = \tau_i(\omega), i = 1, 2, \dots\}$. We decompose Brownian paths $B(t, \omega)$ depending on jump times of $Z(t)$. Let us denote

$$\Delta Z(\tau_i) = Z(\tau_i) - Z(\tau_i-) > 0,$$

and set

$$\widehat{B}^{(\tau_i)}(u) = B(Z(\tau_i-) + u \wedge \Delta Z(\tau_i)) - B(Z(\tau_i-)).$$

Then

$$\widehat{B}^{(\tau_i)} = \{\widehat{B}^{(\tau_i)}(u), 0 \le u < \Delta Z(\tau_i)\} \qquad (4.6)$$

are Brownian motions with the life-times $\Delta Z(\tau_i)$. Enlarging the basic probability space, if necessary, we can introduce independent Brownian motions $B^{(\tau)}$, $\tau = \tau_1, \tau_2, \tau_3, \ldots$, with infinite life time such that (i) $\widehat{B}^{(\tau)}(u) = B^{(\tau)}(u)$ for $0 \le u \le \Delta Z(\tau)$, and (ii) the family $\{B^{(\tau)}, \tau = \tau_1, \tau_2, \tau_3, \ldots\}$ is independent of $\widetilde{B}(t)$ which is the continuous part of the Lévy process $B(Z(t))$ and equal to $\sqrt{\beta}B(t)$ in law. Then

$$p(\tau, \omega) = (\Delta Z(\tau), B^{(\tau)}), \ \ \tau \in D(\omega). \tag{4.7}$$

is a stationary Poisson point process with the characteristic measure $\nu(d\theta)\mathrm{W}(dw)$. Let

$$N((0, s] \times d\theta dw) = \#\{\tau \in D(\omega) : \tau \le s, p(\tau, \omega) \in d\theta dw\}$$

be the counting measure of the point process $p(\tau, \omega)$. Then it is a Poisson random measure with the mean measure $ds\,\nu(d\theta)\mathrm{W}(dw)$, where $\mathrm{W}(dw)$ is the Wiener measure on the space Ω_c of all continuous paths. We will sometimes write $\tau = \tau_i$, for simplicity.

We rewrite equation (4.5) as

$$X(Z(t)) = x + \int_0^t \sigma(u, X(Z(u)))d\widetilde{B}(u) + \beta \int_0^t b(u, X(Z(u)))du$$

$$+ \sum_{0<\tau\le t} \{ \int_0^{\Delta Z(\tau)} \sigma(\tau, X(Z(\tau-) + u))dB^{(\tau)}(u) + \int_0^{\Delta Z(\tau)} b(\tau, X(Z(\tau-) + u))du \}. \tag{4.8}$$

Let (Ω, P) be a probability space, and W be the Wiener measure on the space Ω_c of all continuous paths on $\mathbf{R}^d$, and consider a stochastic differential equation

$$\xi(t) = x + \int_0^t \sigma(s, \xi(u))dw(u) + \int_0^t b(s, \xi(u))du, \tag{4.9}$$

where s is fixed. For fixed s and x, we denote by $\xi_t(s, x, w)$ the unique solution of equation (4.9), and set

$$\overline{\xi}_t(s, x, w) = \xi_t(s, x, w) - x. \tag{4.10}$$

Lemma 4.1. *Let $X(t)$ be the unique solution of equation* (4.4), *and define $X^{(\tau)}(t), t \le \Delta Z(\tau)$, by*

$$X^{(\tau)}(t) = X(Z(\tau-) + t). \tag{4.11}$$

Then $X^{(\tau)}(t)$ *satisfies equation* (4.9) *with* $x = X^{(\tau)}(0) = X(Z(\tau-))$ *and* $w = B^{(\tau)}$.

Proof. In view of equation (4.4), we have, for $t \le \Delta Z(\tau)$,

$$X^{(\tau)}(t) = X(Z(\tau-)+t)$$

$$= x + \int_0^{Z(\tau-)+t} \sigma(Z^{-1}(u), X(u))dB(u) + \int_0^{Z(\tau-)+t} b(Z^{-1}(u), X(u))du,$$

and

$$X^{(\tau)}(0) = X(Z(\tau-))$$

$$= x + \int_0^{Z(\tau-)} \sigma(Z^{-1}(u), X(u))dB(u) + \int_0^{Z(\tau-)} b(Z^{-1}(u), X(u))du.$$

Subtracting, we obtain

$$X^{(\tau)}(t) - X^{(\tau)}(0) = \int_{Z(\tau-)}^{Z(\tau-)+t} \sigma(Z^{-1}(u), X(u))dB(u)$$

$$+ \int_{Z(\tau-)}^{Z(\tau-)+t} b(Z^{-1}(u), X(u))du.$$

Since $Z^{-1}(u) = \tau$, for $u \in (Z(\tau-), Z(\tau)]$, we have, for $t \le \Delta Z(\tau)$,

$$X^{(\tau)}(t) - X^{(\tau)}(0) = \int_0^t \sigma(\tau, X^{(\tau)}(u))dB^{(\tau)}(u) + \int_0^t b(\tau, X^{(\tau)}(u))du, \quad (4.12)$$

on the right-hand of which we have used equation (4.11). This completes the proof.

In particular, equation (4.12) yields

$$X^{(\tau)}(\Delta Z(s)) - X^{(\tau)}(0) = \int_0^{\Delta Z(\tau)} \sigma(\tau, X^{(\tau)}(u))dB^{(\tau)}(u)$$

$$+ \int_0^{\Delta Z(\tau)} b(\tau, X^{(\tau)}(u))du,$$

which implies

$$\overline{\xi}_{\Delta Z(\tau)}(\tau, X^{(\tau)}(0), B^{(\tau)}) = \int_0^{\Delta Z(\tau)} \sigma(\tau, X^{(\tau)}(u))dB^{(\tau)}(u) + \int_0^{\Delta Z(\tau)} b(\tau, X^{(\tau)}(u))du,$$

where $\overline{\xi}_t(s, x, w)$ is defined by (4.10). Therefore, equation (4.8) can be expressed in terms of $\overline{\xi}_{\Delta Z(\tau)}(\tau, X^{(\tau)}(0), B^{(\tau)})$ as

$$X(Z(t)) = x + \int_0^t \sigma(r, X(Z(u)))d\widetilde{B}(u) + \beta\int_0^t b(u, X(u))du + \sum_{0<\tau\leq t} \overline{\xi}_{\Delta Z(\tau)}(\tau, X(Z(\tau\text{-})), B^{(\tau)}), \tag{4.13}$$

where

$$\sum_{0<\tau\leq t} \overline{\xi}_{\Delta Z(\tau)}(\tau, X(Z(\tau\text{-})), B^{(\tau)}) = \int_{(0,t]\times(0,\infty)\times\Omega_c} \overline{\xi}_\theta(s, X(Z(s\text{-})), w)N(dsd\theta dw), \tag{4.14}$$

with the counting measure $N(dsd\theta dw)$ of the Poisson point process $p(\tau, \omega) = (\Delta Z(\tau), B^{(\tau)})$ given in (4.7). Equation (4.13) together with equation (4.14) implies that $Y(t) = X(Z(t))$ satisfies

$$Y(t) = x + \int_0^t \sigma(u, Y(u))\sqrt{\beta}dB(u) + \beta\int_0^t b(u, Y(u))du + \int_{(0,t]\times(0,\infty)\times\Omega_c} \{\xi_\theta(s, Y(s\text{-}), w) - Y(s\text{-})\}N(dsd\theta dw). \tag{4.15}$$

In general, for the case that $\nu((0, \infty)) = \infty$, let $Z(t)$ and $Z_\varepsilon(t)$ be defined by (3.14). Then (3.15), (3.16) and (3.17) hold. Let $X_\varepsilon(t)$ be the unique solution of equation (4.4) with $Z_\varepsilon(t)$. Then $Y_\varepsilon(t) = X_\varepsilon(Z_\varepsilon(t))$ satisfies equation (4.15) with $N_\varepsilon(dsd\theta dw)$, whose mean measure is $ds\nu((\varepsilon, \infty)\cap d\theta)\mathrm{W}(dw)$, that is,

$$Y_\varepsilon(t) = x + \int_0^t \sigma(u, Y_\varepsilon(u))\sqrt{\beta}dB(u) + \beta\int_0^t b(u, Y_\varepsilon(u))du$$

$$+ \int_{(0,t]\times(\varepsilon,\infty)\times\Omega_c} \{\xi_\theta(s, Y_\varepsilon(s-), w) - Y_\varepsilon(s-)\}N(dsd\theta dw). \quad (4.16)$$

Letting $\varepsilon \downarrow 0$, since

$$Y_\varepsilon(t) = X_\varepsilon(Z_\varepsilon(t)) \Rightarrow Y(t) = X(Z(t)), \quad as \ \varepsilon \downarrow 0, \quad (4.17)$$

we have

Theorem 4.1. *Let $Z(t)$ be a subordinator of the form as in equation* (3.1). *Then the time-dependent subordination $Y(t) = X(Z(t))$ of the unique solution $X(t)$ of equation* (4.4) *satisfies*

$$Y(t) = x + \int_0^t \sigma(u, Y(u))\sqrt{\beta}dB(u) + \beta\int_0^t b(u, Y(u))du$$

$$+ \int_{(0,t]\times(0,\infty)\times\Omega_c} \{\xi_\theta(s, Y(s-), w) - Y(s-)\}N(dsd\theta dw), \quad (4.18)$$

where $\xi_t(s, x, w)$ is the unique solution of equation (4.9).

For the case $t_0 \neq 0$, we apply, instead of Lemmas 3.1 and 3.2,

Lemma 4.2. *Let $Z(t)$ be given in* (3.1), *define $Z^{-1}(t)$ by* (2.4), *and denote $\Delta Z(s) = Z(s) - Z(s-)$. Then*

$$\int_{t_0}^{t_0+Z(t-t_0)} f(t_0 + Z^{-1}(u - t_0), g(u))du$$

$$= \beta\int_{t_0}^t f(u, g(t_0 + Z(u - t_0)))du$$

$$+ \sum_{t_0<s\le t} \int_0^{\Delta Z(s-t_0)} f(s, g(t_0 + Z((s - t_0)-) + u))du,$$

for any $\mathbf{R}^d$*-valued continuous function* $g(s)$ *on* $[t_0, \infty)$ *and real-valued continuous function* $f(s, x)$ *on* $[t_0, \infty)\times\mathbf{R}^d$.

Lemma 4.3. *Let* $B(t)$ *be a d-dimensional Brownian motion, and let* $f(s, x)$ *and* $g(s)$, $s \in [t_0, \infty)$, *be as in Lemma* 3.2, *and* $Z(t)$ *be a subordinator of the form as in equation* (3.1). *Then*

$$\int_{t_0}^{t_0+Z(t-t_0)} f(t_0+Z^{-1}(u-t_0), g(u))dB(u-t_0)$$

$$= \int_{t_0}^{t} f(u, g(t_0+Z(u-t_0)))d\widetilde{B}(u-t_0) + \sum_{t_0<s\le t}\int_{Z((s-t_0)-)}^{Z(s-t_0)} f(s, g(t_0+u))dB(u),$$

where $\widetilde{B}(u)$ *is defined by* (3.13) *and equal to* $\sqrt{\beta}B(u)$ *in law.*

Proofs of the lemmas can be carried over in the same way as for Lemmas 3.1 and 3.2. Then, applying Lemmas 4.2 and 4.3, we obtain the general forms of equation (4.8) and equation (4.18) on a time interval $(t_0, t]$. Hence we have

Theorem 4.2. *The time-dependent subordination* $Y_{t_0,x}(t) = X_{t_0,x}(t_0+Z(t-t_0))$ *of the unique solution* $X_{t_0,x}(t)$ *of equation* (4.1) *satisfies a stochastic differential equation with jumps*

$$Y(t) = x + \int_{t_0}^{t} \sigma(u, Y(u))\sqrt{\beta}dB(u-t_0) + \beta\int_{t_0}^{t} b(u, Y(u))du$$

$$+ \int_{(t_0,t]\times(0,\infty)\times\Omega_c} \{\xi_\theta(s, Y(s-), w) - Y(s-)\}N(dsd\theta dw),$$

where $\{\sqrt{\beta}B(t) : t\ge 0\} = \{\widetilde{B}(t) : t\ge 0\}$, *in law, which is the continuous part of the Lévy process* $B(Z(t))$, $\xi_t(s, x, w)$ *is the unique solution of equation* (4.9), *and* $N(dsd\theta dw)$ *is a Poisson random measure with the mean measure* $ds\,\nu(d\theta)W(dw)$, *where* $\nu(d\theta)$ *is the Lévy measure of* $Z(t)$, *and* $W(dw)$ *is the Wiener measure defined on the space* Ω_c *of all continuous sample paths.*

Theorem 4.2 solves the problem of constructing Markov processes with jumps in the case of no scalar potential. To solve the case with potential functions, we shall generalize the method of Kac to the time-dependent subordination.

5. A Formula of Feynman-Kac Type

Let $X_{s,x}(t)$, $a \le s \le t \le b$, be the unique solution of the stochastic differential equation in (4.1) with $t_0 = s$, defined on a probability space $\{\Omega, \mathrm{P}\}$. Let $c(t, x)$ be continuous *a.e.* and bounded above

$$c(t, x) \le c_0 < \infty. \tag{5.1}$$

We define

$$M(s, t) = \exp\left(\int_s^{s+Z(t-s)} c(s + Z^{-1}(u-s), X_{s,x}(u))du\right),$$

taking it for granted that the right-hand side is well-defined, where $Z(t)$ is the same subordinator adopted to define $X_{s,x}(t)$ in (4.1). We set

$$c(t, x) = c^+(t, x) + c^-(t, x),$$

with $c^+ = (c) \vee 0$ and $c^- = (c) \wedge 0$. Then

$$M(s, t) = M^{(+)}(s, t) M^{(-)}(s, t),$$

where

$$M^{(+)}(s, t) = \exp\left(\int_s^{s+Z(t-s)} c^+(s + Z^{-1}(u-s), X_{s,x}(u))du\right),$$

and

$$M^{(-)}(s, t) = \exp\left(\int_s^{s+Z(t-s)} c^-(s + Z^{-1}(u-s), X_{s,x}(u))du\right).$$

Moreover, by Lemma 4.2

$$M^{(+)}(s, t) = M^{(1,+)}(s, t) M^{(2,+)}(s, t), \quad M^{(-)}(s, t) = M^{(1,-)}(s, t) M^{(2,-)}(s, t),$$

with

$$M^{(1,\pm)}(s, t) = \exp\left(\beta \int_s^t c^\pm(r, X_{s,x}(s + Z(r-s)))dr\right),$$

and

$$M^{(2,\pm)}(s, t) = \exp\left(\sum_{s<r\le t} \int_{Z((r-s)-)}^{Z(r-s)} c^\pm(s, X_{s,x}(s+u))du\right),$$

where

$$0 \le M^{(1,-)}(s, t), \ M^{(2,-)}(s, t) \le 1,$$

and, by (5.1),

$$\mathrm{P}[M^{(1,+)}(s,t)] \le e^{\beta c_0(t-s)}.$$

We then assume

$$\mathrm{P}[M^{(2,+)}(s,t)] \le e^{c_1(t-s)}, \tag{5.2}$$

so that $M^{(2,+)}$ is well-defined. We notice that

$$M^{(2,+)}(s,t) \le \exp(c_0 \sum_{0<r\le t-s} (Z(r) - Z(r-))) = \exp(c_0 Z(t-s)),$$

and hence

$$\mathrm{P}[M^{(2,+)}(s,t)] \le \exp((t-s)\int_0^\infty (e^{c_0\theta}-1)\nu(d\theta)).$$

Therefore, a sufficient condition for (5.2) is

$$\int_0^\infty (e^{c_0\theta}-1)\nu(d\theta) = c_1 < \infty.$$

In this case we have (5.2) with this constant $c_1 < \infty$.

Lemma 5.1. (i) *Let* $Y_{s,x}(t) = X_{s,x}(s + Z(t-s))$. *Then*

$$Y_{r,x}(t) = Y_{s,Y_{r,x}(s)}(t), \quad r \le s \le t.$$

(ii) *Denote the functional* $M(s,t)$ *as* $M(s,x,t,\omega)$. *Then*

$$M(r,x,t,\omega) = M(r,x,s,\omega)M(s,Y_{r,x}(s),t,\omega), \quad r \le s \le t.$$

The proof of the lemma is routine and omitted.

For fixed s we define a semi-group $P_t^{(s)}$, $t \ge 0$, by

$$P_t^{(s)}f(x) = \mathrm{W}[f(\xi_t(s,x))\exp(\int_0^t c(s,\xi_r(s,x))dr)], \tag{5.3}$$

with the unique solution $\xi_t(s,x)$ of

$$\xi(t) = x + \int_0^t \sigma(s,\xi(u))dw(u) + \int_0^t b(s,\xi(u))du. \tag{5.4}$$

We assume that the semi-group $P_t^{(s)}$ is a strongly continuous on $C_0(\mathbf{R}^d)$, whose generator is

$$A_s^c = \frac{1}{2}\sum_{i,j=1}^{d} (\sigma\sigma^T)^{ij}(s,x)\frac{\partial^2}{\partial x_i \partial x_j} + \sum_{i=1}^{d} b^i(s,x)\frac{\partial}{\partial x_i} + c(s,x)\mathrm{I}, \tag{5.5}$$

with a core $C_K^\infty(\mathbf{R}^d)$. Moreover, we apply Bochner's subordination with

$$Z^{(d)}(t) = \int_{(0,t]\times(0,\infty)} \theta N(drd\theta),$$

(see (3.14)) to the semi-group $P_t^{(s)}$. Then the subordinate generator is given by

$$\mathbf{M}^{(s)}f(x) = \int_0^\infty \{P_\theta^{(s)}f(x) - f(x)\}\,\nu(d\theta), \tag{5.6}$$

where $\nu(d\theta)$ is the Lévy measure of $Z^{(d)}(t)$, and $C_K^\infty(\mathbf{R}^d)$ is a core of $\mathbf{M}^{(s)}$.

Let us define

$$T(s,x) = \sup\{t > s : \int_s^{s+Z(t-s)} c^-(s + Z^{-1}(u-s), X_{s,x}(u))du) > -\infty\},$$
$$(\sup \phi = s),$$

and set

$$D = \{(s,x) : \mathrm{P}[s < T(s,x)] = 1\}.$$

We define $\mathrm{Q}_{s,t}^c$ by

$$\mathrm{Q}_{s,t}^c f(x) = \mathrm{P}[f(Y_{s,x}(t))M(s,t)], \quad for \ (s,x) \in D. \tag{5.7}$$

We set $\mathrm{Q}_{s,t}^c f(x) = 0$, for $(s,x) \notin D$. Then we have

Theorem 5.1. *Let $c(s,x)$ be any potential function being continuous a.e., and satisfying the conditions in* (5.1) *and* (5.2), *and let $\mathrm{Q}_{s,t}^c f$ be defined by the formula in* (5.7). *Assume $\mathbf{M}^{(s)}f(x)$ is continuous in (s,x). Then*

$$\lim_{t\downarrow s}\frac{\mathrm{Q}_{s,t}^c f(x) - f(x)}{t-s} = \beta A_s^c f(x) + \mathbf{M}^{(s)}f(x), \quad (s,x) \in D,$$

for any $f \in C_K^\infty(\mathbf{R}^d)$.

Proof. To avoid notational complexity we set $s = 0$ and $(0, x) \in D$, and consider

$$Q^c_{0,t}f(x) = P[f(X(Z(t)))\exp(\int_0^{Z(t)} c(Z^{-1}(u), X(u))du)], \tag{5.8}$$

where $X(t)$ is the unique solution of equation (4.4). Let us denote

$$F(t) = f(X(Z(t))),$$

and

$$M(t) = \exp(\int_0^{Z(t)} c(Z^{-1}(u), X(u))du).$$

Let $0 = t_0 < t_1 < t_2 < \ldots < t_n = b$ be a partition of $[0, b]$, and $\delta = \max |t_i - t_{i-1}|$. Then

$$F(b)M(b) - f(x) = \sum_{i=1}^{n} \{F(t_i)M(t_i) - F(t_{i-1})M(t_{i-1})\}$$

$$= \sum_{i=1}^{n} F(t_{i-1})\{M(t_i) - M(t_{i-1})\} + \sum_{i=1}^{n} M(t_i)\{F(t_i) - F(t_{i-1})\}. \tag{5.9}$$

We set $D_\varepsilon(\omega) = \{\tau : \Delta Z(\tau) > \varepsilon\} = \{0 < \tau_1 < \tau_2 < \tau_3 < \ldots\}$, since $\nu((0, \infty))$ may be infinite, and denote by $p_\varepsilon(\tau, \omega) = (\Delta Z_\varepsilon(\tau), B^{(\tau)})$ the point process defined on $D_\varepsilon(\omega)$, where $Z_\varepsilon(t)$ is given by (3.14). Let $X_\varepsilon(t)$ be the solution of equation (4.4) with $Z_\varepsilon(t)$. Let us set

$$F_\varepsilon(t) = f(X_\varepsilon(Z_\varepsilon(t))),$$

and

$$M_\varepsilon(t) = \exp(\int_0^{Z_\varepsilon(t)} c(Z_\varepsilon^{-1}(u), X_\varepsilon(u))du).$$

We have, by Lemma 3.1,

$$\int_0^{Z_\varepsilon(t)} c(Z_\varepsilon^{-1}(u), X_\varepsilon(u))du$$

$$= \beta\int_0^{t} c(u, Y_\varepsilon(u))du + \sum_{0<s\le t} \int_{Z_\varepsilon(s-)}^{Z_\varepsilon(s)} c(s, X_\varepsilon(u))du,$$

where $s = \tau_k$ in the summation and $Y_\varepsilon(u) = X_\varepsilon(Z_\varepsilon(u))$. Hence,

$$M_\varepsilon(t) = M_1(t)M_2(t),$$

with

$$M_1(t) = \exp(\beta\int_0^t c(u, Y_\varepsilon(u))du),$$

and

$$M_2(t) = \exp(\sum_{0<s\le t}\int_{Z_\varepsilon(s-)}^{Z_\varepsilon(s)} c(s, X_\varepsilon(u))du).$$

Then the first summation on the right-hand side of (5.9) with F_ε and M_ε is equal to

$$\sum_{i=1}^{n} F_\varepsilon(t_{i-1})M_2(t_i)\{M_1(t_i) - M_1(t_{i-1})\}$$

$$+ \sum_{i=1}^{n} F_\varepsilon(t_{i-1})M_1(t_{i-1})\{M_2(t_i) - M_2(t_{i-1})\},$$

where the first summation converges, as $\delta \downarrow 0$, to

$$\beta\int_0^t c(u, Y_\varepsilon(u))f(Y_\varepsilon(u-))M_\varepsilon(u)du.$$

The second summation is equal to

$$\sum_{i=1}^{n} F_\varepsilon(t_{i-1})M_1(t_{i-1})M_2(t_{i-1})\{M_2(t_i)/M_2(t_{i-1}) - 1\},$$

and it converges, as $\delta \downarrow 0$, to

$$\sum_{0<s\le t} f(Y_\varepsilon(s-))M_\varepsilon(s-)\{m(\Delta Z_\varepsilon(s)) - 1\},$$

with

$$m(\Delta Z_\varepsilon(s)) = M_\varepsilon(s)/M_\varepsilon(s-), \tag{5.10}$$

where

$$m(\Delta Z_\varepsilon(s)) = \exp(\int_0^{\Delta Z_\varepsilon(s)} c(s, X_\varepsilon(Z_\varepsilon(s-) + u))du). \tag{5.11}$$

Moreover, in view of (4.7), setting $S_\varepsilon = (0, t]\times(\varepsilon, \infty)\times\Omega_c$, we have

$$\sum_{0<s\le t} f(Y_\varepsilon(s-))M_\varepsilon(s-)\{m(\Delta Z_\varepsilon(s)) - 1\}$$

$$= \int_{S_\varepsilon} f(Y_\varepsilon(s-))M_\varepsilon(s-)\{m_s^\theta(Y_\varepsilon(s-)) - 1\}N(dsd\theta dw), \quad (5.12)$$

where

$$m_s^\theta(x) = \exp(\int_0^\theta c(s, \xi_u(s, x, w))du). \quad (5.13)$$

Therefore, we have

$$\lim_{\delta\downarrow 0} \mathrm{P}[\sum_{i=1}^n F_\varepsilon(t_{i-1})\{M_\varepsilon(t_i) - M_\varepsilon(t_{i-1})\}]$$

$$= \beta\mathrm{P}[\int_0^t c(u, Y_\varepsilon(u))f(Y_\varepsilon(u-))M_\varepsilon(u)du]$$

$$+ \mathrm{P}[\int_{S_\varepsilon} f(Y_\varepsilon(s-))M_\varepsilon(s-)\{m_s^\theta(Y_\varepsilon(s-)) - 1\}N(dsd\theta dw)]. \quad (5.14)$$

We now compute the second summation on the right-hand side of (5.9) with F_ε and M_ε.

We first notice that we have, by Theorem 4.1,

$$Y_\varepsilon(t) = X_\varepsilon(Z_\varepsilon(t)) = x + \int_0^t \sigma(u, Y_\varepsilon(u))\sqrt{\beta}dB(u) + \beta\int_0^t b(u, Y_\varepsilon(u))du$$

$$+ \int_{(0,t]\times(\varepsilon,\infty)\times\Omega_c} \bar{\xi}_\theta(s, Y_\varepsilon(s-), w)N(dsd\theta dw),$$

where $\bar{\xi}_r(s, x, w) = \xi_r(s, x, w) - x$.

Then, by Itô (1951) and Kunita-Watanabe (1967),

$$f(Y_\varepsilon(t)) - f(Y_\varepsilon(0)) = \sum_{i=1}^{d} \int_0^t \frac{\partial f}{\partial x^i}(Y_\varepsilon(u))\sigma^{ij}(u, Y_\varepsilon(u))\sqrt{\beta}dB^j(u)$$

$$+ \beta \sum_{i=1}^{d} \int_0^t b^i(u, Y_\varepsilon(u)) \frac{\partial f}{\partial x^i}(Y_\varepsilon(u))du$$

$$+ \beta \sum_{i,j=1}^{d} \int_0^t (\sigma\sigma^T)^{ij}(u, Y_\varepsilon(u)) \frac{1}{2} \frac{\partial^2 f}{\partial x^i \partial x^j}(Y_\varepsilon(u))du$$

$$+ \int_{(0,t]\times(\varepsilon,\infty)\times\Omega_c} \{f(Y_\varepsilon(s-) + \overline{\xi}_\theta(s, Y_\varepsilon(s-), w)) - f(Y_\varepsilon(s-))\}N(dsd\theta dw),$$

in the last integral of which we apply $Y_\varepsilon(s-) + \overline{\xi}_r(s, Y_\varepsilon(s-), w) = \xi_r(s, Y_\varepsilon(s-), w)$. Therefore, we have

$$\lim_{\delta\downarrow 0} \mathrm{P}[\sum_{i=1}^{n} M_\varepsilon(t_i)\{F_\varepsilon(t_i) - F_\varepsilon(t_{i-1})\}] = \mathrm{P}[\int_0^t M_\varepsilon(s)\beta A_s f(Y_\varepsilon(s))ds]$$

$$+ \mathrm{P}[\int_{S_\varepsilon} M_\varepsilon(s-)m_s^\theta(Y_\varepsilon(s-))\{f(\xi_\theta(s, Y_\varepsilon(s-), w)) - f(Y_\varepsilon(s-))\}N(dsd\theta dw)], \tag{5.15}$$

where we have applied equations (5.10), (5.11) and (5.13).

Combining equations (5.9), (5.14) and (5.15), we have

$$\mathrm{P}[f(X_\varepsilon(Z_\varepsilon(t)))M_\varepsilon(t)] - f(x)$$

$$= \mathrm{P}[\int_0^t M_\varepsilon(s)\beta\{A_s f(Y_\varepsilon(s)) + c(s, Y_\varepsilon(s))f(Y_\varepsilon(s-))\}ds]$$

$$+ \mathrm{P}[\int_0^t dsM_\varepsilon(s-)\int_\varepsilon^\infty \mathrm{W}[f(\xi_\theta(s, Y_\varepsilon(s-))m_s^\theta(Y_\varepsilon(s-)) - f(Y_\varepsilon(s-))]\nu(d\theta)], \tag{5.16}$$

where the second integral is equal to

$$\mathrm{P}[\int_0^t dsM_\varepsilon(s-)\int_0^\infty 1_{(\varepsilon,\infty)}(\theta)\{P_\theta^{(s)}f(Y_\varepsilon(s-)) - f(Y_\varepsilon(s-))\}\nu(d\theta)],$$

with $P_\theta^{(s)} f(x)$ given in (5.3). We notice that $(1 \wedge \theta)\nu(d\theta)$ is a finite measure and

$$\frac{1}{\theta}\{P_\theta^{(s)} f(Y_\varepsilon(s-)) - f(Y_\varepsilon(s-))\}, \quad s < t,$$

is bounded, since it converges, as $\theta \downarrow 0$, to the generator of $P_\theta^{(s)}$. Therefore, making $\varepsilon \downarrow 0$ in equation (5.16), we have, by the dominated convergence theorem,

$$\mathrm{P}[f(X(Z(t)))M(t)] - f(x) = \mathrm{P}[\int_0^t \{\{M(s)\beta A_s^c f(Y(s)) + M(s-)\mathbf{M}^{(s)} f(Y(s-)\} ds], \tag{5.17}$$

where A_s^c is given by (5.5) and $\mathbf{M}^{(s)}$ by (5.6). Equation (5.17) implies

$$\lim_{t \downarrow 0} \frac{\mathrm{Q}_{0,t}^c f(x) - f(x)}{t} = \beta A_0^c f(x) + \mathbf{M}^{(0)} f(x), \quad (0, x) \in D,$$

for $f \in C_K^\infty(\mathbf{R}^d)$, by the dominated convergence theorem.

We can prove the case $s \neq 0$ in the same way. Let $X_{s,x}(t)$ be the unique solution of equation (4.1) with $t_0 = s$, and set

$$F(t) = f(X_{s,x}(s + Z(t - s))), \; and \; M(t) = M(s, t).$$

Then we can repeat the same argument that we have adopted in the case $s = 0$. However, to make it simpler, we can apply the integration by parts formula

$$F(t)M(t) - F(s)M(s) = \int_s^t F(r-)dM(r) + \int_s^t M(r)dF(r), \tag{5.18}$$

and the same approximation argument. We remark that, by Theorem 4.2,

$$Y(t) = Y_{s,x}(t) = x + \int_s^t \sigma(r, Y(r))d\widetilde{B}(r) + \beta \int_s^t b(r, Y(r))dr$$
$$+ \int_{(s,t] \times (0,\infty) \times \Omega_c} \overline{\xi}_\theta(r, Y(r-), w) N(dr d\theta dw),$$

where $\overline{\xi}_\theta(r, x, w) = \xi_\theta(r, x, w) - x$. Moreover, we have, by Lemma 4.2,

$$\int_s^{s+Z(t-s)} c(s+Z^{-1}(r-s), X_{s,x}(r))dr$$

$$= \beta\int_s^t c(u, X_{s,x}(s+Z(u-s)))du + \sum_{s<r\le t}\int_{Z((r-s)-)}^{Z(r-s)} c(r, X_{s,x}(s+u))du$$

$$= \beta\int_s^t c(r, Y(r))dr + \sum_{s<r\le t}\int_{Z((r-s)-)}^{Z(r-s)} c(r, X_{s,x}(s+u))du.$$

Hence, the first integral on the right-hand side of (5.18) is

$$\int_s^t F(r-)dM(r)$$

$$= \int_s^t f(Y(r-))\, d\{\exp(\int_s^{s+Z(r-s)} c(s+Z^{-1}(u-s), X_{s,x}(u))du)\}$$

$$= \beta\int_s^t c(r, Y(r))f(Y(r-))M(r)dr$$

$$+ \sum_{s<r\le t} f(Y(r-))M(r-)\{m(\Delta Z(r-s)) - 1\}$$

where

$$m(\Delta Z(r-s)) = \exp(\int_0^{\Delta Z(r-s)} c(r, X_{s,x}(s+Z((r-s)-)+u))du\ ,$$

and hence

$$= \beta\int_s^t c(r, Y(r))f(Y(r-))M(r)dr$$

$$+ \int_{(s,t]\times(0,\infty)\times\Omega_c} M(r-)f(Y(r-))\{m_r^\theta(Y(r-)) - 1\}N(drd\theta dw), \qquad (5.19)$$

where $m_r^\theta(x)$ is defined by (5.13). The second integral of the right-hand side of (5.18) is

$$\int_s^t M(r)dF(r) = \int_s^t M(r)\beta A_r f(Y(r))dr$$

$$+ \int_{(s,t]\times(0,\infty)\times\Omega_c} M(r-)m_r^{\theta}(Y(r-))\{f(\xi_{\theta}(r, Y(r-), w)) - f(Y(r-))\}N(drd\theta dw)$$

$$+ \int_s^t M(r) \sum_{i,j=1}^{d} \frac{\partial f}{\partial x^i}(Y(r))\sigma^{ij}(r, Y(r))\sqrt{\beta}dB^j(r), \tag{5.20}$$

where $\xi_t(s, x, w)$ is the unique solution of equation (5.4) and $Y(t) = Y_{s,x}(t)$.

Combining (5.19) and (5.20) we have

$$\mathrm{P}[f(X(Z(t)))M(t)] - f(x)$$

$$= \mathrm{P}[\int_s^t M(r)\beta\{A_r f(Y(r)) + c(r, Y(r))f(Y(r-))\}dr]$$

$$+ \mathrm{P}[\int_{(s,t]\times(0,\infty)\times\Omega_c} M(r-)\{f(\xi_{\theta}^r(w))\, m_r^{\theta}(Y(r-)) - f(Y(r-))\}N(drd\theta dw)],$$

$$= \mathrm{P}[\int_s^t M(r)\beta A_r^c f(Y(r))dr] + \mathrm{P}[\int_s^t M(r-)\mathbf{M}^{(r)} f(Y(r-))\, dr],$$

where $Y(t) = Y_{s,x}(t)$. We have thus obtained the general form of equation (5.17) on $(s, t]$, and we can complete the proof.

Let us consider a special case that a subordinator $Z(t)$ has the Lévy measure

$$\nu^{(\kappa)}(d\theta) = \frac{1}{2\sqrt{\pi}} e^{-\kappa^2\theta} \frac{1}{\theta^{3/2}} d\theta, \tag{5.21}$$

with a parameter κ, and potential functions satisfying

$$c(t, x) \le \kappa^2 < \infty. \tag{5.22}$$

Then the condition (5.2) is satisfied. We can then generalize the results in Nagasawa-Tanaka (1998, 1999) as follows.

Theorem 5.2. *Let $X_{s,x}(t)$ be the solution of the stochastic differential equation in (4.1) ($t_0 = s$) with the subordinator $Z(t)$ which is independent of $B(t)$ and has the Lévy measure $\nu^{(\kappa)}(d\theta)$ given in (5.21), and let $c(t, x)$ be any potential function being continuous a.e., and satisfying (5.22). Define evolution operators $Q^c_{s,t}$ by the formula in (5.7). Then it solves*

$$\frac{\partial u}{\partial s} + \beta A^c_s u + M^c_s u = 0, \tag{5.23}$$

where

$$A^c_s = \frac{1}{2} \sum_{i,j=1}^{d} (\sigma\sigma^T)^{ij}(s, x) \frac{\partial^2}{\partial x_i \partial x_j} + \sum_{i=1}^{d} b^i(s, x) \frac{\partial}{\partial x_i} + c(s, x)\mathrm{I},$$

and

$$M^c_s = -\sqrt{-A^c_s + \kappa^2 \mathrm{I}} + \kappa \mathrm{I}.$$

Remark. If $\beta = 0$, equation (5.23) reduces to

$$\frac{\partial u}{\partial s} + M^c_s u = 0. \tag{5.24}$$

This generalizes the results on stochastic differential equations of pure-jumps in Nagasawa-Tanaka (1998, 1999)). Equation (5.24) is the equation of motion for a relativistic spinless quantum particle in an electromagnetic field mentioned in Introduction, cf. Nagasawa (1997) for details.

6. Markov Processes with (Pure) Jumps

Markov processes with jumps determined by the evolution operator $Q^c_{s,t}$ in equation (5.7) can be constructed with the help of the Schrödinger representation. In this section we assume that A_s in (2.3) and A^c_s in (5.5) are given by

$$A_s = \frac{1}{2}\Delta + \sum_{i=1}^{d} b^i(s, x) \frac{\partial}{\partial x_i}, \tag{6.1}$$

and

$$A^c_s = \frac{1}{2}\Delta + \sum_{i=1}^{d} b^i(s, x) \frac{\partial}{\partial x_i} + c(s, x)\mathrm{I}, \tag{6.2}$$

respectively, where Δ is the Laplace-Beltrami operator

$$\Delta = \frac{1}{\sqrt{\sigma_2(x)}} \frac{\partial}{\partial x^i} \left(\sqrt{\sigma_2(x)}\, (\sigma\sigma^T(x))^{ij} \frac{\partial}{\partial x^j}\right),$$

with a positive definite diffusion matrix $\sigma\sigma^T(x)$, where we denote $\sigma_2(t,x) = det(\sigma\sigma^T(x))$. To adopt the operators A_s and A_s^c in (6.1) and (6.2), respectively, we need to replace the drift coefficient $\boldsymbol{b}(t,x)$ in the preceding sections by

$$\boldsymbol{b}^\circ(t,x) = \boldsymbol{b}(t,x) + \boldsymbol{b}_\sigma(t,x),$$

with a correction term

$$\boldsymbol{b}_\sigma(x)^j = \frac{1}{2}\frac{1}{\sqrt{\sigma_2(x)}}\frac{\partial}{\partial x^i}(\sqrt{\sigma_2(x)}\,(\sigma\sigma^T(x))^{ij}).$$

If σ is independent of the space variables, then the correction is not necessary and $\boldsymbol{b}^\circ(t,x) = \boldsymbol{b}(t,x)$.

Let $Q^c_{s,t}$ be the operator defined by (5.7). Then there exists a transition kernel $q^c(s,x\,;t,dy)$ such that

$$Q^c_{s,t}f(x) = \int q^c(s,x\,;t,dy)f(y), \tag{6.3}$$

which obeys the Chapman-Kolmogorov equation but $Q^c_{s,t}1 \neq 1$, namely, the normality condition is not satisfied. To construct the Schrödinger process with jumps we need a prescribed entrance-exit law $\{\widehat{\phi}_a(x),\ \phi_b(x)\}$ satisfying a normality condition

$$\int dx\,\widehat{\phi}_a(x)q^c(a,x\,;b,dy)\phi_b(y) = 1.$$

With the triplet $\{q^c(s,x\,;t,dy),\ \widehat{\phi}_a(x),\ \phi_b(x)\}$ we define a probability measure Q by the Schrödinger representation (cf. Nagasawa (1993, 1997)):

$$Q[f(X_a, X_{t_1}, \dots, X_{t_{n-1}}, X_b)]$$

$$= \int dx_0\widehat{\phi}_a(x_0)q^c(a,x_0\,;t_1,dx_1)q^c(t_1,x_1\,;t_2,dx_2)\cdots$$

$$\cdots\, q^c(t_{n-1},x_{n-1}\,;b,dx_n)\phi_b(x_n)\,f(x_0,x_1,\dots,x_n),$$

where $a < t_1 < \cdots < t_{n-1} < b$ and $f(x_0, x_1, \dots, x_n)$ is any bounded measurable function on the product space $(\mathbf{R}^d)^{n+1}$, $n = 1, 2, \dots$. We thus obtain a Schrödinger process $\{X_t, \mathcal{F}_a^r \vee \mathcal{F}_s^b, a \le r < t < s \le b, Q\}$, where $\mathcal{F}_a^t = \sigma\{X_r\,; r \in [a,t]\}$, and $\{X_t, \mathcal{F}_a^t, a \le t \le b, Q\}$ is a Markov process with jumps by Theorem 4 of Nagasawa (1997) (or Theorem 3.6 of Nagasawa (1993)).

Acknowledgment. This work was partially supported by the Swiss National Foundation (21-29833.90).

References

Bochner, S., (1949): Diffusion equations and stochastic processes. Proc. Nat. Acad. Sci. USA, **35**, 368-370.

Itô, K., (1951): *On stochastic differential equations*. Memoirs of the AMS, **4**. American Math. Soc.

Kunita, H. & Watanabe, S., (1967) On square integrable martingales. Nagoya Math. J. **30**, 209-245.

Nagasawa, M., (1993): *Schrödinger equations and Diffusion Theory*. Birkhäuser Verlag, Basel Boston Berlin.

Nagasawa, M., (1996): Quantum theory, theory of Brownian motions, and relativity theory. Chaos, Solitons and Fractals, **7**, 631-643.

Nagasawa, M., (1997): Time reversal of Markov processes and relativistic quantum theory. Chaos, Solitons and Fractals, **8**, 1711-1772.

Nagasawa, M., Tanaka, H., (1998): Stochastic differential equations of pure-jumps in relativistic quantum theory. Chaos, Solitons and Fractals, 10, No 8, 1265-1280.

Nagasawa, M., Tanaka, H., (1999): The principle of variation for relativistic quantum particles. Preprint.

Sato, K., (1990): Subordination depending on a parameter. Probability Theory and Mathematical Statistics, Proc. Fifth Vilnius Conf. Vol. **2**, 372-382.

Vershik, A., Yor, M., (1995): Multiplicativité du processus gamma et étude asymptotique des lois stables d'indice α, lorsque α tend vers 0. Prepublications du Lab. de probab. de l'université Paris VI, 284 (1995).

Marked Excursions and Random Trees

David G. Hobson

Department of Mathematical Sciences,
University of Bath,
Claverton Down, Bath, BA2 7AY. UK.

1 Introduction

This article is devoted to the study of the properties of a marked Brownian excursion, and an embedding of a branching tree in an excursion with marks. The embedding depends upon the heights of the excursion at the various mark times, and the heights of minima between consecutive marks. The resulting tree is a critical binary tree with an independent branching structure. The construction also gives rise to a natural connection between a Brownian excursion with at least one mark, and a family of Brownian excursions each with exactly one mark.

The relationship between Brownian excursions and trees has been much explored in the literature, a selection of which we note here. Neveu and Pitman [9, 10] (see also Le Gall [5]) considered excursions of height at least h, and embedded a tree in the excursion based on the locations of *h-maxima* and *h-minima*, which are excursion dependent local-extremes (see [9, 10] for definitions.) LeGall [6] has described one fruitful application of the relationship between excursions and trees to the construction of super-processes.

Aldous [1, 2, 3] based a tree on the excursion value at arbitrary times chosen independently of the excursion structure. His tree, the *Brownian continuum random tree* contains an infinite number of 'leaves'. Le Gall [7] considers a tree with a finite number of leaves which correspond to times chosen uniformly over the lifetime of a Brownian excursion. This article has much in common with this last paper.

Le Gall chooses an excursion according to Itô measure, re-weighted by the length σ of the excursion, and imagines putting p points uniformly, and independently, in the interval $[0, \sigma]$. He then defines a construction of a tree based upon the excursion, and in particular the excursion values at these time-points. Here we imagine a Brownian motion and an independent Poisson process marking the time axis. We consider the first Brownian excursion to contain a mark, or equivalently we choose an excursion according to the Itô measure of marked excursions. This first marked excursion contains a random number of marks, but by the properties of the Poisson process, each mark is uniformly distributed over the lifetime of the excursion. Now we have an excursion with a (random) number of identified points and we can define an embedded tree in the manner of Le Gall [7]. We investigate the properties of the resultant tree: since the excursion and the marks have been chosen in a canonically simple fashion, the resulting tree is also particularly simple.

We extend this idea to construct a second tree in which each node of the tree corresponds to a mark in the original excursion, and in the final section use this construction to explain an observation on the Arcsin law.

In this article we consider marked excursions. A related problem involving normalised excursions (ie excursions of unit length) has been considered by Pitman [12]. Pitman derives the joint distribution of the values of a normalised Brownian excursion at times $0 < U_{(1)} < \cdots < U_{(n)} < 1$ and the minima of the process over subintervals $([U_{(i)}, U_{(i+1)}])_{1 \le i \le n-1}$. Here $U_{(1)}, \ldots, U_{(n)}$ are the order statistics of n independent standard uniform random variables. The choice in this paper of excursions which contain a mark introduces a size bias which is exactly the right factor to guarantee the independent branching structure of the embedded tree.

It is a pleasure to thank Jean-Francois Le Gall for discussions and correspondence on this subject, and an anonymous referee whose detailed reading of an earlier version of this manuscript ensured that this version is much clearer.

2 Preliminaries on Trees

2.1 Trees

A tree consists of a finite family of elements ordered into generations. The zeroth generation contains a single parent individual who has a random number of offspring. These offspring form the members of the first generation. Subsequently each member of the kth generation has a random number of offspring who form part of the $(k+1)^{th}$ generation.

Formally, following Neveu [8], define a tree as follows; τ is a tree if τ is a finite subset of $U := \cup_{n=0}^{\infty} \mathbb{N}^n$, where $\mathbb{N}^0 := \{\emptyset\}$, such that

- $\emptyset \in \tau$;
- for $k \ge 1$, if $u = u_1 \ldots u_k \in \tau$ then $u_1 \ldots u_{k-1} \in \tau$;
- for $k \ge 1$, if $u = u_1 \ldots u_k \in \tau$ with $u_k = n$ then $u = u_1 \ldots u_{k-1} m \in \tau$ for $1 \le m < n$.

Let $\mathbb{T}$ be the set of trees, and if $u = u_1 \ldots u_k$ let $|u| = k$.

2.2 Marked trees

A marked tree is of the form

$$\tilde{\tau} = (\tau, (L_u, u \in \tau))$$

where $\tau \in \mathbb{T}$ and $L_u \in \mathbb{R}_+, \forall u \in \tau$. The mark L_u should be interpreted as the branch length or lifetime of the individual associated with element u. Let $\tilde{\mathbb{T}}$ be the space of marked trees.

Note the unfortunate dual use of the word marked. In an excursion a mark refers simply to an identified time. In a tree a mark is a piece of additional information associated with an element. Hopefully no confusion will arise.

2.3 Binary trees

In a binary tree the number of offspring is always either zero or two. Let $\mathbb{T}_b$ be the set of binary trees. Introduce a measure μ on $\mathbb{T}_b$ by setting

$$\mu(\tau) = 2^{-\|\tau\|}$$

where $\|\tau\|$ is the cardinality of τ. The measure ν models a branching process with a binary branching distribution where each individual has zero or two offspring with probability $\frac{1}{2}$. If $p := \|\{u \in \tau : u1 \notin \tau\}\|$ then p is the number of elements of the tree with no descendents and $\|\tau\| = 2p - 1$. Such elements are termed 'leaves' of the tree.

Let $\tilde{\mathbb{T}}_b$ be the space of marked binary trees. For α positive define a measure μ_α on $\tilde{\mathbb{T}}_b$ by setting

$$\mu_\alpha(d\tilde{\tau}) \equiv \mu_\alpha(\tau, dL_u) := \mu(\tau) \prod_{u\in\tau} (\alpha e^{-\alpha L_u}) dL_u.$$

This measure corresponds to a continuous time branching process with offspring distribution the uniform measure on $\{0, 2\}$, and exponential, rate α, lifetimes. In particular, for $u \in \tau$, the law of the subtree rooted at u is the same as the law of the whole tree.

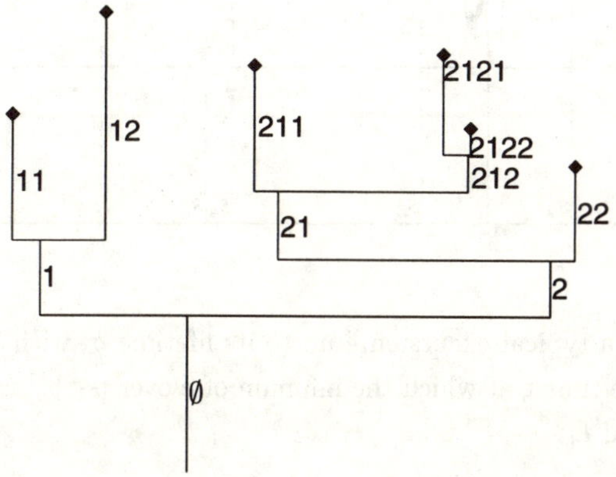

Figure 1: A representation of a marked binary tree. Vertical distances correspond to the branch lengths. Note that the horizontal scale and spacings are not part of the definition of the tree.

2.4 Binary trees and excursions

Define the set E of excursions to be the family of functions $e : \mathbb{R}_+ \mapsto \mathbb{R}_+ \cup \Delta$ with an associated lifetime $\sigma \equiv \sigma(e) \in \mathbb{R}_+$ such that $e(0) = 0$, $\lim_{s\uparrow\sigma} e(s) = 0$, e is continuous on $[0, \sigma)$ and positive on $(0, \sigma)$, and $e(s) = \Delta, \forall s \geq \sigma$. The element Δ is a graveyard state. We will loosely define an excursion by describing it's lifetime σ and a continuous function $e : [0, \sigma) \mapsto \mathbb{R}_+$ with the appropriate properties. In the sequel we will also use the concept of an excursion started from $h > 0$, which is as above except that $e(0) = h$. We will also specify time-points $0 < s_1 < \ldots s_p < \sigma$ during the lifetime of an excursion. These points will be identified with leaves on an associated tree. Let E_p be the space of excursions, with p specified time-points, and let $\tilde{E} \equiv E_1$.

Given an excursion e, for a finite integer p, fix $0 < s_1 < \ldots s_p < \sigma$. Then in each interval $[s_r, s_{r+1}]$ $(r = 1, \ldots p-1)$, there exists a time t_r such that the infimum of e over the interval is attained at t_r. Assume that t_r is unique and in the open interval (s_r, s_{r+1}), and further that the values $e(t_j)$ are distinct. When we consider Brownian excursions these assumptions will be satisfied, almost surely.

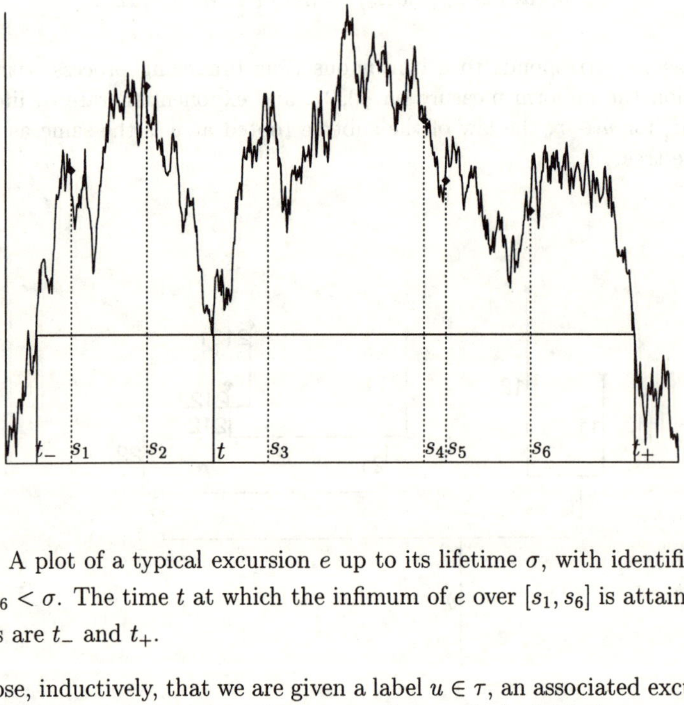

Figure 2: A plot of a typical excursion e up to its lifetime σ, with identified points $s_1 < \ldots s_6 < \sigma$. The time t at which the infimum of e over $[s_1, s_6]$ is attained is also shown, as are t_- and t_+.

Suppose, inductively, that we are given a label $u \in \tau$, an associated excursion e_u, and an ordered set of times $0 < s_1^u < \ldots < s_{p(u)}^u < \sigma(e_u)$. Initially we take $u \equiv \emptyset$, $e_\emptyset \equiv e$, and $(s_1^\emptyset, \ldots, s_{p(u)}^\emptyset) \equiv (s_1, \ldots, s_p)$. Then

- if $p(u) = 1$, set u to be a *leaf*, so that $u1 \notin \tau$ and $u2 \notin \tau$. Let $L_u = e_u(s_1^u)$ and $\tilde{e}_u \equiv e_u$. Note that $\tilde{e}_u \in \tilde{E}$.

- if $p(u) > 1$, let t^u be the time-point between the first mark (at s_1^u) and the last mark (at $s_{p(u)}^u$) at which the excursion e_u attains its minimum value, so that $e_u(t^u) = \min_{s_1^u \leq t \leq s_{p(u)}^u} e_u(t)$. Associate with the label u the length $L_u \equiv e_u(t^u)$. Moreover, set $u1, u2 \in \tau$. We wish to decompose e_u into three excursions; an excursion $\tilde{e}_u$ with exactly one labelled point, and two other excursions e_{u1}, e_{u2}, each with at least one labelled point. Define the times $t_-^u \equiv \sup_{s<s_1^u}\{e_u(s) = e_u(t^u)\}$ and $t_+^u \equiv \inf_{s>s_{p(u)}^u}\{e_u(s) = e_u(t^u)\}$. The three components of the decomposition are then as follows: $e_{u1}(s) := e_u(t_-^u + s) - e_u(t^u)$ is an excursion with lifetime $t^u - t_-^u$; $e_{u2}(s) := e_u(t^u + s) - e_u(t^u)$ an excursion with lifetime $(t_+^u - t^u)$; and $\tilde{e}_u(s)$ which has lifetime $\sigma(e_u) - (t_+^u - t_-^u)$ and is given by $\tilde{e}_u(s) := e_u(s)$ for $s \le t_-^u$ and $\tilde{e}_u(s) = e(t_+^u - t_-^u + s)$, for $s > t_-^u$.

For the label $u \in \tau$ this construction defines a mark L_u and associated excursion $\tilde{e}_u \in \tilde{E}$. Moreover, when $p(u) > 1$ the construction produces two sub-excursions labelled $u1$ and $u2$, with marks $(s_1^{u1}, \ldots, s_{p(u1)}^{u1}) \equiv (s_1^u - t_-^u, \ldots s_j^u - t_-^u)$ and $(s_1^{u2}, \ldots, s_{p(u2)}^{u2}) \equiv (s_{j+1}^u - t^u, \ldots, s_{p(u)}^u - t^u)$ respectively, where $j = \sup\{k : s_k^u < t^u\}$.

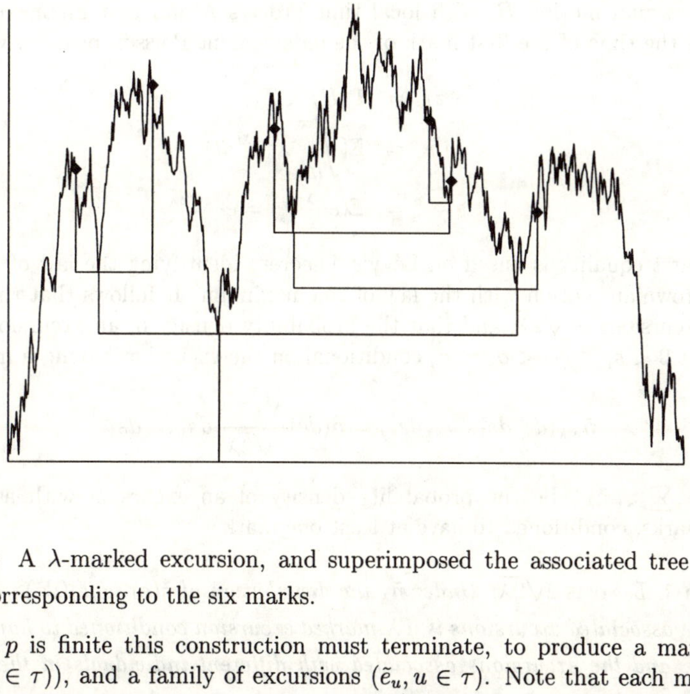

Figure 3: A λ-marked excursion, and superimposed the associated tree with six leaves, corresponding to the six marks.

Since p is finite this construction must terminate, to produce a marked tree $(\tau, (L_u, u \in \tau))$, and a family of excursions $(\tilde{e}_u, u \in \tau)$. Note that each member of this family of excursions has exactly one identified point which corresponds either to one of the original marks, or to one of the split times t_r.

The first aim of this article is to describe the law of the marked tree, and the associated excursions, when the original excursion is chosen to be a marked Brownian excursion.

3 Itô excursions and marked trees

Let B be a Brownian motion with local time process l at zero. Then, as Itô observed, B can be decomposed into excursions away from zero, indexed by l. These excursions form a Poisson process, on the space of excursions (as defined above, but also with the choice of sign), with intensity $n(de)$. By saying that 'e is a Brownian excursion' we mean that e has been chosen according to n.

Suppose further that time is marked by an independent Poisson process with constant intensity λ. For an excursion e, with duration $\sigma = \sigma(e)$, the probability that there are k marks is $(\lambda\sigma)^k e^{-\lambda\sigma}/k!$. By properties of the Poisson process the intensity of Brownian excursions e with k marks at $0 < s_1 < \cdots < s_k < \sigma$ is

$$n_{\lambda,k}(de, ds_1, \ldots, ds_k) = n(de)(\lambda\sigma)^k e^{-\lambda\sigma}\frac{ds_1}{\sigma}\cdots\frac{ds_k}{\sigma}$$

We will say that 'e is a λ-marked excursion' or just 'e is a marked excursion' if e is chosen according to n and the time domain of e contains at least one mark of the Poisson process.

For Brownian motion B, with local time process Λ and first hitting times H, and for T_λ the time of the first mark of the independent Poisson process, we have

$$\begin{aligned}\mathbb{P}(\Lambda_{T_\lambda} > z) &= \mathbb{P}(H_z < T_\lambda)\\ &= \mathbb{E}(\int_{H_z}^{\infty} \lambda e^{-\lambda t} dt)\\ &= \mathbb{E}(e^{-\lambda H_z}) = e^{-z\sqrt{2\lambda}}.\end{aligned} \tag{1}$$

Here the first equality is based on Lévy's Theorem identifying the law of the local time of Brownian motion with the law of the maximum. It follows that the rate of marked excursions is $\sqrt{2\lambda}$, and that the probability density of an excursion e with k marks at $0 < s_1 < \cdots < \sigma_k < \sigma$, conditional on the excursion having a mark, is

$$\tilde{n}_{\lambda,k}(de, ds_1, \ldots, ds_k) = n(de)\frac{\lambda^k e^{-\lambda\sigma}}{\sqrt{2\lambda}} ds_1 \ldots ds_k$$

Let $\tilde{n}_\lambda = \sum_{k\geq 1} \tilde{n}_{\lambda,k}$ be the probability density of an excursion with associated Poisson marks, conditioned to have at least one mark.

Theorem 1 *Let $\alpha \equiv 2\sqrt{2\lambda}$. Under $\tilde{n}_\lambda$ the distribution of $\tilde{\tau}(e)$ is $\mu_\alpha(d\tilde{\tau})$. Moreover each of the associated excursions is a λ-marked excursion conditioned to have exactly one mark, and the excursions associated with different individuals in the tree are independent.*

In the proof of this theorem we need a key lemma on excursions with a single mark.

Lemma 2 *A λ-marked excursion conditioned to have exactly one mark has the following structure; the height Z of the mark has an exponential distribution, rate α, the post-mark process is an independent Brownian motion, started at Z and conditioned to hit zero before being marked, and the pre-mark process is a time reversal of a further independent Brownian motion, again started at Z and again conditioned to hit zero before being marked.*

Proof of Lemma 2
We have that

$$\mathbb{P}(|B_{T_\lambda}| > z) = 2\mathbb{P}(B_{T_\lambda} > z) = \mathbb{P}(H_z < T_\lambda) = e^{-z\sqrt{2\lambda}}$$

(recall (1)) so that the height of the first mark in a Brownian motion has an exponential distribution, rate $\sqrt{2\lambda}$. Further the probability started from z of hitting zero unmarked is $e^{-z\sqrt{2\lambda}}$. Hence, the height of the mark in an excursion conditioned to have exactly one mark has density proportional to

$$(\sqrt{2\lambda}e^{-z\sqrt{2\lambda}})(e^{-z\sqrt{2\lambda}}) \equiv \sqrt{2\lambda}e^{-2\sqrt{2\lambda}z}$$

so that it has an exponential distribution, rate $\alpha = 2\sqrt{2\lambda}$. It also follows that the probability that a λ-marked excursion has exactly one mark is $1/2$.

The other statements follow from the strong Markov property and the invariance of the excursion measure under time reversal (see Rogers and Williams [13, VI.49]). □

Given an excursion e, started from $h > 0$, define $\grave{e}(s) \equiv \inf_{[0,s]} e(u)$. Then by Lévy's Theorem, if e is a Brownian excursion run until it first hits 0, then $((e-\grave{e})(s), s \geq 0)$ is a reflected Brownian motion started at 0 and run until the local time at 0 first reaches h. Further, if e is a Brownian motion conditioned to have no marks, then the excursions of $(e - \grave{e})$ from 0 are Brownian excursions conditioned to have no marks.

Proof of Theorem 1
Let e be a λ-marked Brownian excursion with lifetime σ. Let T_1 be the time of the first mark, and let $\xi \equiv e(T_1)$ be the height of the first mark. Then ξ has an exponential distribution with rate $\sqrt{2\lambda}$ and if $e_T(s) \equiv e(T_1 + s)$ then the process $(e_T - \grave{e}_T)$ is a reflecting Brownian motion, independent of ξ, run until the local time at 0 reaches ξ. For $s \leq \sigma - T_1$ define $Y(s) := (e_T - \grave{e}_T)(\sigma - s)$. The process Y is a reflecting Brownian motion run until it's local time at zero first reaches ξ, but we can think of it as the first part of a reflecting Brownian motion run for all time, and also labelled Y. We are interested in the last marked excursion from 0 (if any) of the process $e-\grave{e}$, this corresponds to the first marked excursion of the time reversed process Y (if that excursion occurs before the local time at zero reaches ξ.)

The first λ-marked excursion of Y occurs when the local time at 0 reaches η where η has an exponential distribution with rate $\sqrt{2\lambda}$ (again recall (1)). The probability that the excursion e has exactly one mark is precisely the probability that $\xi < \eta$, namely one-half, and moreover the law of $\eta \wedge \xi$ is exponential rate $\alpha = 2\sqrt{2\lambda}$.

Consider the case where $\xi < \eta$. Conditional on $\xi < \eta$ the excursion e contains exactly one mark at a height ξ which has an exponential rate α distribution, and we can apply the description in Lemma 2. Consequently, with probability $1/2$, the marked tree consists of just the parent individual, who has lifetime ξ. Note that, conditional on $\xi < \eta$, ξ has an exponential, rate α distribution.

Now consider the case where $\eta < \xi$. The process Y contains excursions which are unmarked, followed by a first marked excursion at local time η, which begins at time t_Y^- and ends at t_Y^+. Now consider these times in terms of the original excursion e. Let $t := \sigma - t_Y^+$ and $t_+ = \sigma - t_Y^-$, and let $t_- := \sup_{u<t}\{e(u) = e(t)\}$. Then t, t_-, t_+

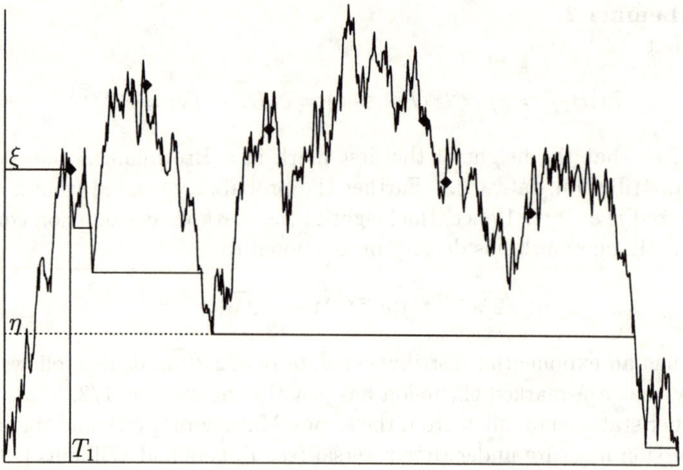

Figure 4: A Brownian excursion e, with first mark at T_1 and the process $\grave{e}$ also plotted.

have the meanings they were given in Section 2 and the excursion e is divided into three excursions, $e_1, e_2, \tilde{e}_\emptyset$ given by $e_1(s) := e(t_- + s) - e(t_-)$ for $0 \le s \le t - t_-$, $e_2(s) := e(t+s) - e(t)$ for $0 \le s \le t_+ - t$ and $\tilde{e}_\emptyset(s) = e(s)$ for $s < t_-$, and $\tilde{e}_\emptyset(s) = e(t_+ - t_- + s)$ for $s \ge t_-$. By construction the time reversal of e_2 is a marked excursion of Y, and, by invariance under time-reversal of n_λ, so is e_2. Moreover, again by the invariance under time-reversal of n_λ, and by the strong Markov property, so is e_1. Finally, by Lemma 2, $\tilde{e}_\emptyset$ is precisely a Brownian excursion with exactly one mark.

Recall that $\mathbb{P}(\eta < \xi) = 1/2$ so that the probability that the excursion contains more than one individual is one half. Then the marked tree has a parent individual (with lifetime η; conditional on $\eta < \xi$, η has an exponential distribution, rate α) who has exactly two offspring. These offspring correspond to the marked excursions e_1 and e_2, which are independent of each other, and of the excursion $\tilde{e}_\emptyset$ associated with the parent.

The theorem follows by induction. □

Remark 3 It follows from the theorem that if we extend the notion of a mark or label on an element of a tree from a lifetime to an excursion with exactly one mark or identified time then we have a correspondence $e \leftrightarrow (\tau, (\tilde{e}_u, u \in \tau))$ where e is a λ-marked Brownian excursion, τ a critical binary tree, and $\tilde{e}_u$ a family of Brownian excursions conditioned to have exactly one mark.

4 Equating marks with leaves

Our goal in this section is to embed a different tree, ρ, in a marked Brownian excursion. This new tree has one node (rather than one leaf) for each mark in the excursion. The motivation for consideration of this new tree is a Ray-Knight interpretation of excursions and local times at different levels. An application is given in the next section.

Suppose, inductively, that we are given an individual labelled $u \in \rho$, an associated excursion e_u, and an ordered set of times $0 < s_1^u < \cdots < s_{p(u)}^u < \sigma(e_u)$. Initially we take $u \equiv \emptyset$, $e_\emptyset \equiv e$, and $(s_1^\emptyset, \ldots, s_{p(u)}^\emptyset) \equiv (s_1, \ldots, s_p)$. Let $\{e_u(s_1^u), \ldots, e_u(s_{p(u)}^u)\}$ be minimised at $h = e_u(s_i^u)$ for some i. Associate with the label u the lifetime $L_u = h$. If $p(u) = 1$ then this individual has no offspring. Otherwise there is at least one excursion above the height h which is marked; label the $d = d(u)$ such marked excursions sequentially $u1, u2, \ldots, ud$. These excursions correspond to direct descendents of u in the tree ρ. For $j = 1, \ldots, d$ let t^{uj} be the start of the jth marked excursion e_{uj} above h, and define new mark times $s_\cdot^{uj} = s_\cdot^u - t^{uj}$ as appropriate. We are now in a position to repeat the construction for these sub-excursions. Since the number of marks p is finite this construction must terminate, to produce a marked tree $(\rho, (L_u, u \in \rho))$

Theorem 4 *Under $\tilde{n}_\lambda$ the distribution of $\rho(e)$ is that of a critical Galton-Watson process with geometric offspring distribution. In particular the measure of a tree ρ is $\nu(\rho) \equiv 2^{-(2\|\rho\|-1)}$. The marked tree has law*

$$\begin{aligned}\nu_\alpha(d\tilde{\rho}) = \nu_\alpha(\rho, dL_u) &= \prod_{u\in\rho} 2^{-(1+d(u))}(1+d(u))\alpha e^{-\alpha L_u}(1-e^{-\alpha L_u})^{d(u)} dL_u \\ &= \nu(\rho)\prod_{u\in\rho}(1+d(u))\alpha e^{-\alpha L_u}(1-e^{-\alpha L_u})^{d(u)} dL_u\end{aligned}$$

where $d(u)$ is the number of offspring of the individual labelled u.

We prove Theorem 4 by defining a surjection from $\tilde{\mathbb{T}}_b$ to $\tilde{\mathbb{T}}$. Given a marked binary tree (with marks corresponding to individual lifetimes) associate with each leaf $u = u_1 \ldots u_k \in \tau$ a total lapsed time $l_u = \sum_{j=0}^k L_{u_1\ldots u_j}$ (the sum of its own lifetime plus the lifetimes of its ancestors).

Let u be the label of the leaf with lowest lapsed time. Let this lapsed time l_u be the lifetime of the parent $\emptyset$ of the marked tree $\tilde{\rho}$. Now consider the number of descendents of the parent individual in the binary tree τ (excluding the individual corresponding to the leaf u), who are alive at the time l_u, see Figure 5. Each of these individuals is the root of a marked binary tree (we consider only that part of the sub-tree above l_u), and to each of these individuals we can associate an offspring of the parent in ρ. In this way we can build inductively a marked tree $\tilde{\rho}$ in which each node, including the leaves, corresponds to a leaf in τ.

Proof of Theorem 4

Most of Theorem 4, and in particular the independent branching structure, follows from Theorem 1, and in particular the fact that the marks have an exponential

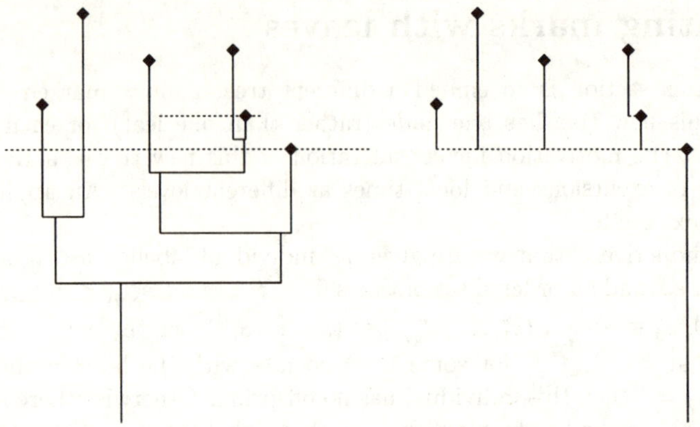

Figure 5: Constructing the general tree from a binary tree. In the new tree the parent has four offspring, the last of whom has one offspring. Again the horizontal scale is unimportant, although it has been inherited from a Brownian excursion. Note that each node in the new tree corresponds to a leaf in the old tree.

distribution, and from the mapping $\tilde{\tau} \mapsto \tilde{\rho}$. The remaining part is to identify the joint law of the lifetime and offspring distribution.

Number of offspring. Each event in the binary tree is with equal probability either a split or a death. The number of offspring of the parent in ρ is exactly the number of splits before the first death in τ (working upwards through the tree) and has a Geometric, parameter $1/2$, distribution.

Each individual alive at this first death is the root of a future tree, by the exponential lifetime property in $\tilde{\tau}$, each of these trees is independent and has the same probabilistic structure as the whole tree.

The law of the lifetime, conditioned on the number of offspring. Conditional on the parent $\emptyset$ having k offspring, the lifetime of $\emptyset$ is the sum of independent exponentials $T_{\alpha}^{\emptyset} + T_{2\alpha}^{\emptyset} + \cdots + T_{(k+1)\alpha}^{\emptyset}$. Here $T_{\alpha}^{\emptyset}$, rate α, is the time until the first event in the binary tree, $T_{2\alpha}^{\emptyset}$, rate 2α, is the subsequent time until the second event, (there now being two individuals in the binary tree), and so on.

By well known facts on the Yule process, or inductively on the number of offspring, the lifetime of the parent, conditioned on k offspring, has density

$$f_k(t) = (k+1)\alpha e^{-\alpha t}(1 - e^{-\alpha t})^k.$$

□

5 Trees and the Arcsin Law

Let B_t be a Brownian motion and let $V_t = (1/t) \int_0^t I_{\{B_s>0\}} ds$ be the proportion of time that Brownian motion has spent positive by time t. Then by Brownian scaling $V_t \stackrel{\mathcal{D}}{=} V$, a random variable independent of t which has the Arcsin distribution:

$$\mathbb{P}(V \leq v) = \int_0^v \frac{du}{\pi\sqrt{u(1-u)}}; \qquad 0 \leq v \leq 1.$$

The original motivation of this study was to explain and understand the observation that, for all integers $k \geq 0$,

$$\mathbb{E}(V^k) = 2^{-2k} \binom{2k}{k} = \mathbb{P}(\text{random walk of length } 2k \text{ is always non-negative}).$$

Since V is a distribution on $[0,1]$ and is completely determined by its moments, an explanation of this observation can be viewed as an alternative proof (though hardly the most direct) of the Arcsin Law.

The final equivalence that we need relates a tree ρ with $m \geq 1$ individuals to an excursion of a simple random walk S with $2m$ steps. Given a tree ρ, define a map $f : \{1, 2, \ldots, 2m-1\} \mapsto \{u : u \in \rho\}$ as follows. Let $f(1) = \emptyset$. Given $f(i) = u = u_1 \ldots u_k$ choose, if possible, the first child v of u which has not already been visited, and let $f(i+1) = v$, otherwise let $f(i+1) = u_1 \ldots u_{k-1}$. Finally let $S_0 = 0$, $S_i = |f(i)| + 1$ and $S_{2m} = \Delta$.

Standard combinatorial properties give that if ρ is a critical branching process with geometric parameter $\frac{1}{2}$ offspring distribution, then the resulting random walk is an excursion of a simple symmetric random walk. This construction, called the contour process or exploration process, is due to Harris [4], see also Le Gall [5].

Proposition 5

$$\mathbb{E}(V^k) = \mathbb{P}(\text{random walk of length } 2k \text{ is always non-negative})$$

Proof.
The proof is based upon the observation that the kth moment of V is the probability that for a Brownian path with k marks, the value of the Brownian motion is positive at each of the mark-times. We combine this remark with Theorem 4 and the bijection between critical Galton-Watson branching processes and excursions of random walks to deduce our result.

Let T_{k+1} be the time of the $(k+1)^{th}$ mark. Consider Brownian motion on $[0, T_{k+1}]$ (with exactly k marks) and $V_{T_{k+1}}$. Conditional on T_{k+1}, the times $T_1 < \cdots < T_k$ are (the order statistics of) k uniform random variables on $[0, T_{k+1}]$. Thus

$$\begin{aligned} \mathbb{E}(V^k) \equiv \mathbb{E}(V^k_{T_{k+1}}) &= \int_0^{T_{k+1}} \cdots \int_0^{T_{k+1}} I_{\{B_{s_1}>0\}} \cdots I_{\{B_{s_1}>0\}} \frac{ds_1}{T_{k+1}} \cdots \frac{ds_k}{T_{k+1}} \\ &= \mathbb{P}(B_{T_i} > 0, 1 \leq i \leq k) \end{aligned}$$

Now consider the whole path of B. Use Theorem 4 to identify the jth marked excursion of B, (with m_j marks), with a jth realisation of a branching tree (with m_j

individuals) and a positive or negative label according as the Brownian excursion is positive or negative. The jth realisation of the branching tree is in one-to-one correspondence with an excursion of a simple symmetric random walk (with $2m_j$ steps). We use the label of the tree to decide if this is a positive or negative excursion of the random walk. Now glueing these excursions together gives us a simple random walk $(S_n, n \geq 0)$. The positivity of $B_{T_1}, \dots B_{T_k}$ is equivalent to the fact that $(S_n \geq 0)$ for every $n \leq 2k$. □

For other consequences of the relationship between Brownian motion sampled uniformly at random times, and simple random walks, see Pitman [11, Corollary 3].

References

[1] ALDOUS, D.J.; The continuum random tree. I., *Annals of Probability*, **19**, 1–28, 1991.

[2] ALDOUS, D.J.; The continuum random tree. II: an overview, In *Proceedings of the Durham Symposium on Stochastic Analysis, 1990*, Editors, Barlow, M.T. and Bingham, N.H.. Cambridge University Press. 23–70, 1991.

[3] ALDOUS, D.J.; The continuum random tree. III., *Annals of Probability*, **21**, 248–289, 1993.

[4] HARRIS, T.E.; First passage and recurrence distributions, *Transactions of the American Mathematical Society*, **73**, 471–486, 1952.

[5] LE GALL, J-F.; Marches aléatoires, mouvement brownien et processus de branchement, *Séminaire de Probabilités*, **XXIII**, 447-464, 1989.

[6] LE GALL, J-F.; Brownian excursions, trees and measure-valued branching processes, *Annals of Probability*, **19**, 1399-1439, 1991.

[7] LE GALL, J-F.; The uniform random tree in a Brownian excursion, *Probability Theory and Related Fields*, **96**, 369–383, 1993.

[8] NEVEU, J; Arbes et processus de Galton-Watson, Annales de l'Institute Henri Poincaré, Série B, **22**, 199-207, 1986.

[9] NEVEU, J. AND PITMAN, J.; Renewal property of the extrema and tree property of a one-dimensional Brownian motion, *Séminaire de Probabilités*, **XXIII**, 239–247, 1989.

[10] NEVEU, J. AND PITMAN, J.; The branching process in a Brownian excursion, *Séminaire de Probabilités*, **XXIII**, 248-257, 1989.

[11] PITMAN, J.; Partition structures derived from Brownian motion and stable subordinators *Bernoulli*, **3**, 79-96, 1997.

[12] PITMAN, J.; Brownian motion, bridge, excursion, and meander characterized by sampling at independent uniform times, *Technical Report No. 545*, Department of Statistics, Berkeley, 1999.

[13] ROGERS, L.C.G. AND WILLIAMS, D.; *Diffusions, Markov processes and Martingales*, Vol. 2, Wiley, Chichester, 1987.

Laws of the iterated logarithm for the Brownian snake

Laurent Serlet

Abstract

We consider the path-valued process (W_s, ζ_s) called the Brownian snake, with lifetime process (ζ_s) a reflected Brownian motion. We first give an estimate of the probability that this process exits a "big" ball. Then we show the following laws of the iterated logarithm for the euclidean norm of the "terminal point" of the Brownian snake:

$$\limsup_{s\uparrow+\infty} \frac{|W_s(\zeta_s)|}{s^{1/4}\,(\log\log s)^{3/4}} = c, \quad \limsup_{s\downarrow 0} \frac{|W_s(\zeta_s)|}{s^{1/4}\,(\log\log(1/s))^{3/4}} = c$$

where $c = 2.3^{-3/4}$.

AMS Classification numbers: 60F10, 60F15, 60G15, 60G17, 60J25

Keywords: Brownian snake, large deviations, law of the iterated logarithm, Borel-Cantelli lemma.

1 Introduction

The Brownian snake is a random process whose values are paths in $\mathbf{R}^d$. This process introduced by Le Gall is closely related to super-Brownian motion, see [Lg1] and is a powerful tool to study the properties of super-Brownian motion, see for example [LP]. The Brownian snake also gives a nice probabilistic representation of the solutions of the partial differential equation $\Delta u = u^2$, see [Lg2] for an introduction.

Let us first recall the definition of the Brownian snake. For a detailed exposition the reader should refer to [Lg1] and [Lg2]. We call stopped path in $\mathbf{R}^d$ a pair (w, ζ) where ζ is a non-negative real number and w is a continuous function from $[0, +\infty)$ to $\mathbf{R}^d$ such that, for every $t \geq \zeta$, $w(t) = w(\zeta)$. We denote by $\mathcal{W}$ the space of all stopped paths in $\mathbf{R}^d$. We call ζ the lifetime of the stopped path (w, ζ). We denote $\hat{w} = w(\zeta)$ the "terminal point" of the path w. We often abuse notation and write only w to designate the stopped path (w,ζ). In this case we use the notation $\zeta(w)$ to designate the liftetime. The space $\mathcal{W}$ is a complete metric space when equipped with the metric

$$\mathrm{dist}((w,\zeta);\,(w',\zeta')) = \sup_{t\geq 0}|w(t)-w'(t)| + |\zeta-\zeta'|.$$

For $x \in \mathbf{R}^d$, we denote $\tilde{x}$ the path with lifetime 0 constantly equal to x.

The Brownian snake $((W_s,\zeta_s);\ s\in[0,+\infty))$ is a strong Markov process with values in $\mathcal{W}$ caracterized under the probability $\mathbf{P}_w$, by the following properties

- $W_0 = w$,
- $(\zeta_s;\ s \in [0,+\infty))$ has the law of a reflected Brownian motion starting from $\zeta(w)$, that is has the law of the absolute value of a linear Brownian motion starting from $\zeta(w)$,
- the conditional law of $(W_s)_{s\geq 0}$ knowing $(\zeta_s)_{s\geq 0}$ is the law of an inhomogeneous Markov process whose transition kernel is described as follows : for $0 \leq s < t$,
  - $W_t(u) = W_s(u)$ for all $u \leq m(s,t) = \inf_{s\leq v\leq t}\zeta_v$,
  - conditionally given $W_s(m(s,t))$, $(W_t(m(s,t)+u),\ u \geq 0)$ is independent of W_s and distributed as a Brownian motion in $\mathbf{R}^d$ starting from $W_s(m(s,t))$ and stopped at time $\zeta_t - m(s,t)$.

For the rest of the paper we will work most of the time under the probability $\mathbf{P} = \mathbf{P}_{\tilde{0}}$.

Occasionally we will need to work with the "excursion measure" of the Brownian snake. Note that, for every $x \in \mathbf{R}^d$ the path $\tilde{x}$ is regular for the Markov process (W_s). Hence we may define the associated excursion measure $\mathbf{N}_x$ (see [Lg2]). Under $\mathbf{N}_x$ the lifetime process $(\zeta(W_t))$ is distributed according to the Itô measure of positive excursions of linear Brownian motion. The conditional distribution of (W_t) knowing $(\zeta(W_t))$ is the same as above. Moreover we suppose that $\mathbf{N}_x$ is normalized so that $\mathbf{N}_x(\sup_{t\geq 0}\zeta(W_t) > h) = 1/2h$ for every $h > 0$.

At the end of this paper we also consider the Brownian snake started at w and stopped at the time σ where its lifetime reaches 0. We will denote $\mathbf{P}^*_w$ the corresponding probability.

A first step for us is to prove the following large deviation result which estimates the probability that the Brownian snake exists a "big" ball before time 1. The notation log refers to natural logarithm.

Theorem 1 *We have*

$$\lim_{A\to+\infty}\frac{1}{A^{4/3}}\log\mathbf{P}\left[\sup_{s\in[0,1]}|\hat{W}_s| \geq A\right] = -c_0$$

with $c_0 = 3\,.2^{-4/3}$.

Then our results follow.

Theorem 2 *Let* $h(s) = s^{1/4}\,(\log\log s)^{3/4}$. *Then,* **P***-almost surely,*

$$\limsup_{s\uparrow+\infty} \frac{|\hat{W}_s|}{h(s)} = c_1$$

with $c_1 = c_0^{-3/4} = 2\,.3^{-3/4}$.

Theorem 3 *Let* $\psi(s) = s^{1/4}\,(\log\log(1/s))^{3/4}$. *Then,* **P***-almost surely,*

$$\limsup_{s\downarrow 0} \frac{|\hat{W}_s|}{\psi(s)} = c_1$$

with, as previously, $c_1 = 2\,.3^{-3/4}$.

Surprisingly a similar law of the iterated logarithm with the same function ψ holds for the so-called iterated Brownian motion, see [CC]. In this model a single Brownian trajectory is described according to a reflected linear Brownian motion whereas in our model a "tree" of Brownian trajectories is described according to a reflected linear Brownian motion.

We will only prove theorem 2. It is easy to adapt this proof to treat the case of theorem 3.

2 A large deviation principle

Our first goal is to prove a large deviation principle concerning the finite-dimensional marginals of $(\varepsilon\,\hat{W}_s, \varepsilon^{2/3}\zeta_s;\ s \in [0,1])$. This has already been done in [Se2] when the considered Brownian snake has a lifetime process which is a normalised Brownian excursion instead of a reflected Brownian motion. The arguments in [Se2] may be adapted to the new setting. The methodology was to describe the law of the $2n$-tuple

$$(\hat{W}_{u_1}, \ldots \hat{W}_{u_n}, \zeta_{u_1}, \ldots, \zeta_{u_n})$$

with $0 < u_1 < \cdots < u_n < 1$ and then analyse the behavior of the densities of this $2n$-tuple when it is appropriately scaled. We know that the two stopped paths W_{u_i} and $W_{u_{i+1}}$ coincide up to time $\inf_{[u_i,u_{i+1}]}\zeta$. We note that, under **P**, there exists a unique m_i such that $\zeta_{m_i} = \inf_{[u_i,u_{i+1}]}\zeta$ and we use the notation $m_i = \text{arginf}([u_i, u_{i+1}], \zeta)$. Thus it is interesting to study the law under **P** of the (4n-2)-tuple

$$(\hat{W}_{u_1}, \ldots, \hat{W}_{u_n}, \zeta_{u_1}, \ldots, \zeta_{u_n}, \zeta_{m_1}, \ldots, \zeta_{m_{n-1}}, \hat{W}_{m_1}, \ldots, \hat{W}_{m_{n-1}}).$$

To describe this law and consequentely to express the large deviation principle, we need some notations related to tree structure introduced in [Se1].

Let $\alpha_1, \ldots, \alpha_{n-1}$ be distinct nonnegative real numbers. We denote by $A(\alpha_1, \ldots, \alpha_{n-1})$ the mapping $a : \{1, \ldots, n-1\} \to \{0, \ldots, n-1\}$ given, for every $i \in \{1, \ldots, n-1\}$, by $a(i) = l$ with

$$\alpha_l < \alpha_i, \quad \forall j \in (l \wedge i, l \vee i),\ \alpha_j > \alpha_i, \quad \alpha_l \text{ as large as possible},$$

if such an integer l exists and $a(i) = 0$ otherwise. We also define a mapping $v : \{1, \ldots, n\} \to \{1, \ldots, n-1\}$ by setting $v(1) = 1$, $v(n) = n-1$ and, for $i \in \{2, \ldots, n-1\}$ $v(i) = i-1$ if $\alpha_{i-1} > \alpha_i$ and $v(i) = i$ otherwise. As it is proved in [Se1], the mapping v is determined by $a = A(\alpha_1, \ldots, \alpha_{n-1})$ so that we use the notation v_a for v. We state the result without proof referring the reader to [Se2] for a detailled proof in a slightly different setting.

Proposition 4 *Let $\sigma = [0 < u_1 < \cdots < u_n < 1]$ be a finite partition of $[0,1]$. Under $\mathbf{P}$, the laws μ_ε of*

$$(\varepsilon \hat{W}_{u_1}, \ldots, \varepsilon \hat{W}_{u_n}, \varepsilon^{2/3} \zeta_{u_1}, \ldots, \varepsilon^{2/3} \zeta_{u_n},$$
$$\varepsilon^{2/3} \zeta_{m_1}, \ldots, \varepsilon^{2/3} \zeta_{m_{n-1}}, \varepsilon \hat{W}_{m_1}, \ldots, \varepsilon \hat{W}_{m_{n-1}})$$

satisfy a large deviation principle with speed $\varepsilon^{-4/3}$ and rate function

$$I_\sigma(y_1, \ldots, y_n, \beta_1, \ldots, \beta_n, \alpha_1, \ldots, \alpha_{n-1}, z_1, \ldots, z_{n-1})$$

$$= \frac{\beta_1^2}{2\,u_1} + \sum_{i=1}^{n-1} \frac{(\beta_i + \beta_{i+1} - 2\alpha_i)^2}{2\,(u_{i+1} - u_i)} + \sum_{i=1}^{n-1} \frac{|z_i - z_{a(i)}|^2}{2\,(\alpha_i - \alpha_{a(i)})} + \sum_{i=1}^{n} \frac{|y_i - z_{v_a(i)}|^2}{2\,(\beta_i - \alpha_{v_a(i)})}$$

if $0 < \alpha_i < \beta_i \wedge \beta_{i+1}$ for every i and $+\infty$ otherwise.

We recall that we have set $z_0 = 0$, $\tilde{0}$ being the "starting point" of the Brownian snake under $\mathbf{P}$. By "large deviation principle" we mean that

- for every U relatively open subset of $(\mathbf{R}^d)^n \times ([0, +\infty))^n \times ([0, +\infty))^{n-1} \times (\mathbf{R}^d)^{n-1}$,
$$\liminf_{\varepsilon \downarrow 0} \varepsilon^{4/3} \log \mu_\varepsilon(U) \geq -\inf_U I_\sigma$$
- for every K relatively closed subset of $(\mathbf{R}^d)^n \times ([0, +\infty))^n \times ([0, +\infty))^{n-1} \times (\mathbf{R}^d)^{n-1}$,
$$\limsup_{\varepsilon \downarrow 0} \varepsilon^{4/3} \log \mu_\varepsilon(K) \leq -\inf_K I_\sigma.$$

3 Probability of exit from a "big ball"

The aim of this section is to prove theorem 1. Let us denote by $\mathcal{A}$ the set of continuous $\mathcal{W}$-valued processes $(W'_s, \zeta'_s)_{s\in[0,1]}$ which have the "snake property" that is $W'_s(u) = W'_t(u)$ for $0 \le s < t \le 1$ and $u \le \inf_{[s,t]} \zeta'$ and which satisfy moreover $\sup_{s\in[0,1]} |\hat{W}'_s| \ge 1$. We have to show that

$$\lim_{\varepsilon\downarrow 0} \varepsilon^{4/3} \log \mathbf{P}\left[((\varepsilon W_s, \varepsilon^{2/3}\zeta_s))_{s\in[0,1]} \in \mathcal{A}\right] = -c_0 \tag{1}$$

We start with two lemmas.

Lemma 5 *For* $0 \le \alpha'_0 < \alpha'_1 < \cdots < \alpha'_N$ *and* $\gamma_1, \ldots, \gamma_N \in (0, +\infty)$ *and* $z'_0, z'_1, \ldots, z'_N \in \mathbf{R}^d$ *we have*

$$\sum_{i=1}^{N} \frac{|z'_i - z'_{i-1}|^2}{\alpha'_i - \alpha'_{i-1}} \ge \frac{|z'_N - z'_0|^2}{\alpha'_N - \alpha'_0}$$

and

$$\sum_{i=1}^{N} \frac{(\alpha'_i - \alpha'_{i-1})^2}{\gamma_i} \ge \frac{(\alpha'_N - \alpha'_0)^2}{\sum_{i=1}^{N} \gamma_i}$$

Proof. The first inequality (and similarly the second) is an easy consequence of Cauchy-Schwarz inequality:

$$|z'_N - z'_0| \le \sum_{i=1}^{N} |z'_i - z'_{i-1}| \le \left(\sum_{i=1}^{N} \frac{|z'_i - z'_{i-1}|^2}{\alpha'_i - \alpha'_{i-1}}\right)^{1/2} \left(\sum_{i=1}^{N} (\alpha'_i - \alpha'_{i-1})\right)^{1/2}$$

Lemma 6 *For* $j \in \{1, \ldots, n\}$ *we have the following upper bound*

$$I_\sigma(y_1, \ldots, y_n, \beta_1, \ldots, \beta_n, \alpha_1, \ldots, \alpha_{n-1}, z_1, \ldots, z_{n-1}) \ge \frac{|y_j|^2}{2\,\beta_j} + \frac{\beta_j^2}{2} \ge 3\,2^{-4/3}\,|y_j|^{2/3}$$

Proof. The second inequality follows from the first one by minimizing over $\beta_j > 0$. For the first one the main argument is the previous lemma. We first note that $\beta_{v_a(j)} + \beta_{v_a(j)+1} - 2\,\alpha_{v_a(j)} \ge \beta_j - \alpha_{v_a(j)}$ and, for every i, $\beta_{a(i)} + \beta_{a(i)+1} - 2\,\alpha_{a(i)} \ge \alpha_i - \alpha_{a(i)}$. This leads us to a lower bound for I_σ. We simply select certain terms in the expression of I_σ. In heuristic terms describing the "tree" formed by the paths $W_{u_1}, \ldots, W_{u_n}$, we start at the leaf y_j and back up to the root $0 \in \mathbf{R}^d$ along the branches of the tree. We denote by p the smallest integer such that $\alpha_{a^p(v_a(j))} = 0$ where $a^p = a \circ \cdots \circ a$. Then

$$I_\sigma \ \ge \ \left[\frac{|y_j - z_{v_a(j)}|^2}{2\,(\beta_j - \alpha_{v_a(j)})} + \frac{|z_{v_a(j)} - z_{a(v_a(j))}|^2}{2\,(\alpha_{v_a(j)} - \alpha_{a(v_a(j))})}\right.$$

$$+ \frac{|z_{a(v_a(j))} - z_{a(a(v_a(j)))}|^2}{2\left(\alpha_{a(v_a(j))} - \alpha_{a(a(v_a(j)))}\right)} + \cdots + \frac{|z_{a^{p-1}(v_a(j))} - z_0|^2}{2\left(\alpha_{a^{p-1}(v_a(j))} - \alpha_{a^p(v_a(j))}\right)}\Bigg]$$
$$+ \Bigg[\frac{\left(\beta_j - \alpha_{v_a(j)}\right)^2}{2\left(u_{v_a(j)+1} - u_{v_a(j)}\right)} + \frac{\left(\alpha_{v_a(j)} - \alpha_{a(v_a(j))}\right)^2}{2\left(u_{a(v_a(j))+1} - u_{a(v_a(j))}\right)}$$
$$+ \cdots + \frac{\left(\alpha_{a^{p-1}(v_a(j))} - \alpha_{a^p(v_a(j))}\right)^2}{2\left(u_{a^p(v_a(j))+1} - u_{a^p(v_a(j))}\right)}\Bigg]$$

Then we use lemma 5. A lower bound of the first quantity in brackets is given by the first inequality of lemma 5. Similarly the second quantity in brackets is dealt with the second inequality of lemma 5. We deduce

$$I_\sigma \geq \frac{|y_j|^2}{2\,\beta_j} + \frac{\beta_j^2}{2}$$

as wanted. This completes the proof of the lemma.

Proof of theorem 1. For $\sigma = [0 < u_1 < \cdots < u_n \leq 1]$ subdivision of $[0,1]$ we denote by π_σ the projection which associate to the $\mathcal{W}$-valued process $(W'_s)_{s\in[0,1]}$ the finite dimensional marginal:

$$(\hat{W'}_{u_1}, \ldots, \hat{W'}_{u_n}, \zeta'_{u_1}, \ldots, \zeta'_{u_n}, \zeta'_{m'_1}, \ldots, \zeta'_{m'_{n-1}}, \hat{W'}_{m'_1}, \ldots, \hat{W'}_{m'_{n-1}})$$

where, as in the previous section, $m'_i = \operatorname{arginf}([u_i, u_{i+1}], \zeta')$. With the notation μ_ε of proposition 4 we have

$$\limsup_{\varepsilon\downarrow 0} \varepsilon^{4/3} \log \mathbf{P}\left[((\varepsilon\, W_s, \varepsilon^{2/3}\zeta_s))_{s\in[0,1]} \in \mathcal{A}\right] \leq \limsup_{\varepsilon\downarrow 0} \varepsilon^{4/3} \log \mu_\varepsilon(\pi_\sigma \mathcal{A})$$

But by proposition 4 the limit on the right hand side is less than minus the infimum of I_σ over $\pi_\sigma\mathcal{A}$.
By lemma 6 the latter quantity is lower than $-c_0\ \sup_{\pi_\sigma\mathcal{A}} \sup_i |W_{u_i}|^{2/3}$ (we recall that c_0 has numerical value $3\,2^{-4/3}$). When the stepsize of σ tend to 0 the previous quantity has the asymptotic upper bound $-c_0$. So we get

$$\limsup_{\varepsilon\downarrow 0} \varepsilon^{4/3} \log \mathbf{P}\left[((\varepsilon\, W_s, \varepsilon^{2/3}\zeta_s))_{s\in[0,1]} \in \mathcal{A}\right] \leq -c_0.$$

Conversely,

$$\begin{aligned}
&\liminf_{\varepsilon\downarrow 0} \varepsilon^{4/3} \log \mathbf{P}\left[((\varepsilon\, W_s, \varepsilon^{2/3}\zeta_s))_{s\in[0,1]} \in \mathcal{A}\right] \\
&\geq \liminf_{\varepsilon\downarrow 0} \varepsilon^{4/3} \log \mathbf{P}\left[\varepsilon\,|\hat{W}_1| \geq 1\right] \\
&\geq -\inf\left\{\frac{\beta_1^2}{2} + \frac{|y_1|^2}{2\,\beta_1};\ |y_1| \geq 1,\ \beta_1 > 0\right\} = -c_0
\end{aligned}$$

Combining the above results on the liminf and the limsup give theorem 1.

4 Proof of the law of the iterated logarithm

Our aim is now to prove theorem 2. We first get the upper bound on the limsup. We take $\lambda > 1$, $c > c_1 = c_0^{-3/4}$ and set

$$A_n^\lambda = \left\{ \sup_{s\in[0,\lambda^n]} |\hat{W}_s| \geq c\,h(\lambda^n) \right\}$$

By the scaling property of the Brownian snake

$$\mathbf{P}[A_n^\lambda] = \mathbf{P}\left[\sup_{s\in[0,1]} |\hat{W}_s| \geq c\,(\log\log(\lambda^n))^{3/4} \right]$$

By theorem 1 we deduce that, for $\varepsilon > 0$ et n large enough

$$\begin{aligned} \mathbf{P}[A_n^\lambda] &\leq \exp -(c_0 - \frac{\varepsilon}{2}) \left[c\,(\log\log(\lambda^n))^{3/4} \right]^{4/3} \\ &\leq \exp -(c_0 - \varepsilon)\, c^{4/3} \log n = \frac{1}{n^{(c_0-\varepsilon)c^{4/3}}} \end{aligned}$$

Since $c_0\, c^{4/3} > 1$, we may choose $\varepsilon > 0$ so that $(c_0 - \varepsilon)\, c^{4/3} > 1$ which then implies $\sum_n \mathbf{P}(A_n^\lambda) < +\infty$. By the Borel-Cantelli lemma we easily deduce that

$$\limsup_{s\uparrow+\infty} \frac{|\hat{W}_s|}{h(s)} \leq c\,\lambda^{1/4}.$$

As this is valid for every $c > c_1$ and every $\lambda > 1$ we have proved that

$$\limsup_{s\uparrow+\infty} |\hat{W}_s|/h(s) \leq c_1.$$

We now pass to the proof of the lower bound on the limsup. We set

$$U_n^\lambda = \left\{ \sup_{s\in[0,\lambda^{n^\alpha}]} |\hat{W}_s| \geq c\,h(\lambda^{n^\alpha}) \right\}.$$

We claim that for $c < c_1$, $\lambda > 1$ and $\alpha > 1$ close enough to 1, we have

$$\sum_n \mathbf{P}[U_n^\lambda] = +\infty. \qquad (2)$$

This is obtained as previously by scaling and use of theorem 1: for $\varepsilon > 0$ et n large enough

$$\begin{aligned} \mathbf{P}[U_n^\lambda] &= \mathbf{P}\left[\sup_{s\in[0,1]} |\hat{W}_s| > c\,(\log\log(\lambda^{n^\alpha}))^{3/4} \right] \\ &\geq \exp -(c_0 + \varepsilon) c^{4/3}\, \alpha \log n = \frac{1}{n^{(c_0+\varepsilon)\,\alpha\, c^{4/3}}}. \end{aligned}$$

Let us admit temporarily that for $c < c_1$, $\alpha > 1$, λ big enough, there exists a constant M such that for all integers $m < n$

$$\mathbf{P}(U_m^\lambda \cap U_n^\lambda) \leq M\, \mathbf{P}(U_m^\lambda)\, \mathbf{P}(U_n^\lambda) \tag{3}$$

Then we may apply to (U_n^λ) a Borel-Cantelli lemma as stated for example in [PS] p.65. This lemma implies that, with positive probability, U_n^λ occurs infinitely often. Hence the event

$$H = \left\{ \limsup_{s\uparrow+\infty} |\hat{W}_s|/h(s) \geq c \right\}$$

occurs with positive probability. But the asymptotic event H satisfies a 0-1 law. Indeed $H \in \sigma\{W_u;\ u \geq d_v\}$ where d_v denotes the smallest zero of the lifetime after time $v \geq 0$. By construction of the Brownian snake this implies that H is independent of $\sigma\{W_u;\ u \leq d_v\}$. If we let v tend to $+\infty$, we see that H is independent of $\sigma\{W_u;\ u \geq 0\}$ hence of himself. Thus $\mathbf{P}(H) = 1$. Since it is valid for every $c < c_1$, we have proved theorem 2.

It remains to prove equation (3). We start with a lemma.

Lemma 7 *There exists a universal constant K_1 such that, for $w \in \mathcal{W}$, $A > \sup_{s\in[0,\zeta]} |w(s)|$,*

$$\mathbf{P}_w^*\left[\sup_{s\in[0,\sigma]} |\hat{W}_s| \geq A\right] \leq K_1 \int_0^\zeta \frac{ds}{(A - |w(s)|)^2}$$

Proof. We use proposition 2.5 of [Lg2]. As in this paper we set $\tilde{\zeta}_s = \inf_{u\in[0,s]} \zeta_u$. We denote by $\{(\alpha_i, \beta_i);\ i \in I\}$ the excursion of $\zeta - \tilde{\zeta}$ away from 0 before time σ (where the lifetime first hits 0). We denote by $\{W^i;\ i \in I\}$ the corresponding path-valued excursions that is

$$W_s^i(u) = W_{(\alpha_i+s)\wedge\beta_i}(\zeta_{\alpha_i} + u).$$

We know that the point measure

$$\sum_{i\in I} \delta_{W^i}(\cdot)$$

is, under $\mathbf{P}_w$, a Poisson point measure with intensity $2\int_0^\zeta ds\, \mathbf{N}_{w(s)}(\cdot)$. We recall from the introduction that $\mathbf{N}_x(\cdot)$ denotes the excursion measure of the Brownian snake away from the path $\tilde{x}$. The probability, under $\mathbf{P}$, that the Brownian snake exits the ball of radius A is the probability that at least one of the excursions W^i starting from $\hat{W}_{\alpha_i}$ does so and thus goes further than $A - |\hat{W}_{\alpha_i}|$ from its origin $\hat{W}_{\alpha_i}$. More precisely :

$$\begin{aligned}
\mathbf{P}^*_w\left[\sup_{s\in[0,\sigma]}|\hat{W}_s|\geq A\right] &= 1-\exp -2\int_0^\zeta ds\,\mathbf{N}_{w(s)}\left(\sup_{s\in[0,\sigma]}|\hat{W}'_s|\geq A\right)\\
&\leq 2\int_0^\zeta ds\,\mathbf{N}_0\left(\sup_{s\in[0,\sigma]}|\hat{W}'_s|\geq A-|w(s)|\right)\\
&\leq 2\int_0^\zeta \frac{ds}{(A-|w(s)|)^2}\,\mathbf{N}_0\left(\sup_{s\in[0,\sigma]}|\hat{W}'_s|\geq 1\right).
\end{aligned}$$

In the last line we have used a scaling argument under the excursion measure cf. [Lg2] proposition 2.3. The proof of the lemma is complete.

Now we come back to the proof that equation (3) holds for $c<c_1$, $\alpha>1$, λ big enough, m large enough and $n>m$. We already know that for $\varepsilon>0$ and n large enough, $\mathbf{P}(U_n^\lambda)\geq 1/n^\beta$ with $\beta=(c_0+\varepsilon)\,c^{4/3}\,\alpha$. We write the following decomposition

$$\begin{aligned}
\mathbf{P}(U_m^\lambda\cap U_n^\lambda) &= \mathbf{P}\left[\sup_{s\in[0,\lambda^{m^\alpha}]}|\hat{W}_s|\geq c\,h(\lambda^{m^\alpha});\ \sup_{s\in[0,\lambda^{n^\alpha}]}|\hat{W}_s|\geq c\,h(\lambda^{n^\alpha})\right]\\
&\leq T_1+T_2+T_3
\end{aligned}$$

where

$$T_1 = \mathbf{P}\left[\sup_{s\in[0,\lambda^{m^\alpha}]}|\hat{W}_s|\geq \frac{c}{2}\,h(\lambda^{n^\alpha})\right] \tag{4}$$

$$T_2 = \mathbf{P}\left[\sup_{s\in[0,\lambda^{m^\alpha}]}\zeta_s\geq \kappa\,\lambda^{m^\alpha/2}\sqrt{\log n}\right] \tag{5}$$

$$\begin{aligned}
T_3 = \mathbf{E}\Bigg[&\mathbf{1}\left(\sup_{s\in[0,\lambda^{m^\alpha}]}|\hat{W}_s|\in[c\,h(\lambda^{m^\alpha}),\frac{c}{2}\,h(\lambda^{m^\alpha})]\right)\mathbf{1}\left(\sup_{s\in[0,\lambda^{m^\alpha}]}\zeta_s\leq\kappa\,\lambda^{m^\alpha/2}\sqrt{\log n}\right)\\
&\times\mathbf{P}_{W_{\lambda^{m^\alpha}}}\left(\sup_{s\in[0,\lambda^{n^\alpha}]}|\hat{W}_s|\geq c\,h(\lambda^{n^\alpha})\right)\Bigg]
\end{aligned} \tag{6}$$

The last term arises after application of the Markov property. Let us start with the first term:

$$\begin{aligned}
T_1 &= \mathbf{P}\left[\sup_{s\in[0,1]}|\hat{W}_s|\geq\frac{c}{2}\frac{h(\lambda^{n^\alpha})}{\lambda^{m^\alpha}}\right]\\
&\leq \mathbf{P}\left[\sup_{s\in[0,1]}|\hat{W}_s|\geq\frac{c}{2}\,\lambda\,(\log\log(\lambda^{n^\alpha})^{3/4}\right]
\end{aligned}$$

$$
\begin{aligned}
&\leq \exp -(c_0 - \varepsilon) \left[\frac{c}{2}\, \lambda\, (\log\log(\lambda^{n^\alpha})^{3/4}\right]^{4/3} \\
&\leq n^{-(c_0-\varepsilon)\left(\frac{c\lambda}{2}\right)^{4/3} \alpha} \\
&\leq m^{-\beta}\, n^{-\beta}
\end{aligned}
$$

for large enough $m < n$, as soon λ is chosen so large that

$$
(c_0 - \varepsilon)\left(\frac{c\lambda}{2}\right)^{4/3} \alpha \geq 2\beta = 2\,(c_0 + \varepsilon)\, c^{4/3}\, \alpha.
$$

For the second term we use scaling and a well known large deviation result for Brownian motion:

$$
\begin{aligned}
T_2 &= \mathbf{P}\left[\sup_{s\in[0,1]} \zeta_s \geq \kappa\sqrt{\log n}\right] \\
&\leq \exp\left(-\frac{\kappa^2}{2}\log n\right) \leq m^{-\beta}\, n^{-\beta}
\end{aligned}
$$

as soon as $\kappa^2 \geq 4\beta$. For T_3 we first notice that

$$
\begin{aligned}
&\mathbf{P}_{W_{\lambda m^\alpha}}\left(\sup_{s\in[0,\lambda^{n^\alpha}]} |\hat{W}_s| \geq c\, h(\lambda^{n^\alpha})\right) \\
&\leq \mathbf{P}^*_{W_{\lambda m^\alpha}}\left(\sup_{s\in[0,\sigma]} |\hat{W}'_s| \geq c\, h(\lambda^{n^\alpha})\right) + \mathbf{P}\left(\sup_{s\in[0,\lambda^{n^\alpha}]} |\hat{W}_s| \geq c\, h(\lambda^{n^\alpha})\right)
\end{aligned}
$$

The aim is to show that these two quantities are bounded by $M\,\mathbf{P}(U_n^\lambda)$. The second one is precisely equal to $\mathbf{P}(U_n^\lambda)$. For the first one we use lemma 7:

$$
\mathbf{P}^*_{W_{\lambda m^\alpha}}\left(\sup_{s\in[0,\sigma]} |\hat{W}'_s| \geq c\, h(\lambda^{n^\alpha})\right) \leq K_1 \int_0^{\zeta_{\lambda m^\alpha}} \frac{ds}{(c\, h(\lambda^{n^\alpha}) - |W_{\lambda m^\alpha}(s)|)^2}
$$

Let us recall that in this computation we have $\sup_{s\in[0,\lambda^{m^\alpha}]} |\hat{W}_s| \leq \frac{c}{2}\, h(\lambda^{n^\alpha})$ and thus for every s, $|W_{\lambda m^\alpha}(s)| \leq \frac{c}{2}\, h(\lambda^{n^\alpha})$. We also have $\zeta_{\lambda m^\alpha} \leq \kappa\, \lambda^{\frac{m^\alpha}{2}} \sqrt{\log n}$. We deduce

$$
\begin{aligned}
\mathbf{P}^*_{W_{\lambda m^\alpha}}\left(\sup_{s\in[0,\sigma]} |\hat{W}'_s| \geq c\, h(\lambda^{n^\alpha})\right) &\leq \frac{K_1\, \zeta_{\lambda m^\alpha}}{(\frac{c}{2}\, h(\lambda^{n^\alpha}))^2} \\
&\leq \frac{4\, K_1\, \kappa}{c^2\, \alpha}\, \frac{\lambda^{\frac{m^\alpha}{2}} \sqrt{\log n}}{\lambda^{\frac{n^\alpha}{2}} (\log n)^{3/2}} \\
&\leq C\, \frac{\lambda^{\frac{(n-1)^\alpha}{2}}}{\lambda^{\frac{n^\alpha}{2}}} \leq \frac{C'}{\lambda^{C''\, n^{\alpha-1}}} \leq \frac{1}{n^\beta}
\end{aligned}
$$

for n large enough. Substituting this result in the definition of T_3 give the sought-after bound, as for T_1 and T_2 and we conclude that inequality (3) is true.

Acknowledgments. This work originates in a stimulating conversation with J.F. Le Gall and O. Zeitouni.

References.

Bl Blumenthal R.M. *Excursions of Markov processes.* Birkhauser, Boston.

CC Csaki E., Csorgo M., Foldes A., Revesz P. (1995) Global Strassen-type theorems for iterated Brownian motions. *Stoc. Proc. App.* **59**, 321-341.

Lg1 Le Gall J.F. (1993) A class of path-valued Markov processes and its applications to super-processes. *Probab. Th. Rel. Fields* **95**, 25-46.

Lg2 Le Gall J.F. (1994) A path-valued Markov process and its connections with partial differential equations. *Proc. First European Congress of Mathematics*, vol. II, p.185-212. Birkhauser, Boston.

LP Le Gall J.F., Perkins E.A. (1993) The Hausdorff measure of the support of two dimensional super-Brownian motion. *Ann. Probab.* **23** (4).

PS Port S.C., Stone C.J. (1978) *Brownian motion and classical potential theory*, Academic Press, New-York.

RY Revuz D., Yor M. (1994) *Continuous martingales and Brownian motion*, second edition. Springer-Verlag, Heidelberg.

Se1 Serlet L. (1995) On the Hausdorff measure of multiple points and collision points of super-Brownian motion. *Stochastics and Stoc. Rep.* **54**, 169-198.

Se2 Serlet L. (1997) A large deviation principle for the Brownian snake. *Stoc. proc. appl.* **67** 101-115.

Laurent Serlet
Université René Descartes
UFR Math-Info; 45 rue des Saints Pères
75270 PARIS CEDEX 06 FRANCE
E-mail: serlet@math-info.univ-paris5.fr

On the Onsager-Machlup functional for elliptic diffusion processes

Mireille CAPITAINE
University of Toulouse

Abstract

In this paper, we show, following K. Hara and Y. Takahashi [7], how the stochastic Stokes theorem and the Kunita-Watanabe theorem on orthogonal martingales may be used to produce a general and easy computation of the Onsager-Machlup functional of an elliptic diffusion process, even for large classes of norms in Wiener space.

1 Introduction

In their paper [7], K. Hara and Y. Takahashi present a rather simple and efficient computation of the Onsager-Machlup functional of an elliptic diffusion process for the supremum norm. The first task of this work is to present this simple approach and to show how it extends to various families of norms on Wiener space. In particular, this approach to the Onsager-Machlup functional is rather elementary when the diffusion matrix is the identity matrix and may be applied to the cases of the L^2-norm and L^p-norms with $2 < p < 4$ yielding some new results.

First we introduce the problem of the Onsager-Machlup functional for an elliptic diffusion process. Denote by $\mathcal{C}_0\left([0,1];\mathbb{R}^d\right)$ the space of $\mathbb{R}^d$-valued, continuous functions on $[0,1]$ vanishing at the origin, by $\|\cdot\|_\infty$ the supremum norm on $[0,1]$ and by $\mathcal{B}$ the Borel σ-field of $\left(\mathcal{C}_0\left([0,1];\mathbb{R}^d\right), \|\cdot\|_\infty\right)$. Let P be the Wiener measure defined on $\mathcal{B}$ and $\omega = (\omega_1, \ldots, \omega_d)$ be the canonical Brownian motion on the Wiener space $\left(\mathcal{C}_0\left([0,1];\mathbb{R}^d\right), \mathcal{B}, P\right)$. Let $X(t)$ be the elliptic diffusion process solution of the stochastic differential equation

$$dX(t) = \sigma\left(X(t)\right) d\omega(t) + b\left(X(t)\right) dt, \; X(0) = x_0, \; X(t) \in \mathbb{R}^m,$$

where σ is an $m \times d$ matrix of smooth vector fields on $\mathbb{R}^m$, b is a smooth vector field on $\mathbb{R}^m$ and x_0 belongs to $\mathbb{R}^m$. A Riemannian structure is naturally induced by the diffusion coefficients on $\mathbb{R}^m$, such that the generator of the diffusion is $\frac{1}{2}\Delta_M + f$ where Δ_M is the Laplace-Beltrami operator and f is a vector field. To be more precise, equip $\mathbb{R}^m$ with the metric $g = (\sigma\sigma^*)^{-1}$ and the Levi-Cevita connection (we denote by σ^* the transpose of the matrix σ). It is quite natural to consider X as the diffusion on the Riemannian manifold $M = (\mathbb{R}^m, g)$, with generator $\frac{1}{2}\Delta_M + f$, where for every $1 \leq i \leq m$,

$$f_i(x) = b_i(x) + \frac{1}{2}\sum_{l,j=1}^{m} (\sigma\sigma^*)_{lj}(x) \left\{ {}_{l\,j}^{\,i} \right\},$$

denoting by $\left\{ {}_{l\,j}^{\,i} \right\}$ the Christoffel symbol:

$$\left\{ {}_{l\,j}^{\,i} \right\} = \frac{-1}{2}\sum_{q=1}^{m} \left\{ \frac{\partial}{\partial x_l}(\sigma\sigma^*)^{-1}_{qj} + \frac{\partial}{\partial x_j}(\sigma\sigma^*)^{-1}_{lq} - \frac{\partial}{\partial x_q}(\sigma\sigma^*)^{-1}_{lj} \right\} (\sigma\sigma^*)_{iq}.$$

Recall that the Laplace-Beltrami operator is given by

$$\Delta_M F = \sum_{i,j=1}^{m} (\sigma\sigma^*)_{ij} \frac{\partial^2 F}{\partial x_i \partial x_j} - \sum_{i,j,k=1}^{m} (\sigma\sigma^*)_{ij} \left\{ {}_{i\,j}^{\,k} \right\} \frac{\partial F}{\partial x_k}.$$

Let $\rho(x,y)$ be the Riemannian distance on M and let $\|\cdot\|$ be a measurable norm on a subspace of $\mathcal{C}_0([0,1],\mathbb{R})$. Let Φ and Ψ be two smooth M-valued functions on $[0,1]$, starting at x_0. Denote by $\rho(X,\Phi)$ the $\mathbb{R}$-valued function on $[0,1]$ defined by $\rho(X,\Phi) : t \to \rho(X(t), \Phi(t))$. If the limit

$$\lim_{\epsilon \to 0} \frac{P(\|\rho(X,\Phi)\| < \epsilon)}{P(\|\rho(X,\Psi)\| < \epsilon)}$$

exists and can be written as

$$\exp\left[\int_0^1 L\left(\Phi(s), \dot{\Phi}(s)\right) ds - \int_0^1 L\left(\Psi(s), \dot{\Psi}(s)\right) ds\right],$$

such a function L on the tangent bundle TM is called the Onsager-Machlup function of X for the norm $\|\cdot\|$.

Y. Takahashi and S. Watanabe proved in [15] that, if one chooses on $\mathcal{C}_0([0,1];\mathbb{R})$ the supremum norm, the Onsager-Machlup function on the tangent bundle TM is given by

$$L(p,v) = -\frac{1}{2}\|f(p) - v\|_p^2 - \frac{1}{2}\operatorname{div} f(p) + \frac{1}{12}R(p),$$

where, for every p in M, $\|\cdot\|_p$ denotes the Riemannian norm on the tangent space T_pM at p, $\operatorname{div} f(p)$ is the divergence of f at p and $R(p)$ is the scalar curvature at p. They used probabilistic techniques such as Girsanov's formula, stochastic Stokes' theorem (see Lemma 2 below) and the Kunita-Watanabe theorem on orthogonal martingales (see Lemma 3 below). In [5], T. Fujita and S. Kotani obtained the same result by a purely analytical approach. In [4], we proved that this expression of the Onsager-Machlup function is still valid for a large class of norms on $\mathcal{C}_0([0,1];\mathbb{R})$, including in particular Hölder norms $\|\cdot\|_\alpha$ with $0 < \alpha < \frac{1}{2}$. One key idea in [15] consisted in a Besselization technique to come down to small balls of a Bessel process, using Girsanov's transformation. But as the Besselizing drift was singular, the proof was quite complex. Recently, K. Hara has found a new Besselizing drift which is smooth (see [6] and [7]). This makes the proof of [15] simpler and this shorter proof has been presented in [7]. As a consequence of the regularity of the drift, the stochastic Stokes theorem may be applied in greater generality and the proofs do not require anymore bounds on small balls probabilities. In particular, we observe here that this new approach is still valid for large classes of norms on Wiener space, simplifying the early reasonings of [4].

The method of K. Hara and Y. Takahashi is presented in section 3 of this paper. Section 2 consists of a collection of basic lemmas that will be used throughout this work. In the last section, we discuss the case of constant diffusion coefficients and compare recent results for various norms on Wiener space in this context.

2 Fundamental lemmas

Lemma 1 *(see [8]) Let $I_1, \ldots, I_n$ be n random variables on a probability space $(\Omega, \mathcal{B}, P)$. Let $\{A_\epsilon\}_{0<\epsilon<1}$ be a family of events in $\mathcal{B}$. Let $a_1, \ldots, a_n$ be n real numbers. If, for every real number c and every $1 \leq i \leq n$,*

$$\limsup_{\epsilon \to 0} E\left(\exp\{cI_i\} \middle| A_\epsilon\right) \leq \exp(ca_i),$$

then

$$\lim_{\epsilon \to 0} E\left(\exp\left\{\sum_{i=1}^n I_i\right\} \middle| A_\epsilon\right) = \exp\left(\sum_{i=1}^n a_i\right).$$

The proofs presented in this paper are based on the following two lemmas.

Lemma 2 *(Stochastic Stokes' theorem, see [15] or [7]) Let $Y = (Y_1, \ldots, Y_m)$ be a continuous semimartingale starting at zero and let β be a space-time 1-form defined by*

$$\beta = \sum_{i=1}^m \beta_i(s,x)dx_i.$$

Let us define

$$\bar{\beta}_i(s,x) = \int_0^1 \beta_i(s,ux)du,$$

$$S^{ij}(s) = \int_0^s Y_i(u)dY_j(u) - Y_j(u)dY_i(u).$$

Then, we have the following identity (where $\circ$ stands for the Stratonovitch integral) :

$$\begin{aligned}\sum_{i=1}^m \int_0^1 \beta_i(s,Y(s)) \circ dY_i(s) &= \frac{1}{2}\sum_{i,j=1}^m \int_0^1 \left(\frac{\partial \bar{\beta}_j}{\partial x_i} - \frac{\partial \bar{\beta}_i}{\partial x_j}\right)(s,Y(s)) \circ dS^{ij}(s) \\ &\quad - \sum_{i=1}^m \int_0^1 \frac{\partial \bar{\beta}_i}{\partial s}(s,Y(s))\, Y_i(s)ds + \sum_{i=1}^m \bar{\beta}_i(1,Y(1))\, Y_i(1).\end{aligned}$$

To prove Lemma 2, one can apply the classical Stokes theorem for the 1-form β and the random surface $\{(t, uP(t)), 0 \leq t \leq 1, 0 \leq u \leq 1\}$, where P is every polygonal line approximating the sample path Y. Then, the formula follows from the fact that the Stratonovitch integral is the limit of the line integral along polygonal lines.

Lemma 3 *Let $(\Omega, \mathcal{F}, P)$ be a complete probability space and $\{\mathcal{F}_t\}_{0 \leq t \leq 1}$ be a filtration which satisfies the usual conditions. Let $(Z(t))_{0 \leq t \leq 1}$ and $(M(t))_{0 \leq t \leq 1}$ be two continuous square integrable $\mathcal{F}_t$-martingales such that Z has the predictable representation property and $\langle Z, M\rangle = 0$. Let $\mathcal{G} = \sigma\{Z(s), 0 \leq s \leq 1\}$ be the σ-field generated by Z.*

Let A in $\mathcal{G}$ such that $P(A) > 0$. Let $(F(t))_{0\leq t\leq 1}$ be a continuous $\mathcal{F}_t$-adapted process. Let assume that there exists $C > 0$ such that, for every ω in A and every s in $[0,1]$, $\langle M\rangle_1(\omega) \leq C$ and $|F(s)(\omega)| \leq C$. Then,

$$E\left[\exp\left\{\int_0^1 F(s)dM(s) - \frac{1}{2}\int_0^1 F^2(s)d\langle M\rangle_s\right\} \,\middle|\, A\right] = 1,$$

and consequently

$$E\left[\exp\left\{\int_0^1 F(s)dM(s)\right\} \,\middle|\, A\right] \leq \left\{E\left[\exp\left\{2\int_0^1 F^2(s)d\langle M\rangle_s\right\} \,\middle|\, A\right]\right\}^{\frac{1}{2}}.$$

Proof: Define on $(\Omega, \mathcal{F})$ the probability measure $dQ = \frac{1}{P(A)}\mathbb{1}_A dP$ and denote by E_Q the expectation relative to Q. Since A belongs to $\mathcal{G}$, there exists a process ϕ adapted to the filtration generated by Z such that

$$\mathbb{1}_A = P(A) + \int_0^1 \phi(t)dZ(t).$$

Using the fact that M and $M\int_0^\cdot \phi(t)dZ(t)$ are martingales, we obtain that, for all $0 \leq s < t \leq 1$ and every bounded $\mathcal{F}_s$-measurable variable ξ,

$$\begin{aligned} E_Q(M(t)\xi) &= \frac{1}{P(A)}E\left[M(t)\xi\left(P(A) + \int_0^t \phi(u)dZ(u)\right)\right] \\ &= E(M(s)\xi) + \frac{1}{P(A)}E\left(M(s)\xi\int_0^s \phi(u)dZ(u)\right) \\ &= E_Q(M(s)\xi). \end{aligned}$$

We conclude that M is a square integrable martingale under Q. Moreover, the stochastic integrals of F with respect to M under P and under Q are obviously the same Q-almost everywhere. Similarly, the quadratic variations of M under P and under Q are equal Q-almost everywhere. Now, we have

$$E_Q\left[\exp\left(\frac{1}{2}\int_0^1 F^2(s)d\langle M\rangle_s\right)\right] \leq e^{\frac{1}{2}C^3}.$$

Novikov's criterion (see [12]) allows us to conclude that $(N(t))_{0\leq t\leq 1}$, defined by

$$N_t = \exp\left\{\int_0^t F(s)dM(s) - \frac{1}{2}\int_0^t F^2(s)d\langle M\rangle_s\right\},$$

is a martingale under Q and in particular

$$E\left[\exp\left\{\int_0^1 F(s)dM(s) - \frac{1}{2}\int_0^1 F^2(s)d\langle M\rangle_s\right\} \,\middle|\, A\right] = 1,$$

which yields the first assertion of Lemma 3.

Now, write that

$$\int_0^1 F(s)\,dM(s) = \left\{\int_0^1 F(s)\,dM(s) - \int_0^1 F^2(s)\,d\langle M\rangle_s\right\} + \int_0^1 F^2(s)\,d\langle M\rangle_s.$$

Then, using the Cauchy-Schwarz inequality and applying the first point of this lemma to $2F$, we obtain, for every $\epsilon > 0$,

$$E\left[\exp\left\{\int_0^1 F(s)dM(s)\right\} \,\middle|\, A\right] \leq \left\{E\left[\exp\left\{2\int_0^1 F^2(s)d\langle M\rangle_s\right\} \,\middle|\, A\right]\right\}^{\frac{1}{2}}.$$

The proof of Lemma 3 is thus complete.

3 The general case

Let $\|\cdot\|$ be a measurable norm on a subspace of $\mathcal{C}_0([0,1],\mathbb{R})$.

3.1 Theorem

Theorem 1 *Let $X(t)$ be the elliptic diffusion process which is the solution of the stochastic differential equation*

$$dX(t) = \sigma(X(t))\,d\omega(t) + b(X(t))\,dt,\ X(0) = x_0,\ X(t) \in \mathbb{R}^m,$$

where σ is an $m \times d$ matrix of smooth vector fields on $\mathbb{R}^m$, b is a smooth vector field on $\mathbb{R}^m$ and x_0 belongs to $\mathbb{R}^m$. Assume that the norm $\|\cdot\|$ under consideration on $\mathcal{C}_0([0,1],\mathbb{R})$ dominates the supremum norm. Then, the Onsager-Machlup function L is given by

$$L(p,v) = -\frac{1}{2}\|f(p) - v\|_p^2 - \frac{1}{2}divf(p) + \frac{1}{12}R(p).$$

3.2 Reduction of the problem

Let us consider a norm $\|\cdot\|$ on $\mathcal{C}_0([0,1],\mathbb{R})$ which dominates the supremum norm. We will reduce the problem of the Onsager-Machlup function for this norm $\|\cdot\|$ to an evaluation of a conditional exponential moment with regard to small balls of a Bessel process. The approach we will follow here is the one by Y. Takahashi and S. Watanabe in [15] (where they study the case of the supremum norm). However, the smooth Besselizing drift γ we will use has been found by K. Hara in [6] and is different from the drift used by Y. Takahashi and S. Watanabe in [15], which is singular at the origin. Let Φ be a smooth M-valued function on $[0,1]$, starting at x_0.

3.2.1 Introduction of a system of normal coordinates along the curve Φ

On the product manifold $[0,1] \times M$, let $\tilde{\Phi}$ be the curve: $t \in [0,1] \to (t,\Phi(t))$. Let introduce a coordinate system in a neighborhood U of $\tilde{\Phi}$ as follows. Let us choose an orthonormal basis $e^0 = \{e_1^0,\ldots,e_m^0\}$ in the tangent space $T_{\Phi(0)}M$ at $\Phi(0)$. For every $t > 0$, let $e^t = \{e_1^t,\ldots,e_m^t\}$ be the orthonormal basis in $T_{\Phi(t)}M$ obtained as the parallel translate of e^0 along the curve Φ. There exists a neighborhood U of $\tilde{\Phi}$ in $[0,1] \times M$ such that the mapping $(t,x) \in U \to (t,x_1,\ldots,x_m) \in [0,1] \times \mathbb{R}^m$, where $x = \exp_{\Phi(t)}(\sum_{i=1}^m x_i e_i^t)$, is well defined ($s \to \exp_q(sv)$ is the geodesic c with initial conditions $c(0) = q$ and $c'(0) = v$). $\Theta : (t,x) \to (t,x_1,\ldots,x_m)$ is a diffeomorphism of U onto some neighborhood V of the curve $t \to (t,0)$ in $[0,1] \times \mathbb{R}^m$, and for each fixed t, $x \to (x_1,\ldots,x_m)$ is the normal coordinate system N_t in a neighborhood of $\Phi(t)$ with respect to the frame e^t. Denote respectively by $g_{ij}(t,x)$, $g^{ij}(t,x)$, $\Gamma_{ij}^k(t,x)$ and $f_i(t,x)$, the components in the normal coordinates system N_t of the metric, its inverse, the Christoffel symbols and the vector field f. Denote by $\left(\dot{\Phi}_i(t)\right)_{1\le i\le m}$ the components of the tangent vector $\dot{\Phi}_t$ in the frame e^t. Finally, let us consider the space-time process $(t,X(t))$. Its generator is $\frac{\partial}{\partial t} + \frac{1}{2}\Delta_M + f$. This differential operator on $U \subset [0,1] \times M$ is transformed by the above diffeomorphism to the differential operator $\frac{\partial}{\partial t} + \frac{1}{2}\sum_{i,j=1}^m g^{ij}(t,x)\frac{\partial^2}{\partial x_i \partial x_j} + \sum_{i=1}^m \tilde{b}_i(t,x)\frac{\partial}{\partial x_i}$ on $V \subset [0,1] \times \mathbb{R}^m$, where $\tilde{b}_i(t,x) = f_i(t,x) - \dot{\Phi}_i(t) + \epsilon_i(t,x) - \frac{1}{2}\sum_{l,q=1}^m g^{lq}(t,x)\Gamma_{lq}^i(t,x)$, where $\epsilon_i(t,x)$ is a smooth function satisfying $\sup_{0\le t\le 1}|\epsilon_i(t,x)| = O\left(|x|^2\right)$ (cf Lemma 1.2 [5]). Let $T = \inf\{t \ge 0 : (t, X(t)) \notin V\}$ and $\left(t\wedge T, \tilde{X}(t\wedge T)\right) = \Theta(t\wedge T, X(t\wedge T))$. Denoting by $|\cdot|$ the Euclidean norm on $\mathbb{R}^m$, we have for ϵ small enough,

$$P(\|\rho(X,\Phi)\| < \epsilon) = P\left(\left\| |\tilde{X}| \right\| < \epsilon\right).$$

Moreover, $\tilde{X}$ is given as the solution of the stochastic differential equation

$$d\tilde{X}(t) = \tilde{\sigma}\left(t, \tilde{X}(t)\right) d\tilde{B}(t) + \tilde{b}\left(t, \tilde{X}(t)\right) dt,\ \tilde{X}(0) = 0,\ \tilde{X}(t) \in \mathbb{R}^m,$$

where $\tilde{\sigma}(t,x) = \left(\tilde{\sigma}^{ij}(t,x)\right)_{\substack{1\le i\le m\\ 1\le j\le m}}$ is the square root of the inverse of the metric $g(t,x)$ in the normal coordinates N_t, and $\tilde{B}$ is a Brownian motion on $\mathbb{R}^m$.

3.2.2 Use of Girsanov's transformation

For each fixed t in $[0,1]$, since the coordinates system is the normal coordinates system N_t, we have by Gauss lemma, for every $1 \le i \le m$,

$$\sum_{j=1}^m g^{ij}(t,x)x_j = x_i, \tag{1}$$

and thus

$$\sum_{j=1}^m \tilde{\sigma}^{ij}(t,x)x_j = x_i.$$

Let us define for every $1 \le i \le m$,

$$\gamma_i(t,x) = \frac{1}{2}\sum_{j=1}^m \frac{\partial g^{ij}}{\partial x_j}(t,x).$$

Differentiating both sides of (1), we obtain that γ satisfies

$$\sum_{i=1}^m \left(1 - g^{ii}(t,x)\right) = 2\sum_{j=1}^m \gamma_j(t,x)x_j.$$

Let $Y(t) = (Y_1(t), \ldots, Y_m(t))$ be the solution of the stochastic differential equation on $[0,1]$:

$$dY(t) = \tilde{\sigma}(t, Y(t))\, d\tilde{B}(t) + \gamma(t, Y(t))\, dt,\ Y(0) = 0,\ Y(t) \in \mathbb{R}^m.$$

We get

$$\begin{aligned} d|Y(t)|^2 &= 2\sum_{i=1}^m Y_i(t)dY_i(t) + \sum_{i=1}^m d\langle Y_i\rangle_t \\ &= 2\sum_{i,k=1}^m Y_i(t)\tilde{\sigma}^{ik}(t, Y(t))\, d\tilde{B}_k(t) + 2\sum_{i=1}^m Y_i(t)\gamma_i(t, Y(t))\, dt + \sum_{i=1}^m g^{ii}(t, Y(t))\, dt \\ &= 2\sum_{k=1}^m Y_k(t)d\tilde{B}_k(t) + m dt, \end{aligned}$$

since $\sum_{i=1}^m Y_i(t)\tilde{\sigma}^{ik}(t,Y(t)) = Y_k(t)$ and $\sum_{i=1}^m (1-g^{ii}(t,Y(t))) = 2\sum_{j=1}^m \gamma_j(t,Y(t))Y_j(t)$. Define $B_t = \sum_{k=1}^m \int_0^t \frac{Y_k(s)}{|Y(s)|} d\tilde{B}_k(s)$. B is a Brownian motion on $\mathbb{R}$ and, as $d|Y(t)|^2 = 2|Y(t)|\,dB(t) + mdt$, we thus get that $|Y(t)|$ is an m-dimensional Bessel process. Using Girsanov's transformation, we obtain that for every $\epsilon > 0$,

$$P\left(\left\| \left|\tilde{X}\right| \right\| < \epsilon\right) = E\left(\exp\left\{\sum_{i,j=1}^m \int_0^1 \tilde{\sigma}_{ij}(t,Y(t))\,\delta^j(t,Y(t))\,d\tilde{B}_i(t) - \frac{1}{2}\sum_{i,j=1}^m \int_0^1 g_{ij}(t,Y(t))\,\delta^i(t,Y(t))\,\delta^j(t,Y(t))\,dt\right\} : \ \|\,|Y|\,\| < \epsilon\right)$$

where $\delta^i(t,x) = \tilde{b}_i(t,x) - \gamma_i(t,x)$ and $\tilde{\sigma}_{ij}(t,x)$ denote the components in the normal coordinate system N_t of the inverse matrix of $\tilde{\sigma}(t,x)$ (i.e the components of the square-root of $g(t,x)$). Thus, we get

$$\begin{aligned} P\left(\left\| \left|\tilde{X}\right| \right\| < \epsilon\right) &= E\left(\exp\left\{\sum_{i,j=1}^m \int_0^1 \tilde{\sigma}_{ij}(t,Y(t))\,\delta^j(t,Y(t))\,d\tilde{B}_i(t)\right.\right. \\ &\quad \left.\left. - \frac{1}{2}\sum_{i,j=1}^m \int_0^1 g_{ij}(t,Y(t))\,\delta^i(t,Y(t))\,\delta^j(t,Y(t))\,dt\right\} \,\Big|\, \|\,|Y|\,\| < \epsilon\right) \\ &\quad \times P\left(\|\,|Y|\,\| < \epsilon\right). \end{aligned}$$

If we succeed in evaluating the limit of $\frac{P(\|\rho(X,\Phi)\|<\epsilon)}{P(\|\,|Y|\,\|<\epsilon)}$, when ϵ tends to zero, for every smooth path Φ starting at x_0, we will obviously be able to evaluate the limit of $\frac{P(\|\rho(X,\Phi)\|<\epsilon)}{P(\|\rho(X,\Psi)\|<\epsilon)}$, when ϵ tends to zero, for all smooth paths Φ and Ψ starting at x_0. Since, for ϵ small enough,

$$\frac{P\left(\|\rho(X,\Phi)\| < \epsilon\right)}{P\left(\|\,|Y|\,\| < \epsilon\right)} = E\left(\exp\mathcal{I} \,\Big|\, \|\,|Y|\,\| < \epsilon\right)$$

where

$$\mathcal{I} = \sum_{i,j=1}^m \int_0^1 \tilde{\sigma}_{ij}(t,Y(t))\,\delta^j(t,Y(t))\,d\tilde{B}_i(t) - \frac{1}{2}\sum_{i,j=1}^m \int_0^1 g_{ij}(t,Y(t))\,\delta^i(t,Y(t))\,\delta^j(t,Y(t))\,dt,$$

the computation of the Onsager-Machlup functional therefore consists in the asymptotic evaluation of a conditional exponential moment. Thanks to Lemma 1, it suffices to handle the conditional exponential moments of each term appearing in the linear expression of $\mathcal{I}$.

By making the change of drift with respect to the smooth function γ, we will be able to make a general and unique study for every natural norm on the Wiener space.

3.3 Control of the different conditional exponential moments

According to the second theorem of Elie Cartan (see [11]), for each fixed t, since the system of coordinates is the normal coordinates systen , we have a Taylor development of $g_{ij}(t,x)$ at zero (and therefore of $\tilde{\sigma}_{ij}(t,x)$, $\tilde{\sigma}^{ij}(\ ,x)$ and $\Gamma^i_{jk}(t,x)$) where the coefficients are universal polynomials in the successive covariant derivatives of the curvature tensor at $\Phi(t)$. All the $O(|x|^q)$'s, $q \geq 0$, which will appear in the developments of the functions of the form $h(t,x)$ will be uniform in t in $[0,1]$. We will use the following developments (see [5], [6], [7], [15]):

$$g^{ij}(t,x) = \delta_{ij} + O(|x|^2),$$

$$g_{ij}(t,x) = \delta_{ij} + O(|x|^2),$$

$$\gamma_i(t,x) = -\frac{1}{6}\sum_{j=1}^{m} R_{ij}(t,0)x_j + O(|x|^2),$$

$$\tilde{b}_i(t,x) = f_i(t,0) - \dot{\Phi}_i(t) + \sum_{j=1}^{m}\left(\frac{\partial f_i}{\partial x_j}(t,0) - \frac{1}{3}R_{ij}(t,0)\right)x_j + O(|x|^2),$$

$$\delta^i = f_i(t,0) - \dot{\Phi}_i(t) + \sum_{j=1}^{m}\left(\frac{\partial f_i}{\partial x_j}(t,0) - \frac{1}{6}R_{ij}(t,0)\right)x_j + O(|x|^2),$$

where $R_{ij}(t,x)$ are the components of the Ricci tensor in N_t. We get

$$g_{ij}(t,Y(t))\,\delta^i(t,Y(t))\,\delta^j(t,Y(t)) = \delta_{ij}\left(f_i(t,0) - \dot{\Phi}_i(t)\right)^2 + O(|Y(t)|).$$

Then we see easily that, for every $1 \leq i \leq m$, every $1 \leq j \leq m$ and every real c,

$$\limsup_{\epsilon\to 0} E\left(\exp\left\{-\frac{c}{2}\int_0^1 g_{ij}(t,Y(t))\,\delta^i(t,Y(t))\,\delta^j(t,Y(t))\,dt\right\}\Big|\, \|\,|Y|\,\| < \epsilon\right)$$
$$\leq \exp\left\{-\frac{c}{2}\delta_{ij}\int_0^1\left(f_i(t,0) - \dot{\Phi}_i(t)\right)^2 dt\right\}.$$

By Lemma 1, we can deduce that, for every real c,

$$\limsup_{\epsilon\to 0} E\left(\exp\left\{-\frac{c}{2}\sum_{i,j=1}^{m}\int_0^1 g_{ij}(t,Y(t))\,\delta^i(t,Y(t))\,\delta^j(t,Y(t))\,dt\right\}\Big|\, \|\,|Y|\,\| < \epsilon\right)$$
$$\leq \exp\left\{-\frac{c}{2}\sum_{i=1}^{m}\int_0^1\left(f_i(t,0) - \dot{\Phi}_i(t)\right)^2 dt\right\}.$$

Now, write that

$$\sum_{i,j=1}^{m}\int_0^1 \tilde{\sigma}_{ij}(t,Y(t))\,\delta^j(t,Y(t))\,d\tilde{B}_i(t) = \sum_{l,j=1}^{m}\int_0^1 g_{jl}(t,Y(t))\,\delta^j(t,Y(t))\,dY_l(t)$$
$$-\sum_{l,j=1}^{m}\int_0^1 g_{jl}(t,Y(t))\,\delta^j(t,Y(t))\,\gamma_l(t,Y(t))\,dt.$$

Since $g_{jl}(t,Y(t))\,\delta^j(t,Y(t))\,\gamma_l(t,Y(t)) = O(|Y(t)|)$, for every $1 \leq j \leq m$ and every $1 \leq l \leq m$, we easily get that, for every real c,

$$\limsup_{\epsilon\to 0} E\left(\exp\left\{c\int_0^1 g_{jl}(t,Y(t))\,\delta^j(t,Y(t))\,\gamma_l(t,Y(t))\,dt\right\}\Big|\, \|\,|Y|\,\| < \epsilon\right) \leq 1.$$

Similarly, write that

$$\int_0^1 g_{jl}(t,Y(t))\,\delta^j(t,Y(t))\,dY_l(t) = \int_0^1 g_{jl}(t,Y(t))\,\delta^j(t,Y(t))\circ dY_l(t)$$
$$-\frac{1}{2}\langle g_{jl}(\cdot,Y(\cdot))\,\delta^j(\cdot,Y(\cdot)),Y_l(\cdot)\rangle_1.$$

We have

$$d\langle g_{jl}(\cdot,Y(\cdot))\,\delta^j(\cdot,Y(\cdot)),Y_l(\cdot)\rangle_t = \sum_{1\le k,q\le m} \tilde{\sigma}^{qk}(t,Y(t))\,\tilde{\sigma}^{lk}(t,Y(t))\,\frac{\partial(g_{jl}\delta^j)}{\partial x_q}(t,Y(t))\,dt$$

$$= \sum_{1\le q\le m} g^{lq}(t,Y(t))\,\frac{\partial(g_{jl}\delta^j)}{\partial x_q}(t,Y(t))\,dt$$

$$= \delta_{jl}\left\{\frac{\partial f_j}{\partial x_j}(t,0) - \frac{1}{6}R_{jj}(t,0)\right\}dt + O(|Y(t)|)dt.$$

Then, we see easily that for every real number c,

$$\limsup_{\epsilon\to 0} E\left(\exp\left\{-\frac{c}{2}\langle g_{jl}(\cdot,Y(\cdot))\,\delta^j(\cdot,Y(\cdot)),Y_l(\cdot)\rangle_1\right\}\Big|\,\|\,|Y|\,\|<\epsilon\right)$$

$$\le \exp\left\{-\frac{c}{2}\delta_{jl}\int_0^1\left\{\frac{\partial f_j}{\partial x_j}(t,0)-\frac{1}{6}R_{jj}(t,0)\right\}dt\right\}.$$

Consider now $\int_0^1 g_{jl}(t,Y(t))\,\delta^j(t,Y(t))\circ dY_l(t)$. Applying Lemma 2, we get

$$\int_0^1 g_{jl}(t,Y(t))\,\delta^j(t,Y(t))\circ dY_l(t) = \sum_{q=1}^m\int_0^1 \frac{\partial\overline{(g_{jl}\delta^j)}}{\partial x_q}(s,Y(s))\circ dS^{ql}(s)$$
$$-\int_0^1\frac{\partial\overline{(g_{jl}\delta^j)}}{\partial s}(s,Y(s))\,Y_l(s)ds$$
$$+\overline{(g_{jl}\delta^j)}(1,Y(1))\,Y_l(1),$$

where

$$\overline{(g_{jl}\delta^j)}(s,x) = \int_0^1 (g_{jl}\delta^j)(s,ux)du,$$

$$S^{ql}(s) = \int_0^s Y_q(u)dY_l(u) - Y_l(u)dY_q(u).$$

Since moreover

$$d\langle\frac{\partial\overline{(g_{jl}\delta^j)}}{\partial x_q}(\cdot,Y(\cdot)),S^{ql}\rangle_t = \sum_{k=1}^m\frac{\partial^2\overline{(g_{jl}\delta^j)}}{\partial x_k\partial x_q}(t,Y(t))\left(g^{kl}(t,Y(t))Y_q(t)-g^{kq}(t,Y(t))Y_l(t)\right)dt$$
$$= O(|Y(t)|)dt$$

we get that

$$\int_0^1(g_{jl}\delta^j)(t,Y(t))\circ dY_l(t) = \sum_{q=1}^m\int_0^1\frac{\partial\overline{(g_{jl}\delta^j)}}{\partial x_q}(t,Y(t))\,dS^{ql}(t)+\int_0^1 O(|Y(t)|)dt+O(|Y(1)|),$$

and the control of the conditional exponential moments of $\int_0^1(g_{jl}\delta^j)(t,Y(t))\circ dY_l(t)$ is equivalent to the control of those of $\sum_{q=1}^m\int_0^1\frac{\partial\overline{(g_{jl}\delta^j)}}{\partial x_q}(t,Y(t))\,dS^{ql}(t)$.

Let M^{ql} be the martingale part of S^{ql}:

$$M^{ql}(t) = \sum_{k=1}^m\int_0^t Y_q(u)\tilde{\sigma}^{lk}(u,Y(u))\,d\tilde{B}_k(u) - \sum_{k=1}^m\int_0^t Y_l(u)\tilde{\sigma}^{qk}(u,Y(u))\,d\tilde{B}_k(u).$$

We have

$$\int_0^1 \frac{\partial \overline{(g_{jl}\delta^j)}}{\partial x_q}(t,Y(t))\,dS^{ql}(t) = \int_0^1 \frac{\partial \overline{(g_{jl}\delta^j)}}{\partial x_q}(t,Y(t))\,dM^{ql}(t) + \int_0^1 \frac{\partial \overline{(g_{jl}\delta^j)}}{\partial x_q}(t,Y(t))\left\{Y_q(t)\gamma_l(t,Y(t)) - Y_l(t)\gamma_q(t,Y(t))\right\}dt.$$

We see easily that, for every real number c,

$$\limsup_{\epsilon\to 0} E\left(\exp\left\{c\int_0^1 \frac{\partial \overline{(g_{jl}\delta^j)}}{\partial x_q}(t,Y(t))\left\{Y_q(t)\gamma_l(t,Y(t)) - Y_l(t)\gamma_q(t,Y(t))\right\}dt\right\}\Big|\, \|\,|Y|\,\| < \epsilon\right) \le 1.$$

Recall that $d|Y(t)|^2 = 2|Y(t)|\,dB(t) + m\,dt$, where $B_t = \sum_{k=1}^m \int_0^t \frac{Y_k(s)}{|Y(s)|}d\tilde{B}_k(s)$. The σ-field generated by B, $\sigma\{B(s), 0\le s\le 1\}$, is the same as the one generated by $|Y|$, $\sigma\{|Y(s)|, 0\le s\le 1\}$; moreover

$$\langle B, M^{ql}\rangle_t = \sum_{k=1}^m \int_0^t \frac{Y_k(s)}{|Y(s)|}\left(Y_q(s)\tilde{\sigma}^{lk}(s,Y(s)) - Y_l(s)\tilde{\sigma}^{qk}(s,Y(s))\right)ds,$$

and since $\sum_{k=1}^m Y_k(s)\tilde{\sigma}^{qk}(s,Y(s)) = Y_q(s)$ and $\sum_{k=1}^m Y_k(s)\tilde{\sigma}^{lk}(s,Y(s)) = Y_l(s)$, we obtain

$$\langle B, M^{ql}\rangle_t = \int_0^t \frac{Y_q(s)Y_l(s) - Y_q(s)Y_l(s)}{|Y(s)|}ds = 0.$$

Thus, by Lemma 3, we get for every real number c and every $0<\epsilon<1$,

$$E\left(\exp\left\{c\int_0^1 \frac{\partial \overline{(g_{jl}\delta^j)}}{\partial x_q}(t,Y(t))\,dM^{ql}(t)\right\}\Big|\, \|\,|Y|\,\| < \epsilon\right) \le \left\{E\left(\exp\left\{2c^2\int_0^1\left(\frac{\partial \overline{(g_{jl}\delta^j)}}{\partial x_q}(t,Y(t))\right)^2 d\langle M^{ql}\rangle_t\right\}\Big|\, \|\,|Y|\,\| < \epsilon\right)\right\}^{\frac{1}{2}}.$$

Moreover, as

$$d\langle M^{ql}\rangle_t = \sum_{k=1}^m \left(Y_l(t)\tilde{\sigma}^{qk}(t,Y(t)) - Y_q(t)\tilde{\sigma}^{lk}(t,Y(t))\right)^2 dt = O\left(|Y(t)|^2\right)dt,$$

we have

$$\limsup_{\epsilon\to 0} E\left(\exp\left\{c\int_0^1 \frac{\partial \overline{(g_{jl}\delta^j)}}{\partial x_q}(t,Y(t))\,dM^{ql}(t)\right\}\Big|\, \|\,|Y|\,\| < \epsilon\right) \le 1.$$

By Lemma 1, we deduce that, for every real number c,

$$\limsup_{\epsilon\to 0} E\left[\exp\left\{c\sum_{i,j=1}^m \int_0^1 \tilde{\sigma}_{ij}(t,Y(t))\,\delta^j(t,Y(t))\,d\tilde{B}_i(t)\right\}\Big|\, \|\,|Y|\,\| < \epsilon\right] \le \exp\left\{-\frac{c}{2}\sum_{j=1}^m\int_0^1\left\{\frac{\partial f_j}{\partial x_j}(t,0) - \frac{1}{6}R_{jj}(t,0)\right\}dt\right\}.$$

Using once more Lemma 1, the proof of Theorem 1 is thus complete.

4 When σ is the identity matrix

If $m = d$ and if for each x in $\mathbb{R}^d$, $\sigma(x)$ is the identity matrix, the Riemannian structure induced by the diffusion is the Euclidean one. Denote by $|\cdot|$ the Euclidean norm on $\mathbb{R}^d$. So, Theorem 1 gives the limit, when ϵ tends to zero, of $\frac{P(\| |X-\Phi| \|<\epsilon)}{P(\| |X-\Psi| \|<\epsilon)}$, for every smooth functions Φ and Ψ and every norm $\|\cdot\|$ on $\mathcal{C}_0([0,1],\mathbb{R})$ which dominates the supremum norm. Actually, studies on the Onsager-Machlup functional in the case $\sigma := I_d$ used to consider the problem in a more general framework and to investigate the limit, when ϵ tends to zero, of $\frac{P(\|X-\Phi\|<\epsilon)}{P(\|X-\Psi\|<\epsilon)}$, for every functions Φ and Ψ such that $\Phi - x_0$ and $\Psi - x_0$ belong to the Cameron-Martin space $\mathcal{H}^d$, and every norm $\|\cdot\|$ on $\mathcal{C}_0\left([0,1],\mathbb{R}^d\right)$. That is to say in particular, the norm under consideration is not assumed a priori of the form: $\|\cdot\| = \| |\cdot| \|'$ where $\|\cdot\|'$ is a norm on $\mathcal{C}_0([0,1],\mathbb{R})$. In this context, we noticed that the preceding techniques (the stochastic Stokes theorem and the Kunita-Watanabe theorem) give a very quick proof of the following theorem whose framework is a little more general that the one of Theorem 1 for $\sigma := I_d$.

4.1 Theorem

Theorem 2 *Let $X(t)$ be the diffusion process which is the solution of the stochastic differential equation*

$$dX(t) = b(X(t))\,dt + d\omega(t),\ \ X(0) = x_0,\ \ X(t) \in \mathbb{R}^d,$$

where x_0 belongs to $\mathbb{R}^d$ and b is a $\mathbb{R}^d$-valued function on $\mathbb{R}^d$ of class C^2, bounded, such that all its derivatives are bounded and its second derivatives are Lipschitz continuous. Let $\|\cdot\|$ be a measurable norm defined on a subspace F of the Wiener space $\mathcal{C}_0\left([0,1];\mathbb{R}^d\right)$ to which ω and $X-\Phi$ belong (for every Φ in $x_0+\mathcal{H}^d$), such that $(F,\|\cdot\|)$ is separable. If the norm $\|\cdot\|$ dominates the L^2-norm and is such that the random variable $\|\omega\|$ is measurable with respect to the σ-field $\sigma\{|\omega(s)|, 0 \le s \le 1\}$ (where $|\cdot|$ denotes the Euclidean norm on $\mathbb{R}^d$), then the Onsager-Machlup functional of X for the norm $\|\cdot\|$ exists and is given by

$$L\left(\Phi,\dot{\Phi}\right) = -\frac{1}{2}\sum_{i=1}^{d}\left|\dot{\Phi}_i - b_i(\Phi)\right|^2 - \frac{1}{2}\sum_{i=1}^{d}\frac{\partial b_i}{\partial x_i}(\Phi).$$

4.2 Proof

We will need one more lemma.

Lemma 4 *(see [13], [9], [2]) Let f be a deterministic function in $L^2[0,1]$. Define $I_i(f) = \int_0^1 f(t)\,d\omega_i(t)$. If the norm $\|\cdot\|$ dominates the L^1-norm then*

$$\lim_{\epsilon\to 0} E\left(\exp\{|I_i(f)|\} \mid \|\omega\| < \epsilon\right) = 1.$$

Using Girsanov's transformation (see [8]), we obtain, for every $\epsilon > 0$,

$$P(\|X-\Phi\|<\epsilon) = \exp\left(-\frac{1}{2}\int_0^1 \left|\dot{\Phi}(s) - b(\Phi(s))\right|^2 ds\right)$$
$$\times E\Big(\exp\Big\{\frac{1}{2}\int_0^1 |b(\Phi(s))|^2 ds - \frac{1}{2}\int_0^1 |b(\omega(s)+\Phi(s))|^2 ds$$
$$-\sum_{i=1}^d \int_0^1 b_i(\Phi(s))\, d\Phi_i(s) + \sum_{i=1}^d \int_0^1 b_i(\omega(s)+\Phi(s))\, d\Phi_i(s)$$
$$-\sum_{i=1}^d \int_0^1 \dot{\Phi}_i(s)\, d\omega_i(s)$$
$$+\sum_{i=1}^d \int_0^1 b_i(\omega(s)+\Phi(s))\, d\omega_i(s)\Big\} : \|\omega\|<\epsilon\Big).$$

Therefore

$$\frac{P(\|X-\Phi\|<\epsilon)}{P(\|\omega\|<\epsilon)} = \exp\left(-\frac{1}{2}\int_0^1 \left|\dot{\Phi}(s) - b(\Phi(s))\right|^2 ds\right) E_\epsilon,$$

where $E_\epsilon = E\Big(\exp\Big\{\frac{1}{2}\int_0^1 |b(\Phi(s))|^2 ds - \frac{1}{2}\int_0^1 |b(\omega(s)+\Phi(s))|^2 ds$

$$-\sum_{i=1}^d \int_0^1 b_i(\Phi(s))\, d\Phi_i(s) + \sum_{i=1}^d \int_0^1 b_i(\omega(s)+\Phi(s))\, d\Phi_i(s)$$
$$-\sum_{i=1}^d \int_0^1 \dot{\Phi}_i(s)\, d\omega_i(s) + \sum_{i=1}^d \int_0^1 b_i(\omega(s)+\Phi(s))\, d\omega_i(s)\Big\}\Big|\|\omega\|<\epsilon\Big).$$

The computation of the Onsager-Machlup functional therefore consists in the asymptotic evaluation of E_ϵ. According to Lemma 1, it suffices to handle the conditional exponential moments of each term of the sum inside the exponential map.

By using the fact that b is Lipschitz continuous and bounded and that $\|\cdot\|$ dominates the L^2-norm, we easily get for every real number c,

$$\limsup_{\epsilon\to 0} E\Big(\exp\Big\{c\Big(\frac{1}{2}\int_0^1 |b(\Phi(s))|^2 ds$$
$$-\frac{1}{2}\int_0^1 |b(\omega(s)+\Phi(s))|^2 ds\Big)\Big\}\Big|\|\omega\|<\epsilon\Big) \le 1$$

and for every $1 \le i \le d$,

$$\limsup_{\epsilon\to 0} E\left(\exp\left\{c\left(\int_0^1 (b_i(\omega(s)+\Phi(s)) - b_i(\Phi(s)))\, d\Phi_i(s)\right)\right\}\Big|\|\omega\|<\epsilon\right) \le 1.$$

As an immediate consequence of Lemma 4, we have for every $1 \le i \le d$ and for every real number c,

$$\limsup_{\epsilon\to 0} E\left(\exp\left\{c\left(\int_0^1 \dot{\Phi}_i(s)\, d\omega_i(s)\right)\right\}\Big|\|\omega\|<\epsilon\right) \le 1.$$

We are left with the control of the exponential moments of $\int_0^1 b_i(\omega(s)+\Phi(s))\, d\omega_i(s)$ which is the main interest of this proof. We can write

$$\int_0^1 b_i(\omega(s)+\Phi(s))\, d\omega_i(s) = \int_0^1 b_i(\omega(s)+\Phi(s)) \circ d\omega_i(s) - \frac{1}{2}\int_0^1 \frac{\partial b_i}{\partial x_i}(\omega(s)+\Phi(s))ds.$$

Now, developping $\frac{\partial b_i}{\partial x_i}(\omega(s) + \Phi(s))$ at the point $\Phi(s)$ up to the order 1, we get

$$\int_0^1 \frac{\partial b_i}{\partial x_i}(\omega(s) + \Phi(s))ds = \int_0^1 \frac{\partial b_i}{\partial x_i}(\Phi(s))ds + \int_0^1 \Psi_i(s,\omega)ds.$$

Since the derivatives of order 2 of b_i are bounded and since the norm $\|\cdot\|$ dominates the L^1-norm, there is $C > 0$ such that

$$\int_0^1 \Psi_i(s,\omega)ds \le C\epsilon$$

on the event $\{\|\omega\| < \epsilon\}$. Thus we get for every $1 \le i \le d$ and for every real number c,

$$\limsup_{\epsilon\to 0} E\left(\exp\left\{c\left(-\frac{1}{2}\int_0^1 \frac{\partial b_i}{\partial x_i}(\omega(s) + \Phi(s)))ds\right)\right\}\right) \le \exp\left\{c\left(-\frac{1}{2}\int_0^1 \frac{\partial b_i}{\partial x_i}(\Phi(s))ds\right)\right\}.$$

Let us now analyze $\int_0^1 b_i(\omega(s) + \Phi(s)) \circ d\omega_i(s)$. By using an approximation of Φ by smooth functions on $[0,1]$, one can easily check that Lemma 2 is still valid for $\beta(s,x) = b(x + \Phi(s))$. So, we get

$$\begin{aligned}\int_0^1 b_i(\omega(s) + \Phi(s)) \circ d\omega_i(s) &= \sum_{j=1}^d \int_0^1 \frac{\partial \bar{b}_i}{\partial x_j}(s,\omega(s)) \circ dA^{ji}(s) \\ &\quad - \int_0^1 \frac{\partial \bar{b}_i}{\partial s}(s,\omega(s))\,\omega_i(s)ds + \bar{b}_i(1,\omega(1))\,\omega_i(1),\end{aligned}$$

where

$$\bar{b}_i(s,x) = \int_0^1 b_i(ux + \Phi(s))du,$$

$$A^{ji}(s) = \int_0^s \omega_j(u)d\omega_i(u) - \omega_i(u)d\omega_j(u).$$

Since b is bounded, we get by Lemma 4 that for every $1 \le i \le d$ and every real number c,

$$\limsup_{\epsilon\to 0} E\left(\exp\left\{c\bar{b}_i(1,\omega(1))\,\omega_i(1)\right\}\Big|\|\omega\| < \epsilon\right) \le 1.$$

Since the derivatives of b are bounded and since the norm $\|\cdot\|$ dominates the L^2-norm, we easily get for every $1 \le i \le d$ and every real number c,

$$\limsup_{\epsilon\to 0} E\left(\exp\left\{-c\int_0^1 \frac{\partial \bar{b}_i}{\partial s}(s,\omega(s))\,\omega_i(s)ds\right\}\Big|\|\omega\| < \epsilon\right) \le 1.$$

Now, rewrite

$$\begin{aligned}\int_0^1 \frac{\partial \bar{b}_i}{\partial x_j}(s,\omega(s)) \circ dA^{ji}(s) &= \int_0^1 \frac{\partial \bar{b}_i}{\partial x_j}(s,\omega(s))\,dA^{ji}(s) - \frac{1}{2}\int_0^1 \frac{\partial^2 \bar{b}_i}{\partial x_j^2}(s,\omega(s))\,\omega_i(s)ds \\ &\quad + \frac{1}{2}\int_0^1 \frac{\partial^2 \bar{b}_i}{\partial x_i \partial x_j}(s,\omega(s))\,\omega_j(s)ds.\end{aligned}$$

Similarly, for every $1 \le i \le d$ and every real number c,

$$\limsup_{\epsilon\to 0} E\left(\exp\left\{\frac{c}{2}\left(\int_0^1 \frac{\partial^2 \bar{b}_i}{\partial x_i \partial x_j}(s,\omega(s))\,\omega_j(s)ds - \int_0^1 \frac{\partial^2 \bar{b}_i}{\partial x_j^2}(s,\omega(s))\,\omega_i(s)ds\right)\right\}\Big| \|\omega\| < \epsilon\right) \leq 1.$$

Define for every $0 \leq s \leq 1$, the Brownian motion $Z_s = \sum_{k=1}^{d} \int_0^s \frac{\omega_k(t)}{|\omega(t)|} d\omega_k(t)$. A^{ji} and Z are orthogonal since $\langle A^{ji}, Z\rangle_t = 0$, for every $0 \leq t \leq 1$. The filtration generated by Z, $\mathcal{G} = \sigma\{Z_s, 0 \leq s \leq 1\}$, is also the one generated by $|\omega|$, $\sigma\{|\omega(s)|, 0 \leq s \leq 1\}$. Therefore, provided that the norm under consideration is a measurable variable with respect to the radial process, the event $\{\|\omega\| < \epsilon\}$ belongs to $\mathcal{G}$ for every $0 < \epsilon < 1$, and by Lemma 3, we get for every real number c and all $1 \leq i, j \leq d$,

$$E\left[\exp\left\{c \int_0^1 \frac{\partial \bar{b}_i}{\partial x_j}(s,\omega(s))\, dA^{ji}(s)\right\}\Big| \|\omega\| < \epsilon\right]$$
$$\leq \left\{E\left[\exp\left\{2c^2 \int_0^1 \left(\frac{\partial \bar{b}_i}{\partial x_j}(s,\omega(s))\right)^2 \left(\omega_i^2(s) + \omega_j^2(s)\right) ds\right\}\Big| \|\omega\| < \epsilon\right]\right\}^{\frac{1}{2}}.$$

Therefore, for every real number c and all $1 \leq i, j \leq d$,

$$\limsup_{\epsilon\to 0} E\left[\exp\left\{c \int_0^1 \frac{\partial \bar{b}_i}{\partial x_j}(s,\omega(s))\, dA^{ji}(s)\right\}\Big| \|\omega\| < \epsilon\right] \leq 1$$

Now, Lemma 1 allows us to conclude that

$$\lim_{\epsilon\to 0} \frac{P(\|X - \Phi\| < \epsilon)}{P(\|\omega\| < \epsilon)} = \exp\left(-\frac{1}{2}\int_0^1 \left|\dot{\Phi}(s) - b(\Phi(s))\right|^2 ds - \frac{1}{2}\sum_{i=1}^{d} \int_0^1 \frac{\partial b_i}{\partial x_i}(\Phi(s))ds\right),$$

and the proof of Theorem 2 is complete.

In the last section, we briefly compare what we obtained to earlier, as well as more recent, results in this framework.

4.3 Previous results

It was proved in [8] that, if one chooses on $\mathcal{C}_0\left([0,1]; \mathbb{R}^d\right)$ the supremum norm

$$\|\omega\|_\infty = \sup_{t\in[0,1]} |\omega(t)|, \tag{2}$$

and if Φ is of class $\mathcal{C}^2$, the Onsager-Machlup functional is given by

$$L\left(\Phi, \dot{\Phi}\right) = -\frac{1}{2}\sum_{i=1}^{d} \left|\dot{\Phi}_i - b_i(\Phi)\right|^2 - \frac{1}{2}\sum_{i=1}^{d} \frac{\partial b_i}{\partial x_i}(\Phi).$$

This case enters the setting of Theorem 1 or Theorem 2. L.A. Shepp and O. Zeitouni proved in [13] that the result still holds for every norm which is equivalent to (2) and if $\Phi - x_0$ only belongs to $\mathcal{H}^d$. Moreover they showed in [14] that this expression is still valid for other norms, in particular for L^p-norms, $p \geq 4$, and Hölder norms $\|\cdot\|_\alpha$ with $0 < \alpha < \frac{1}{3}$ in the case $d \geq 2$ and with $0 < \alpha < \frac{1}{2}$ in the case $d = 1$. They deal more generally with a class of completely convex norms and obtain the Onsager-Machlup functional for some norms which do not enter the context of Theorem 1 or Theorem 2. In [3], we extended this result to a large class of natural norms on Wiener space, including in particular Hölder norms for every $0 < \alpha < \frac{1}{2}$; our approach closely follows [14] but we have to use versions of the norms which are rotationaly invariant on the range of the Brownian paths. Nevertheless, the context in [3] is still a bit more general than the one of Theorem 2. Recently, O. Zeitouni and T. Lyons showed in [10] that this geometric property may actually be relaxed in the case of Hölder norms. On the other hand, using different approaches, O. Zeitouni in [16] and E. Mayer-Wolf and O. Zeitouni in [11], obtained the Onsager-Machlup functional for the L^2-norm.

In [8], [13], [14], [3], [10] the authors use Girsanov's transformation to come down to Brownian small balls (as it was done in the proof of Theorem 2). The difficulty is then to evaluate conditional exponential moments of the stochastic integral $\int_0^1 \langle b(\omega(s) + \Phi(s)), d\omega(s)\rangle$. The usual method consists in a Taylor development of $b(\omega(s) + \Phi(s))$ at the point $\Phi(s)$; the minimum order of this development is fixed by the probability of the small balls relative to the norm under consideration. The computation of the Onsager-Machlup functional therefore consists in the asymptotic evaluation of conditional exponential moments of the stochastic integrals appearing in the development.

Now, as we saw in the proof of Theorem 2, the stochastic Stokes theorem and the Kunita-Watanabe theorem (techniques which were naturally used in the case of diffusion processes on manifolds) give immediately the asymptotic evaluation of the exponential moments of $\int_0^1 \langle b(\omega(s) + \Phi(s)), d\omega(s)\rangle$, provided that the norm under consideration on $\mathcal{C}_0\left([0,1];\mathbb{R}^d\right)$ is such that the random variable $\|\omega\|$ is measurable with respect to the σ-field $\sigma\{|\omega(s)|, 0 \le s \le 1\}$. Using this approach, we have to assume that b is of class $\mathcal{C}^2$, bounded, such that all its derivatives are bounded and its second derivatives are Lipschitz continuous. This smoothness assumption on b is independent of the norm under consideration whereas in the previous methods it is imposed by the order of the requisite Taylor development and thus depends on the norm. At last, this new proof has the noteworthy advantage of including the case of the L^2-norm as well as the case of the L^p-norms with $2 < p < 4$ (which did not seem to have been considered up to now). Indeed, in the previous methods, the techniques used to handle some stochastic integrals in which appear first derivatives of b, only hold in the case of a norm which dominates the L^4-norm.

References

[1] Berger, M., Gauduchon, P., Mazet, E., Le spectre d'une variété Riemannienne, Lecture Notes in Math., 194 (1971), Springer-Verlag, Berlin.

[2] C. Borell, "A note on Gauss measures which agree on small balls", *Annal. Inst. Henri Poincaré*, Vol. XIII, n 3, 1977, 231-238.

[3] Capitaine, M., " Onsager-Machlup functional for some smooth norms on Wiener space" *Prob. Th. Rel. Fields*, Vol. 102, 1995, 189-201.

[4] Capitaine, M., "Fonctionnelle d'Onsager-Machlup pour un processus de diffusion-Inégalités de Sobolev logarithmiques sur l'espace des chemins", Thèse de l'université Paul Sabatier (1996).

[5] Fujita, T. and Kotani, S., "The Onsager-Machlup function for diffusion processes", *J. Math. Kyoto Univ.*, 22-1 (1982), 115-130.

[6] Hara, K., "Wiener functionals associated with joint distributions of exit time and position from small geodesic balls", *Annals of Probability*, 24 (1996), 825-837.

[7] Hara, K. and Takahashi, Y., "Lagrangian for pinned diffusion process", Itô's Stochastic Calculus and Probability Theory, Springer (1996), 117-128.

[8] Ikeda, N. and Watanabe, S., Stochastic Differential Equations and Diffusion Processes, North-Holland, 1981.

[9] Ledoux, M., Isoperimetry and Gaussian analysis, Ecole d'Eté de Probabilités de Saint-Flour 1994, Lecture Notes in Math. 1648, 165-294 (1996), Springer-Verlag.

[10] Lyons, T. and Zeitouni, O., "Conditional exponential moments for iterated Wiener integrals, with application to Onsager-Machlup functionals", Ann. Probability, to appear.

[11] E. Mayer-Wolf et O. Zeitouni, "The Probability of small gaussian ellipsoids and associated conditional moments", *Annals of Probability*, Vol. 21, n 1, (1993), 14-24.

[12] Revuz, D. and Yor, M., Continuous Martingales and Brownian Motion, Springer-Verlag.

[13] Shepp, L.A. and Zeitouni, O., "A note on conditional exponential moments and the Onsager Machlup functional", *Annals of Probability*, 20 (1992), 652-654.

[14] Shepp, L.A. and Zeitouni, O., "Exponential estimates for convex norms and some applications", Barcelona Seminar on Stochastic Analysis, St Feliu de Guixols, 1991, Progress in Prob. vol. 32, Birkhäuser Verlag. Basel. Boston. Berlin, 1993. 203-215.

[15] Takahashi, Y. and Watanabe, S., "The probability functional (Onsager-Machlup function) of diffusion processes", Stochastic Integrals (D. Williams, Ed.), Lecture Notes in Math., 851 (1981), 433-463. Springer, Berlin-New-York.

[16] O. Zeitouni, "On the Onsager Machlup functional of diffusion processes around non C^2 curves", *Annals of Probability*, 17 (1989), 1037-1054.

Laboratoire de Statistique et Probabilités
Université Paul Sabatier
118, route de Narbonne
31062 Toulouse Cedex, France
capitain@cict.fr

A Unified Approach to Several Inequalities for Gaussian and Diffusion Measures

Yaozhong Hu
Department of Mathematics, University of Kansas
405 Snow Hall, Lawrence, KS 66045-2142

Abstract

This paper presents a simple unified approach to several inequalities for Gaussian and diffusion measures. They include hypercontractive inequalities, logarithmic Sobolev inequalities, FKG inequalities, and correlation inequalities.

1 Introduction

This paper is concerned with several inequalities associated with Gaussian measures and measures generated by diffusions. We are mainly interested in hypercontractive inequalities, Poincaré inequalities (spectral gap inequalities), logarithmic Sobolev inequalities, FKG inequalities, and correlation inequalities. A unified approach is developed to produce all these inequalities at the same time. This approach is the so-called semigroup approach [9], [10], [6]. However, one usually obtains hypercontractive inequalities from logarithmic Sobolev inequalities. An interesting point of this paper is to illustrate that the hypercontractive inequalities can be obtained in a somewhat simpler way. We prove hypercontractivity by differentiating an auxiliary function once and (while if we differentiate it twice we obtain the logarithmic Sobolev inequalities). This approach is natural since it is well-known that the logarithmic Sobolev inequality is the infinitesimal form of hypercontractivity. If we are only interested in Gaussian measures, our approach for hypercontractivity is as simple as the famous Neveu approach ([11], see also [7]).

2 Hypercontractivity for Diffusions

Let $(E, \mathcal{E}, \mu)$ be a measure space and let $L^2(\mu)$ be the set of all square integrable functions from E to $\mathbb{R}$. Let L be the Markov generator associated with a continuous Markov semigroup P_t in $L^2(\mu)$. Denote $\mu(f) = \int_E f(x)\mu(dx)$.

Assumptions: *μ is invariant with respect to L and $(P_t)_{t\geq 0}$. There is a nice algebra $\mathcal{A}$ of bounded functions on E which is dense in the L^2 domain of L, stable by L, by P_t and by the action of composition with C^∞ real functions which are 0 at 0. Let also $\mathcal{A}$ be dense in $L^p(\mu)$ for any $1 < p < \infty$. P_t is ergodic in the sense that for any f, $P_t f \to \mu(f)$ μ-almost surely.*

Following P.-A. Meyer, we introduce the so-called "carré du champ" operator Γ as the symmetric bilinear form on $\mathcal{A}$ defined by

$$\Gamma(f,g) = \frac{1}{2}\{L(fg) - fLg - gLf\}\,, \quad \forall\, f,g \in \mathcal{A}\,. \tag{2.1}$$

Denote $\Gamma(f) = \Gamma(f,f)$. It is well-known that Γ is positive-definite.

We shall consider a Markov semigroup whose generator is a diffusion in the following sense:

For every C^∞ function ψ on $\mathbb{R}^n$, and for every finite family $F = (f_1, \cdots, f_n)$, where $f_1, \cdots, f_n \in \mathcal{A}$,

$$L\psi(F) = \sum_{i=1}^{n} \frac{\partial\psi}{\partial x_i}(f_1, \cdots, f_n) L(f_i) + \sum_{i,j=1}^{n} \frac{\partial^2\psi}{\partial x_i \partial x_j}(f_1, \cdots, f_n)\Gamma(f_i, f_j)\,. \tag{2.2}$$

The iterated "carré du champ" operator is defined as

$$\Gamma_2(f,g) := \frac{1}{2}\{L\Gamma(f,g) - \Gamma(f, Lg) - \Gamma(g, Lf)\}\,,\ \forall f,g \in \mathcal{A}\,. \tag{2.3}$$

Let $\mathrm{Rg}(f)$ denote the closure of the range of f.

The main resut of this work is the following statement.

Theorem 2.1 *Let R, τ be fixed real numbers. Let ϕ and ψ be C^∞ functions on an open interval $J \subset \mathbb{R}$. If $\Gamma_2(f) \geq R\Gamma(f)$, $\forall f \in \mathcal{A}$, and if $\phi(x) \geq 0\,, \psi(x) \geq 0\,, \phi''(x) \leq 0\,, \psi''(x) \leq 0$, and*

$$\left(e^{-R\tau}\phi'(x)\psi'(y)\right)^2 \leq \phi(x)\phi''(x)\psi(y)\psi''(y)\,, \quad \forall\, x,y \in J\,, \tag{2.4}$$

then for all f, $g \in \mathcal{A}$ such that $\mathrm{Rg}(P_t(f)) \subset J$, $\mathrm{Rg}(P_t(g)) \subset J$,

$$\mu\{\psi(g)\ P_\tau(\phi(f))\} \leq \phi(\mu(f))\psi(\mu(g))\,. \tag{2.5}$$

Proof It is easy to see that for every such ϕ, and $f \in \mathcal{A}$ with $\mathrm{Rg}(f) \subset J$, $\phi(f)$ is well-defined and (2.2) holds. Denote $F_t = P_t f$ and $G_t = P_t g$. For any $T > 0$, introduce an auxiliary function of t,

$$h_t := P_{T-t}\{[P_\tau(\phi(F_t))]\psi(G_t)\}\,, \quad 0 \leq t \leq T\,.$$

Let us compute its derivative. Let $L_t = L - \frac{\partial}{\partial t}$. By (2.2) it is easy to check that for any f_t, g_t defined on $\mathbb{R}_+ \times E$,

$$L_t\phi(f_t) = \phi''(f_t)\Gamma(f_t)\,, \quad L_t(f_t g_t) = f_t L_t g_t + g_t L_t f + 2\Gamma(f_t, g_t), \tag{2.6}$$

Set $\Phi = \phi(F_t)$, $\Psi = \psi(G_t)$, $\Phi' = \phi'(F_t)$, $\Psi' = \psi'(G_t)$, $\Phi'' = \phi''(F_t)$, and $\Psi'' = \psi''(G_t)$. By (2.6),

$$\begin{aligned} \frac{dh_t}{dt} &= -P_{T-t}\left\{L_t\left[(P_\tau(\Phi))\Psi\right]\right\} \\ &= -P_{T-t}\left\{L_t\left(P_\tau(\Phi)\right)\Psi + P_\tau(\Phi)L_t\left(\Psi\right) + 2\Gamma(P_\tau(\Phi)\,,\Psi)\right\} \\ &= -P_{T-t}\left\{P_\tau\left[\Phi''\Gamma(F_t)\right]\Psi + P_\tau[\Phi]\Psi''\Gamma(G_t) + 2\Gamma(P_\tau(\Phi)\,,\Psi)\right\}\,. \quad (2.7) \end{aligned}$$

Since Γ is positive definite, we have that

$$|\Gamma(P_\tau(\Phi),\Psi)|^2 \leq \Gamma(P_\tau(\Phi))\Gamma(\Psi)\,.$$

It is proved in [1] (page 149) that if $\Gamma_2(f) \geq R\Gamma(f)$ for all $f \in \mathcal{A}$, then

$$\sqrt{\Gamma(P_T f)} \leq e^{-RT} P_T\sqrt{\Gamma(f)}\,, \quad \forall\, f \in \mathcal{A}\,. \quad (2.8)$$

Thus we have

$$\begin{aligned} |\Gamma(P_\tau(\Phi),\Psi)| &\leq \sqrt{P_\tau[\Gamma(\Phi)]}\sqrt{\Gamma(\Psi)} \leq e^{-R\tau}P_\tau\sqrt{\Gamma(\Phi)}\sqrt{\Gamma(\Psi)} \\ &= e^{-R\tau}P_\tau\left[|\Phi'|\sqrt{\Gamma(F_t)}\right]|\Psi'|\sqrt{\Gamma(G_t)}\,. \end{aligned}$$

The last inequality follows from $\Gamma(\phi(f)) = \phi'(f)^2\Gamma(f)$, which is an easy consequence of (2.1) and (2.2). Let us then introduce the operator $\tilde{P}_\tau$ acting on $\mathcal{A}\times\mathcal{A}$ by $\tilde{P}_\tau(f\times g) = P_\tau(f)(x)g(x)$ (where $(f\times g)(x,y) = f(x)g(y)\,,\ x,y \in E$). Since P_τ is positivity preserving, the same is true for $\tilde{P}_\tau$. Then we have

$$\begin{aligned} \frac{dh_t}{dt} \geq\ & -P_{T-t}\{\tilde{P}_\tau[\Phi''\Gamma(F_t)\times\Psi + \Phi\times\Psi''\Gamma(G_t) \\ & +2e^{-R\tau}|\Phi'||\Psi'|\sqrt{\Gamma(F_t)}\times\sqrt{\Gamma(G_t)}]\}\,. \end{aligned} \quad (2.9)$$

If (2.4) holds, then $\frac{dh_t}{dt} \geq 0$. Thus $h_0 \leq h_T$. Computing h_T and h_0, we obtain the inequality

$$P_T\left\{(P_\tau(\phi(f)))\psi(g)\right\} \leq P_\tau(\phi(P_T(f)))\psi(P_T g)\,. \quad (2.10)$$

Letting $T \to \infty$, the theorem is proven. □

This theorem is a generalization of the hypercontractivity theorem. Set $||f||_p^p = \mu(|f|^p)$.

Corollary 2.2 *Let $p > 1$ and $q > 1$. If $\Gamma_2(f) \geq R\Gamma(f)$ and if $e^{-2R\tau} \leq \dfrac{p-1}{q-1}$, then $||P_\tau f||_q \leq ||f||_p$, $\forall f \in \mathcal{A}$.*

Proof Let us take $J_\varepsilon = (-\varepsilon/2, +\infty)$, $\phi_\varepsilon(x) = |x+\varepsilon|^{1/p}$, and $\psi_\varepsilon(y) = |y+\varepsilon|^{1/q'}$ with $1/q + 1/q' = 1$. Therefore (2.4) is true for ϕ_ε and ψ_ε if and only if $e^{-2R\tau} \leq (p-1)/(q-1)$. Thus (2.5) holds for ϕ_ε and ψ_ε. Letting $\varepsilon \to 0$, we obtain that for all $f,g \in \mathcal{A}$ and $f,g \geq 0$,

$$\mu\left\{P_\tau(f^{1/p})g^{1/q'}\right\} \leq (\mu(f))^{1/p}(\mu(g))^{1/q'}\,.$$

In the above inequality replacing f by $(f^2+\varepsilon)^{p/2}-\varepsilon^{p/2}$ and g by $(g^2+\varepsilon)^{q'/2}-\varepsilon^{q'/2}$ and then letting $\varepsilon \to 0$, we get

$$\mu((P_\tau|f|)|g|) \leq (\mu(|f|^p))^{1/p}(\mu(|g|^{q'}))^{1/q'}, \quad \forall\ f, g \in \mathcal{A}.$$

The corollary is thus established. □

3 Logarithmic Sobolev Inequalities

It is not so easy to compute the second derivative of h_t introduced in the previous section. We shall compute it in the next section for a Gaussian measure. Here we take $\psi(x) = x$ and $\tau = 0$ and compute h_t''. In this case we have $\frac{dh_t}{dt} = P_{T-t}\{\phi''\Gamma(F)\}$. Differentiating this identity once more with respect to t and using (2.2) for $\Psi(x,y) = \phi''(x)y$ and $F_1 = F$ and $F_2 = \Gamma(F)$, we obtain

$$\begin{aligned}
\frac{d^2h_t}{dt^2} &= P_{T-t}\left\{L_t\left[\phi''\Gamma(F)\right]\right\} \\
&= P_{T-t}\left\{\phi^{(4)}\Gamma(F)^2 + \phi'' L_t\Gamma(F) + 2\phi^{(3)}\Gamma(F,\Gamma(F))\right\} \\
&= P_{T-t}\left\{\phi^{(4)}\Gamma(F)^2 + 2\phi''\Gamma_2(F) + 2\phi^{(3)}\Gamma(F,\Gamma(F))\right\} \\
&= P_{T-t}\left\{\mathrm{Tr}(AB) + 2R\phi''\Gamma(F)\right\},
\end{aligned}$$

where $A = \begin{pmatrix} \phi''/2 & \phi^{(3)} \\ \phi^{(3)} & \phi^{(4)} \end{pmatrix}$ and $B = \begin{pmatrix} 4\left[\Gamma_2(F) - R\Gamma(F)\right] & \Gamma(F,\Gamma(F)) \\ \Gamma(F,\Gamma(F)) & \Gamma(F)^2 \end{pmatrix}$. B is positive definite by [1], p.149. If (3.2) below holds, A is also positive definite. Thus we have

$$\frac{d^2h_t}{dt^2} \geq 2RP_{T-t}\left[\phi''(F)\Gamma(F)\right] = 2R\frac{dh_t}{dt}. \tag{3.1}$$

This implies that the derivative of $h_t - h_t'\frac{1-e^{-2Rt}}{2R}$ is non-negative. Hence

$$h_T - h_0 \leq h_T'(1-e^{-2RT})/(2R).$$

Thus we have

Theorem 3.1 *Let ϕ be a C^∞ function on some open interval $J \subset \mathbb{R}$ satisfying*

$$\phi''(x) \geq 0, \quad \phi^{(4)}(x) \geq 0, \quad 2(\phi^{(3)}(x))^2 \leq \phi''(x)\phi^{(4)}(x), \quad \forall x \in J. \tag{3.2}$$

Assume that $\Gamma_2(f) \geq R\Gamma(f)$, $\forall f \in \mathcal{A}$. Then for every $f \in \mathcal{A}$ such that $\mathrm{Rg}(P_t f) \subset J$,

$$P_T(\phi(f)) \leq \phi(P_T(f)) + \frac{1-e^{-2RT}}{2R}P_T(\phi''(f)\Gamma(f)), \quad \forall\, T > 0. \tag{3.3}$$

Letting $T \to \infty$, we have

$$\mu(\phi(f)) \leq \phi(\mu(f)) + \frac{1}{2R}\mu(\phi''(f)\Gamma(f)). \tag{3.4}$$

Remark 1 *In (3.4) letting $\phi(x) = x^2$, one gets the Poincaré inequality and letting $\phi(x) = x \log x$ (using the argument of Corollary 2.2), one gets the logarithmic Sobolev inequality, see [6]. For the Ornstein-Uhlenbeck semigroup, (2.8) is obvious so that the preceding proof of hypercontractivity for Gaussian measures is indeed rather simple.*

4 Gaussian Measures

For a Gaussian measure, we may also use the heat semigroup [8] instead of the Ornstein-Uhlenbeck semigroup. Let μ be the standard Gaussian measure on $\mathbb{R}^d$. Let P_t be heat kernel semigroup associated with the standard Laplacian: $\frac{\partial P_t}{\partial t} = \Delta P_t$ (We omit the factor 1/2 for simplicity). Let $\Delta_t = \Delta - \frac{\partial}{\partial t}$. Denote by $\langle \cdot, \cdot \rangle$ the inner product of two vectors or the Hilbert-Schmidt product of two matrices. The same computation rule (2.2) applies to Δ_t and moreover, for any functions f and g on $\mathbb{R}^d$ (in this section, f and g always denote C^∞ functions with compact supports except otherwise stated), if $F_t = P_t f$ and $G_t = P_t g$ as before,

$$\begin{aligned} \Delta_t \langle \nabla F_t , \nabla G_t \rangle &= \langle \nabla(\Delta_t F_t) , \nabla G_t \rangle + \langle \nabla F_t , \nabla(\Delta_t G_t) \rangle + \langle \nabla^2 F_t , \nabla^2 G_t \rangle \\ &= 2 \langle \nabla^2 F_t , \nabla^2 G_t \rangle . \end{aligned} \tag{4.1}$$

Fix $T \geq 0$. Consider $h_t = P_{T-t} \{\phi(F_t)\psi(G_t)\}$, $0 \leq t \leq T$. Then it is easy to see that

$$\begin{aligned} \frac{dh_t}{dt} &= -P_{T-t}\{\phi''(F_t)\psi|\nabla F_t|^2 + \phi(F_t)\psi''(G_t)|\nabla G_t|^2 \\ &\quad + \phi'(F_t)\psi'(G_t)\langle \nabla F_t , \nabla G_t \rangle\} . \end{aligned}$$

If we take $\phi(x) = \psi(x) = x$, and if $\nabla f \geq 0$ and $\nabla g \geq 0$ component-wise, then we see $h_t' \geq 0$. Thus $h_T \geq h_0$. Thus we obtain in this way the classical FKG inequality for Gaussian measures.

Proposition 4.1 *If $\nabla f \geq 0$, $\nabla g \geq 0$ component-wise, then*

$$\mu(f)\mu(g) \leq \mu(fg) .$$

Now, compute the second derivative of h_t for general ϕ and ψ. We have that

$$\begin{aligned} \frac{d^2 h_t}{dt^2} &= P_{T-t}\Big\{\Delta_t \left(\phi''(F_t)\psi(G_t)|\nabla F_t|^2\right) + \Delta_t \left(\phi(F_t)\psi''(G_t)|\nabla G_t|^2\right) \\ &\quad + 2\Delta_t \left(\phi'(F_t)\psi'(G_t)\langle \nabla F_t , \nabla G_t \rangle\right)\Big\} . \end{aligned} \tag{4.2}$$

Apply then (2.2) to $\Psi = \phi''(x_1)\psi(x_2)x_3$ and use (4.1) to compute $\Delta_t |\nabla F_t|^2$. It follows that

$$\begin{aligned} \Delta_t \left(\phi''(F_t)\psi(G_t)|\nabla F_t|^2\right) &= 2\phi''\psi\langle \nabla^2 F_t, \nabla^2 F_t \rangle + \phi^{(4)}\psi|\nabla F_t|^4 \\ &\quad + \phi''\psi''|\nabla F_t|^2|\nabla G_t|^2 + 2\phi^{(3)}\psi'\langle \nabla F_t , \nabla G_t \rangle|\nabla F_t|^2 \\ &\quad + 4\phi^{(3)}\psi\langle \nabla F_t \otimes \nabla F_t , \nabla^2 F_t \rangle \\ &\quad + 4\phi''\psi'\langle \nabla F_t \otimes \nabla G_t , \nabla^2 F_t \rangle . \end{aligned}$$

In a similar way we can compute the other terms in (4.2). If we introduce $X^T = [X_1, X_2, X_3, X_4, X_5]$, where

$$X_1 = \nabla F_t \otimes \nabla F_t\,, \quad X_2 = \nabla G_t \otimes \nabla G_t\,, \quad X_3 = \nabla F_t \otimes \nabla G_t\,,$$

$$X_4 = \nabla^2 F_t\,, \quad X_5 = \nabla^2 G_t\,,$$

then we can write

$$\frac{d^2 h_t}{dt^2} = \langle X\,, AX\rangle\,, \tag{4.3}$$

where $A = B + C$, B is a 5×5 matrix with $B_{45} = B_{54} = 2\phi'\psi'$ and the other elements 0 and C is defined as

$$C = \begin{pmatrix} \phi^{(4)}\psi & 2\phi''\psi'' & 2\phi^{(3)}\psi' & 2\phi^{(3)}\psi & 2\phi''\psi' \\ 2\phi''\psi'' & \phi\psi^{(4)} & 2\phi'\psi^{(3)} & 2\phi'\psi'' & 2\phi\psi^{(3)} \\ 2\phi^{(3)}\psi' & 2\phi'\psi^{(3)} & 2\phi''\psi'' & 4\phi''\psi' & 4\phi'\psi'' \\ 2\phi^{(3)}\psi & 2\phi'\psi'' & 4\phi''\psi' & 2\phi''\psi & 0 \\ 2\phi''\psi' & 2\phi\psi^{(3)} & 4\phi'\psi'' & 0 & 2\phi\psi'' \end{pmatrix}\,. \tag{4.4}$$

If A is positive definite, then $h_t'' \geq 0$. Namely, $h_0 \leq h_T - h_0'$. Thus we proved

Theorem 4.2 *If ϕ and ψ are C^∞ functions on some open interval $J \subset \mathbb{R}$ such that A is positive definite on J, then for all f, g such that $\mathrm{Rg}(P_t(f)) \subset J$, $\mathrm{Rg}(P_t(g)) \subset J$,*

$$\begin{aligned} \mu(\phi(f)\psi(g)) \;\leq\; & \phi(\mu(f))\psi(\mu(g)) + \mu\Big\{\phi''(f)\psi(g)|\nabla f|^2 + \phi(f)\psi''(g)|\nabla g|^2 \\ & +2\phi'(f)\psi'(g)\langle\nabla f\,, \nabla g\rangle\Big\}\,. \end{aligned} \tag{4.5}$$

Remark 2 *1) Take $\phi(x) = \psi(x) = x$. If f and g are convex functions and if $\mu(\nabla f) = 0$ and $\mu(\nabla g) = 0$, then (4.5) implies the following correlation inequality [8]:*

$$\mu(fg) \geq \mu(f)\mu(g)\,.$$

2) Let $\psi(x) \equiv 1$ and let ϕ satisfy (3.2). Then it is easy to check that A is positive definite. Thus (4.5) implies

$$\mu(\phi(f)) \leq \phi(\mu(f)) + \mu(\phi''(f)|\nabla f|^2)\,. \tag{4.6}$$

As we mentioned, this implies the Poincaré and the logarithmic Sobolev inequalities. In fact, if $\phi(x) = x^2$, then ϕ satisfies (3.2). (4.6) becomes the Poincaré inequality

$$\mu(|f|^2) \leq \mu(f)^2 + \mu(|\nabla f|^2)\,.$$

If $\phi(x) = x\log^+ x$, then ϕ satisfies (3.2) (using the argument of Corollary 2.2). (4.6) implies the logarithmic Sobolev inequality

$$\mu(f\log f) \leq \mu(f)\log\mu(f) + \mu(\frac{1}{f}|\nabla f|^2)\,, \quad \forall f \geq 0\,.$$

References

[1] Bakry, D. Transformations de Riesz pour les semigroupes symétriques. Seconde partie: Etude sous la condition $\Gamma_2 \geq 0$, Séminaire de probabilités, XIX, 1983/84, 145-174, Lecture Notes in Math., 1123, Springer, Berlin-New York, 1985.

[2] Bakry, D. L'hypercontractivité et son utilisation en théorie des semigroupes. Lectures on probability theory (Saint-Flour, 1992), 1–114, Lecture Notes in Math., 1581, Springer, Berlin, 1994.

[3] Bakry, D.; Emery, M. Diffusions hypercontractives. Séminaire de probabilités, XIX, 1983/84, 177–206, Lecture Notes in Math., 1123, Springer, Berlin-New York, 1985.

[4] Bakry, D.; Ledoux, M. Lévy-Gromov's isoperimetric inequality for an infinite-dimensional diffusion generator. Invent. math. 123 (1996), 259–281.

[5] Davies, E. B.; Gross, L.; Simon, B. Hypercontractivity: a bibliographic review. Ideas and methods in quantum and statistical physics (Oslo, 1988), 370–389, Cambridge Univ. Press, Cambridge, 1992.

[6] Driver, B. K.; and Hu, Y.Z. On heat kernel logarithmic Sobolev inequalities. Stochastic analysis and applications (Powys, 1995), 189–200, World Sci. Publishing, River Edge, NJ, 1996.

[7] Dellacherie, C.; Meyer, P.-A.; Maisonneuve, B. Probabilités et Potentiel. V. Hermann, Paris, 1992.

[8] Hu, Y.Z. Itô-Wiener chaos expansion with exact residual and correlation, variance inequalities. J. Theoret. Probab. 10 (1997), no. 4, 835–848.

[9] Ledoux, M. L'algèbre de Lie des gradients itérés d'un générateur markovien – Développements de moyennes et entropies. Ann. Sci. École Norm. Sup. 28 (1995), no. 4, 435–460.

[10] Ledoux, M. Semigroup proofs of the isoperimetric inequality in Euclidean and Gauss space. Bull. Sci. Math. 118 (1994), 485–510.

[11] Neveu, J. Sur l'espérance conditionnelle par rapport à un mouvement brownien. Ann. Inst. H. Poincaré Sect. B (N.S.) 12 (1976), 105–109.

Note: More complete references can be found in [2], [5].

Trous spectraux pour certains algorithmes de Métropolis sur $\mathbb{R}$

Laurent Miclo et Cyril Roberto

Laboratoire de Statistique et Probabilités, UMR C5583
Université Paul Sabatier et CNRS
118, route de Narbonne
31 062 Toulouse cedex
email : miclo@cict.fr, roberto@cict.fr

RÉSUMÉ :

Sur $\mathbb{R}$ considérons le noyau markovien $Q(x, dy) = \mathbb{1}_{]x-\frac{1}{2}, x+\frac{1}{2}[}(y)\, dy$ réversible par rapport à la mesure de Lebesgue dy. A partir de celui-ci on effectue une transformation de type Métropolis pour obtenir un noyau $P(x, dy)$ réversible par rapport à la loi de densité proportionnelle à $e^{-U(x)}$ relativement à la mesure de Lebesgue dx où U est une fonction convexe régulière vérifiant $\lim_{x\to\pm\infty} U(x) = +\infty$ et une autre hypothèse de croissance.

Le but de cette note est de prouver l'existence d'un trou spectral pour l'opérateur positif $\mathrm{Id} - P$ (où Id désigne l'identité), la difficulté résidant surtout dans la non-compacité du modèle.

Nous introduirons tout d'abord plus en détail les objets utilisés avant de réduire le problème et de le résoudre, nous utiliserons principalement une technique de multiflots.

1 Introduction

L'existence d'un trou spectral permet d'avoir des évaluations quantitatives de la vitesse de convergence vers l'équilibre d'algorithmes stochastiques comme par exemple ceux de Métropolis. Ce sont ces estimations qui motivent notre étude.

Plus précisément, on se donne sur $\mathbb{R}$ le noyau markovien $Q(x, dy) = \mathbb{1}_{]x-\frac{1}{2}, x+\frac{1}{2}[}(y)\, dy$ réversible par rapport à la mesure de Lebesgue dy et U une fonction convexe vérifiant $\lim_{x\to\pm\infty} U(x) = +\infty$.

Sous cette hypothèse et par convexité, on montre facilement qu'il existe deux réels a et α strictement positifs tels que pour tout $|x| \geq a$,

$$U(x) \geq \alpha|x| \quad \text{et} \quad |U'(x)| \geq \alpha \,. \tag{1}$$

On définit la mesure

$$\mu(dx) = K^{-1} e^{-U(x)} dx \quad \text{avec} \quad K = \int_{\mathbb{R}} e^{-U(x)} dx$$

Lecture Notes in Mathematics, Vol. 1709 "Séminaire de Probabilités XXXIII"
ISBN 3-540-66342-8

During the processing of Vol. XXXIII (LNM 1709) all author's addresses were mistakenly deleted. Springer-Verlag apologizes to the authors and readers; please put this correction sheet in your copy of volume XXXIII. (The addresses of some authors who have moved since have been updated)

Prof. Marc Arnaudon
Laboratoire de Mathématiques
Université de Poitiers
BP 179
86860 Futuroscope Cedex
France

Mrs. Samia Beghdadi-Sakrani
Laboratoire de Probabilités
Université Pierre et Marie Curie
4, Place Jussieu
75252 Paris cedex 05
France

Dr. Nacereddine Belili
LAMS, Site Colbert
Université de Rouen
76821 Mont-Saint-Aignan
France

Prof. Michel Benaïm
Département de Mathématiques
Université Cergy-Pontoise
95031 Neuville sur Oise
France

Dr. Abdellatif Bentaleb
Département de Mathématiques
Faculté des Sciences de Méknès
Université My Ismail
B.P. 4010, Beni M'Hamed
Méknès, Maroc

Dr. Werner Brannath
Institut f. Medizinische Statistik
und Dokumentation
der Universität Wien
Schwarzspanierstr. 17
1210 Wien
Austria

Prof. Olivier Catoni
Laboratoire de Probabilités
Université Pierre et Marie Curie
4, Place Jussieu
75252 Paris cedex 05
France

Mr. Karl Chrétien
Appartment 24
77 rue de Koechelin
68200 Mulhouse
France

Prof. Freddy Delbaen
Dept. für Mathematik
ETH Zürich
Rämistr. 101
8092 Zürich
Schweiz

Prof. Bernard De Meyer
ESSTIN
2, rue Jean Lamour
54500 Vandoeuvre-les-Nancy
France

Prof. Lester E. Dubins
Dept. of Mathematics
University of California
Berkeley, CA 94720
U.S.A.

Dr. Nathalie Eisenbaum
Laboratoire de Probabilités
4, Place Jussieu
75252 Paris Cedex 05
France

Dr. Aziz Es-Sahib
Département Génie Mathématique
Insa de Rouen
Place E. Blondel
76131 Mont-Saint-Aignan Cedex
France

Dr. Peter Grandits
Vienna University of Technology
Financial and Actuarial Mathematics
Wiedner Hauptstr. 8-10/107-5
1040 Vienna
Austria

Prof. Henri Heinich
Département Génie Mathématique
Insa de Rouen
Place E. Blondel
76131 Mont-Saint-Aignan Cedex
France

Dr. Jan Kallsen
Institut für Mathematische Stochastik
der Universität Freiburg
Eckerstr. 1
79104 Freiburg i. Br.
Germany

Mr. David Kurtz
Mathématiques
7 rue René Descartes
67084 Strasbourg Cedex
France

Prof. Bernard Maisonneuve
IMSS
UPMF
B.P. 47 X
38040 Grenoble
France

Prof. Michal Morayne
Institute of Mathematics
University of Wroclaw
Pl. Grunwaldzki 2/4
50-384 Wroclaw
Poland

Prof. Jim Pitman
Dept. of Statistics
University of California
367 Evans Hall No. 3860
Berkeley, CA 94720-3860
USA

Prof. Maurizio Pratelli
Dipartimento di Matematica
Università di Pisa
Via Buonarotti 2
56127 Pisa
Italy

Prof. Walter Schachermayer
Vienna University of Technology
Financial and Actuarial Mathematic
Wiedner Hauptstr. 8-10/107-5
1040 Vienna
Austria

Prof. Krzysztof Tabisz
Institute of Mathematics
University of Wroclaw
Pl. Grunwaldzki 2/4
50-384 Wroclaw
Poland

Prof. Koichiro Takaoka
Faculty of Commerce
Hitotsubashi University
Kunitachi City, Tokyo 186-8601
Japan

Prof. Hideatsu Tsukahara
Dept. of Economics
Seijo University
6-1-20 Seijo
Tokyo 157-8511
Japan

Dr. Jon Warren
Dept. of Statistics
University of Warwick
Coventry CV4 7AL
United Kingdom

Prof. Shinzo Watanabe
Division of Mathematics
Kyoto University
Kyoto 606-8502
Japan

Erratum: Page IV of Séminaire de Probabilités XXXIII,
Michel Ledoux' e-mail address should be ledoux@cict.fr

$K < \infty$ par définition de a, et on construit par un procédé de type Métropolis le noyau

$$P(x,dy) = e^{-(U(y)-U(x))_+}\,Q(x,dy) + a_x\delta_x(dy) = f_x(y)\,dy + a_x\delta_x(dy)$$

où

$$x_+ \stackrel{\text{déf.}}{=} \max(x,0),\quad a_x = 1 - \int_{\mathbb{R}} f_x(y)\,dy \quad \text{et} \quad f_x(y) = \begin{cases} e^{-(U(y)-U(x))_+} & \text{si } |y-x| \leq \frac{1}{2} \\ 0 \text{ sinon} \end{cases}.$$

Par construction, P est réversible par rapport à μ et est irréductible. On pourra donc se servir du semi-groupe markovien $(e^{t(P-\mathrm{Id})})_{t\geq 0}$ pour approximer μ.

On suppose alors de plus sur U qu'il existe une constante $C > 1$ telle que :

$$(H) \qquad \begin{cases} \forall x \in [a,\infty[& U'(x+1) \leq CU'(x) \\ \forall x \in]-\infty,-a] & |U'(x-1)| \leq C|U'(x)| \end{cases}$$

On remarque que l'hypothèse (H) n'est pas très restrictive car tout polynôme convient. En particulier, la loi gaussienne rentre dans le cadre de notre étude. En revanche, des fonctions "trop rapides" comme e^{x^2} ne conviennent pas. On peut également noter que cette hypothèse est technique et que l'intuition nous invite à penser qu'elle n'est pas nécessaire.

Le but de cette note est de montrer que sous ces hypothèses, le trou spectral de $\mathrm{Id} - P$ est strictement positif :

$$\lambda \stackrel{\text{déf.}}{=} \inf_{f\in\mathbb{L}^2(\mu)\setminus\mathrm{Vect}(\mathbb{1})} \frac{\mu(f(\mathrm{Id}-P)f)}{\mu((f-\mu(f))^2} > 0$$

Par ce biais, nous pourrons évaluer la vitesse de convergence de $e^{t(P-\mathrm{Id})}$ pour t grand vers μ car nous savons que cette convergence est exponentiellement rapide dans $\mathbb{L}^2(\mu)$, de taux λ.

Nous montrerons tout d'abord dans les deux sections suivantes qu'il existe un trou spectral sur $]-\infty,-a]$, $[-a-1,a+1]$ et $[a,\infty[$ pour le noyau P et la mesure μ restreints à ces différents intervalles, où a est défini plus haut. Par symétrie autour de 0, il nous suffira de traiter les cas $[-a-1,a+1]$ et $[a,\infty[$. Nous commencerons par nous intéresser à l'intervalle $[a,\infty[$ qui représente la plus grosse difficulté, on y développera une technique de multiflots due à Sinclair (voir [12]), nous nous placerons ensuite sur $[-a-1,a+1]$ où l'existence d'un trou spectral strictement positif est relativement aisée à obtenir. Dans la quatrième section, nous recollerons enfin les morceaux pour obtenir le résultat général sur $\mathbb{R}$ tout entier.

Puis dans une dernière section, nous discuterons des propriétés similaires dans le cas où l'on s'intéresse plutôt aux chaînes de Markov en temps discret admettant P pour noyaux de probabilités de transition.

Comme nous l'a pertinemment rappelé le rapporteur, il existe en temps discret d'autres manières d'obtenir une convergence exponentiellement rapide vers la probabilité invariante. Par exemple si P était quasi-compact (cf. [9]), on saurait qu'uniformément en $x \in \mathbb{R}$, la variation totale $\|P^n(x,\cdot)-\mu\|_{\mathrm{vt}}$ converge vers 0 pour n grand, mais ceci n'est pas possible ici (considérer des points initiaux x très grands). Une approche alternative consiste à trouver une fonction mesurable de Lyapounov, $V \geq 1$, qui satisfait en dehors d'un compact $P(V) \leq \alpha V$, avec $0 \leq \alpha < 1$, car ceci assure tout d'abord que $V \in \mathbb{L}^1(\mu)$ puis qu'il existe deux constantes $K > 0$ et $0 < \widetilde{\rho} < 1$ telles que

$$(2) \qquad \forall\, x_0 \in \mathbb{R}, \qquad \sup_{\{f\in\mathbb{L}^1(\mu)\,:\,|f|\leq V\}} |P^n(f)(x_0) - \mu(f)| \;\leq\; K\widetilde{\rho}^n V(x_0)$$

(voir le chapitre 16 du livre de Meyn et Tweedie [6], en notant que dans notre situation les ensembles compacts sont des « petite sets »).

De telles applications V existent pour les cas traités dans cet article (car on peut toujours prendre $V(\cdot) = \exp(\alpha\,|\cdot|\,/2)$, avec α la constante intervenant dans (1), ceci fournissant d'ailleurs des exemples où $V \notin \mathbb{L}^2(\mu)$, en considérant $U(\cdot) = \alpha\,|\cdot|$), toutefois l'existence d'un trou spectral permet d'obtenir des résultats de convergence uniforme en un sens différent. Par exemple on peut remplacer dans (2) l'ensemble sur lequel est considéré le maximum par

$$\{f \in \mathbb{L}^2(\mu)\,:\, f(x_0) \leq K_1,\, \mu(f^2) \leq K_2\}$$

où $K_1, K_2 \geq 0$ sont deux constantes (pour plus de précisions, voir la fin de l'article). Cependant l'importance de cette uniformité (et le fait que le trou spectral est relativement stable par perturbations bornées de U sur tout $\mathbb{R}$) apparaît plutôt pour des chaînes de Markov inhomogènes en temps. Mais ce gain se paye, car l'étude du trou spectral est plus délicate que l'obtention d'une bonne fonction de Lyapounov V.

D'autre part la dimension 1 est certainement trop restrictive, cependant notons que le résultat présenté s'étend immédiatement par comparaison et tensorisation sur $\mathbb{R}^n$, $n \geq 2$, pour des potentiels de la forme

$$\forall\, x = (x_1, \cdots, x_n) \in \mathbb{R}^n, \qquad U(x) \;=\; U_1(x_1) + \cdots + U_n(x_n)$$

où les U_i, $1 \leq i \leq n$ satisfont les hypothèses considérées ici, en utilisant les inégalités $(U(y) - U(x))_+ \leq \sum_{1\leq i\leq n}(U_i(y_i) - U_i(x_i))_+$, valables pour tous $x = (x_1, \cdots, x_n)$, $y = (y_1, \cdots, y_n) \in \mathbb{R}^n$ (on peut d'ailleurs s'interroger sur un résultat plus général pour des fonctions $U : \mathbb{R}^n \to \mathbb{R}$ satisfaisant une bonne condition de stricte convexité en dehors d'un compact, à l'instar des travaux connus pour les diffusions, voir [2] ou [1]). Mais nous pensons que le cas réel donne une bonne illustration de l'utilité des multiflots pour palier la non-bornitude des modèles.

2 Existence du trou spectral sur $[a, \infty[$

On restreint le noyau P et la mesure μ à $[a, \infty[$ pour obtenir le noyau

$$P_1(x, dy) = f_x(y)\, \mathbb{I}_{[a,\infty[}(y)\, dy + a_{1,x}\delta_x(dy)$$

avec

$$a_{1,x} = 1 - \int_a^\infty f_x(y)\, dy\ ,$$

et la mesure

$$\mu_1(dx) = K_1^{-1} e^{-U(x)}\, dx \quad \text{où} \quad K_1 = \int_a^\infty e^{-U(x)}\, dx.$$

On a toujours P_1 réversible par rapport à la mesure μ_1, on veut montrer :

$$\lambda_1 \stackrel{\text{déf.}}{=} \inf_{f\in\mathbb{L}^2(\mu_1)\setminus\mathrm{Vect}(\mathbf{I})} \frac{\mu_1(f(\mathrm{Id}-P_1)f)}{\mu_1((f-\mu_1(f))^2} > 0 .$$

Il n'existe que peu de résultats donnant des estimations satisfaisantes du trou spectral d'un noyau évoluant sur un ensemble continu, sauf si celui-ci provient d'un semi-groupe de diffusion considéré par exemple au temps 1. Ainsi pour cette étude, on a commencé par chercher à faire des comparaisons avec le processus d'Ornstein-Uhlenbeck, mais sans véritablement y arriver. Rosenthal (voir [10, exemple 5.4 pages 69-71]) s'intéresse par exemple au cas $U(x) = x^2$ (cas gaussien), par ses méthodes, il obtient une borne du trou spectral sur $[-M, M]$ qui explose lorsque M tend vers l'infini. La difficulté provient essentiellement du caractère non compact de $\mathbb{R}$ (ou de l'intervalle $[a,\infty[$).

Aussi, nous allons chercher, par discrétisation, à nous ramener à un noyau sur $\mathbb{N}$ où les résultats sont plus nombreux. Le noyau sur $[a,\infty[$ (et donc le trou spectral correspondant sur $[a,\infty[$) sera vu comme une limite en un certain sens de noyaux sur $\mathbb{N}$. L'idée directrice du choix de la discrétisation est principalement le souci de simplifier les transitions tout en gardant des comportements proches du noyau initial. Proche étant pris au sens où les trous spectraux sont comparables.

2.1 Réduction du problème

On effectue une discrétisation non homogène de $[a,\infty[$ *:*

Par convexité et par construction de a, il existe une suite croissante $(x_k)_{k\in\mathbb{N}}$ de $[a,\infty[$, définie par

$$x_0 = a \quad \text{et} \quad U(x_{k+1}) = U(x_k) + \delta \quad \text{où} \quad \delta \in]0,1[\quad \text{est fixé.}$$

Par convexité à nouveau, $\delta_k \stackrel{\text{déf.}}{=} x_{k+1} - x_k$ est une suite décroissante et $x_k \stackrel{k\to\infty}{\longrightarrow} \infty$.

On introduit alors

$$\text{la tribu} \quad \mathcal{A}_\delta = \sigma(I_k \stackrel{\text{déf.}}{=} [x_k, x_{k+1}[\,, k \in \mathbb{N}) ,$$

$$\text{la mesure} \quad \tilde{\mu}_\delta(i) = \mu_1(I_i) , \text{ pour } i \in \mathbb{N} ,$$

(3) $$\text{et la probabilité de transition} \quad \tilde{P}_\delta(i,j) = \frac{1}{\tilde{\mu}_\delta(i)} \int_{I_i\times I_j} P_1(x,dy)\mu_1(dx) .$$

On se ramène ainsi à $\mathbb{N}$, car en posant pour $\delta > 0$,

(4) $$\tilde{\lambda}_\delta \stackrel{\text{déf.}}{=} \inf_{f\in\mathbb{L}^2(\tilde{\mu}_\delta)\setminus\mathrm{Vect}(\mathbf{I})} \frac{\tilde{\mu}_\delta(f(\mathrm{Id}-\tilde{P}_\delta)f)}{\tilde{\mu}_\delta((f-\tilde{\mu}_\delta(f))^2)}$$

il est bien connu que $\lim_{n\to\infty} \tilde{\lambda}_{2^{-n}} = \lambda_1$ (ceci découle du fait que les tribus $\mathcal{A}_{2^{-n}}$ convergent en croissant pour $n \in \mathbb{N}$ grand, vers la tribu borélienne de $[a,+\infty[$, pour plus de détails voir par exemple [7]).

En vue d'estimer $\tilde{\lambda}_\delta$, cherchons à maîtriser $\tilde{\mu}_\delta$ et $\tilde{P}_\delta$:
Par construction de $(x_k)_{k\in\mathbb{N}}$ et par définition de δ_k,

$$K_1^{-1}(x_{k+1}-x_k)e^{-U(x_{k+1})} \;\leq\; \tilde{\mu}_\delta(k) = K_1^{-1}\int_{x_k}^{x_{k+1}} e^{-U(x)}dx \;\leq\; K_1^{-1}(x_{k+1}-x_k)e^{-U(x_k)}$$

d'où

$$K_1^{-1}e^{-1}\delta_k e^{-U(x_k)} \;\leq\; \tilde{\mu}_\delta(k) \leq K_1^{-1}\delta_k e^{-U(x_k)} \tag{5}$$

On pose alors

$$\boxed{\hat{\mu}_\delta(k) = Z_\delta^{-1}\delta_k e^{-U(x_k)}}$$

où Z_δ est la constante de renormalisation (qui a priori dépend très peu de $\delta > 0$, car elle vérifie $K_1 \leq Z \leq K_1 e$), et on note que $\hat{\mu}_\delta(k)\#\tilde{\mu}_\delta(k)$, où $\#$ signifie un encadrement à des constantes indépendantes du niveau d'approximation δ près. Par un calcul analogue, on montre que $\tilde{P}_\delta(i,j)\#\delta_j e^{-(U(x_j)-U(x_i))_+}$, pour i et j vérifiant $|i-j| \leq \lfloor \frac{1}{2\delta_{i\wedge j}} - 1 \rfloor$, on pose donc

$$\boxed{\hat{P}_\delta(i,j) = \begin{cases} \delta_j e^{-(U(x_j)-U(x_i))_+}, & \text{si } i \text{ et } j \text{ vérifient } 0 < |i-j| \leq \lfloor \frac{1}{2\delta_{i\wedge j}} - 1 \rfloor \\ 1 - \sum_{k\neq i} \hat{P}_\delta(i,k), & \text{si } i = j \\ 0 & \text{sinon} \end{cases}}$$

où $\lfloor\ \rfloor$ désigne la partie entière.

Par décroissance de δ_k, la condition $|i-j| \leq \lfloor \frac{1}{2\delta_{i\wedge j}} - 1 \rfloor$ assure que $x_{j+1} - x_i \leq \frac{1}{2}$ et $x_{i+1} - x_j \leq \frac{1}{2}$. Ce qui permet de réaliser plus précisément que pour tous $i \neq j \in \mathbb{N}$, on a $\tilde{P}_\delta(i,j) \geq e^{-2}\hat{P}_\delta(i,j)$. Tenant compte de (5) et de résultats généraux de comparaison (voir par exemple [11]), on obtient que

$$\tilde{\lambda}_\delta \geq e^{-3}\hat{\lambda}_\delta$$

où

$$\hat{\lambda}_\delta \stackrel{\text{déf.}}{=} \inf_{f\in\mathbb{L}^2(\hat{\mu}_\delta)\setminus\text{Vect}(\mathbb{1})} \frac{\hat{\mu}_\delta(f(\text{Id} - \hat{P}_\delta)f)}{\hat{\mu}_\delta(f - \hat{\mu}_\delta(f))^2}$$

Ce dernier résultat et (4) montrent que le trou spectral λ_1 sera strictement positif si nous arrivons à minorer $\hat{\lambda}_\delta$ indépendamment de δ (c'est l'objet de la section 2.2).

Afin de simplifier les notations, introduisons les fonctions $j_{\min}$ et $j_{\max}$ de $\mathbb{N}$ dans $\mathbb{N}$ définies par :

$$\text{pour tout } i,\ \ \hat{P}_\delta(i,j) \neq 0 \text{ si et seulement si } j \in \{j_{\min}(i), \ldots, j_{\max}(i)\}$$

On vérifie facilement que $i - j_{\min}(i)$ est croissante et $j_{\min}(j_{\max}(i)) = j_{\max}(j_{\min}(i)) = i$. On remarque enfin qu'on connaît explicitement $j_{\max}(i)$ car $j_{\max}(i) - i = \lfloor \frac{1}{2\delta_i} - 1 \rfloor$ et que $j_{\min}(i)$ n'est défini qu'implicitement.

Au cours de la démonstration, nous aurons besoin des résultats suivants relatifs à ces fonctions :

Lemme 1 *Si C désigne la constante définie dans l'hypothèse (H), pour δ assez petit et pour tout i,*

$$j_{\max}(i) - i - 1 \leq 2C(i - j_{\min}(i)).$$

Lemme 2 *Pour δ assez petit et pour tout i,*

$$\frac{1}{i - j_{\min}(i)} \leq 4C\delta_i.$$

Demonstration du lemme 1 :
Soit $j = j_{\min}(i)$, d'après le théorème des accroissements finis, il existe $x_i' \in [x_i, x_{i+1}]$ et $x_j' \in [x_j, x_{j+1}]$ tels que

$$\delta_i = \frac{\delta}{U'(x_i')} \quad \text{et} \quad \delta_j = \frac{\delta}{U'(x_j')}.$$

D'après l'hypothèse (H) on a

$$\begin{aligned} U'(x_i') &\leq CU'(x_j') \\ \Rightarrow \quad \frac{1}{2\delta_i} &\leq C\frac{1}{2\delta_j}\,. \end{aligned}$$

Par ailleurs $U' \geq \alpha$ sur $[a, \infty[$ d'après (1), d'où

$$\frac{1}{2\delta_j} \geq \frac{\alpha}{2\delta} > 4 \quad \text{pour } \delta \text{ petit.}$$

Ainsi

$$\begin{aligned} \frac{1}{2\delta_i} &\leq 2C(\frac{1}{2\delta_j} - 2) \\ \Rightarrow \quad \frac{1}{2\delta_i} - 1 - 1 &\leq 2C(\frac{1}{2\delta_j} - 1 - 1) \\ \Rightarrow \quad \lfloor\frac{1}{2\delta_i} - 1\rfloor - 1 &\leq 2C\lfloor\frac{1}{2\delta_j} - 1\rfloor \end{aligned}$$

ce qui donne le résultat puisque par définition

$$\lfloor\frac{1}{2\delta_i} - 1\rfloor = j_{\max}(i) - i \quad \text{et} \quad \lfloor\frac{1}{2\delta_j} - 1\rfloor = i - j_{\min}(i)\,.$$

Démonstration du lemme 2 :
On pose ici aussi $j = j_{\min}(i)$.
D'après la démonstration précédente on a le résultat intermédiaire

$$\frac{1}{2\delta_i} \leq 2C(\frac{1}{2\delta_j} - 2)$$

donc

$$\frac{1}{2\delta_i} \leq 2C\lfloor\frac{1}{2\delta_j} - 1\rfloor$$

il vient

$$\begin{aligned} \frac{1}{i - j_{\min}(i)} = \lfloor\frac{1}{2\delta_j} - 1\rfloor^{-1} &\leq 2C\frac{1}{\frac{1}{2\delta_i}} \\ &\leq 4C\delta_i\,. \end{aligned}$$

2.2 Minoration de $\hat{\lambda}_\delta$

Pour montrer que $\hat{\lambda}_\delta$ est minorée par une constante indépendante de δ, on va utiliser un résultat de Sinclair (voir [12, section 4]).

Sur $\mathbb{N}$ on construit un graphe où les sommets sont les points et $e = (u, v)$ est une arête si et seulement si $\hat{\mu}_\delta(u)\hat{P}_\delta(u,v) > 0$. Pour tout x différent de y, on note Γ_{xy} l'ensemble des chemins joignant x à y. Sur Γ_{xy}, on choisit une probabilité ϕ_{xy}, $\phi_{xy}(\gamma_{xy})$ représente la probabilité de choisir le chemin γ_{xy} parmi l'ensemble des chemins joignant x à y. On notera souvent γ au lieu de γ_{xy} pour ne pas alourdir les notations.

Sous cette construction, la borne de Sinclair est

$$\eta_\delta(\phi) = \sup_{e=(u,v)} \eta_\delta(\phi)(e) = \sup_{e=(u,v)} \frac{1}{\hat{\mu}_\delta(u)\hat{P}_\delta(u,v)} \sum_{x,y\,:\,\gamma_{xy}\ni e} \hat{\mu}_\delta(x)\hat{\mu}_\delta(y)\phi_{xy}(\gamma) \,.$$

Le théorème s'énonce alors :

$$\hat{\lambda}_\delta \geq \frac{1}{8\eta_\delta^2(\phi)}$$

La clef de la démonstration repose sur le choix de ϕ :

Soit $x < y$ (si $x > y$ on prendra le chemin construit à partir de celui allant de y à x à l'envers, ce qui est permis par réversibilité). Partant de y, on choisit y_1 tel que $y - j_{\min}(y) \leq y_1 < y$ avec probabilité $p_y \overset{\text{déf.}}{=} \frac{1}{y - j_{\min}(y)}$ puis on choisit y_2 de manière équiprobable, $y_1 - j_{\min}(y_1) \leq y_2 < y_1$, avec probabilité p_{y_1} et ainsi de suite jusqu'à dépasser x, $y_{n+1} \leq x < y_n$, on pose alors $y_{n+1} = x$. Le chemin ainsi construit est formé des arêtes $(x, y_n), (y_n, y_{n-1}), \ldots, (y_1, y)$.

On va majorer explicitement $\eta_\delta(\phi)$. Pour cela, on peut par construction se restreindre aux arêtes $e = (u, v)$ avec $u < v$. Soit $e = (u, v)$ une telle arête fixée $(u < v)$, on doit calculer en premier lieu

$$\text{(6)} \qquad \frac{1}{\hat{\mu}_\delta(u)\hat{P}_\delta(u,v)} = \frac{1}{\delta_u \delta_v} e^{U(x_v)}$$

et en deuxième lieu, la somme

$$\sum_{x,y\,:\,\gamma_{xy}\ni e} \hat{\mu}_\delta(x)\hat{\mu}_\delta(y)\phi_{xy}(\gamma) = \sum_{x \leq u} \hat{\mu}_\delta(x) \sum_{y \geq v} \hat{\mu}_\delta(y) \sum_{\gamma_{xy}\ni e} \phi_{xy}(\gamma)$$

Le dernier terme $\sum_{\gamma_{xy}\ni e} \phi_{xy}(\gamma)$ est à comprendre comme la somme, à x et y fixés, sur tous les chemins $\gamma \in \Gamma_{xy}$ tels que $\gamma \ni e$.

D'après le choix de ϕ, on doit distinguer les cas $x < u$ et $x = u$:

On pose

$$\begin{aligned}
\sum_{x,y\,:\,\gamma_{xy}\ni e} \hat{\mu}_\delta(x)\hat{\mu}_\delta(y)\phi_{xy}(\gamma) = \quad & \sum_{x=u} \hat{\mu}_\delta(x) \sum_{y \geq v} \hat{\mu}_\delta(y) \sum_{\gamma_{xy}\ni e} \phi_{xy}(\gamma) \\
& + \sum_{x<u} \hat{\mu}_\delta(x) \sum_{y \geq v} \hat{\mu}_\delta(y) \sum_{\gamma_{xy}\ni e} \phi_{xy}(\gamma) \\
= \quad & Q_1 + Q_2 + Q_3 + Q_4
\end{aligned}$$

avec

$$
\begin{aligned}
Q_1 &= \hat{\mu}_\delta(u)\hat{\mu}_\delta(v)\sum_{\gamma_{uv}\ni e}\phi_{uv}(\gamma)\\
Q_2 &= \hat{\mu}_\delta(u)\sum_{y>v}\hat{\mu}_\delta(y)\sum_{\gamma_{uy}\ni e}\phi_{uy}(\gamma)\\
Q_3 &= \hat{\mu}_\delta(v)\sum_{x<u}\hat{\mu}_\delta(x)\sum_{\gamma_{xv}\ni e}\phi_{xv}(\gamma)\\
Q_4 &= \sum_{x<u}\hat{\mu}_\delta(x)\sum_{y>v}\hat{\mu}_\delta(y)\sum_{\gamma_{xy}\ni e}\phi_{xy}(\gamma)
\end{aligned}
$$

Considérons chacun des cas séparément :

Pour Q_4 :

On doit s'intéresser à $\sum_{\gamma_{xy}\ni e}\phi_{xy}(\gamma)$ qui représente, à $x < u$ et $y > v$ fixés, la probabilité globale (sous $\phi_{x,y}$) d'emprunter l'arête $e = (u, v)$, elle aussi préalablement fixée, en allant de y à x. Soit m le premier point par lequel le chemin passe entre v et $j_{\max}(v)$ (à partir de ce point, le chemin peut directement sauter en v). A partir de m, le chemin va passer par n points $x_1, x_2, \ldots, x_n$ puis par v et par u et enfin continuer vers x. Pour se rendre de m à x_1 puis de x_1 à $x_2, \ldots$, on a une probabilité $p_m, p_{x_1}, \ldots$ A chaque étape les probabilités vérifient par croissance de la fonction $\mathrm{id} - j_{\min}$, $p_m \le p_{x_1} \le \cdots \le p_v$.

Ainsi on a, à x et y fixés, en sommant sur les chemins γ joignant x à y

$$
\sum_{\gamma_{xy}\ni e}\phi_{xy}(\gamma) \le \sum_{n=0}^{m-v-1} C^n_{m-v-1}(p_v)^{n+1}p_v\ , \tag{7}
$$

où C^n_{m-v-1} est le nombre de choix de n points entre m et v, p_v^{n+1} est une majoration de la probabilité de passer par ces n points puis par v et où le dernier p_v représente la probabilité de choisir u une fois en v.

Or, d'après le lemme 1, $j_{\max}(v) - v - 1 \le \frac{2C}{p_v}$, donc

$$
\sum_{n=0}^{m-v-1} C^n_{m-v-1}(p_v)^{n+1}p_v = (1+p_v)^{m-v-1}(p_v)^2 \le (1+p_v)^{j_{\max}(v)-v-1}(p_v)^2 \le e^{2C}(p_v)^2 \tag{8}
$$

Pour finir d'estimer Q_4, il faut calculer

$$
\sum_{y>v}\hat{\mu}_\delta(y) = \sum_{y>v}\delta_y e^{-U(x_y)} \le e\int_{x_{v+1}}^{\infty} e^{-U(x)}dx\ .
$$

(l'inégalité provient de (5)).

Il vient

$$
\begin{aligned}
\sum_{y>v}\hat{\mu}_\delta(y) &\le e\,e^{-U(x_{v+1})}\int_0^\infty e^{-(U(x+x_{v+1})-U(x_v))}dx\\
&\le e\,\frac{e^{-U(x_v)}}{\alpha} \qquad (9)
\end{aligned}
$$

car par convexité et par l'hypothèse (1),

$$U(x+x_{v+1})-U(x_{v+1})\geq\frac{1}{\alpha}(x+x_{v+1}-x_{v+1})\quad U'\geq\frac{1}{\alpha}\quad\text{et}\quad U(x_{v+1})\geq U(x_v).$$

En regroupant les calculs de (7), (8) et (9) et en majorant grossièrement $\sum_{x<u}\hat{\mu}_\delta(x)$ par $\sum_{x\in\mathbb{N}}\hat{\mu}_\delta(x)=1$, on obtient

(10) $$Q_4\leq e^{-U(x_v)}\frac{e}{\alpha}e^{2C}{p_v}^2$$

Pour Q_3 :

On majore le terme $\sum_{\gamma_{xv}\ni e}\phi_{xv}(\gamma)$ par p_v car dans Q_3, on a $x<u$, le chemin doit directement sauter de v en u (avec probabilité p_v) puis se rendre en x. De même que précédemment on majore grossièrement $\sum_{x<u}\hat{\mu}_\delta(x)$ par 1, il vient

(11) $$Q_3\leq\delta_v e^{-U(x_v)}p_v$$

Pour Q_2 :

On doit s'intéresser à $\sum_{\gamma_{uy}\ni e}\phi_{uy}(\gamma)$. Ici, $x=u$, le chemin se rend de y à v puis saute directement de v à $u=x$. En reprenant les calculs du cas Q_4, les majorations (7) et (8) restent valables sinon le dernier p_v qui ici n'apparaît plus :

$$\sum_{\gamma_{uy}\ni e}\phi_{uy}(\gamma)\leq\sum_{n=0}^{m-v-1}C^n_{m-v-1}(p_v)^{n+1}\leq e^{2C}p_v\,.$$

Ce dernier calcul et (9) nous donne

(12) $$Q_2\leq\delta_u e^{-U(x_u)}\frac{e}{\alpha}e^{-U(x_v)}e^{2C}p_v$$

Pour Q_1 :

On a

$$\sum_{\gamma_{uv}\ni e}\phi_{uv}(\gamma)=\phi_{uv}(\gamma)\leq 1$$

où γ est ici l'unique chemin joignant u et v en passant directement par l'arête $e=(u,v)$.

Ainsi

(13) $$Q_1\leq\hat{\mu}_\delta(u)\hat{\mu}_\delta(v)=\delta_u\delta_v e^{-U(x_u)}e^{-U(x_v)}$$

Avant de regrouper les calculs, remarquons que d'après le lemme 2,

$$p_v=\frac{1}{v-j_{\min}(v)}\leq 4C\delta_v$$

Si on pose $K'=\sup_{x\in\mathbb{R}}(e^{-U(x)})$ ($K'<\infty$ par hypothèses sur U), d'après (10), (11), (12), (13) et (6), on a

(14) $$\eta_\delta(\phi)(e)\leq K'+\frac{4CK'e}{\alpha}e^{2C}+4C+\frac{e}{\alpha}e^{2C}(4C)^2\,.$$

où on a utilisé $\frac{\delta_v}{\delta_u}\leq 1$ par croissance de δ_k.

Ainsi, $\eta_\delta(\phi)(e)$ est bornée par une constante indépendante de u, v donc de $e = (u, v)$ et de δ, en prenant le sup sur les arêtes et en passant à la limite en δ, on a l'existence du trou spectral sur $[a, \infty[$,

$$\lambda_1 > 0.$$

On remarque que la constante trouvée dans (14) tend vers l'infini lorsque C tend vers l'infini, ainsi nos résultats ne sont pas généralisables par cette démonstration à des densités par exemple de la forme $e^{-e^{x^2}}$ où $U(x) = e^{x^2}$ ne vérifie pas l'hypothèse (H). Cependant, l'intuition nous laisse penser que plus U croit rapidement vers l'infini en l'infini et plus le trou spectral a des chances d'être grand, aussi, l'hypothèse (H) n'est-elle pas très naturelle.

3 Existence du trou spectral sur $[-a-1, a+1]$

Comme précédemment, on restreint le noyau P et la mesure μ à $[-a-1, a+1]$, pour obtenir le noyau

$$P_2(x, dy) = f_x(y)\, \mathbb{1}_{[-a-1,a+1]}(y)\, dy + a_{2,x}\delta_x(dy)$$

où

$$a_{2,x} = 1 - \int_{-a-1}^{a+1} f_x(y)\, dy$$

et la mesure

$$\mu_2(dx) = K_2^{-1} e^{-U(x)}\, dx \quad \text{avec } K_2 = \int_{-a-1}^{a+1} e^{-U(x)}\, dx\ .$$

Pour montrer que le noyau P_2, réversible par rapport à la mesure μ_2 sur $[-a-1, a+1]$, admet un trou spectral non nul, on va utiliser des arguments très simples de comparaison et d'analyse de Fourier. On aurait également pu reprendre une méthode due à Rosenthal ([10, théorème 5]), qui généralise sur un ensemble continu la borne de Sinclair exposée dans la section précédente.

Soit $f \in \mathbb{L}^2(\mu)$, on cherche donc à minorer

$$\begin{aligned}
&\frac{\int_{[-1-a,1+a]^2} \mu_2(dx)P_2(x,dy)(f(y)-f(x))^2}{\int_{[-1-a,1+a]} \mu_2(dx)(f(x)-\mu_2(f))^2} \\
&\geq \frac{\int_{[-1-a,1+a]^2} \mu_2(dx)P_2(x,dy)(f(y)-f(x))^2}{\int_{[-1-a,1+a]} \mu_2(dx)(f(x)-\int_{[-1-a,1+a]} f(y)\,dy/(2(1+a)))^2} \\
&\geq \exp(-A)\frac{\int_{[-1-a,1+a]^2} \mathbf{1}_{[x-1/2,x+1/2]}(y)(f(y)-f(y))^2\,dxdy}{\int_{[-1-a,1+a]} \left(f(x)-\int_{[-1-a,1+a]} f(y)\,dy/(2(1+a))\right)^2 dx}
\end{aligned}$$

avec $A = \max_{[-1-a,1+a]} U - \min_{[-1-a,1+a]} U$.

Soit $\widetilde{f}$ la fonction définie sur $T = \mathbb{R}/\mathbb{Z}$ (aussi identifié avec $[0, 1[$) par

$$\forall\, 0 \leq t < 1, \qquad \widetilde{f}(t) \;=\; \begin{cases} f(-1-a+4t(a+1)) & , \text{si } 0 \leq t \leq 1/2 \\ \widetilde{f}(1-t) & , \text{si } 1/2 < t < 1 \end{cases}$$

En considérant séparément les interactions entre des couples de points de T de part et d'autre de $1/2$ et de 0 (qui est identifié avec 1 dans T), il apparaît sans difficulté que

$$\begin{aligned}&\int_{[-1-a,1+a]^2} \mathbf{I}_{[x-1/2,x+1/2]}(y)(f(y)-f(x))^2\,dxdy\\ &\geq\ 4(1+a)^2\int_{T^2}\mathbf{I}_{[x-1/(8(1+a)),x+1/(8(1+a))]}(y)(\widetilde{f}(y)-\widetilde{f}(x))^2\,\widetilde{\mu}(dx)\widetilde{\mu}(dy)\end{aligned}$$

où $\widetilde{\mu}$ est la mesure de Lebesgue sur T, et

$$\begin{aligned}&\int_{[-1-a,1+a]}\left(f(x)-\int_{[-1-a,1+a]}f(y)\,dy/(2(1+a))\right)^2 dx\\ &=\ 4(1+a)\int_T(\widetilde{f}(x)-\widetilde{\mu}(f))^2\,\widetilde{\mu}(dx)\end{aligned}$$

Ainsi le trou spectral de P_2 dans $\mathbb{L}^2(\mu_2)$ est minoré par celui de $\widetilde{P}$ dans $\mathbb{L}^2(\widetilde{\mu})$, multiplié par $\exp(-A)/8$, où $\widetilde{P}$ est le noyau de convolution défini sur le tore T par

$$\forall\ x\in T,\qquad \widetilde{P}(x,dy)\ =\ 4(1+a)\mathbf{I}_{[x-1/(8(1+a)),x+1/(8(1+a))]}(y)\,\widetilde{\mu}(dy)$$

le segment ci-dessus étant toujours compris modulo 1.

Or il est facile d'obtenir par une décomposition en série de Fourier (voir [7]) une minoration du trou spectral $\lambda(\widetilde{P})$ de $\widetilde{P}$

$$\lambda(\widetilde{P})\ \geq\ 1-\frac{|\exp(i4\pi/(8(1+a))-1|}{4\pi/(8(1+a))}>0$$

d'où le résultat annoncé.

4 Existence du trou spectral sur $\mathbb{R}$ ou comment recoller les morceaux

Afin de montrer l'existence du trou spectral sur $\mathbb{R}$, on va chercher à se ramener à 5 points. On introduit pour cela les intervalles

$$I_1=]-\infty,-a],\ \ I_2=[-a-1,a+1],\ \ I_3=[a,\infty[,$$

et une partition de $\mathbb{R}$ en 5 ensembles

$$A_1=]-\infty,-a-1],\ \ A_2=]-a-1,-a],\ \ A_3=]-a,a[,\ \ A_4=[a,a+1[,\ \ A_5=[a+1,\infty[$$

qui seront représentés par les 5 points auxquels on veut se ramener.

Sur chaque I_i on dispose du noyau $P_i(x,dy)=f_x(y)\mathbb{I}_{I_i}(y)dy+a_{i,x}\delta_x(dy)$ réversible par rapport à la mesure $\mu_i(dx)=K_i^{-1}\mu(dx)$ avec $K_i=\int_{I_i}\mu(dx)$. D'après les sections précédentes, sur chaque I_i, le noyau P_i admet un trou spectral $\lambda_i>0$ sous μ_i, on a donc pour tout $i=1$ à 3, les inégalités

$$(15)\qquad\qquad \mu_i(f(\mathrm{Id}-P_i)f)\geq\lambda_i\mu_i(f-\mu_i(f))^2\ .$$

Le recollement des morceaux s'effectue en plusieurs étapes. La première d'entre elles consiste à construire une nouvelle forme à partir de laquelle on se ramènera à 5 points (deuxième étape). On finit la démonstration en montrant que notre nouveau noyau sur 5 points admet un trou spectral, on utilisera pour cela un résultat de Cheeger.

Première étape : construction d'une nouvelle forme

Il est bien connu que la forme de Dirichlet $\mathcal{E}_{P,\mu}(f,f) \stackrel{\text{déf.}}{=} \mu(f(\mathrm{Id}-P)f)$ s'écrit sous la forme $\mathcal{E}_{P,\mu}(f,f) = \frac{1}{2}\int_{\mathbb{R}} d\mu(x) \int_{\mathbb{R}} P(x,dy)(f(x)-f(y))^2$, on montre alors facilement

$$\begin{aligned}
\mathcal{E}(f,f) &= \frac{1}{2}\int_{\mathbb{R}} d\mu(x) \int_{\mathbb{R}} P(x,dy)(f(x)-f(y))^2 \\
&= \frac{1}{2}\sum_{i=1}^{5} \int_{A_i} d\mu(x) \int_{\mathbb{R}} P(x,dy)(f(x)-f(y))^2 \\
&\geq \frac{1}{4}\sum_{i=1}^{3} \int_{I_i\times I_i} d\mu(x)P(x,dy)(f(x)-f(y))^2 \\
&= \frac{1}{4}\sum_{i=1}^{3} \int_{I_i\times I_i} K_i d\mu_i(x)P_i(x,dy)(f(x)-f(y))^2 \\
&= \frac{1}{2}\sum_{i=1}^{3} K_i \mathcal{E}_{P_i,\mu_i}(f_i,f_i) \quad \text{où } f_i = f\mathbb{1}_{I_i} \\
&\geq \frac{1}{2}\sum_{i=1}^{3} K_i\lambda_i\mu_i((f_i-\mu_i(f_i))^2)
\end{aligned}$$

Pour la troisième inégalité, on compte au plus deux fois chaque passage de A_i à A_j $(j \neq i)$, pour la dernière inégalité, on utilise l'inégalité (15) trois fois.

On construit alors la nouvelle forme

$$\tilde{\mathcal{E}}(f,f) \stackrel{\text{déf.}}{=} \frac{1}{2}\sum_{i=1}^{3} K_i\lambda_i\mu_i(f_i-\mu_i(f_i))^2 .$$

Il existe un générateur de sauts L, vu comme un noyau d'intensités de transition $L(x,dy)$, tel que

$$\tilde{\mathcal{E}}(f,f) = -\mu(fLf).$$

Un calcul élémentaire où l'on décompose $\tilde{\mathcal{E}}(f,f)$ en $\sum_{i,j=1}^{5}\int_{A_i}\int_{A_j}\mu(dx)L(x,dy)f(y)$, et une identification nous donnent

$$\begin{aligned}
&L(x,\mathbb{1}_{\mathbb{R}\setminus\{x\}}(y)dy) = \\
&\quad \mathbb{1}_{A_1}(x)\frac{1}{2}\lambda_1\mu_1(dy) + \mathbb{1}_{A_2}(x)\left[\frac{1}{2}\lambda_1\mu_1(dy)\mathbb{1}_{A_1}(y) + \left[\frac{1}{2}\lambda_1\mu_1(dy)+\frac{1}{2}\lambda_2\mu_2(dy)\right]\mathbb{1}_{A_2}(y)\right. \\
&\quad \left.+\frac{1}{2}\lambda_2\mu_2(dy)\mathbb{1}_{A_3}(y)\right] + \mathbb{1}_{A_3}(x)\frac{1}{2}\lambda_2\mu_2(dy) + \mathbb{1}_{A_4}(x)\left[\frac{1}{2}\lambda_2\mu_2(dy)\mathbb{1}_{A_3}(y)\right. \\
&\quad \left.+\left[\frac{1}{2}\lambda_2\mu_2(dy)+\frac{1}{2}\lambda_3\mu_3(dy)\right]\mathbb{1}_{A_4}(y) + \frac{1}{2}\lambda_3\mu_3(dy)\mathbb{1}_{A_5}(y)\right] + \mathbb{1}_{A_5}(x)\frac{1}{2}\lambda_3\mu_3(dy)
\end{aligned}$$

et

$$L(x,\{x\}) = -L(x,\mathbb{R}\backslash\{x\})$$

Le générateur s'écrit

$$L(x,dy) = L(x,\mathbb{I}_{\mathbb{R}\backslash\{x\}}(y)dy) + L(x,\{x\})\delta_x(dy)$$

En résumé, on a montré

$$\mathcal{E}_{P,\mu}(f,f) \geq \tilde{\mathcal{E}}(f,f)\,. \tag{16}$$

On va désormais travailler avec cette nouvelle forme.

Deuxième étape : on se ramène à 5 points

Soit maintenant $\mathcal{F} = \sigma(A_1, A_2, \ldots, A_5)$ et $F = \mu(f|\mathcal{F})$ (si $x \in A_i$, $F(x) = \frac{\mu(f\mathbf{1}_{A_i})}{\mu(A_i)}$). Notons également, pour $1 \leq i \leq 5$, $F_i = \mathbb{I}_{A_i}(f - F)$ et $l_i = -L(x,\{x\}) > 0$, qui est une quantité ne dépendant pas du choix du point $x \in A_i$.

On calcule alors que

$$\widetilde{\mathcal{E}}(f,f) \;=\; \widetilde{\mathcal{E}}(F,F) + \sum_{1\leq i\leq 5} l_i\mu(F_i^2)$$

et que

$$\mu((f-\mu(f))^2) \;=\; \mu((F-\mu(F))^2) + \sum_{1\leq i\leq 5} \mu(F_i^2)$$

Il apparaît ainsi que

$$\lambda \geq \inf_{f\in\mathbb{L}^2(\mu)\backslash\mathrm{Vect}(\mathbf{1})} \frac{\widetilde{\mathcal{E}}(f,f)}{\mu(f-\mu(f))^2} \;=\; \lambda_{\check{L}} \wedge \min_{1\leq i\leq 5} l_i$$

où

$$\lambda_{\check{L}} \stackrel{\text{déf.}}{=} \inf_{f\in\mathbb{L}^2(\mathcal{F},\mu)\backslash\mathrm{Vect}(\mathbf{1})} \frac{\widetilde{\mathcal{E}}(f,f)}{\mu(f-\mu(f))^2}$$

Ceci nous ramène au calcul du trou specral pour un processus irréductible à valeurs dans un ensemble à 5 points (d'autant plus qu'en considérant l'indicatrice de A_i, pour $1 \leq i \leq 5$, on se persuade facilement que $l_i \geq \lambda_{\check{L}}$), et on va rappeler dans l'étape suivante une manière de minorer cette quantité par un nombre strictement positif, ce qui nous permettra de conclure.

Troisième étape : existence du trou spectral sur les 5 points

Sur les 5 points $A_1, A_2, \ldots, A_5$, on dispose d'une matrice $\check{L}$, 5×5, tridiagonale, symétrique dans $\mathbb{L}^2(\check{\mu})$ où $\check{\mu}$ est la mesure définie sur A_i par $\check{\mu}(A_i) = \mu(A_i)$ pour tout $i = 1$ à 5 . Les coefficients de cette matrice nous sont donnés explicitement par L.

Il est bien connu que ce noyau d'intensités irréductible admet un trou spectral. Pour l'évaluer, on a choisit d'utiliser la constante de Cheeger (voir [3], [5], ou [8])

$$I(\check{L},\check{\mu}) \stackrel{\text{déf.}}{=} \inf_{\substack{0<\check{\mu}(A)\leq\frac{1}{2}\\ A \text{ connexe}\\ A^c \text{ connexe}}} \frac{\check{\mu}(\mathbb{I}_A\check{L}\mathbb{I}_{A^c})}{\check{\mu}(A)}$$

où la connexité est celle relative à la matrice d'incidence de $\check{L}$. On sait que dans le cas d'un espace d'état à 5 points, le trou spectral $\lambda_{\check{L}}$ du noyau $\check{L}$ associé à la mesure $\check{\mu}$ est directement lié à la constante de Cheeger par la relation

$$2I(\check{L},\check{\mu}) \geq \lambda_{\check{L}} \geq \frac{I(\check{L},\check{\mu})}{5}$$

La connexité de A et A^c réduit l'étude à

$$A = [A_1, A_i] \text{ et } A^c = [A_{i+1}, A_5] \quad \text{ou} \quad A^c = [A_1, A_i] \text{ et } A = [A_{i+1}, A_5]$$

On en déduit immédiatement

$$I(\check{L},\check{\mu}) \geq \min_{i,j\,:\,\check{L}(A_i,A_j)>0} \check{L}(A_i,A_j) = \min_{i,j\,:\,\mu(\mathbf{I}_{A_i}L(\mathbf{I}_{A_j}))>0} \frac{\mu(\mathbf{I}_{A_i}L(\mathbf{I}_{A_j}))}{\mu(A_i)}$$

5 Temps discret

Comme on l'a déjà mentionné, le trou spectral λ fournit une vitesse de convergence vers l'équilibre μ du processus de Markov de sauts associé au semi-groupe qui admet $P - \mathrm{Id}$ comme générateur. Mais on peut aussi s'intéresser en temps discret à la chaîne de Markov qui admet P pour noyau de probabilités de transition. Il est bien connu (cf. par exemple [4]) que pour une telle chaîne, une vitesse de convergence exponentiellement rapide en le temps (dans $\mathbb{L}^2(\mu)$) est notamment impliquée par l'existence d'un trou spectral $\widetilde{\lambda}$ non nul pour $\mathrm{Id} - PP^*$ (où P^* est l'adjoint de P dans $\mathbb{L}^2(\mu)$), cette constante permettant d'expliciter une minoration du taux : plus précisément, pour tout $f \in \mathbb{L}^2(\mu)$, on a

$$\|P^n(f) - \mu(f)\|_2 \leq \rho^n \sqrt{\mu(f^2) - \mu(f)^2}$$

avec $\rho = \sqrt{1-\widetilde{\lambda}}$ et où $\|\cdot\|_2$ désigne la norme dans $\mathbb{L}^2(\mu)$.

Par réversibilité, on a ici $P^* = P$, et $\widetilde{\lambda}$ est donc le trou spectral de $\mathrm{Id} - P^2$. Sans vouloir non plus présenter une minoration quantitative, on va se contenter d'indiquer le résultat suivant :

Proposition 3 *Sous les hypothèses précédentes, on est assuré que* $\widetilde{\lambda} > 0$.

En effet, remarquons que pour $x \geq a$, on a

$$\begin{aligned} a_x &= 1 - \int_{[x-1/2,x+1/2]} \exp(-(U(y)-U(x))_+)\,dy \\ &\geq \frac{1}{2} - \int_{[x,x+1/2]} \exp(-\alpha(y-x))\,dy \\ &= \frac{1}{\alpha}(\exp(-\alpha/2) - 1 + \alpha/2) \end{aligned}$$

De même pour $x \leq -a$, ce qui nous conduit à poser

$$\eta \stackrel{\text{déf.}}{=} \frac{1}{\alpha}(\exp(-\alpha/2) - 1 + \alpha/2) > 0$$

Comme dans la section 2, considérons R_1 la restriction du noyau P^2 à $[a, +\infty[$, et rappelons que P_1 et μ_1 désignent respectivement la restriction de P et la renormalisation de la restriction de μ à cet intervalle. Pour tout $f \in \mathbb{L}^2(\mu)$, on constate immédiatement que

$$\begin{aligned}\int \mu_1(x)R_1(x,dy)(f(y)-f(x))^2 &\geq \int \mu_1(x)P_1^2(x,dy)(f(y)-f(x))^2 \\ &\geq \eta \int \mu_1(x)P_1(x,dy)(f(y)-f(x))^2\end{aligned}$$

ainsi d'après les calculs de la section 2, le couple réversible (R_1, μ_1) admet un trou spectral strictement positif.

De la même manière, on traite la restriction de P^2 à $]-\infty, -a]$. Le fait que la restriction de P^2 à $[-1-a, 1+a]$ admette également un trou spectral se prouve comme dans la section 3, et on conclut de manière identique aux considérations de la section précédente, à l'existence d'un trou spectral $\widetilde{\lambda}$ strictement positif.

Voyons maintenant comment à partir de la stricte positivité de $\widetilde{\lambda}$, on peut obtenir les estimées indiquées à la fin de l'introduction :

Soit $x_0 \in \mathbb{R}$ fixé, et notons m la probabilité admettant la densité $\widetilde{f}_{x_0}(\cdot) \stackrel{\text{déf.}}{=} K f_{x_0}(\cdot)$ $\exp(U(\cdot))/(1-a_{x_0})$ (sous-entendu par rapport à μ).

Pour $n \geq 0$, la probabilité mP^n admet alors pour densité $P^{*n}(\widetilde{f}_{x_0}) = P^n(\widetilde{f}_{x_0})$.

Soit une fonction $g \in \mathbb{L}^2(\mu)$ telle que $\mu(g) = 0$, il apparaît que pour tout $n \geq 0$,

$$P^n(x_0, g) = \sum_{1\leq k\leq n} a_{x_0}^{n-k}(1-a_{x_0})(mP^{k-1})(g) + a_{x_0}^n g(x_0)$$

or d'après les considérations précédentes, pour $k \geq 1$,

$$\begin{aligned}(mP^{k-1})(g) &= \mu(P^{k-1}(\widetilde{f}_{x_0})g) \\ &= \mu[(P^{k-1}(\widetilde{f}_{x_0}) - \mu(\widetilde{f}_{x_0}))g] \\ &\leq \sqrt{\mu[(P^{k-1}(\widetilde{f}_{x_0}) - \mu(\widetilde{f}_{x_0}))^2]}\sqrt{\mu(g^2)} \\ &\leq \rho^{k-1}\sqrt{\mu(\widetilde{f}_{x_0}^2) - \mu(\widetilde{f}_{x_0})^2}\sqrt{\mu(g^2)} \\ &\leq \rho^{k-1}\left\|\widetilde{f}_{x_0}\right\|_2 \|g\|_2\end{aligned}$$

Notons par ailleurs que

$$\mu(\widetilde{f}_{x_0}^2) = \frac{K}{(1-a_{x_0})^2}\int_{[x_0-1/2,x_0+1/2]} \exp(-U(x) + 2U(y) - 2(U(y)-U(x))_+)\,dy$$

ainsi il apparaît facilement que pour $x_0 \geq 1 + a$,

$$\mu(\widetilde{f}_{x_0}^2) \leq \frac{K}{(1-a_{x_0})^2}\exp(U(x_0))$$

Il en découle qu'il existe une constante $A \geq K$ telle que

$$\forall\, x_0 \in \mathbb{R}, \qquad \mu(\widetilde{f}_{x_0}^2) \leq \frac{A}{(1-a_{x_0})^2}\exp(U(x_0))$$

d'où en fait

$$\begin{aligned} |P^n(x_0,g)| &\leq A\exp(U(x_0)) \sum_{1\leq k\leq n} a_{x_0}^{n-k}\rho^k \|g\|_2 + a_{x_0}^n g(x_0) \\ &\leq \begin{cases} A\exp(U(x_0))a_{x_0}\dfrac{a_{x_0}^n - \rho^n}{a_{x_0}-\rho}\|g\|_2 + a_{x_0}^n g(x_0) & \text{, si } \rho \neq a_{x_0} \\ (An\exp(U(x_0))\|g\|_2 + \rho g(x_0))\rho^n & \text{, si } \rho = a_{x_0} \end{cases} \end{aligned}$$

ce qui permet d'établir le résultat annoncé, en remplaçant g par $g - \mu(g)$, pour $g \in \mathbb{L}^2(\mu)$ quelconque.

Un autre avantage du trou spectral par rapport aux choix d'une fonction de Lyapounov est que le taux de convergence est plus explicite, ici $a_{x_0} \vee \rho \leq (1/2) \vee \rho$.

Ainsi par exemple avec $U : \mathbb{R} \ni x \mapsto x^2/2$, en itérant les noyaux P on finit par approximer la loi gaussienne standard relativement vite.

Références

[1] D. Bakry. L'hypercontractivité et son utilisation en théorie des semigroupes. In P. Bernard, editor, *Lectures on Probability Theory. Ecole d'Eté de Probabilités de Saint-Flour XXII-1992*, Lecture Notes in Mathematics 1581. Springer-Verlag, 1994.

[2] H.J Brascamp and E.H. Lieb. On extensions of the Brunn-Minkowski and Prékopa-Leindler theorems, including inequalities for log concave functions, and with an application to the diffusion equation. *Journal of Functional Analysis*, 22:366–389, 1976.

[3] Cheeger J., *A lower bound for smallest eigenvalue of the Laplacian*, Problems in Analysis, Symposium in honor of S. Bochner, Princeton University Press, (1970), 195-199.

[4] Fill J.A., *Eigenvalue bounds on convergence to stationarity for nonreversible Markov chains, with an application to the exclusion process*, The Annals of Applied Probability, (1991) 1 (1), 62-87.

[5] Lawler G. and Sokal A., *Bounds on the L^2 spectrum for Markov chains and Markov processes: a generalization of Cheeger inequality*, Transactions of the American Mathematical Society, (1988) 309 (2), 557-580.

[6] S.P. Meyn and R.L. Tweedie. *Markov Chains and Stochastic Stability*. Springer-Verlag, 1993.

[7] Miclo L., *Trous spectraux à basse température : un contre-exemple à un comportement asymptotique escompté*, Séminaire de Probabilités XXXII, Lecture Notes in Mathematics 1686, Springer-Verlag, (1998), 36-55.

[8] Miclo L., *Une variante de l'inégalité de Cheeger pour les chaînes de Markov finies*, ESAIM : P&S, URL : http://www.emath.fr/ps/, (1998) 2, 1-21.

[9] J. Neveu. *Bases mathématiques du calcul des probabilités*. Masson, 1970.

[10] Rosenthal J.S., *Markov chain convergence : from finite to infinite*, Stochastic Process and their applications, (1996) 62, 55-72.

[11] L. Saloff-Coste. Lectures on finite Markov chains. In P. Bernard, editor, *Lectures on Probability Theory and Statistics. Ecole d'Eté de Probabilités de Saint-Flour XXVI-1996*, Lecture Notes in Mathematics 1665. Springer-Verlag, Berlin, 1997.

[12] Sinclair A., *Improved bounds for mixing rates of Markov chains and multicommodity flow*, Combinatories, proba. comput., (1992) 1, 351-370.

Comportement asymptotique des fonctions harmoniques sur les arbres

Frédéric Mouton

Résumé. On considère ici une marche aléatoire sur un arbre infini ayant des probabilités de transition "raisonnables", c'est-à-dire bornées entre deux constantes strictement comprises entre 0 et 1/2. Pour une fonction harmonique sur cet arbre, on montre que les notions de convergence radiale, bornitude radiale, finitude de l'énergie radiale et les notions stochastiques correspondantes sont toutes équivalentes en presque tout point du bord à l'infini. La démonstration est inspirée de celle d'un résultat analogue de l'auteur sur les variétés riemanniennes de courbure négative pincée.

Abstract. Considering a random walk on an infinite tree, we suppose that transition probabilities are "reasonnable", *id est* that they are bounded between two constants taken in $(0, 1/2)$. It is shown that, for a given harmonic function on the tree, properties of radial convergence, radial boundedness, finiteness of radial energy and corresponding stochastic notions are all equivalent at almost each point of the geometric boundary. The idea of the proof comes from an analoguous result on Riemannian manifolds of pinched negative curvature due to the author.

Introduction

L'étude des propriétés radiales ou non-tangentielles des fonctions harmoniques remonte aux résultats de Fatou [11] du début du siècle : une fonction harmonique positive sur le disque unité de $\mathbf{R}^2$ admet en presque tout point du bord une limite radiale (resp. non-tangentielle). On s'est ensuite aperçu progressivement de la grande généralité de cet énoncé puisqu'il reste valide sur des espaces très divers : demi-espace euclidien, ouverts lipschitziens, espace hyperbolique, variétés riemanniennes simplement connexes de courbure négative pincée [3, 1], groupe libre [10, 8], arbres [7], graphes hyperboliques au sens de Gromov [2], pour n'en citer que quelques uns.

Une autre question a été de savoir ce qu'on pouvait dire lorsque la fonction harmonique n'était plus supposée positive, ce qui a entraîné la recherche de critères pour que la fonction admette une limite radiale ou non-tangentielle en un point donné du bord. Le résultat global de Fatou n'impliquant de convergence qu'en presque tout point du bord, on a recherché des critères ponctuels qui soient valides en presque tout point du bord. Les deux critères qui nous intéressent ici sont celui de bornitude (radiale ou non-tangentielle) et de finitude de l'*intégrale d'aire* (aussi appelée intégrale de Lusin). Le critère de bornitude est

---

* *Mots-clés* : fonctions harmoniques — arbres — théorème de Fatou — marches aléatoires.
† *Classification math.* : 05C05, 31C20, 31C35, 60J15, 60J50.

dû à I.I. Privalov [18] en dimension 2 et à A.P. Calderón [6] dans le demi-espace. Celui de l'intégrale d'aire est dû d'une part à J. Marcinkiewicz et A. Zygmund [15] et d'autre part à D.C. Spencer [19] en dimension 2, et à A.P. Calderón [5] et E.M. Stein [20] dans le demi-espace.

Contrairement au théorème de Fatou, ces critères ponctuels ont été assez peu développés dans d'autres cadres que le demi-espace euclidien. Cette étude est cependant intéressante comme on peut le voir sur l'exemple de l'espace hyperbolique réel, où plusieurs des notions mises en jeu trouvent une interprétation géométrique très satisfaisante : le bord qui devient le bord "à l'infini" (ou bord géométrique), les rayons qui deviennent des géodésiques, les cônes non-tangentiels qui deviennent des voisinages tubulaires de géodésiques et, surtout, l'intégrale d'aire qui reçoit une expression plus simple et naturelle, puisqu'elle devient l'intégrale d'un gradient au carré, c'est-à-dire une énergie [16]. Les premiers travaux en ce sens sont ceux d'A. Korányi et R.B. Putz sur les espaces symétriques [12, 13], qui utilisent la structure algébrique du groupe d'isométrie de l'espace considéré. Plus récemment, nous avons établi ces critères ponctuels pour les variétés riemanniennes simplement connexes de courbure négative pincée [16], en nous inspirant d'une démonstration des critères euclidiens due à J. Brossard [4] et utilisant le mouvement brownien.

Il semblait alors naturel de voir ce qu'on pouvait dire dans le cas discret, c'est-à-dire lorsqu'on regarde les fonctions harmoniques sur un graphe. Le cas le plus simple est celui des arbres, qui sont de bons analogues discrets des variétés de courbure négative, et c'est celui que nous traitons ici. Nous pensions *a priori* que les arguments du cas continu s'adapteraient sans problème. Cela a été le cas pour les arguments géométriques, qui sont même considérablement simplifiés par la structure d'arbre. Il n'en a pas été de même pour les arguments probabilistes car les martingales discrètes sont d'un maniement plus délicat que les martingales continues. En ce qui concerne la théorie du potentiel, la démonstration dans le cas des variétés utilisait la connaissance du bord de Martin et des inégalités de Harnack à l'infini [3, 1]. Dans le cas présent, bien qu'on ait à notre disposition les résultats théoriques d'A. Ancona [1, 2], il semble plus naturel d'utiliser les résultats antérieurs de P. Cartier sur le bord de Martin des arbres [7]. Le principe de Harnack à l'infini est alors remplacé par des estimations reposant sur la structure d'arbre et la propriété forte de Markov, ce qui permet d'avoir des démonstrations assez dépouillées et des majorations explicites.

Par ailleurs, la lecture de cet article nécessite assez peu de prérequis, l'essentiel étant d'avoir (ou de se forger au cours du texte) une idée intuitive de la propriété forte de Markov. En particulier, il n'est nulle part fait recours à la géométrie différentielle, ce qui devrait permettre de faire connaître les techniques de démonstration à un public plus large que dans le cas des variétés. À part les résultats de P. Cartier, dont nous rappelons en détail (mais sans démonstrations) ce qui nous servira, tout est démontré. Cela devrait permettre à cet article de se suffire à lui-même. Le lecteur plus exigeant pourra se référer sans problème à l'article de P. Cartier qui ne demande aucune connaissance préalable.

La structure de cet article est la suivante. Dans la section 1, nous introduisons les concepts de base de graphe, de marche aléatoire et de fonction harmonique. Le lemme 2 illustre la subtilité du cas discret en ce qui concerne les martingales. Dans la section 2, on rappelle les principaux résultats de P. Cartier. Dans la section 3, nous rappelons la méthode de Doob pour conditionner la marche à sortir en un point donné du bord (qui est aussi exposée dans l'article de P. Cartier). La section 4 introduit l'hypothèse d'uniformité sur la marche aléatoire et expose le résultat principal. Ce résultat est ensuite démontré dans les sections 6 et 7, après avoir tiré des conséquences de l'hypothèse d'uniformité dans la section 5.

1 Marche aléatoire sur un graphe

Après avoir défini les notions de graphe et de marche aléatoire, nous introduisons la fonction de Green en rappelant la dichotomie "récurrence-transience". Nous définissons ensuite les fonctions harmoniques et montrons un premier résultat sur leurs différentes propriétés asymptotiques le long de la marche aléatoire.

1.1 Définitions

On définit un *graphe non-orienté* $\mathcal{G} = (S, A)$ par son ensemble de *sommets* S et son ensemble d'*arêtes* A, une arête étant une partie à deux éléments de S (remarquons que cette définition exclut les "boucles" et les "arêtes multiples"). On dira que deux sommets x et y sont *voisins* s'ils sont reliés par une arête ($\{x, y\} \in A$), ce qu'on notera $x \sim y$. Une suite $x = x_0, x_1, x_2, \ldots, x_n = y$ telle que $x_i \sim x_{i+1}$ pour tout $0 \leq i \leq n-1$ est appelée un *chemin* de *longueur* n joignant x à y.

Le graphe $\mathcal{G}$ sera supposé *localement fini* (tout sommet a un nombre fini de voisins) et *connexe* (deux points quelconques sont toujours joignables par un chemin). La *distance* $d(x, y)$ entre deux points x et y est alors la longueur minimum des chemins joignant x à y. Un chemin réalisant ce minimum est appelé un *segment géodésique.*

Une *marche aléatoire* sur le graphe $\mathcal{G}$ est alors une chaîne de Markov sur S dont les probabilités de transition $p(x, y)$ (probabilité de passer de x à y en une étape) sont non nulles si et seulement si x et y sont deux sommets voisins. On a alors, pour un sommet x fixé, $\sum_{y \sim x} p(x, y) = 1$. La marche aléatoire est donnée par la famille de variables aléatoires $(X_n)_{n \in \mathbf{N}}$ où X_n est la "position" au temps n. Sa loi est la donnée d'une famille $(P_x)_{x \in S}$ où P_x est la probabilité obtenue lorsque le point de départ de la marche est x. On note $\mathcal{F}_n$ la tribu engendrée par les variables aléatoires X_i, $i \leq n$ (c'est la tribu du passé au sens large : elle "représente" les informations connues jusqu'à l'instant n compris). Si T est un temps d'arrêt, on note $\mathcal{F}_T$ la tribu qu'il définit (informations connues jusqu'à l'instant T). La propriété forte de Markov traduit le fait que tout ce qui se passe après un temps d'arrêt ne dépend pas du passé au sens strict, mais uniquement de la situation au temps T.

Pour énoncer précisément cette propriété, on introduit l'*espace des chemins* : on sait qu'on peut choisir comme univers sous-jacent l'espace Ω des chemins infinis (indexés par $\mathbf{N}$), muni de la tribu "trace" sur Ω de la tribu produit dénombrable des $\mathcal{P}(S)$. On introduit aussi l'*opérateur de décalage* Θ de Ω dans lui-même qui, à un chemin ω, fait correspondre le chemin $\Theta(\omega) : n \longmapsto \omega(n+1)$. Pour un temps d'arrêt presque sûrement fini T on note alors Θ^T l'application qui, à un chemin ω, fait correspondre $(\Theta^T)(\omega) = \Theta^{T(\omega)}(\omega) : n \longmapsto \omega(n+T(\omega))$. On peut alors énoncer la propriété :

Théorème 1 *(Propriété forte de Markov)*
Soient F une variable aléatoire positive et T un temps d'arrêt presque sûrement fini. Alors

$$E_x[F \circ \Theta^T | \mathcal{F}_T] = u_F(X_T) \text{ où l'on pose } u_F(y) = E_y[F].$$

1.2 Fonctions de Green

On définit $p_n(x,y) = P_x[X_n = y]$, la probabilité de passer de x à y en exactement n coups ($n \geq 1$). On a évidemment $p_1 = p$. La *fonction de Green* G est alors définie sur $S \times S$ par $G(x,y) = \sum_{n=0}^{\infty} p_n(x,y)$. Si on note $L(y) = \sum_{n=0}^{\infty} \mathbf{1}_{(X_n=y)}$ le *temps de séjour* en y, on a $G(x,y) = E_x[L(y)]$ par le théorème de Fubini.

On peut exprimer $G(x,x)$ en fonction de p_x, probabilité de revenir au moins une fois en x en partant de x. Pour cela on partitionne l'univers Ω avec les événements "revenir exactement n fois en x", qu'on note Ω_n, pour $0 \leq n \leq \infty$. Pour $n \neq \infty$, on obtient $P_x[\Omega_n] = p_x^n(1-p_x)$ en utilisant la propriété forte de Markov aux temps de k-ième retour en x pour $1 \leq k \leq n$. Pour $n = \infty$, la propriété forte de Markov au temps de premier retour en x donne $P_x[\Omega_\infty] = p_x \cdot P_x[\Omega_\infty]$. On distingue alors deux cas.

Si $p_x = 1$, les Ω_n sont de probabilité nulle pour tout $n \neq \infty$, donc leur réunion (dénombrable) aussi et son complémentaire Ω_∞ est presque sûr : on revient presque sûrement une infinité de fois en x et x est dit *récurrent*. Le temps de séjour en x est alors presque toujours infini et $G(x,x) = \infty$.

Si $p_x < 1$, la relation avec $P_x[\Omega_\infty]$ prouve que cette probabilité est nulle : on ne revient presque jamais une infinité de fois en x et x est dit *transient*. On a alors $G(x,x) = \sum_{n=0}^{\infty}(n+1)p_x^n(1-p_x) = \frac{1}{1-p_x}$.

Pour exprimer $G(x,y)$, on introduit $H(x,y)$, probabilité partant de x d'atteindre y. Remarquons que H ne s'annule pas, d'après l'hypothèse de connexité, et que $H(x,y) = \sum_{n=0}^{\infty} q_n(x,y)$, où $q_n(x,y)$ est la probabilité, partant de x, d'atteindre pour la première fois y au temps n. En utilisant la propriété forte de Markov au temps τ_y de première atteinte de y, on obtient :

$$G(x,y) = H(x,y)G(y,y) \tag{1}$$

Sous l'hypothèse de connexité du graphe, il est classique que tous les points sont de même nature (récurrence ou transience) : si un point est récurrent, alors tous les points le sont et la fonction de Green n'a que des valeurs infinies. Dans le cas contraire, tous les points sont transients et la fonction de Green n'a que des valeurs finies. On dira que la marche aléatoire est *récurrente* ou *transiente*. Remarquons qu'une marche sur un graphe fini est forcément récurrente.

Plus généralement, si U est une partie de S, on définit sur $U \times U$ la *fonction de Green* $G_U(x,y)$ de U comme l'espérance, partant de x, du temps de séjour en y avant de sortir de U. Le théorème de Fubini donne aisément la formule suivante.

Soient φ une fonction positive sur U et $x \in U$. En notant τ le temps de sortie de U, on a

$$E_x\left[\sum_{k=0}^{\tau-1} \varphi(X_k)\right] = \sum_{y \in U} \varphi(y) G_U(x,y). \tag{2}$$

1.3 Fonctions harmoniques

On associe à la marche aléatoire un opérateur *laplacien* défini sur l'espace $\mathbf{R}^S$ des fonctions sur l'ensemble S des sommets par

$$(\Delta f)(x) = E_x[f(X_1)] - f(x) = \left(\sum_{y \sim x} p(x,y) f(y)\right) - f(x).$$

Une fonction est dite *harmonique* si son laplacien est nul. Par exemple, en utilisant la propriété de Markov au temps 1 puis en prenant les espérances, on voit facilement que, pour $y \in S$ fixé, la fonction $H(\cdot, y)$ est harmonique sur $S \setminus \{y\}$. Il en est alors de même pour $G(\cdot, y)$ d'après la formule 1. Le lien entre la marche aléatoire et le laplacien s'exprime par la propriété suivante, qui découle de la propriété de Markov.

Lemme 1 *(Propriété de martingale)*
Pour tout point x et toute fonction f, les variables aléatoires

$$M_n = f(X_n) - \sum_{k=0}^{n-1} \Delta f(X_k)$$

forment une $(\mathcal{F}_n)$-martingale pour la probabilité P_x. En particulier, si f est harmonique, $(f(X_n))$ est une martingale.

Nous étudierons par la suite le comportement des fonctions harmoniques le long de certaines trajectoires de la marche aléatoire. Pour cela, nous introduisons ici quelques définitions.

Soit u une fonction harmonique fixée. On appelle *énergie aléatoire* de u la variable aléatoire

$$J^*(u) = \sum_{k=0}^{\infty} \left(\Delta u^2\right)(X_k).$$

Cette définition permet de se passer de l'introduction du gradient [16] et elle est bien valide car la série ci-dessus est à termes positifs. En effet, pour tout $x \in S$,

$$\Delta u^2(x) = \sum_{y \sim x} p(x,y)(u^2(y) - u^2(x))$$

$$= \sum_{y\sim x} p(x,y)\left[(u(y)-u(x))^2 + 2u(x)(u(y)-u(x))\right]$$
$$= \sum_{y\sim x} p(x,y)(u(y)-u(x))^2,$$

car la fonction u est harmonique.

On définit alors les trois événements correspondant au comportement de u le long des trajectoires lorsque n tend vers ∞ : convergence, bornitude et finitude de l'énergie, c'est-à-dire

$$\mathcal{L}^{**} = \{\omega \in \Omega |\text{ la suite } (u(X_n(\omega)) \text{ a une limite finie}\},$$
$$\mathcal{N}^{**} = \{\omega \in \Omega |\text{ la suite } (u(X_n(\omega)) \text{ est bornée}\},$$
$$\mathcal{J}^{**} = \{\omega \in \Omega | J^*(u)(\omega) < \infty\}.$$

Dans le cas d'un graphe quelconque, on a les inclusions suivantes :

Lemme 2 *Pour toute fonction harmonique u (et tout point x), on a*

$$\mathcal{J}^{**} \tilde{\subset} \mathcal{L}^{**} \subset \mathcal{N}^{**}$$

le première inclusion étant P_x-presque sûre.

La deuxième inclusion est triviale car toute suite convergente est bornée. Montrons la première : pour cela, il suffit de montrer que pour tout $N \in \mathbf{N}$, l'ensemble $\mathcal{J}_N^{**} = \{\omega \in \Omega | J^*(u)(\omega) \leq N\}$ est presque inclus dans $\mathcal{L}^{**}$, le résultat cherché en découlant par réunion dénombrable. Fixons alors $N \in \mathbf{N}$ et notons τ_N le premier temps n pour lequel $\sum_{k=0}^{n-1} (\Delta u^2)(X_k) > N$ en convenant que $\tau_N = \infty$ lorsque cela n'arrive pas. Pour $n \in \mathbf{N}$, l'événement $\{\tau_N > n+1\}$ étant égal à $\{\sum_{k=0}^{n} (\Delta u^2)(X_k) \leq N\}$ (la somme étant à termes positifs), il est $\mathcal{F}_n$-mesurable. L'événement complémentaire $\{\tau_N - 1 \leq n\}$ est donc $\mathcal{F}_n$-mesurable et $\tau'_N = \tau_N - 1$ est un temps d'arrêt. Remarquons que $\sum_{k=0}^{\tau'_N-1} (\Delta u^2)(X_k) \leq N$ et que $\tau'_N = \infty$ si et seulement si $J^*(u) \leq N$. Stoppons alors la martingale $M_n = u^2(X_n) - \sum_{k=0}^{n-1} \Delta u^2(X_k)$ au temps d'arrêt τ'_N.

$$M_{n\wedge\tau'_N} \geq - \sum_{k=0}^{(n\wedge\tau'_N)-1} \Delta u^2(X_k) \geq -N.$$

Cette martingale étant minorée par une constante, elle est presque sûrement convergente. Sur l'événement $\mathcal{J}_N^{**}$, d'une part $\tau'_N = \infty$, donc $M_n = M_{n\wedge\tau'_N}$ est presque sûrement convergente, et d'autre part $\sum_{k=0}^{\infty} \Delta u^2(X_k)$ converge, donc $u^2(X_n)$ a presque sûrement une limite finie. Pour se ramener à la fonction u, on utilise une astuce classique : la fonction $v = u + 1$ est aussi une fonction harmonique et $\Delta v^2 = \Delta(u^2 + 2u + 1) = \Delta u^2$ car $2u + 1$ est harmonique. Ainsi on peut appliquer à v ce qui précède et, sur l'événement $\mathcal{J}_N^{**}$ (qui est le même pour u et v), $v^2(X_n)$ a presque sûrement une limite finie. Comme $v^2 = u^2 + 2u + 1$, si v^2 et u^2 ont une limite finie, alors u en a une. Ainsi $\mathcal{J}_N^{**} \overset{\sim}{\subset} \mathcal{L}^{**}$, ce qui achève la preuve du lemme.

Avant de passer au cas des arbres, faisons quelques remarques. Les inclusions ci-dessus semblent strictes en général et il devrait être possible de construire un contre-exemple. Dans le cas que nous allons considérer, ce seront au contraire des égalités presque sûres, ce qui sera montré indirectement comme une conséquence du théorème principal. En termes de martingales, $J^*(u)$ correspond à "une" variation quadratique de la martingale $(u(X_n))$ (il y a en effet plusieurs définitions de la variation quadratique dans le cas discret) et, dans ce langage, le résultat ci-dessus doit être connu.

2 Cas d'un arbre, bords à l'infini

On décrit dans cette section des résultats sur les arbres connus depuis les travaux de P. Cartier [7]. On commence par définir les arbres, puis on introduit les différents bords à l'infini : le bord géométrique, le bord de Martin et le bord de Poisson, ainsi que les résultats associés et les liens entre ces différents bords.

2.1 Définitions

On se place sur le graphe $\mathcal{G} = (S, A)$ de la section précédente (non orienté, localement fini et connexe). On dit que deux chemins sont *élémentairement homotopes* si on peut passer de l'un à l'autre en supprimant un "aller-retour", c'est-à-dire remplacer $\ldots, x, y, x, \ldots$ par $\ldots, x, \ldots$ et on appelle *homotopie* la relation d'équivalence engendrée sur l'ensemble des chemins. On appelle *lacet* un chemin qui finit à l'endroit où il a commencé et *lacet trivial* un lacet réduit à un point. Le graphe $\mathcal{G}$ est appelé un *arbre* s'il est *simplement connexe*, c'est-à-dire si tout lacet est homotope à un lacet trivial.

Dans toute la suite, $\mathcal{G}$ sera un **arbre**, qu'on supposera de plus **transient** (donc infini).

On déduit aisément de la définition d'un arbre qu'il y a unicité du segment géodésique joignant x à y et que tout chemin de x à y passe par tous les points de ce segment géodésique. Autrement dit, pour une partie de S, la connexité et la convexité sont deux notions équivalentes. Si maintenant on choisit un point z sur le segment géodésique joignant x à y, on est donc obligé de passer par z lorsqu'on va de x à y et cela donne les formules suivantes, grâce à la propriété forte de Markov :

$$H(x,y) = H(x,z)H(z,y) \tag{3}$$

$$G(x,y) = H(x,z)G(z,y) \tag{4}$$

(La fonction H définie ici est égale à la fonction F de l'article de P. Cartier sauf sur la diagonale de $S \times S$.)

Remarquons enfin que, si U est une partie connexe (donc convexe par ce qui précède) de S et $x \in S$, il existe un unique point de U situé à distance minimale de x, qu'on appelle la *projection* de x sur U. De plus, si un chemin partant de x rencontre U, il le fait pour la première fois à l'endroit de cette projection.

2.2 Bord géométrique

Le bord géométrique de l'arbre est défini à l'aide des rayons géodésiques. Une *géodésique* est un chemin doublement infini (indexé par $\mathbf{Z}$) dont tous les sous-chemins finis sont des segments géodésiques. Un *rayon géodésique* est un chemin simplement infini (indexé par $\mathbf{N}$) dont tous les sous-chemins finis sont des segments géodésiques. Deux rayons géodésiques γ et δ sont dits *asymptotes* s'ils restent à distance bornée l'un de l'autre, c'est-à-dire si la suite $(d(\gamma(n), \delta(n)))_{n\in\mathbf{N}}$ est bornée. Dans le cas d'un arbre, cela signifie qu'à une translation près, les rayons géodésiques coïncident à partir d'un certain rang. La relation ainsi définie est une relation d'équivalence et on appelle *bord géométrique* de l'arbre l'ensemble des rayons géodésiques quotienté par cette relation d'équivalence, qu'on note ∂S. Dans le cas d'un arbre, il est facile de voir qu'étant donné un point $x \in S$, chaque point du bord géométrique est représenté par un unique rayon géodésique partant de x.

Nous nous limiterons donc à considérer les rayons géodésiques qui partent d'un *point base* donné o, **qui est fixé pour toute la suite**.

Si θ est un point du bord géométrique ∂S, on note alors γ_θ le rayon géodésique partant de o qui représente θ. On dira que γ_θ *pointe vers* θ. Par abus de langage, on notera encore γ_θ l'image du rayon géodésique, c'est-à-dire l'ensemble $\{\gamma_\theta(n), n \in \mathbf{N}\}$.

On munit $S \cup \partial S$ de la topologie des cônes : si on fixe un segment géodésique partant de o, on considère tous les rayons géodésiques qui débutent par ce segment et on appelle *cône* la réunion de ces rayons géodésiques privée du segment initial moins son extrémité, à laquelle on a rajouté les points à l'infini correspondant. La topologie engendrée par les cônes et les points de S fournit alors une compactification de S, appelée *compactification géométrique*. Dire qu'un point y est "proche" de θ signifie que la partie initiale commune entre γ_θ et le segment géodésique (oy) est "grande".

2.3 Bord de Martin

Le bord de Martin est défini à l'aide des *noyaux de Green normalisés* $K_y = \frac{G(\cdot,y)}{G(o,y)}$ définis pour $y \in S$. Ces noyaux sont harmoniques sauf en y et on cherche à envoyer cette "singularité" à l'infini : on considère les suites (y_n) telles que (K_{y_n}) converge simplement vers une fonction harmonique sur S (remarquons que cette fonction harmonique est positive). Deux telles suites sont dites équivalentes si la fonction harmonique limite est la même. Le *bord de Martin* est obtenu en quotientant l'ensemble de ces suites par cette relation d'équivalence. En identifiant S et l'ensemble de noyaux de Green normalisés (par $y \longmapsto K_y$) et le bord de

Martin et l'ensemble des fonctions harmoniques limites, puis en mettant sur la réunion de ces deux ensembles la topologie de la convergence simple, on obtient une compactification de S, appelée *compactification de Martin.* Avec cette topologie, les suites précédentes convergent vers le point du bord de Martin qu'elles définissent. Remarquons qu'en général, il faut définir le bord de Martin à l'aide de la convergence uniforme sur tout compact, qui coïncide ici avec la convergence simple.

En vue de comparer le bord de Martin et le bord géométrique, il est naturel de regarder si les rayons géodésiques définissent un point du bord de Martin, c'est-à-dire si la suite $(\frac{G(\cdot,\gamma_\theta(n))}{G(o,\gamma_\theta(n))})$ converge simplement vers une fonction harmonique. Soit $x \in S$ et y la projection de x sur γ_θ. Pour tout point $z \in \gamma_\theta$ situé entre y et θ, la formule 4 donne $G(o,z) = H(o,y)G(y,z)$ et $G(x,z) = H(x,y)G(y,z)$. Ainsi la suite $(\frac{G(x,\gamma_\theta(n))}{G(o,\gamma_\theta(n))})$ est stationnaire et converge vers $\frac{H(x,y)}{H(o,y)}$ où y est la projection de x sur γ_θ. On voit facilement que la limite simple ainsi obtenue est harmonique et on l'appelle le *noyau de Martin* K_θ de θ. Cela fournit un plongement du bord géométrique dans le bord de Martin.

Le résultat principal de P. Cartier est qu'il n'y a pas d'autres points dans le bord de Martin :

Théorème 2 *Pour un arbre transient, les compactifications de Martin et géométrique coïncident. De plus, tous les points de la frontière de Martin sont extrémaux.*

Il en découle d'après la théorie de Martin qu'on peut représenter les fonctions harmoniques positives par des mesures positives sur le bord à l'infini :

Corollaire 1 *(Représentation des fonctions harmoniques positives)*
La formule

$$\forall x \in S,\ u(x) = \int_{\partial S} K_\theta(x)\nu(d\theta).$$

établit une bijection entre l'ensemble des fonctions harmoniques positives u et l'ensemble des mesures boréliennes positives ν sur le bord ∂S.

2.4 Bord de Poisson

En appliquant la théorie de Martin à la fonction constante égale à 1, on obtient des résultats sur la marche aléatoire :

Théorème 3 *Soit x un point quelconque de S. Alors, P_x-p.s., la suite (X_n) converge vers un point $X_\infty \in \partial S$. La loi de X_∞ (loi de sortie de la marche aléatoire), notée μ_x et appelée mesure harmonique partant de x, est une mesure sur ∂S. On obtient ainsi une famille de mesures $\mu = (\mu_x)_{x \in S}$ sur ∂S qui sont absolument continues les unes par rapport aux autres et dont la dérivée de Radon-Nykodim est donnée par le noyau de Martin :*

$$\frac{d\mu_x}{d\mu_o}(\theta) = K_\theta(x).$$

Ces mesures harmoniques ont les mêmes ensembles négligeables ce qui permet de définir la μ-négligeabilité et l'espace $L^\infty(\partial S, \mu)$. Leur support commun (ensemble des points de sortie de la marche aléatoire) est appelé *bord de Poisson.* C'est la partie du bord de Martin qui sert à représenter les fonctions harmoniques positives bornées et il permet aussi de représenter celles qui ne sont pas positives :

Théorème 4 *(Représentation des fonctions harmoniques bornées)*
La formule

$$\forall x \in S,\ u(x) = \int_{\partial S} f(\theta)\mu_x(d\theta) = E_x\left[f(X_\infty)\right]$$

établit une bijection entre l'ensemble des fonctions harmoniques bornées u et l'ensemble des fonctions $f \in L^\infty(\partial S, \mu)$.

Nous verrons plus tard comment obtenir f comme "limite" (géométrique ou stochastique) de la fonction u.

Par ailleurs, les mesures harmoniques vérifient une propriété analogue aux formules 3 et 4 sur les fonctions H et G. Pour l'énoncer, introduisons une définition : si $y \in S$, on appelle *ombre* de y sur le bord l'ensemble

$$A_o(y) = \{\theta \in \partial S | y \in \gamma_\theta\},$$

obtenu en mettant une "source lumineuse" au point o. On montre alors aisément, grâce à l'unicité des segments géodésiques et à la propriété forte de Markov, le résultat suivant :

Lemme 3 *Si $E \subset \partial S$ est un borélien inclus dans l'ombre $A_o(y)$ de y, alors*

$$\mu_o(E) = H(o, y)\mu_y(E).$$

3 Conditionnement

La méthode des h-processus de Doob permet de construire, pour $x \in S$ et $\theta \in \partial S$, une probabilité P_x^θ, probabilité partant de x et *sachant qu'on sort en θ.* Cette construction est faite dans l'article de Cartier [7] et est identique à celle du cas continu [9, 16]. La probabilité ainsi construite vérifie (et c'est presque sa définition) la propriété suivante :

Proposition 1 *Soient T un temps d'arrêt presque sûrement fini et F une variable aléatoire positive $\mathcal{F}_T$-mesurable. On a alors la formule :*

$$E_x^\theta[F] = \frac{1}{K_\theta(x)} E_x\left[F \cdot K_\theta(X_T)\right].$$

De plus, la marche vérifie encore la **propriété forte de Markov** pour P_x^θ et on a les deux propriétés suivantes :

Proposition 2 *(Désintégration)*
Pour toute variable aléatoire positive F,

$$E_x[F] = \int_{\partial S} E_x^\theta[F]\mu_x(d\theta).$$

La deuxième nécessite une définition : un événement A sera dit *asymptotique* s'il est invariant par l'opérateur de décalage Θ défini dans la sous-section 1.1 (c'est-à-dire si $\mathbf{1}_{(A)} \circ \Theta = \mathbf{1}_{(A)}$). On a alors le résultat suivant :

Proposition 3 *(Loi 0–1 asymptotique)*
Si l'événement A est asymptotique, alors, pour tout $\theta \in \partial S$, l'application $x \longmapsto P_x^\theta[A]$ est constante sur S et vaut 0 ou 1.

Les événements $\mathcal{L}^{**}$, $\mathcal{N}^{**}$ et $\mathcal{J}^{**}$ définis dans la section 1.3 par le comportement d'une fonction harmonique u le long de la marche aléatoire sont clairement asymptotiques. La loi 0–1 asymptotique affirme que les quantités $P_x^\theta(\mathcal{L}^{**})$, $P_x^\theta(\mathcal{N}^{**})$ et $P_x^\theta(\mathcal{J}^{**})$ prennent les valeurs 0 ou 1, cela indépendamment du point base x. On définit alors les ensembles (indépendants de x) suivants :

$$\mathcal{L}^* = \left\{\theta \in \partial S | P_x^\theta(\mathcal{L}^{**}) = 1\right\},$$

$$\mathcal{N}^* = \left\{\theta \in \partial S | P_x^\theta(\mathcal{N}^{**}) = 1\right\},$$

$$\mathcal{J}^* = \left\{\theta \in \partial S | P_x^\theta(\mathcal{J}^{**}) = 1\right\},$$

Ce sont les ensembles de points θ du bord tels que u converge, est bornée ou a une énergie finie le long des trajectoires aléatoires "conditionnées" à sortir au point θ. Nous dirons alors qu'il y a *convergence stochastique*, *bornitude stochastique* ou *finitude de l'énergie stochastique* de u en θ. De plus, pour $\theta \in \mathcal{L}^*$, les événements $\{\omega \in \Omega | \lim_{n\to+\infty} u(X_n) \leq R\}$ sont asymptotiques, donc la variable aléatoire $\lim_{n\to+\infty} u(X_n)$ est P_x^θ-p.s. constante (cette constante ne dépendant pas de x) et cette valeur est appelée la *limite stochastique* de u en θ.

En appliquant la formule de désintégration ci-dessus au lemme 2, on obtient le résultat suivant :

Lemme 4 *Pour toute fonction harmonique u, on a*

$$\mathcal{J}^* \mathrel{\tilde{\subset}} \mathcal{L}^* \subset \mathcal{N}^*$$

le première inclusion étant μ-presque sûre (μ mesure harmonique).

Nous introduisons dans la section suivante une hypothèse d'uniformité sur la marche aléatoire. Une conséquence immédiate du résultat principal (théorème 5) est que, sous cette hypothèse, les deux inclusions ci-dessus sont en fait des égalités presque sûres.

4 Résultat principal

Les résultats géométriques dont on s'inspire ici [16] concernent les variétés riemanniennes dont la courbure est "pincée" entre deux constantes strictement négatives. L'analogue discret de la courbure négative est obtenu en prenant un arbre. Cela permet, entre autres, d'avoir un comportement similaire des géodésiques : écartement de deux géodésiques, unicité du segment géodésique entre deux points, bord à l'infini. Pour traduire les bornes uniformes sur la courbure, on regarde le nombre d'arêtes partant de chaque point ainsi que leurs poids respectifs : beaucoup d'arêtes avec des poids faibles signifient que la courbure est très négative. On se convainc alors qu'une bonne hypothèse, d'ailleurs déjà utilisée dans les travaux de A. Korányi, M.A. Picardello et M.H. Taibleson [14] et M.A. Picardello et W. Woess [17], est la suivante :

$$(\mathcal{H}) \qquad \exists \varepsilon > 0, \exists \eta > 0, \forall x \sim y, \varepsilon \leq p(x,y) \leq \frac{1}{2} - \eta.$$

Cette hypothèse assure en particulier que le nombre de voisins d'un sommet est minoré par trois et majoré uniformément sur S. De plus, elle entraîne que l'arbre est transient, comme conséquence immédiate du lemme 6 qu'on verra plus loin.

Pour pouvoir énoncer le résultat principal il nous reste à définir les notions "radiales" correspondant aux notions stochastiques déjà introduites mais définies par le comportement de la fonction sur les rayons géodésiques. Soit u une fonction harmonique. On définit son énergie radiale par

$$J_\theta(u) = \sum_{k=0}^{\infty} \Delta u^2(\gamma_\theta(k)).$$

Nous dirons qu'il y a *convergence radiale*, *bornitude radiale* ou *finitude de l'énergie radiale* de u en θ selon que $(u(\gamma_\theta(n)))_n$ converge, qu'elle est bornée ou que $J_\theta(u) < +\infty$. Nous allons démontrer le théorème suivant :

Théorème 5 *Soient $\mathcal{G} = (S, A)$ un arbre muni d'une marche aléatoire vérifiant l'hypothèse d'uniformité $(\mathcal{H})$ et u une fonction harmonique.*

Alors les notions de convergence radiale, bornitude radiale, finitude de l'énergie radiale, convergence stochastique, bornitude stochastique et finitude de l'énergie stochastique de u coïncident pour μ-presque tout point du bord.

Pour plus de commodité, nous noterons $\mathcal{L}$, $\mathcal{N}$ et $\mathcal{J}$, les ensembles de points θ du bord pour lesquels il y a respectivement convergence radiale, bornitude radiale ou finitude de l'énergie radiale de u en θ. Il nous faut montrer que les ensembles $\mathcal{L}$, $\mathcal{N}$, $\mathcal{J}$, $\mathcal{L}^*$, $\mathcal{N}^*$ et $\mathcal{J}^*$ sont égaux μ-presque partout. D'après le lemme 4, on sait déjà que $\mathcal{J}^* \tilde{\subset} \mathcal{L}^* \subset \mathcal{N}^*$ et il est clair que $\mathcal{L} \subset \mathcal{N}$, puisqu'une suite convergente est bornée.

Par ailleurs, puisqu'on est sur un arbre, un chemin partant de o et qui tend vers un point θ du bord passe forcément par tous les points du segment géodésique γ_θ. Ainsi, si $X_0 = o$ et $X_\infty = \theta$, alors la suite $(\gamma_\theta(n))$ est une suite extraite de la suite (X_n) et on a les inclusions $\mathcal{L}^* \subset \mathcal{L}$, $\mathcal{N}^* \subset \mathcal{N}$, et $\mathcal{J}^* \subset \mathcal{J}$. Cela entraîne aussi qu'en cas de convergences stochastique et radiale, les limites sont les mêmes.

En résumé, on a donc les inclusions suivantes :

$$\begin{array}{ccccc} \mathcal{J}^* & \tilde{\subset} & \mathcal{L}^* & \subset & \mathcal{N}^* \\ \cap & & \cap & & \cap \\ \mathcal{J} & & \mathcal{L} & \subset & \mathcal{N} \end{array}$$

et il suffit de montrer que $\mathcal{N} \tilde{\subset} \mathcal{J}$ et $\mathcal{J} \tilde{\subset} \mathcal{J}^*$ pour achever la preuve du théorème, ce qui sera fait dans les sections 6 et 7. Pour cela, nous aurons besoin de traduire l'hypothèse $(\mathcal{H})$, ce qui sera fait dans la section 5 où nous verrons plusieurs conséquences de cette hypothèse.

Nous aurons aussi besoin de savoir construire la fonction à l'infini qui "représente" une fonction harmonique bornée (théorème 4). Il découle des résultats de P. Cartier qu'elle est limite de la fonction harmonique :

Lemme 5 *Toute fonction harmonique bornée converge radialement et stochastiquement en μ-presque tout point du bord et l'unique fonction $f \in L^\infty(\partial S, \mu)$ telle que*

$$\forall x \in S,\ u(x) = \int_{\partial S} f(\theta)\mu_x(d\theta) = E_x\left[f(X_\infty)\right]$$

est μ-presque partout limite radiale et stochastique de u.

5 Conséquences de l'hypothèse d'uniformité

Dans toute la suite, on supposera que la marche aléatoire sur l'**arbre** $\mathcal{G}$ vérifie l'**hypothèse** $(\mathcal{H})$.

Cela permet alors de majorer uniformément les fonctions H et G :

Lemme 6

$$\exists C_H\left(=\frac{\frac{1}{2}-\eta}{\frac{1}{2}+\eta}\right) < 1, \forall x \neq y, H(x,y) \leq C_H.$$

$$\exists C_G\left(=\frac{1}{1-C_H}\right), \forall x,y, G(x,y) \leq C_G.$$

Une démonstration de ce résultat figure dans l'article de M.A. Picardello et W. Woess déjà cité [17]. Nous en donnons ici une autre, due à C. Leuridan (communication personnelle).

D'après les formules 3 et 4, il suffit de montrer la première inégalité lorsque x et y sont voisins et la deuxième lorsque $x = y$ (car H prend des valeurs entre 0 et 1).

Soient alors x et y deux sommets voisins. La majoration de $H(x,y)$ repose sur le fait qu'à tout instant, la probabilité de s'éloigner de y est supérieure ou égale à $\frac{1}{2}+\eta$. Cela découle de l'hypothèse $(\mathcal{H})$ puisqu'en tout point différent de y, il y a une seule arête qui se rapproche de y.

La méthode de M.A. Picardello et W. Woess consiste à comparer la situation à celle d'une chaîne de Markov unidimensionnelle sur $\mathbf{N}$ telle qu'en un point différent de 0, la probabilité d'augmenter de 1 vaille $\frac{1}{2}+\eta$ et celle de décroître de 1 vaille $\frac{1}{2}-\eta$. Dans ce cas particulier, il est possible de calculer explicitement la probabilité d'atteindre 0 en partant de 1. La comparaison repose alors sur une récurrence. On pourrait aussi faire cette comparaison à l'aide du formalisme des réseaux électriques.

Nous utilisons ici une méthode plus probabiliste qui nous a été indiquée par C. Leuridan et qui consiste à faire cette comparaison par le biais d'une surmartingale judicieuse, obtenue en regardant les fonctions harmoniques sur $\mathbf{N}^*$ du cas unidimensionnel ci-dessus : on pose

$$M_n = \left(\frac{\frac{1}{2}-\eta}{\frac{1}{2}+\eta}\right)^{d(X_n,y)}$$

et on a alors :

$$E_x[M_{n+1}|\mathcal{F}_n] = M_n \cdot E_x\left[\left(\frac{\frac{1}{2}-\eta}{\frac{1}{2}+\eta}\right)^{d(X_{n+1},y)-d(X_n,y)} |\mathcal{F}_n\right].$$

Or, sachant que $\{X_0 = x_0; X_1 = x_1; \ldots; X_n = x_n\}$, $d(X_{n+1},y)-d(X_n,y)$ vaut 1 avec une probabilité supérieure ou égale à $\frac{1}{2}+\eta$ et -1 avec une probabilité inférieure ou égale à $\frac{1}{2}-\eta$, donc (en remarquant que $\frac{\frac{1}{2}-\eta}{\frac{1}{2}+\eta} \leq \frac{\frac{1}{2}+\eta}{\frac{1}{2}-\eta}$)

$$E_x\left[\left(\frac{\frac{1}{2}-\eta}{\frac{1}{2}+\eta}\right)^{d(X_{n+1},y)-d(X_n,y)} |\mathcal{F}_n\right] \leq \left(\frac{1}{2}+\eta\right)\frac{\frac{1}{2}-\eta}{\frac{1}{2}+\eta} + \left(\frac{1}{2}-\eta\right)\frac{\frac{1}{2}+\eta}{\frac{1}{2}-\eta} = 1$$

et la suite (M_n) est une surmartingale. Elle est positive, majorée par 1 et converge presque sûrement vers 0. Soit τ_y le temps d'atteinte de y. On a $M_{\tau_y} = \mathbf{1}_{(\tau_y<\infty)}$ et le théorème d'arrêt de Doob donne (x et y étant voisins) :

$$H(x,y) = P_x[\tau_y < \infty] = E_x[M_{\tau_y}] \leq E_x[M_0] = \frac{\frac{1}{2}-\eta}{\frac{1}{2}+\eta}.$$

Pour un sommet x, on a vu dans la sous-section 1.2 que $G(x,x) = \frac{1}{1-p_x}$. Or p_x, probabilité de revenir au moins une fois en x en partant de x, est la moyenne des $H(y,x)$ pour y voisin de x, donc $p_x \leq C_H$, ce qui prouve la majoration de G et achève la preuve du lemme.

Ces majorations permettent de démontrer trois résultats qui seront très utiles par la suite.

Corollaire 2 *("Lemme d'évitement")*
Soit U une partie connexe de S (au sens de l'introduction). Alors

$$P_x\,[\textit{la marche rencontre } U] \leq (C_H)^{d(x,U)}.$$

Notons y la projection de x sur U (U est connexe). On sait qu'on est forcé de passer par y pour rencontrer U lorsqu'on part de x (sous-section 2.1), donc P_x [la marche rencontre U] $= H(x,y)$. Si on note $x = x_0, x_1, \ldots, x_n = y$ les points du segment géodésique joignant x à y, on a (par la formule 3)

$$H(x,y) = H(x_0,x_1)H(x_1,x_2)\cdots H(x_{n-1},x_n) \leq (C_H)^n.$$

Comme $n = d(x,y) = d(x,U)$, le lemme est prouvé.

Pour énoncer les deux résultats suivants, introduisons une notation : si E est une partie du bord, on notera

$$\Gamma(E) = \bigcup_{\theta \in E} \gamma_\theta$$

la réunion des rayons géodésiques pointant vers E. On a alors :

Corollaire 3 *("Lemme de la fin")*
Soit E un borélien du bord. Alors, pour μ-presque tout $\theta \in E$, la marche aléatoire finit P_o^θ-presque sûrement sa vie dans $\Gamma(E)$.

Notons v la fonction harmonique bornée définie par

$$v(x) = \mu_x(E) = P_x\left[X_\infty \in E\right]$$

à l'aide du théorème 4 de représentation des fonctions harmoniques bornées. Si $x \notin \Gamma(E)$, alors, par le corollaire précédent ($\Gamma(E)$ étant connexe),

$$P_x\left[\text{la marche rencontre } \Gamma(E)\right] \leq C_H,$$

donc *a fortiori* $v(x) \leq C_H$, puisqu'il faut rentrer dans $\Gamma(E)$ pour finir dans E. Ainsi, si $v(x) > C_H$, alors $x \in \Gamma(E)$. Par ailleurs, d'après le lemme 5, pour μ-presque tout $\theta \in E$, P_o^θ-presque sûrement, $(v(X_n))$ tend vers 1, donc $X_n \in E$ à partir d'un certain rang par ce qui précède, ce qu'il fallait démontrer.

Et en notant $\widetilde{\gamma_\theta} = \{x \in S | d(x,\gamma_\theta) \leq 1\}$ le rayon géodésique *épaissi*, on a aussi :

Corollaire 4 *("Lemme des pointes")*
Soit E un borélien du bord. Alors, pour μ-presque tout $\theta \in E$, $\Gamma(E)$ contient une "pointe" de $\widetilde{\gamma_\theta}$, c'est-à-dire contient $\widetilde{\gamma_\theta}$ sauf un ensemble fini.

Avec les notations de la démonstration précédente, on sait encore, par le lemme 5, que pour μ-presque tout $\theta \in E$, v converge radialement en θ vers 1. Soit θ un tel point du bord : $v(x)$ tend vers 1 lorsque x tend vers θ en restant dans γ_θ. Comme $v \leq 1$ et que, par l'hypothèse $(\mathcal{H})$, les poids des voisins d'un point sont minorés uniformément, la propriété de la moyenne entraîne qu'il y a aussi convergence de $v(x)$ vers 1 lorsque x tend vers θ en restant dans $\widetilde{\gamma_\theta}$. Comme dans la démonstration précédente, si $v(x) > C_H$, alors $x \in \Gamma(E)$, ce qui conclut.

Nous pouvons maintenant commencer la démonstration proprement dite du théorème.

6 La bornitude radiale implique la finitude de l'énergie radiale

On montre ici que $\mathcal{N} \tilde{\subset} \mathcal{J}$. Pour cela, on introduit l'ensemble

$$\widetilde{\mathcal{N}} = \left\{ \theta \in \partial S \,\middle|\, \sup_{\widetilde{\gamma}_\theta} |u| < +\infty \right\}$$

qui est clairement inclus dans $\mathcal{N}$ ($\widetilde{\gamma}_\theta$ est le rayon géodésique épaissi défini dans la section 5). Réciproquement, le "lemme des pointes" (corollaire 4) permet de montrer que $\mathcal{N} \tilde{\subset} \widetilde{\mathcal{N}}$. Pour cela on écrit $\mathcal{N} = \bigcup_{N \in \mathbf{N}} \mathcal{N}_N$, où

$$\mathcal{N}_N = \left\{ \theta \in \partial S \,\middle|\, \sup_{\gamma_\theta} |u| \leq N \right\}.$$

La réunion étant dénombrable, il nous suffit alors de montrer que pour tout N, $\mathcal{N}_N \tilde{\subset} \widetilde{\mathcal{N}}$. Cependant, le "lemme des pointes" affirme que, pour μ-presque tout $\theta \in \mathcal{N}_N$, $\Gamma(\mathcal{N}_N)$ contient $\widetilde{\gamma}_\theta$ sauf un ensemble fini. Pour un tel θ, u est donc bornée sur $\widetilde{\gamma}_\theta$ puisque, par définition de $\mathcal{N}_N$, $|u| \leq N$ sur $\Gamma(\mathcal{N}_N)$. Ainsi $\mathcal{N}_N \tilde{\subset} \widetilde{\mathcal{N}}$ pour tout N et $\mathcal{N} \tilde{\subset} \widetilde{\mathcal{N}}$, donc ces deux ensembles sont égaux μ-presque partout.

Il nous suffit donc de montrer ici que $\widetilde{\mathcal{N}} \tilde{\subset} \mathcal{J}$. On écrit de nouveau que $\widetilde{\mathcal{N}} = \bigcup_{N \in \mathbf{N}} \widetilde{\mathcal{N}}_N$, où

$$\widetilde{\mathcal{N}}_N = \left\{ \theta \in \partial S \,\middle|\, \sup_{\widetilde{\gamma}_\theta} |u| \leq N \right\}.$$

et il nous suffit de montrer que pour tout N, $\widetilde{\mathcal{N}}_N \tilde{\subset} \mathcal{J}$.

Soit maintenant $N \in \mathbf{N}$ fixé. On pose $\Gamma = \Gamma(\widetilde{\mathcal{N}}_N)$, réunion des rayons géodésiques pointant sur $\widetilde{\mathcal{N}}_N$ et on note τ le temps de sortie de Γ. On sait que

$$M_n = u^2(X_n) - \sum_{k=0}^{n-1} \Delta u^2(X_k)$$

est une martingale d'après le lemme 1. En appliquant le théorème d'arrêt de Doob au temps d'arrêt borné $\tau \wedge n$, on obtient

$$E_o\left[M_{\tau \wedge n}\right] = E_o[M_0] = u^2(o) \geq 0,$$

donc

$$E_o\left[\sum_{k=0}^{\tau \wedge n - 1} \Delta u^2(X_k)\right] \leq E_o\left[u^2(X_{\tau \wedge n})\right].$$

Or $X_{\tau \wedge n}$ est à distance au plus 1 de Γ, donc d'un rayon géodésique γ_θ où $\theta \in \widetilde{\mathcal{N}}_N$. Ainsi $X_{\tau \wedge n} \in \widetilde{\gamma}_\theta$ et, par définition de $\widetilde{\mathcal{N}}_N$, $|u(X_{\tau \wedge n})| \leq N$, donc

$$E_o\left[\sum_{k=0}^{\tau \wedge n - 1} \Delta u^2(X_k)\right] \leq N^2.$$

En faisant tendre n vers ∞, on obtient, par convergence monotone (remarquer que $\Delta u^2 \geq 0$),

$$E_o\left[\sum_{k=0}^{\tau-1}\Delta u^2(X_k)\right] \leq N^2 < +\infty$$

et par la formule de désintégration (proposition 2), on obtient que, pour μ-presque tout $\theta \in \partial S$,

$$E_o^\theta\left[\sum_{k=0}^{\tau-1}\Delta u^2(X_k)\right] < +\infty.$$

Lorsque $X_\infty = \theta$ et $\tau = \infty$, la suite $(\gamma_\theta(n))$ étant une suite extraite de (X_n) et les séries étant à termes positifs,

$$J_\theta(u) = \sum_{k=0}^{\infty}\Delta u^2(\gamma_\theta(k)) \leq \sum_{k=0}^{\infty}\Delta u^2(X_k) = \sum_{k=0}^{\tau-1}\Delta u^2(X_k),$$

donc, en multipliant par $\mathbf{1}_{(\tau=\infty)}$ et en prenant les espérances conditionnelles en θ, on obtient, pour tout $\theta \in \partial S$,

$$J_\theta(u)\cdot P_o^\theta[\tau=\infty] \leq E_o^\theta\left[\mathbf{1}_{(\tau=\infty)}\cdot\sum_{k=0}^{\tau-1}\Delta u^2(X_k)\right] \leq E_o^\theta\left[\sum_{k=0}^{\tau-1}\Delta u^2(X_k)\right].$$

Ainsi, pour μ-presque tout $\theta \in \partial S$,

$$J_\theta(u)\cdot P_o^\theta[\tau=\infty] < +\infty.$$

Pour conclure, il suffit donc de montrer que, pour μ-presque tout $\theta \in \widetilde{\mathcal{N}}_N$, $P_o^\theta[\tau=\infty] > 0$. Par le "lemme de la fin" (corollaire 3), il suffit de montrer cela pour les θ tels que la marche aléatoire finit P_o^θ-presque sûrement sa vie dans Γ. Soit θ un tel point. On a

$$\lim_{n\to\infty} P_o^\theta\left[\forall k \geq n, X_k \in \Gamma\right] = P_o^\theta\left[\text{la marche finit dans } \Gamma\right] = 1.$$

Soit alors n tel que

$$P_o^\theta\left[\forall k \geq n, X_k \in \Gamma\right] > 0.$$

D'après la propriété de Markov,

$$P_o^\theta\left[\forall k \geq n, X_k \in \Gamma\right] = E_o^\theta\left[P_o^\theta\left[\forall k \geq n, X_k \in \Gamma | \mathcal{F}_n\right]\right] = E_o^\theta\left[\varphi(X_n)\right]$$

où $\varphi(x) = P_x^\theta\left[\forall k, X_k \in \Gamma\right] = P_x^\theta[\tau=\infty]$. Comme $E_o^\theta\left[\varphi(X_n)\right] > 0$, la fonction φ n'est pas nulle. Il existe donc un point x tel que $P_x^\theta[\tau=\infty] > 0$. Or

$$P_o^\theta[\tau=\infty] \geq P_o^\theta\left[(X_n) \text{ rencontre } x \text{ et } \tau=\infty\right]$$

et, en appliquant la propriété forte de Markov au temps d'atteinte de x, cette dernière probabilité est égale à

$$P_o^\theta\left[(X_n) \text{ rencontre } x \text{ avant de sortir de } \Gamma\right]\cdot P_x^\theta[\tau=\infty].$$

Par ailleurs, en appliquant la proposition 1 au temps d'arrêt $T = \tau_x \wedge \tau \wedge n$ (où τ_x est le temps d'atteinte de x) et à $F = \mathbf{1}_{(\tau_x \leq \tau \wedge n)} = \mathbf{1}_{(X_T = x)}$, on obtient (puisque $K_\theta(o) = 1$) :

$$P_o^\theta[\tau_x \leq \tau \wedge n] = E_o[\mathbf{1}_{(\tau_x \leq \tau \wedge n)} \cdot K_\theta(X_T)] = K_\theta(x) \cdot P_o[\tau_x \leq \tau \wedge n],$$

ce qui donne par convergence monotone, en faisant tendre n vers ∞,

$$\begin{aligned} &P_o^\theta\left[(X_n) \text{ rencontre } x \text{ avant de sortir de } \Gamma\right] \\ &\qquad = K_\theta(x) \cdot P_o\left[(X_n) \text{ rencontre } x \text{ avant de sortir de } \Gamma\right]. \end{aligned}$$

Cette quantité n'est pas nulle car le noyau de Martin est strictement positif et Γ est connexe (car étoilé) donc il contient au moins un chemin de o à x et il y a une probabilité non nulle de suivre ce chemin. Ainsi $P_o^\theta[\tau = \infty] > 0$, ce qui achève la preuve de l'inclusion $\mathcal{N} \,\tilde{\subset}\, \mathcal{J}$.

7 La finitude de l'énergie radiale implique la finitude de l'énergie aléatoire

On montre ici que $\mathcal{J} \,\tilde{\subset}\, \mathcal{J}^*$. De la même manière que dans la section précédente, on introduit les ensembles $\mathcal{J}_N = \{\theta \in \partial S | J_\theta(u) \leq N\}$ et il suffit de montrer que, pour tout N, $\mathcal{J}_N \,\tilde{\subset}\, \mathcal{J}^*$. On introduit aussi la fonction harmonique bornée $v(x) = \mu_x(\mathcal{J}_N) = P_x\left[X_\infty \in \mathcal{J}_N\right]$ définie à l'aide du théorème 4 de représentation des fonctions harmoniques bornées. Enfin, on définit la quantité

$$I = E_o\left[\sum_{k=0}^{\infty} \Delta u^2(X_k) \cdot \mathbf{1}_{(v \geq \alpha)}(X_k)\right],$$

avec $\alpha = \frac{1+C_H}{2}$ où $C_H < 1$ est la constante du lemme 6.

Remarquons qu'il nous suffit de montrer que $I < +\infty$ pour avoir le résultat voulu. En effet, dans ce cas, la formule de désintégration (proposition 2) assure pour μ-presque tout $\theta \in \partial S$ que

$$E_o^\theta\left[\sum_{k=0}^{\infty} \Delta u^2(X_k) \cdot \mathbf{1}_{(v \geq \alpha)}(X_k)\right] < +\infty$$

donc, P_o^θ-presque sûrement,

$$\sum_{k=0}^{\infty} \Delta u^2(X_k) \cdot \mathbf{1}_{(v \geq \alpha)}(X_k) < +\infty.$$

Par ailleurs, la fonction v converge stochastiquement en μ-presque tout point θ du bord vers $\mathbf{1}_{(\mathcal{J}_N)}(\theta)$ d'après le lemme 5. Ainsi, μ-presque tout point $\theta \in \mathcal{J}_N$ vérifie que, P_o^θ-presque sûrement, la série précédente converge et $\lim_{k\to\infty} v(X_k) = 1$ donc $\sum_{k=0}^{\infty} \Delta u^2(X_k) < +\infty$. Un tel θ est donc dans $\mathcal{J}^*$, donc $\mathcal{J}_N \,\tilde{\subset}\, \mathcal{J}^*$.

Il nous reste donc à prouver que $I < +\infty$. Pour cela, réécrivons I à l'aide de la formule 2 appliquée avec $U = S$:

$$I = \sum_{y \in S} \Delta u^2(y) \mathbf{1}_{(v \geq \alpha)}(y) G(o,y) = \sum_{y \in \{v \geq \alpha\}} \Delta u^2(y) G(o,y). \tag{5}$$

Puis, en intégrant l'énergie (bornée par N) sur $\mathcal{J}_N$ et en appliquant le théorème de Fubini, on obtient une quantité ressemblante, dont on sait qu'elle est finie :

$$N \geq \int_{\mathcal{J}_N} \left(\sum_{k=0}^{\infty} \Delta u^2(\gamma_\theta(k)) \right) \mu_o(d\theta) = \sum_{y \in \Gamma(\mathcal{J}_N)} \Delta u^2(y) \mu_o(\mathcal{J}_N \cap A_o(y)), \tag{6}$$

où $A_o(y) = \{\theta \in \partial S | y \in \gamma_\theta\}$ est comme précédemment l'ombre de y sur le bord. Il ne reste plus qu'à comparer les quantités 5 et 6, ce qu'on fait à l'aide du lemme suivant qui entraîne clairement la finitude cherchée.

Lemme 7 *On a les deux propriétés suivantes :*

1. $\{v \geq \alpha\} \subset \Gamma(\mathcal{J}_N)$;
2. $\forall y \in \{v \geq \alpha\}, G(o,y) \leq 2(C_G)^2 \mu_o(\mathcal{J}_N \cap A_o(y))$.

Montrons alors ce lemme. Remarquons que, par définition de l'ombre de y, $y \notin \Gamma(\partial S \setminus A_o(y))$. En appliquant le "lemme d'évitement" (corollaire 2), on a

$$\mu_y(\partial S \setminus A_o(y)) \leq P_y \,[\text{la marche rencontre } \Gamma(\partial S \setminus A_o(y))] \leq C_H,$$

donc

$$v(y) = \mu_y(\mathcal{J}_N) \leq \mu_y(\mathcal{J}_N \cap A_o(y)) + \mu_y(\partial S \setminus A_o(y)) \leq \mu_y(\mathcal{J}_N \cap A_o(y)) + C_H.$$

Pour $y \in \{v \geq \alpha\}$, on a alors

$$\mu_y(\mathcal{J}_N \cap A_o(y)) \geq v(y) - C_H \geq \alpha - C_H = \frac{1 - C_H}{2} > 0.$$

Cela a deux conséquences. D'une part, $\mathcal{J}_N \cap A_o(y)$ est non vide donc, par définition de l'ombre, $y \in \Gamma(\mathcal{J}_N)$, ce qui prouve la première propriété. D'autre part, en appliquant la formule 1, le lemme 3 et le lemme 6, on obtient :

$$\frac{\mu_o(\mathcal{J}_N \cap A_o(y))}{G(o,y)} = \frac{H(o,y)\mu_y(\mathcal{J}_N \cap A_o(y))}{H(o,y)G(y,y)} \geq \frac{1 - C_H}{2C_G} = \frac{1}{2(C_G)^2},$$

ce qui prouve la deuxième propriété et achève la démonstration de l'inclusion $\mathcal{J} \stackrel{\sim}{\subset} \mathcal{J}^*$.

Remerciements

Nous tenons à remercier ici l'auditoire du *séminaire d'Analyse et de Probabilités* de Grenoble pour l'interaction profitable qu'il a su créer durant nos exposés et tout particulièrement Christophe Leuridan pour l'intérêt qu'il a apporté à ce travail.

Références

[1] Alano Ancona. Negatively curved manifolds, elliptic operators and the Martin boundary. *Ann. of Math.*, 125:495–536, 1987.

[2] Alano Ancona. Positive harmonic functions and hyperbolicity. In J. Kràl et al., editor, *Potential Theory, Surveys and Problems*. Springer Lect. Notes in Math. 1344, Berlin, 1988.

[3] M.T. Anderson and R. Schoen. Positive harmonic functions on complete manifolds of negative curvature. *Ann. of Math.*, 121:429–461, 1985.

[4] Jean Brossard. Comportement non-tangentiel et comportement brownien des fonctions harmoniques dans un demi-espace. Démonstration probabiliste d'un théorème de Calderon et Stein. *Séminaire de Probabilités, Université de Strasbourg*, XII:378–397, 1978.

[5] A.P. Calderón. On a theorem of Marcinkiewicz and Zygmund. *Trans. of A.M.S.*, 68:55–61, 1950.

[6] A.P. Calderón. On the behaviour of harmonic functions at the boundary. *Trans. of A.M.S.*, 68:47–54, 1950.

[7] P. Cartier. Fonctions harmoniques sur un arbre. In *Symposia Mathematica*, volume IX, pages 203–270. Academic Press, London and New-York, 1972.

[8] Yves Derriennic. Marche aléatoire sur le groupe libre et frontière de Martin. *Z. Wahrscheinlichkeitstheorie verw. Gebiete*, 32:261–276, 1975.

[9] Richard Durrett. *Brownian Motion and Martingales in Analysis*. Wadsworth Advanced Books & Software, 1984.

[10] E.B. Dynkin and M.B. Malyutov. Random walks on groups with a finite number of generators. *Soviet Math. Dokl.*, 2:399–402, 1961.

[11] Pierre Fatou. Séries trigonométriques et séries de Taylor. *Acta Math.*, 30:335–400, 1906.

[12] A. Korányi and R.B. Putz. Local Fatou theorem and area theorem for symmetric spaces of rank one. *Trans. Amer. Math. Soc.*, 224:157–168, 1976.

[13] A. Korányi and R.B. Putz. An area theorem for products of symmetric spaces of rank one. *Bull. Sc. math.*, 105:3–16, 1981.

[14] Adam Korányi, Massimo A. Picardello, and Mitchell H. Taibleson. Hardy spaces on non-homogeneous trees. In *Symposia Mathematica*, volume XXIX, pages 205–254. Academic Press, London and New-York, 1987.

[15] J. Marcinkiewicz and A. Zygmund. A theorem of Lusin. *Duke Math. J.*, 4:473–485, 1938.

[16] Frédéric Mouton. Comportement asymptotique des fonctions harmoniques en courbure négative. *Comment. Math. Helvetici*, 70:475–505, 1995.

[17] Massimo A. Picardello and Wolfgang Woess. Finite truncations of random walks on trees. In *Symposia Mathematica*, volume XXIX, pages 255–265. Academic Press, London and New-York, 1987.

[18] I.I. Privalov. Sur les fonctions conjuguées. *Bull. Soc. Math. France*, pages 100–103, 1916.

[19] D.C. Spencer. A function theoric identity. *Amer. J. Math.*, 65:147–160, 1943.

[20] E.M. Stein. On the theory of harmonic functions of several variables II. *Acta Math.*, 106:137–174, 1961.

Université de Grenoble I
Institut Fourier
UMR 5582 CNRS-UJF
UFR de Mathématiques
B.P. 74
38402 SAINT-MARTIN D'HÈRES CEDEX (France)
Email: `mouton@fourier.ujf-grenoble.fr`

Asymptotic estimates for the first hitting time of fluctuating additive functionals of Brownian motion

Y. Isozaki S. Kotani

1 Introduction

In [3], we obtained the following estimates for the first hitting time of the integrated Brownian motion: Let $B(t)$ be the linear Brownian motion started at 0. It holds with some explicit constant $k > 0$

$$(1.1)\qquad P\left[\int_0^u B(s)ds < r \text{ for all } 0 \le u \le t\right] \sim kr^{1/6}t^{-1/4} \text{ as } r^{1/6}t^{-1/4} \to 0,$$

which is a refinement of Sinai's estimates[12].

The above formula as well as the other ones follow systematically from the theorem in [3]: Let $(X(t), Y(t))$ be the Kolmogorov diffusion ([5]).

$$(1.2)\qquad Y(t) = y + B(t), \qquad X(t) = x + \int_0^t Y(s)ds.$$

Let T be the first hitting time to the positive y-axis:

$$(1.3)\qquad T = \inf\{t \ge 0; X(t) = 0, Y(t) \ge 0\}.$$

Hence $Y(T)$ is the hitting place on the positive y-axis. We denote by $E_{(x,y)}$ and $P_{(x,y)}$ the expectation and the probability measure for this diffusion respectively.

Theorem *([3]) For $\mu, \kappa \ge 0$ and $x \le 0$, $y \in \mathbb{R}$ it holds*

$$(1.4)\quad 1 - E_{(\bar\sigma x, \bar\sigma^{1/3}y)}\left[\exp\left\{-\sigma\mu T - \sqrt{\sigma}\sqrt{2\kappa}Y(T)\right\}\right] \sim \tilde K(x,y)\bar\sigma^{1/6}K(\kappa,\mu)\sigma^{1/4}$$

as $\bar\sigma^{1/6}\sigma^{1/4}$ tends to 0, where

$$K(\kappa,\mu) = \frac{3(\sqrt{2\kappa} + \sqrt{2\mu})\Gamma\left(\frac{1}{3}\right) 3^{1/3}}{\sqrt{\pi}\sqrt{\sqrt{2\kappa} + 2\sqrt{2\mu}}\Gamma\left(\frac{1}{6}\right) 2^{1/6}}$$

and

$$\tilde K(x,y) = \frac{|x|^{5/6}e^{-2(y^+)^3/9|x|}}{\Gamma\left(\frac{1}{3}\right)}\int_0^\infty dt e^{-t}\left(|x|t + 2|y^-|^3/9\right)^{1/6}\left(|x|t + 2(y^+)^3/9\right)^{-5/6}.$$

The proof depends heavily on a formula obtained by McKean[8].

We considered in [4] a generalization for this problem. We redefine $(X(t), Y(t))$, the odd additive functional, as

$$(1.5)\qquad Y(t) = y + B(t), \qquad X(t) = x + \int_0^t |Y(s)|^\alpha \operatorname{sgn}(Y(s))ds.$$

and we retain the notations T, $E_{(x,y)}$ and $P_{(x,y)}$. In [4], we were able to prove some weaker estimates:

Theorem *([4]) For $\alpha \geq 0$, $\nu := 1/(\alpha+2)$, $x \leq 0$ and $y = 0$, there exist positive constants $k'(\alpha)$, $k''(\alpha)$ such that*

$$(1.6)\qquad k'(\alpha)|x|^{\nu/2}t^{-1/4} < P_{(x,0)}[T > t] < k''(\alpha)|x|^{\nu/2}t^{-1/4}$$

for all small $|x|^{\nu/2}t^{-1/4}$.

The present paper proves the existence of the limit value for $|x|^{-\nu/2}t^{1/4}P_{(x,0)}[T > t]$, and more generally, we obtain similar results for some additive fuctionals that are not odd, or symmetric. We shall observe that the exponent $-1/4$ of time parameter in the above theorems varies between 0 and $-1/2$ in accordance with the skewness of additive functionals.

There are at least two approaches for our problem: the analytical one using Krein's spectral theory of strings(cf. Kotani–Watanabe[6]) and the probabilistic one based on the excursion theory, among which we mainly take the latter course.

Acknowledgement. The authors would thank M. Yor for helpful discussions.

2 The main theorem

In the remainder of this paper, almost all quantities depend on the parameter $\alpha > -1$ and $\bar{c} > 0$ without any mentioning. Let V be a function on the real line which is positive on $(0, \infty)$ and negative on $(-\infty, 0)$.

$$(2.7)\qquad V(x) = x^\alpha \text{ for } x > 0; V(0) = 0; V(x) = -|x|^\alpha/\bar{c} \text{ for } x < 0.$$

We define a diffusion $(X(t), Y(t))$ on $\mathbb{R}^2$ in a similar way and denote it by the same symbol:

$$(2.8)\qquad Y(t) = y + B(t), \qquad X(t) = x + \int_0^t V(Y(s))ds.$$

We denote by $E_{(x,y)}$ and $P_{(x,y)}$ the expectation and the probability measure for the diffusion started at $(x, y) \in \mathbb{R}^2$. Let T be the first hitting time to the positive y-axis as usual. Let T_0^Y be the first hitting time to x-axis:

$$(2.9)\qquad T_0^Y = \inf\{t \geq 0; Y(t) = 0\}$$

and for $\kappa, \lambda, \mu \geq 0$, $x \leq 0$, $y \in \mathbb{R}$ define $u_0(x, y) \equiv u_0(x, y; \mu)$ by

$$(2.10)\qquad u_0(x, y) = E_{(x,y)}[\exp(-\mu T)]$$

and more generally $u(x,y) \equiv u(x,y;\kappa,\lambda,\mu)$ by

$$u(x,y) = E_{(x,y)}\left[\exp\left\{-\mu T - \lambda X(T_0^Y \circ \theta_T) - \kappa(T_0^Y \circ \theta_T - T)\right\}\right] \tag{2.11}$$
$$\equiv E_{(x,y)}[\exp\{-\mu T\} F(\lambda V + \kappa; Y(T))] \tag{2.12}$$

here θ_t is the usual shift operator on the path space and the function $F(\lambda V + \kappa; z)$ is the unique bounded solution of $\frac{1}{2}F''(z) = (\lambda V(z) + \kappa)F(z)$ on $(0,\infty)$ with $F(0) = 1$. It is clear that $0 \le u(x,y) \le 1$, $u(0,0) = 1$ and $u_0(0,y) = 1$ for $y > 0$.

Theorem 1 *Define positive numbers $0 < \nu < 1$, $0 < \rho < 1$ by $\nu = 1/(\alpha+2)$ and $\bar{c}^{\nu} \sin \pi\nu(1-\rho) = \sin \pi\nu\rho$. Then for $\kappa, \lambda, \mu \ge 0$ there exists a positive constant $C(\kappa,\lambda,\mu)$ such that it holds*

$$1 - u(x,0;\sigma\kappa,\sigma^{1/2\nu}\lambda,\sigma\mu) \sim |x|^{\nu\rho} C(\kappa,\lambda,\mu)\sigma^{\rho/2} \tag{2.13}$$

as $|x|^{\nu\rho}\sigma^{\rho/2}$ tends to 0.

Corollary 1 *It holds that*

$$1 - u_0(x,0;\mu) \sim C(0,0,1)|x|^{\nu\rho}\mu^{\rho/2} \tag{2.14}$$

as $|x|^{\nu\rho}\mu^{\rho/2}$ tends to 0, in other words,

$$P\left[\int_0^s V(B(u))du < |x| \text{for all } 0 \le s \le t\right] \sim \frac{C(0,0,1)}{\Gamma(1-\rho/2)}|x|^{\nu\rho}t^{-\rho/2} \tag{2.15}$$

as $|x|^{\nu\rho}t^{-\rho/2}$ tends to 0.

We have, more generally, the following theorem.

Theorem 2 *There exist a positive constant $\tilde{C}(x,y)$ such that, for $\kappa,\lambda,\mu \ge 0$, $x \le 0$ and $y \in \mathbb{R}$, it holds that*

$$1 - u(\bar{\sigma}x, \bar{\sigma}^{1/\nu}y; \sigma\kappa, \sigma^{1/2\nu}\lambda, \sigma\mu) \sim \tilde{C}(x,y)\bar{\sigma}^{\nu\rho}C(\kappa,\lambda,\mu)\sigma^{\rho/2} \tag{2.16}$$

for positive $\sigma, \bar{\sigma}$ such that $\bar{\sigma}^{\nu\rho}\sigma^{\rho/2}$ tends to 0, where $C(\kappa,\lambda,\mu)$ is the same as in Theorem 1 and $\tilde{C}(x,y)$ is given by

$$\tilde{C}(x,y) = \frac{|x|^{1-\nu+\nu\rho}\exp\left\{-2\nu^2(y^+)^{1/\nu}/|x|\right\}}{\Gamma(\nu)} \int_0^\infty dte^{-t}\left(|x|t + \frac{2\nu^2}{\bar{c}}|y^-|^{1/\nu}\right)^{\nu\rho}\left(|x|t + 2\nu^2(y^+)^{1/\nu}\right)^{-1+\nu-\nu\rho}.$$

Remark 1. The function u has the following scaling property: for any $c > 0$

$$\begin{aligned} u(x,y;\kappa,\lambda,\mu) &\equiv u(c^{1/\nu}x, cy; c^{-2}\kappa, c^{-1/\nu}\lambda, c^{-2}\mu) \\ &\equiv E_{(c^{1/\nu}x,cy)}\left[\exp\left\{-c^{-2}\mu T\right\} F(\lambda V + \kappa; c^{-1}Y(T))\right] \end{aligned}$$

and the theorems are stated accordingly.

Remark 2. The distribution of $Y(T)$ under $P_{(0,y)}$ is known explicitly by Rogers-Williams[10], see also McGill[7]: For $y < 0$,

$$P_{(0,y)}[Y(T) \in d\eta] = \frac{\sin \pi\nu\rho}{\pi\nu\bar{c}^{\nu\rho}} |y|^{\rho} \eta^{1/\nu-1-\rho} \frac{d\eta}{\bar{c}^{-1}|y|^{1/\nu} + \eta^{1/\nu}}, \text{ on } \{\eta > 0\}. \tag{2.17}$$

Their methods do not seem to cover, however, the cases involving the stopping time T.

Remark 3. We denote by $\tau(t)$ the inverse of the local time of Y at 0. It is well known that $\int_0^{\tau(t)} V(B_u)du$ is a stable process with index ν and it holds

$$P\left[\int_0^{\tau(s)} V(B_u)du < |x| \text{ for all } s \le t\right] \sim \text{const } |x|^{\nu\rho} t^{-\rho} \tag{2.18}$$

as $|x|^{\nu\rho}t^{-\rho}$ tends to 0. See e.g. Bertoin[2]. This result has the same order as our Corollary 1 in the space variable $|x|$, but differs in the time variable t.

Remark 4. Note also that ρ is equal to the probability $P[\int_0^{\tau(t)} V(B_u)du > 0]$ independent of t, which can be proved using the result by Zolotarev[13].

3 Proof of Theorem 1

We denote by $L(t)$ the local time at 0 of $Y(T)$: $L_t = \lim_{\varepsilon\downarrow 0} \frac{1}{2\varepsilon}\int_0^t 1_{(-\varepsilon,\varepsilon)}(Y(u))du$ and by τ_t or $\tau(t)$ the right continuous inverse of L_t: $\tau_t \equiv \tau(t) = \inf\{u > 0;\ L_u > t\}$. Let n^+ and n^- be the Itô measure for positive and negative excursions respectively, and set $n = n^+ + n^-$.

We denote a general excursion by $\varepsilon = (\varepsilon_t; t \ge 0)$, its lifetime by $\zeta = \zeta(\varepsilon)$ and define a random time for $x \le 0$,

$$T(\varepsilon, x) = \inf\left\{0 \le t \le \zeta; x + \int_0^t V(\varepsilon_s)ds \ge 0\right\}. \tag{3.19}$$

We set $T(\varepsilon, x) = \zeta$ if there is no such t. It follows, through calculations of the Lévy measure of $X(\tau_t)$, that

$$\begin{aligned} n^+\left[1 - \exp\left\{-\lambda\int_0^\zeta V(\varepsilon_s)ds\right\}\right] &= \frac{\nu^{2\nu-1}2^\nu\Gamma(1-\nu)}{\Gamma(\nu)}\lambda^\nu, \\ n^-\left[1 - \exp\left\{\lambda\int_0^\zeta V(\varepsilon_s)ds\right\}\right] &= \frac{\nu^{2\nu-1}2^\nu\Gamma(1-\nu)}{\Gamma(\nu)}(\lambda/\bar{c})^\nu \end{aligned}$$

for positive λ and that

$$n^+\left[\int_0^\zeta V(\varepsilon_s)ds > \xi\right] = \frac{\nu^{2\nu-1}2^\nu}{\Gamma(\nu)}\xi^{-\nu}, \tag{3.20}$$

$$n^-\left[\int_0^\zeta V(\varepsilon_s)ds < -\xi\right] = \frac{\nu^{2\nu-1}2^\nu}{\Gamma(\nu)}(\bar{c}\xi)^{-\nu} \tag{3.21}$$

for positive ξ.

We have an integral equation for $u(x, 0)$.

Lemma 1 *We extend u for positive x by $u(x,0)=1$. Then it holds for $x<0$*

$$n\left[u\left(x+\int_0^\zeta V(\varepsilon_s)ds,0\right)-u(x,0)\right]$$

$$(3.22)\qquad = n\left[u\left(x+\int_0^\zeta V(\varepsilon_s)ds,0\right)\{1-e^{-\mu T(\varepsilon,x)}F(\lambda V+\kappa;\varepsilon(T(\varepsilon,x)))\}\right].$$

Proof. Let $F(z)=F(\lambda V+\kappa;z)$. Define $a\vee b=\max(a,b)$, $a\wedge b=\min(a,b)$ and

$$M(t)=u(X(\tau_t\wedge T),Y(\tau_t\wedge T))e^{-\mu(\tau(t)\wedge T))}$$

then $u(x,0)=E_{(x,0)}[M(t)]$ holds for any $t\geq 0$ and $x\leq 0$.

If $T\leq\tau_{t-}$ then $M(t)-M(t-)=0$.

If $\tau_{t-}<T\leq\tau_t$ then τ_t is the first hitting time of 0 by Y after T, i.e., $\tau_t=T_0^Y\circ\theta_T$. In this case,

$$M(t)-M(t-)=e^{-\mu T}F(Y(T))-e^{-\mu\tau(t-)}u(X(\tau_{t-}),0)$$

and $T-\tau_{t-}=T(\varepsilon,X(\tau_{t-}))$, here ε denotes the excursion started at τ_{t-} and ended at τ_t: $\varepsilon_s=Y(s+\tau_{t-})$, $s<\tau_t-\tau_{t--}$.

Finally if $\tau_t<T$ then

$$M(t)-M(t-)=e^{-\mu\tau(t)}u((X(\tau_t),0)-e^{-\mu\tau(t-)}u((X(\tau_{t-}),0)$$

and $\tau_t-\tau_{t-}=\zeta(\varepsilon)$. The master formula of excursion theory(cf. Revuz-Yor[11] page 439) tells us

$$\begin{aligned}&E_{(x,0)}\left[M(s)-M(0)\right]=\int_0^s dtE_{(x,0)}\Big[\\ &e^{-\mu\tau(t-)}n^+\left[e^{-\mu T(\varepsilon,X(\tau(t-)))}F(\varepsilon(T(\varepsilon,X(\tau_{t-}))))-u((X(\tau_{t-}),0);T(\varepsilon,X(\tau_{t-}))<\zeta\right]\\ &+e^{-\mu\tau(t-)}n\left[e^{-\mu\zeta}u\left(X(\tau_{t-})+\int_0^\zeta V(\varepsilon_s)ds,0\right)-u(X(\tau_{t-}),0);T(\varepsilon,X(\tau_{t-}))=\zeta\right].\end{aligned}$$

Recalling $u(x,0)=E_{(x,0)}\left[M(s)\right]$, we know that the integrand of the right hand side is identically null.

Since $X(\tau_t)$ is a ν-stable Lévy process, the paths are right continuous and the transition density decays as t goes to 0 uniformly outside any neighbohood of $X(0)$. The proof is hence complete if we show

$$\begin{aligned}&n^+\ \left[e^{-\mu T(\varepsilon,x)}F(\varepsilon(T(\varepsilon,x)))-u(x,0);T(\varepsilon,x)<\zeta\right]\\ &\quad+n\left[e^{-\mu\zeta}u\left(x+\int_0^\zeta V(\varepsilon_s)ds,0\right)-u(x,0);T(\varepsilon,x)=\zeta\right],\end{aligned}$$

which coincides with

$$n\left[u\left(x+\int_0^\zeta V(\varepsilon_s)ds,0\right)e^{-\mu T(\varepsilon,x)}F(\varepsilon(T(\varepsilon,x)))-u(x,0)\right],$$

is continuous on $\{x<0\}$ and its absolute value is dominated by an integrable function plus a constant. We need the following lemma.

Lemma 2 *The function $u(x,0)$ is infinitely differentiable on $\{x<0\}$ and $\frac{\partial u}{\partial x}$ is positive.*

Moreover, if $\alpha \geq 0$, $\nu \leq 1/2$ then $\frac{\partial^2 u}{\partial x^2}$ is positive, in particular $\frac{\partial u}{\partial x} = o(1/|x|)$ as $x \to -\infty$. If $-1<\alpha<0$, $1/2<\nu<1$ then $\frac{\partial u}{\partial x} = O(1/|x|)$ as $x \to -\infty$.

Remark. It can be proved for any $m>0$ and $n>0$, $\frac{\partial^n u}{\partial x^n} = O(|x|^{-m})$. However the statemant above is sufficient for our purpose.

Proof. Let $F(z) = F(\lambda V + \kappa; z)$. By the scaling property it holds that

$$u(x,0;\kappa,\lambda,\mu) = E_{(-1,0)}[e^{-|x|^{2\nu}\mu T} F(|x|^{\nu} Y(T))].$$

Since $F(z)$ decays exponentially as $z \to \infty$, the differentiation inside the expectation can be justified. Hence

$$\begin{aligned}\frac{\partial u}{\partial x}(x,0;\kappa,\lambda,\mu) &= E_{(-1,0)}\Big[2\nu|x|^{2\nu-1}\mu T e^{-|x|^{2\nu}\mu T} F(|x|^{\nu}Y(T)) \\ &\quad + e^{-|x|^{2\nu}\mu T}\nu|x|^{\nu-1}Y(T)(-F'(|x|^{\nu}Y(T))\Big].\end{aligned}$$

Here $-F'(z)$ is a positive decreasing function. The integrand is obviously positive and if $2\nu-1 \leq 0$ it is strictly decreasing in $|x|$. If $\nu > 1/2$, we use again the scaling property:

$$\frac{\partial u}{\partial x}(x,0;\kappa,\lambda,\mu) = \frac{1}{|x|}E_{(x,0)}\left[2\nu\mu T e^{-\mu T}F(Y(T)) + e^{-\mu T}\nu Y(T)(-F'(Y(T))\right].$$

The integrand is a bounded function of two variables T and $Y(T)$. □

End of the proof of Lemma 1. The difference between

$$n\left[u\left\{x + \int_0^{\zeta} V(\varepsilon_s)ds, 0\right\} e^{-\mu T(\varepsilon,x)} F(\varepsilon(T(\varepsilon,x))) - u(x,0)\right]$$

and $n\left[u(x + \int_0^{\zeta} V(\varepsilon_s)ds, 0) - u(x,0)\right]$ is bounded since it is dominated by

$$\begin{aligned} & n\left[1 - e^{-\mu T(\varepsilon,x)}F(\varepsilon(T(\varepsilon,x)))\right] \\ \equiv\ & n\left[1 - \exp\left\{-\mu T(\varepsilon,x) - \kappa(\zeta - T(\varepsilon,x)) - \lambda\int_{T(\varepsilon,x)}^{\zeta} V(\varepsilon_s)ds\right\}\right],\end{aligned}$$

which is also bounded by

$$\begin{aligned} & n\left[1 - \exp\left\{-(\mu\vee\kappa)\zeta - \lambda\int_0^{\zeta} V(\varepsilon_s)\vee 0 ds\right\}\right] \\ <\ & n[1 - \exp\{-(\mu\vee\kappa)\zeta\}] + n\left[1 - \exp\left\{-\lambda\int_0^{\zeta} V(\varepsilon_s)\vee 0 ds\right\}\right] < \infty.\end{aligned}$$

We divide $n\left[u(x + \int_0^{\zeta} V(\varepsilon_s)ds, 0) - u(x,0)\right]$ into two parts.

$n\left[|u(x+\int_0^\zeta V(\varepsilon_s)ds,0)-u(x,0)|;|\int_0^\zeta V(\varepsilon_s)ds|>1\right]$ is bounded because $0\le u\le 1$ and $n\left[|\int_0^\zeta V(\varepsilon_s)ds|>1\right]<\infty$ by (3.20) and (3.21). Integrating by parts,

$$\begin{aligned} & n\left[\left|u\left\{x+\int_0^\zeta V(\varepsilon_s)ds,0\right\}-u(x,0)\right|;\left|\int_0^\zeta V(\varepsilon_s)ds\right|<1\right] \\ = & \int_0^1 d\xi\frac{\partial u}{\partial x}(x+\xi,0)n^+\left[\xi<\int_0^\zeta V(\varepsilon_s)ds<1\right] \\ & -\int_{-1}^0 d\xi\frac{\partial u}{\partial x}(x+\xi,0)n^-\left[-1<\int_0^\zeta V(\varepsilon_s)ds<\xi\right], \end{aligned}$$

which is integrable, since it is a convolution of two integrable functions $\frac{\partial u}{\partial x}$ and $n^\pm[\xi<|\int_0^\zeta V(\varepsilon_s)ds|]$.

The continuity also follows using the above arguments since $T(\varepsilon,x)$ and $\varepsilon(T(\varepsilon,x))$ are continuous in x. □

Putting the explicit value of $n^\pm[\xi<|\int_0^\zeta V(\varepsilon_s)ds|]$ into the left side of Lemma 1, we have

$$\begin{aligned} & n\left[u(x+\int_0^\zeta V(\varepsilon_s)ds,0)-u(x,0)\right] \\ = & \frac{\nu^{2\nu-1}2^\nu}{\Gamma(\nu)}|x|^{-\nu}\Big(\{1-u(x,0)\}-\nu\int_0^1|1-t|^{-\nu-1}(\{1-u(xt,0)\}-\{1-u(x,0)\})dt \\ & -\nu\int_1^\infty\frac{|1-t|^{-\nu-1}}{\bar{c}^\nu}(\{1-u(xt,0)\}-\{1-u(x,0)\})dt\Big) \end{aligned}$$

The integral transform on this right side can be inverted.

Lemma 3 *For $v\in C^1((-\infty,0))$ such that $\frac{dv}{dx}$ is integrable, define $Lv(x)\in C((-\infty,0))$ by*

$$Lv(x)=\nu\int_0^1|1-t|^{-\nu-1}(v(xt)-v(x))dt+\nu\int_1^\infty\frac{|1-t|^{-\nu-1}}{\bar{c}^\nu}(v(xt)-v(x))dt.$$

If $v(x)-Lv(x)=f(x)$ then it holds

$$v(x)=\int_{-\infty}^0\frac{dt}{|t|}f(t)G(-\frac{|t|}{|x|}) \tag{3.23}$$

with a function $G(b)$ defined by

$$\begin{aligned} G(b) &= \tilde{G}(-\log(-b)), \quad b<0, \\ \tilde{G}(\xi) &= \lim_{A\to+\infty}\int_{-A}^A\frac{e^{-i\xi x}}{2\pi r(ix)}dx, \quad \xi\in\mathbb{R} \\ r(z) &= \frac{1}{\Gamma(\nu)\sin\pi\nu\rho}\Gamma(1-z)\Gamma(\nu+z)\sin\pi(\nu\rho+z), \quad z\in\mathbb{C} \end{aligned}$$

and with $\rho\in(0,1)$ defined by $\bar{c}^\nu=\dfrac{\sin\pi\nu\rho}{\sin\pi\nu(1-\rho)}$.

Moreover, if $\int_{-\infty}^{0} |x|^{-1-\nu\rho}|f(x)|dx < \infty$ *then*

$$\lim_{x\to -0} \frac{v(x)}{|x|^{\nu\rho}} = \frac{\Gamma(\nu)\sin\pi\nu\rho}{\pi\nu\rho\Gamma(\nu\rho)\Gamma(\nu-\nu\rho)} \int_{-\infty}^{0} |x|^{-1-\nu\rho} f(x)dx. \tag{3.24}$$

Remark. The Markov process associated to L turns into a Lévy process by taking the logarithm. This property enables us to calculate $\tilde{G}(\xi)$ and $r(z)$ explicitly. We prove this lemma at the end of this section.

Proof of Theorem 1. We set

$$f(x) = \frac{\Gamma(\nu)|x|^{\nu}}{\nu^{2\nu-1}2^{\nu}} n\left[u\left(x + \int_0^{\zeta} V(\varepsilon_s)ds, 0\right)(1 - e^{-\mu T(\varepsilon,x)}F(\varepsilon(T(\varepsilon,x))))\right]$$

for $x < 0$. It is obvious that $f(x)$ is positive everywhere and continuous. As we saw in the proof of Lemma 2, $n[1 - e^{-\mu T(\varepsilon,x)}F(\varepsilon(T(\varepsilon,x)))]$ is bounded, hence $f(x) = O(|x|^{\nu})$ as x tends to 0.

We have also

$$f(x) = \frac{\Gamma(\nu)|x|^{\nu}}{\nu^{2\nu-1}2^{\nu}} n\left[u\left(x + \int_0^{\zeta} V(\varepsilon_s)ds, 0\right) - u(x,0)\right].$$

By integration by parts, $n\left[|u(x + \int_0^{\zeta} V(\varepsilon_s)ds, 0) - u(x,0)|; \frac{|x|}{2} > |\int_0^{\zeta} V(\varepsilon_s)ds|\right]$ is dominated by const $\int_{-|x|/2}^{|x|/2} d\xi|\xi|^{-\nu}\frac{\partial u}{\partial x}(x+\xi,0)$. It is shown in Lemma 2 that $\frac{\partial u}{\partial x} = O(1/|x|)$ as $x \to -\infty$, which implies

$$n\left[\left|u\left(x + \int_0^{\zeta} V(\varepsilon_s)ds, 0\right) - u(x,0)\right|; \frac{|x|}{2} > \left|\int_0^{\zeta} V(\varepsilon_s)ds\right|\right] = O(|x|^{-\nu}).$$

Finally, $n\left[|u(x + \int_0^{\zeta} V(\varepsilon_s)ds, 0) - u(x,0)|; \frac{|x|}{2} < |\int_0^{\zeta} V(\varepsilon_s)ds|\right]$ is easily dominated by $n\left[\frac{|x|}{2} < |\int_0^{\zeta} V(\varepsilon_s)ds|\right] = O(|x|^{-\nu})$ since $0 \le u(x,0) \le 1$ for every x.

Therefore we have shown that $f(x) = O(1)$ as $x \to -\infty$, hence the integrability of $f(x)$ with respect to $|x|^{-1-\nu\rho}dx$ and the existence of the limit value for $(1 - u(x,0))/|x|^{\nu\rho}$ as $x \to -0$.

The statement of the theorem follows from the scaling property of u: for any $c > 0$,

$$u(x, y; c^2\kappa, c^{1/\nu}\lambda, c^2\mu) = u(c^{1/\nu}x, cy; \kappa, \lambda, \mu).$$

Setting $y = 0$, $c = \sqrt{\sigma}$, $u(x, 0; \sigma\kappa, \sigma^{1/2\nu}\lambda, \sigma\mu)$ is equal to $u(\sigma^{1/2\nu}x, 0; \kappa, \lambda, \mu)$, which satisfies $1 - u(\sigma^{1/2\nu}x, 0; \kappa, \lambda, \mu) \sim$ const $|\sigma^{1/2\nu}x|^{\nu\rho}$ as $\sigma^{1/2\nu}|x|$ tends to 0. □

Proof of Lemma 3. Define the functions $\tilde{v}$ and $\tilde{f}$ on $\mathbb{R}$ by $\tilde{v}(x) = v(-\exp(-x))$ and $\tilde{f}(x) = f(-\exp(-x))$. Define the integral operators $\tilde{L}$ and $\tilde{L}'$ by

$$\begin{aligned}\tilde{L}g(x) \; := \; & \nu \int_0^{\infty} |1 - e^{-X}|^{-\nu-1}e^{-X}\{g(x+X) - g(x)\}\,dX \\ & + \frac{\nu}{\bar{c}^{\nu}} \int_{-\infty}^{0} |1 - e^{-X}|^{-\nu-1}e^{-X}\{g(x+X) - g(x)\}\,dX\end{aligned} \tag{3.25}$$

$$\begin{aligned}\tilde{L}'g(x) \; := \; & \frac{\nu}{\bar{c}^{\nu}} \int_0^{\infty} |1 - e^{X}|^{-\nu-1}e^{X}\{g(x+X) - g(x)\}\,dX \\ & + \nu \int_{-\infty}^{0} |1 - e^{X}|^{-\nu-1}e^{X}\{g(x+X) - g(x)\}\,dX\end{aligned} \tag{3.26}$$

for $g \in C^1(\mathbb{R} \to \mathbb{C})$ with integrable $\frac{dg}{dx}$. It is obvious that $\tilde{L}\tilde{v}(x) \equiv Lv(-e^{-x})$.

To prove the lemma, it is sufficient to show that

$$\tilde{v}(x) = \int_{-\infty}^{\infty} \tilde{f}(y)\tilde{G}(y-x)dy. \tag{3.27}$$

We can show, by an standard argument, that $\tilde{f}$, $\tilde{v}$ and $\tilde{L}\tilde{v}$ belong to $\mathcal{S}'(\mathbb{R})$ and for any $\phi \in \mathcal{S}(\mathbb{R})$ it holds

$$(\tilde{v} - \tilde{L}\tilde{v})(\mathcal{F}\phi) = \tilde{v}((1-\tilde{L}')\mathcal{F}\phi) \tag{3.28}$$

It is elementary but tedious to verify that

$$\begin{aligned} 1 - r(i\xi) &= \frac{\nu}{\bar{c}^{\nu}} \int_0^{\infty} |1-e^X|^{-\nu-1}e^X \left\{e^{-i\xi X} - 1\right\} dX \\ &\quad + \nu \int_{-\infty}^{0} |1-e^X|^{-\nu-1}e^X \left\{e^{-i\xi X} - 1\right\} dX, \end{aligned} \tag{3.29}$$

$$(1-\tilde{L}')\mathcal{F}\phi(x) = \mathcal{F}[\phi(\xi)r(i\xi)](x), \quad \phi \in \mathcal{S} \tag{3.30}$$

and that the function $\frac{1}{r(i\xi)}$ on $\mathbb{R}$ is infinitely diffenrentiable and

$$r(-i\xi) = \overline{r(i\xi)}, \qquad \frac{1}{r(i\xi)} = \frac{\text{const}}{x^{\nu}} + O(x^{-1-\nu}) \text{ as } x \to \infty. \tag{3.31}$$

We next show for any $\chi \in \mathcal{S}(\mathbb{R})$

$$\mathcal{F}\left[\frac{\chi(x)}{r(ix)}\right] = \mathcal{F}\chi * \tilde{G}. \tag{3.32}$$

We start with

$$\mathcal{F}\left[\chi(x)\frac{1_{[-A,A]}(x)}{r(ix)}\right] = \frac{1}{\sqrt{2\pi}}\mathcal{F}\chi * \mathcal{F}\left[\frac{1_{[-A,A]}(x)}{r(ix)}\right], \quad A > 0. \tag{3.33}$$

It is clear that the left side of (3.33) converges to the left side of (3.32) as A tends to ∞.

The difference between $\tilde{G}(y)$ and $\frac{1}{\sqrt{2\pi}}\mathcal{F}[\frac{1_{[-A,A]}(x)}{r(ix)}](y)$ is dominated by $c_0 + c_1|y|^{\nu-1}$. To see this, it is sufficient to estimate $\int_A^{\infty} \frac{\exp(-iyx)}{r(ix)}dx$ for positive A.

By (3.31), $|\int_A^{\infty} \frac{\exp(-iyx)}{r(ix)}dx|$ is less than $c_2 \int_A^{\infty} \frac{|\exp(-iyx)|}{x^{1+\nu}}dx + c_3|\int_A^{\infty} \frac{\exp(-iyx)}{x^{\nu}}dx|$ with some positive constants c_2, c_3. $\int_A^{\infty} \frac{\exp(-iyx)}{x^{\nu}}dx = |y|^{\nu-1}\int_{|y|A}^{\infty} \frac{\exp(-i(\operatorname{sgn} y)x)}{x^{\nu}}dx$ is dominated by $M|y|^{\nu-1}$ with

$$M = \sup_{\xi>0}\left|\int_{\xi}^{\infty} \frac{\exp(-ix)}{x^{\nu}}dx\right|.$$

Since $\frac{1}{\sqrt{2\pi}}\mathcal{F}[\frac{1_{[-A,A]}(x)}{r(ix)}](y)$ converges to $\tilde{G}(y)$ for fixed $y > 0$, the right side of (3.33) also converges to the right side of (3.32) as A tends to ∞. Hence we have established the equation (3.32).

We now show (3.27). Set $\chi(x) = \phi(x)r(ix) \in \mathcal{S}(\mathbb{R})$. $\tilde{v}(\mathcal{F}\chi)$ is equal to $\tilde{v}((1-\tilde{L}')\mathcal{F}\phi)$ by (3.30), which is further equal to the left side of (3.28). Since $\tilde{v} - \tilde{L}\tilde{v} = \tilde{f}$, we have

$$\tilde{v}(\mathcal{F}\chi) = \tilde{f}(\mathcal{F}\phi).$$

Here $\mathcal{F}\phi$ is equal to the left side of (3.32). Hence we have

$$\begin{aligned}\tilde{v}(\mathcal{F}\chi) &= \tilde{f}(\mathcal{F}\chi * \tilde{G}) \\ &= \int_{-\infty}^{\infty} dy \tilde{f}(y) \int_{-\infty}^{\infty} d\xi \mathcal{F}\chi(\xi)\tilde{G}(y-\xi) \\ &= \int_{-\infty}^{\infty} d\xi \mathcal{F}\chi(\xi) \int_{-\infty}^{\infty} dy \tilde{f}(y)\tilde{G}(y-\xi).\end{aligned}$$

The both sides of (3.27) are continuous and bounded, and coincide in $\mathcal{S}'(\mathbb{R})$, hence they also coincide in $C_b(\mathbb{R})$.

If, moreover, $f(x)$ is integrable with respect to $|x|^{-1-\nu\rho}dx$ on the negative half line, then $\int_{-\infty}^{0} \frac{dt}{|t|} f(t) \frac{G(-|t|/|x|)}{|x|^{\nu\rho}}$ converges to the right side of (3.24) as x tends to -0 because of the following asymptotics:

$$\begin{aligned}G(b) &\sim \frac{\Gamma(\nu)\sin \pi\nu\rho}{\pi\nu\rho\Gamma(\nu\rho)\Gamma(\nu-\nu\rho)}|b|^{-\nu\rho} \text{ as } b \to -\infty, \\ G(b) &\sim \frac{\Gamma(\nu)\sin \pi\nu\rho}{\pi\nu(1-\rho)\Gamma(\nu\rho)\Gamma(\nu-\nu\rho)}|b|^{1-\nu\rho} \text{ as } b \to -0, \\ G(b) &\sim O(|b+1|^{\nu-1}) \text{ as } b \to -1.\end{aligned}$$

□

4 Proof of Theorem 2.

By the scaling property of Brownian motion, we have for positive c

$$u^c(x,y) := u(c^{1/\nu}x, cy; \kappa, \lambda, \mu) = u(x, y; c^2\kappa, c^{1/\nu}\lambda, c^2\mu).$$

In the previous section it is established with some constant $C > 0$

(4.34) $$1 - u^c(x,0) \sim Cc^\rho |x|^{\nu\rho} \text{ as } c \to +0$$

while in this section we prove

(4.35) $$1 - u^c(x,y) \sim Cc^\rho \tilde{C}(x,y) \text{ as } c \to +0$$

for fixed $x \le 0$, $y \in \mathbb{R}$.

4.1 The case of the starting point (x, y) in the third quadrant.

Let $Y_0 = y < 0$. In this case Y_t is negative until the hitting time T_0^Y. Applying the optional sampling theorem to the martingale $F(\lambda V_-; |Y_t|) \exp\left\{\lambda \int_0^t V(Y(s))ds\right\}$, $\lambda > 0$, we obtain

$$E_{(0,y)}[\exp\{\lambda X(T_0^Y)\}] = F(\lambda V_-; |y|),$$

where $F(\lambda V_-; z)$ is the unique bounded solution of $\frac{1}{2}F''(z) = \frac{\lambda}{c} z^\alpha F(z)$ on $\{z > 0\}$ with $F(0) = 1$.

The function $F(\lambda V_-; z)$ is expressed in terms of modified Bessel functions:

$$F(\lambda V_-; z) = \frac{2\nu^\nu}{\Gamma(\nu)}(2\lambda/\bar{c})^{\nu/2}\sqrt{z}K_\nu(2\nu z^{1/2\nu}(2\lambda/\bar{c})^{1/2}).$$

Here $\nu = 1/(2+\alpha)$ as usual. Using the formula (2.13.42) in Oberhettinger-Badii [9], we can invert the Laplace transform to obtain

$$\text{(4.36)} \quad E_{(0,y)}[X(T_0^Y) \in d\xi] = \frac{\nu^{2\nu}2^\nu|y|}{\Gamma(\nu)\bar{c}^\nu|\xi|^{1+\nu}} \exp\left\{-\frac{2\nu^2|y|^{1/\nu}}{\bar{c}|\xi|}\right\} d\xi \text{ on } \{\xi < 0\}.$$

It is obvious that the law of $X(T_0^Y)$ under $P_{(x,y)}$ is identical to that of $x + X(T_0^Y)$ under $P_{(0,y)}$.

By the strong Markov property of $(X(t), Y(t))$,

$$\begin{aligned} 1 - u^c(x,y) &= 1 - E_{(x,y)}[u^c(X(T_0^Y), 0)\exp\{-c^2\mu T_0^Y\}] \\ &= E_{(x,y)}[1 - u^c(X(T_0^Y), 0)] + O(E[1 - \exp\{-c^2\mu T_0^Y\}]). \end{aligned}$$

We see from (4.35) that $\dfrac{1-u^c(x,0)}{c^\rho}$ is dominated by $C'|x|^{\nu\rho}$ with some constant C', and it is well known that $E[1-\exp\{-c^2\mu T_0^Y\}] = 1 - \exp\{-\sqrt{2\mu}c|y|\} = O(c)$.

Combining this with the integrability of $|x + X(T_0^Y)|^{\nu\rho}$ we know

$$\lim_{c\to+0} \frac{1-u^c(x,y)}{Cc^\rho} = E_{(0,y)}\left[|x + X(T_0^Y)|^{\nu\rho}\right].$$

Putting (4.36) into the right hand side,

$$\begin{aligned} \tilde{C}(x,y) &= E_{(0,y)}\left[|x + X(T_0^Y)|^{\nu\rho}\right] \\ &= \int_0^\infty d\xi\,(|x| + \xi)^{\nu\rho}\frac{\nu^{2\nu}2^\nu|y|}{\Gamma(\nu)\bar{c}^\nu|\xi|^{1+\nu}}\exp\left\{-\frac{2\nu^2|y|^{1/\nu}}{\bar{c}|\xi|}\right\}. \end{aligned}$$

Replacing $\frac{2\nu^2|y|^{1/\nu}}{\bar{c}|\xi|}$ by t, we obtain

$$\tilde{C}(x,y) = \Gamma(\nu)^{-1}\int_0^\infty dt e^{-t}\left(|x|t + \frac{2\nu^2|y|^{1/\nu}}{\bar{c}}\right)^{\nu\rho} t^{-1+\nu-\nu\rho}. \qquad \square$$

4.2 The case of the starting point (x, y) in the second quadrant.

The function u^c satisfies in the left half plain $\{x < 0\}$ the differential equation

$$\frac{1}{2}\frac{\partial^2 u^c}{\partial y^2} + V(y)\frac{\partial u^c}{\partial y} = c^2\mu u^c$$

with the boundary condition on the positive y-axis:

$$u^c(0,y) = F(c^{1/\nu}\lambda V + c^2\kappa; y) \equiv F(\lambda V + \kappa; cy), \qquad y > 0,\ c > 0.$$

Let $U_c(y) = \int_{-\infty}^{0} dx e^{zx} u^c(x,y)$, $z \geq 0$. It follows from Theorem 1

$$(4.37) \qquad 1/z - U_c(0) = Cc^{\rho}\Gamma(1+\nu\rho)z^{-1-\nu\rho} \text{ as } c \to 0.$$

An integration by parts shows

$$\frac{1}{2}U_c''(y) = (zy^{\alpha} + c^2\mu)U^c(y) - y^{\alpha}F(\lambda V + \kappa; cy), \qquad y > 0.$$

Let $\phi_c(y)$, $\psi_c(y)$, $F_c(y)$ be the solutions of the equation $\frac{1}{2}f''(y) = (zy^{\alpha} + c^2\mu)f(y)$ on $(0,\infty)$ determined by the following conditions:

$$\begin{aligned} &\phi_c(0) = 1, \quad \phi'_c(0) = 0 \\ &\psi_c(0) = 0, \quad \psi'_c(0) = 1 \\ &F_c(0) = 1, \quad F_c(y) \text{ is bounded, i.e., } F_c(y) = F(zV + c^2\mu; y). \end{aligned}$$

Let $\phi_0(y)$, $\psi_0(y)$, $F_0(y)$ be the solutions of $\frac{1}{2}f''(y) = zy^{\alpha}f(y)$ normalized similarly.

We have by the method of variation of constants that

$$\begin{aligned} U_c(y) &= U_c(0)F_c(y) + 2F_c(y)\int_0^y \psi_c(\xi)\xi^{\alpha}F(\lambda V + \kappa; c\xi)d\xi \\ &\quad +2\psi_c(y)\int_y^{\infty} F_c(\xi)\xi^{\alpha}F(\lambda V + \kappa; c\xi)d\xi. \end{aligned}$$

Since $F(\lambda V + \kappa; c\xi)$ is a convex decreasing function it holds the inequality $0 < 1 - F(\lambda V + \kappa; c\xi) < \left|\frac{dF}{d\xi}(\lambda V + \kappa; 0)\right| c\xi$. Hence we have, for each fixed $y > 0$,

$$\begin{aligned} \int_0^y \psi_c(\xi)\xi^{\alpha}F(\lambda V + \kappa; c\xi)d\xi &= \int_0^y \psi_c(\xi)\xi^{\alpha}d\xi + O(c) \\ \int_y^{\infty} F_c(\xi)\xi^{\alpha}F(\lambda V + \kappa; c\xi)d\xi &= \int_y^{\infty} F_c(\xi)\xi^{\alpha}d\xi + O(c) \end{aligned}$$

as c tends to 0. Noting the differential equation of ψ_c and F_c we obtain

$$\begin{aligned} &2F_c(y)\int_0^y \psi_c(\xi)\xi^{\alpha}F(\lambda V + \kappa; c\xi)d\xi + 2\psi_c(y)\int_y^{\infty} F_c(\xi)\xi^{\alpha}F(\lambda V + \kappa; c\xi)d\xi \\ &= \frac{F_c(y)(\psi_c'(y) - 1) - \psi_c(y)F_c'(y)}{z} + O(c) = \frac{1 - F_c(y)}{z} + O(c). \end{aligned}$$

We need to prove that, for each fixed $y > 0$, $F_c(y) - F_0(y) = O(c)$ as $c \to 0$. By the Feynmann-Kac formula, $F_c(y) \equiv F(zV + c^2\mu; y)$ is the same as $E_y[\exp(-\int_0^{T_0}(zV(B_s) + c^2\mu)ds)]$. Here T_0 is the first hitting time to 0 by a standard Brownian motion B_s. Now it is clear that $0 < F_0(y) - F_c(y) = E_y[\exp(-\int_0^{T_0}(zV(B_s)ds)\,(1 - \exp(-c^2\mu T_0))] < E_y[1 - \exp(-c^2\mu T_0)] = 1 - \exp(-cy\sqrt{2\mu}) = O(c)$.

Combining these with (4.37) we have

$$(4.38) \qquad 1/z - U_c(y) = Cc^{\rho}\Gamma(1+\nu\rho)F(zV; y)z^{-1-\nu\rho} + O(c) \text{ as } c \to +0.$$

We can conclude by a standard argument that

$$1 - u^c(x,y) \sim Cc^{\rho}\tilde{C}(x,y) \text{ as } c \to +0$$

with

$$\int_0^{\infty} e^{-zx} dx \tilde{C}(x,y) = \Gamma(1+\nu\rho)F(zV;y)z^{-1-\nu\rho}.$$

Since $F(zV;y) = \frac{2\nu^{\nu}}{\Gamma(\nu)}(2z)^{\nu/2}\sqrt{y}K_{\nu}(2\nu y^{1/2\nu}(2z)^{1/2})$, we can invert the Laplace transform (see Oberhettinger-Badii [9] (13.45)) to obtain

$$\tilde{C}(x,y) = \frac{\Gamma(1+\nu\rho)|x|^{1/2+\nu\rho-\nu/2}}{\Gamma(\nu)2^{(1-\nu)/2}\nu^{1-\nu}y^{(1-\nu)/2\nu}} \exp\left\{-\frac{\nu^2 y^{1/\nu}}{|x|}\right\} W_{\frac{\nu}{2}-\frac{1}{2}-\nu\rho,\frac{\nu}{2}}(2\nu^2 y^{1/\nu}/|x|)$$

where $W_{\kappa,\mu}(z)$ is a Whittaker function defined by(see Abramowitz-Stegun [1] (13.1.33) and (13.2.5))

$$W_{\kappa,\mu}(z) = \frac{z^{1/2+\mu}e^{-z/2}}{\Gamma(1/2+\mu-\kappa)} \int_0^{\infty} dt e^{-zt} t^{-1/2+\mu-\kappa}(1+t)^{\mu+\kappa-1/2}.$$

Replacing $2\nu^2 y^{1/\nu}t/|x|$ by t, we obtain

$$\tilde{C}(x,y) = \frac{|x|^{1-\nu+2\nu\rho}\exp\left\{-2\nu^2 y^{1/\nu}/|x|\right\}}{\Gamma(\nu)} \int_0^{\infty} dt e^{-t} t^{\nu\rho}\left(|x|t + 2\nu^2 y^{1/\nu}\right)^{-1+\nu-\nu\rho}. \qquad \square$$

References

[1] M. Abramowitz, I. A. Stegun, *A Handbook of mathematical functions,* Dover, New York, 1964.

[2] J. Bertoin, *Lévy processes*, Cambridge Univ. Press, Cambridge, 1997.

[3] Y. Isozaki, S. Watanabe, An asymptotic formula for the Kolmogorov diffusion and a refinement of Sinai's estimates for the integral of Brownian motion, Proc. Japan Acad., **70A**, (1994), pp. 271–276.

[4] Y. Isozaki, Asymptotic estimates for the distribution of additive functionals of Brownian motion by the Wiener-Hopf factorization method, J. Math. Kyoto Univ., **36**, (1996), pp. 211-227.

[5] A. N. Kolmogorov, Zuffälige Bewegungen, Ann. Math. II., **35** (1934), pp. 116–117.

[6] S. Kotani, S. Watanabe, Krein's spectral theory of strings and generalized diffusion processes, *Functional Analysis in Markov Porcesses, ed. M. Fukushima*, Lecture Notes in Mathematics **923**, pp. 235–259, Springer-Verlag, Berlin, 1982.

[7] P. McGill, Wiener-Hopf factorization of Brownian motion, Prob. Th. Rel. Fields, **83**, (1989), pp. 355–389.

[8] H. P. McKean,Jr., A winding problem for a resonator driven by a white noise, J. Math. Kyoto Univ., **2** (1963), pp. 227–235.

[9] F. Oberhettinger, L. Badii, *Tables of Laplace Transforms,* Springer-Verlag, Berlin, 1973.

[10] L. C. G. Rogers, D. Williams, A differential equation in Wiener-Hopf theory, *Stochastic analysis and applications, ed. A. Truman, D. Williams*, Lecture Notes in Mathematics **1095**, pp. 187–199, Springer-Verlag, Berlin, 1984.

[11] D. Revuz, M. Yor, *Continuous martingales and Brownian motion*, Springer-Verlag, Berlin, 1991.

[12] Ya. G. Sinai, Distribution of some functionals of the integral of a random walk, Theor. Math. Phys., **90** (1992), pp. 219–241.

[13] V. M. Zolotarev, Mellin-Stieltjes transforms in probability theory, Theor. Prob. Appl., **2** (1957), pp. 433–460.

Y. Isozaki, Department of Mathematics, Graduate School of Science, Osaka University, Toyonaka 560-0043, Japan, E-mail: yasuki@math.sci.osaka-u.ac.jp
S. Kotani, Department of Mathematics, Graduate School of Science, Osaka University, Toyonaka 560-0043, Japan, E-mail: kotani@math.sci.osaka-u.ac.jp

Monotonicity Property for a Class of Semilinear Partial Differential Equations.

Siva Athreya*

Abstract

We establish a monotonicity property in the space variable for the solutions of an initial boundary value problem concerned with the parabolic partial differential equation connected with super-Brownian motion.

1. Introduction and main result

The "hot spots" conjecture of J. Rauch has been analyzed in certain planar domains D by R. Bañuelos and K. Burdzy [BB99] using probabilistic methods. In that paper, they synchronously couple two reflected Brownian motions and establish some monotonicity properties for solutions of the heat equation in D.

In this note we show that by applying similar coupling techniques to super-reflected Brownian motion one can prove a monotonicity property for solutions of a class of semilinear elliptic partial differential equations connected with super-Brownian motion. The result follows easily via an application of the existing machinery developed in the field of super-processes.

The purpose of this note is to enunciate the ease with which the probabilistic argument shown in [BB99] can be extended to provide a non-trivial result for solutions of certain semilinear partial differential equations. To the best of our knowledge the result presented in this note is new in the field of semilinear partial differential equations.

We consider solutions $u : \mathbb{R}_+ \times D \to \mathbb{R}_+$ of the following initial boundary value problem:

$$\frac{\partial u}{\partial t}(t,x) = \frac{1}{2}\Delta u(t,x) + \Phi(u(t,x)), \quad x \in D, \quad t > 0, \tag{1}$$

$$u(0,x) = \phi(x), \quad x \in D, \tag{2}$$

$$\frac{\partial u}{\partial n}(t,x) = 0, \quad x \in \partial D, \quad t > 0. \tag{3}$$

Here D is a bounded connected subset of $\mathbb{R}^2$, $\phi \in C^1$, and $\Phi : \mathbb{R}_+ \to \mathbb{R}_+$ is a continuously differentiable function of the form

$$\Phi(\lambda) = a_1\lambda - b_1\lambda^2 + \int_0^\infty (1 - \exp(-\lambda u) - \lambda u)\nu(du), \tag{4}$$

*Research supported in part by NSERC operating grant and Pacific Institute for the Mathematical Sciences

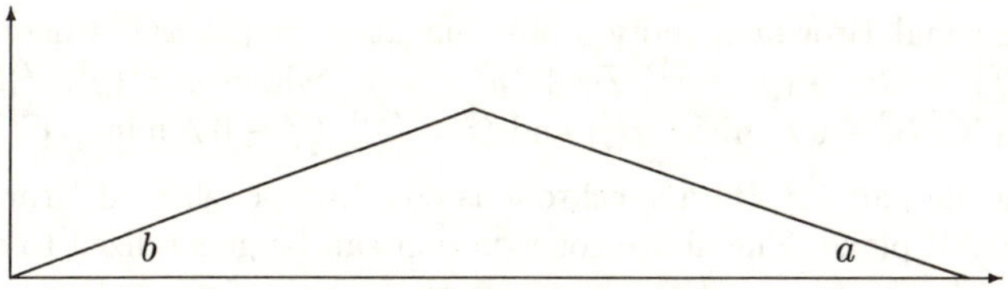

Figure 1: Obtuse triangle

where $a_1 \in \mathbb{R}, b_1 \geq 0$ and $\nu(du)$ is a regular Borel measure in $\mathbb{R}_+$, such that $\int_0^\infty u \wedge u^2 \nu(du) < \infty$. It is well known that solutions to this initial boundary value problem exist and are unique.

The partial differential equations that arise when $\Phi(\lambda) = -\lambda^2$(choose $b_1 = 1$ and $\nu(du) = 0$) and $\Phi(\lambda) = -\lambda^{1+\beta}, 0 < \beta < 1$ (choose $b_1 = 0$ and $\nu(du) = c_1 du/u^{2+\beta}, c_1 > 0$) are connected with binary and β-branching super-Brownian motion. The path properties of the process and various analytical properties of the partial differential equations have been extensively studied ([DIP89], [LG95], [Dyn91]).

Our main result is concerned with the direction of the gradient of $u(t,x)$ in obtuse triangles. We consider an obtuse triangle D, with the longest side of the triangle lying on the horizontal axis. The triangle lies in the first quadrant and one of its vertices is at the origin. The smaller sides of the triangle form angles a and b with the horizontal axis, with $a \in (-\frac{\pi}{2}, 0)$ and $b \in (0, \frac{\pi}{2})$ (See Figure 1). Let $\angle \nabla_x u(t,x)$ be the angle formed by the gradient $\nabla_x u(t,x)$ with the horizontal axis. For the remainder of this article (unless stated otherwise) D will denote this obtuse triangle.

Theorem 1. *Suppose that $u(0,x)$ is C^1 and $c < \angle \nabla_x u(0,x) < d$ for all $x \in D$, where $c > b - \frac{\pi}{2}$ and $d < \frac{\pi}{2} + a$. Then for every t and x we have*

$$min(a,c) \leq \angle \nabla_x u(t,x) \leq max(b,d).$$

The main idea of the proof is to construct a synchronous coupling of historical reflected Brownian motions in D. We use the same method as in [Kle89]. The final step uses the log-Laplace functional of super-reflected Brownian motion to obtain the monotonicity property for the solutions to (1).

Notation: We shall denote $x \in \mathbb{R}^2$ as $x = (x^1, x^2)$, where each $x^i \in \mathbb{R}$ (real numbers). For any Polish space G, $y \in G$, measurable function $f : G \to R$ and a measure m on G, we define $\langle f, m \rangle = \int_G f(y) dm(y)$ and δ_y as the dirac measure at the point y. For $d \geq 1$, $\mathcal{B}_d$ will denote the Borel σ-field on $\mathbb{R}^d$, $C_d = C([0,\infty), \mathbb{R}^d)$, $C_d^t = \{y \in C_d : y = y(\cdot \wedge t)\}$, $\mathcal{C}_d$ will denote the Borel σ-field of C_d, $M_F(C_d)$ the set of all finite measures on C_d, and $M_F(C_d^t) = \{m \in M_F(C_d) : y(\cdot \wedge t) = y \;\; m \;\; \text{a.e. } y\}$. For $z, w \in C_d$, we define

$$(z/s/w)(u) = \begin{cases} z(u) & \text{if } u < s, \\ w(u-s) & \text{if } u \geq s. \end{cases}$$

2. Synchronous coupling of historical reflected Brownian motions

First we provide a brief construction of "synchronous coupling" of reflected Brownian motions. We refer the reader to [BB99] for further details. Let $B_t = (B_t^1, B_t^2)$ be a two dimensional Brownian motion starting at $x = (x^1, x^2)$, where $x^2 > 0$ and $C_t = (C_t^1, C_t^2) = (B_t^1 + (y^1 - x^1), B_t^2 + (y^2 - x^2))$, where $y = (y^1, y^2)$ with $y^2 > 0$. Define $\xi_t^x = (B_t^1, B_t^2 - 0 \wedge \min_{s \leq t} B_s^2)$ and $\xi_t^y = (C_t^1, C_t^2 - 0 \wedge \min_{s \leq t} C_s^2)$.

We shall call the pair (ξ_t^x, ξ_t^y) a synchronous coupling of reflected Brownian motions in the upper half-plane. The above construction can be generalized to any polygonal domain $D \subset \mathbb{R}^2$. The construction gives us for every pair of $x, y \in D$, a pair of reflected Brownian motions (ξ_t^x, ξ_t^y) starting at (x, y), such that $\xi_t^x - \xi_t^y$ remains constant in the time periods when both processes are in the interior of the domain.

We proceed to define a historical process H_t on $D \times D$. Let ξ_t^x, ξ_t^y be spatially coupled reflected Brownian motions as in [BB99], starting at x and y in D. Then $\xi_t = (\xi_t^x, \xi_t^y)$ is a continuous Markov process taking values in $D \times D$. The path valued process $\bar{\xi}_t = (\xi^x(.\wedge t), \xi^y(.\wedge t))$ will be the motion process for H_t and Φ (as in 4) will describe its branching mechanism. Applying Theorem 2.2.3. in [DP91] to the process $(\bar{\xi}, \Phi)$, we see that the measure valued (ξ, Φ)-historical process H_t exists. The semi-group $Q_{r,t}$ of H is determined by

$$Q_{r,t}(\exp(-\langle m, \psi \rangle)) = \exp(-\langle m, V_{r,t}(\psi) \rangle),$$

where $V_{r,t}(\psi(z)) = \langle \delta_z, V_{r,t}(\psi) \rangle$ is the unique solution of

$$V_{r,t}(\psi(z)) = P_{r,y(r)}(\psi(z/r/\bar{\xi}_{t-r})) + \int_0^{t-r} P_{r,y(r)}(\Phi(V_{s+r,t}(z/r/\bar{\xi}_s)))ds, \tag{5}$$

where ψ is a bounded positive measurable function, $m \in M_F(C_4)^r$ and $z \in C_4^r$, $r \leq t$.

Since the branching mechanism is spatially homogeneous, the measures $H_t^1(dx) = H_t(dx \times C_2)$ and $H_t^2(dy) = H_t(C_2 \times dy)$ are the historical processes associated with ξ_t^x and ξ_t^y. We call the pair $\{H_t^1, H_t^2\}$ as synchronously coupled historical Brownian motion.

Let $\Pi_t(y) = y(t)$ be the coordinate map in C_2. Keeping the particle picture in mind, $X_t = H_t^1 \Pi_t^{-1}$ and $Y_t = H_t^2 \Pi_t^{-1}$ are 2-dimensional super-reflected Brownian motions in $\mathbb{R}^2$ with the branching mechanism Φ and the motion process ξ^x and ξ^y respectively. This is verified in Theorem 2.2.4 [DP91]. Let ϕ be a positive bounded measurable function in D. The log-Laplace functional of X_t with $X_0 = \mu$ (see [Fit88]) is given by

$$\langle \mu, V_t \phi \rangle = -\log E_\mu(\exp(-\langle X_t, \phi \rangle)), \quad \mu \in M_F(D), \tag{6}$$

where $v(t,x) = \langle \delta_x, V_t \phi \rangle = V_t \phi(x)$ is the the unique solution of

$$v(t,x) = S_t(\phi(x)) + \int_0^t S_t(\Phi(v(t-u, \cdot))(x) du. \tag{7}$$

As ϕ and Φ are continuously differentiable, by Theorem 1.5 (page 187) in [Paz83], we conclude that $v(t,x)$ is a solution to (1), with initial conditions (2). As v solves (7) and the fact that S_t is the semi-group of reflected Brownian motion imply that v satisfies the Neumann boundary conditions (3).

3. Proof of Theorem 1

Let x, y in D be chosen so that they satisfy

1. $x^1 < y^1$,
2. they are on a line K inside the domain D, such that

$$\max(b,d) - \frac{\pi}{2} \leq \angle K \leq \min(a,c) + \frac{\pi}{2}. \tag{8}$$

Let H_t^1 and H_t^2 be the coupled historical reflected Brownian motion described in the previous section, with H_t starting from $m = \delta_{(\bar{x},\bar{y})}$. Take any (w,z) in the support of the historical measure H_t and let $K(s)$ be the line joining the points $w(s)$ and $z(s)$. The sides of the obtuse triangle are not perpendicular to each other; this and the fact that $w(s)$ and $z(s)$ are "typical" Brownian paths ensures that $K(s)$ is defined in a unique way for all s a.s. and that we will never have $w(s) = z(s)$. The direction of K either remains constant or approaches the direction of the side which is currently reflecting one of the paths [BB99].

For all s, the angle $\angle K(s)$ can never leave the interval $[\max(b,d) - \frac{\pi}{2}, \min(a,c) + \frac{\pi}{2}]$. We always have

$$w(s)^1 < z(s)^1, \quad \text{for all } \ s \geq 0. \tag{9}$$

Let $\phi^1(x,y) = \phi(x)$ and $\phi^2(x,y) = \phi(y)$. Since every path in the support of H starts at (x,y) and ϕ satisfies the hypothesis of the theorem, by (9) we know that $\phi^1(w(s),z(s)) < \phi^2(w(s),z(s))$, for all $s \geq 0$. Using the relationship established earlier between super-reflected Brownian motions and solutions to (1), we may deduce

$$\begin{aligned} u(t,x) &= -\log E(\exp(-\langle X_t, \phi\rangle)) \\ &= -\log E(\exp(-\int H_t^1(dy_1)\phi(y_1(t)))) \\ &= -\log E(\exp(-\int H_t(dy_1,dy_2)\phi^1(y_1(t),y_2(t)))) \\ &\leq -\log E(\exp(-\int H_t(dy_1,dy_2)\phi^2(y_1(t),y_2(t)))) \\ &= -\log E(\exp(-\int H_t^2(dy_2)\phi(y_2(t)))) \\ &= -\log E(\exp(-\langle Y_t, \phi\rangle)) = u(t,y). \end{aligned} \tag{10}$$

Hence the solution $u(t,x)$ is monotonically increasing in x^1 for all $(x^1,x^2) \in K \cap D$. Since this is true for every line K satisfying (8), the gradient of $u(t,x)$ must satisfy the condition stated in the theorem. □

4. Remarks

The assumption that D is a triangle plays no role in the arguments described in the proof of Theorem 1. The only property D needs to satisfy is that, if two reflected Brownian motions in the domain are synchronous coupled, then the left particle will stay left of the other particle for all time t. We refer the reader to [BB99] or [Ath98] for examples of certain polygonal and non-convex domains D in $\mathbb{R}^2$ with the above property.

We also wish to point out that the special form of Φ, which enabled the particle representation of the partial differential equation (1), was crucial for the result to hold true. The Feynman-Kacs representation of solutions to the partial differential equation do not yield monotonicity properties of the solution. Hence there does not seem to be an obvious probabilistic method to extend the above results to other partial differential equations.

Acknowledgments : I would like to thank my thesis advisor, Prof. K. Burdzy for suggesting this problem to me and for all the discussions that went into resolving the technical details.

References

[Ath98] Siva Athreya. Probability and semilinear partial differential equations. *Ph.D. thesis*, 1998.

[BB99] R. Bañuelos and K. Burdzy. On the hot-spots conjecture of J. Rauch. *J. Func. Anal.*, 164:1-33, 1999.

[DIP89] D. Dawson, I. Iscoe, and E. Perkins. Super Brownian motion: path properties and hitting probabilities. *Probability Theory and Related Fields*, 83:135–206, 1989.

[DP91] D. Dawson and E. Perkins. Historical processes. *Memoirs of the American Mathematical Society*, 454, 1991.

[Dyn91] E.B. Dynkin. A probabilistic approach to one class of non-linear differential equations. *Probability Theory and Related Fields*, 89:89–115, 1991.

[Fit88] P. Fitzsimmons. Construction and regularity of measure-valued branching processes. *Israel J. Math.*, 64:337–361, 1988.

[Kle89] A. Klenke. Multiple scale analysis of clusters in spatial branching models. *Annals of Probability*, 83:135–206, 1989.

[LG95] J.F. Le Gall. Brownian snake and partial differential equations. *Probability Theory and Related Fields*, 102:393–432, 1995.

[Paz83] A. Pazy. *Semigroups of Linear Operators and Applications to Partial Differential Equations.* Springer-Verlag, New York, 1983.

Siva Athreya
1984, Mathematics Road
Department of Mathematics, The University of British Columbia
Vancouver, B.C., Canada V6T 1Z2.
Email: athreya@math.ubc.ca

Fast Sets and Points for Fractional Brownian Motion

By

Davar Khoshnevisan* & Zhan Shi
University of Utah & Université Paris VI

Summary. In their classic paper, S. Orey and S.J. Taylor compute the Hausdorff dimension of the set of points at which the law of the iterated logarithm fails for Brownian motion. By introducing "fast sets", we describe a converse to this problem for fractional Brownian motion. Our result is in the form of a limit theorem. From this, we can deduce refinements to the aforementioned dimension result of Orey and Taylor as well as the work of R. Kaufman. This is achieved via establishing relations between stochastic co-dimension of a set and its Hausdorff dimension along the lines suggested by a theorem of S.J. Taylor.

Keywords and Phrases. Fast point, fast set, fractional Brownian motion, Hausdorff dimension.

A.M.S. 1991 Subject Classification. 60G15, 60G17, 60J65.

§1. Introduction

Suppose $W \triangleq (W(t); t \geqslant 0)$ is standard one–dimensional Brownian motion starting at 0. Continuity properties of the process W form a large part of classical probability theory. In particular, we mention A. Khintchine's law of the iterated logarithm (see, for example, [21, Theorem II.1.9]): for each $t \geqslant 0$, there exists a null set $\mathcal{N}_1(t)$ such that for all $\omega \notin \mathcal{N}_1(t)$,

$$\limsup_{h \to 0^+} \frac{|W(t+h) - W(t)|}{\sqrt{h \ln\ln(1/h)}} = \sqrt{2}. \tag{1.1}$$

Later on, P. Lévy showed that $\cup_{t \geqslant 0} \mathcal{N}_1(t)$ is not a null set. Indeed, he showed the existence of a null set $\mathcal{N}_2$ outside which

$$\limsup_{h \to 0^+} \sup_{0 \leqslant r \leqslant 1} \frac{|W(r+h) - W(r)|}{\sqrt{h \ln(1/h)}} = \sqrt{2}. \tag{1.2}$$

* Research supported by a grant from the National Security Agency

See [13, p. 168] or [21, Theorem I.2.7], for example. It was observed in [18] that the limsup is actually a limit. Further results in this direction can be found in [2, p. 18]. The apparent discrepancy between (1.1) and (1.2) led S. Orey and S.J. Taylor to further study the so-called fast (or rapid) points of W. To describe this work, for all $\lambda > 0$, define $\mathrm{F}_1(\lambda)$ to be the collection of all times $t \geqslant 0$ at which

$$\limsup_{h \to 0^+} \frac{|W(t+h) - W(t)|}{\sqrt{h \ln(1/h)}} \geqslant \lambda\sqrt{2}.$$

The main result of [18] is that with probability one,

$$\dim\left(\mathrm{F}_1(\lambda)\right) = 1 - \lambda^2. \tag{1.3}$$

One can think of this as the multi-fractal analysis of white noise. Above and throughout, "dim(A)" refers to the Hausdorff dimension of A. Furthermore, whenever $\dim(A)$ is (strictly) negative, we really mean $A = \varnothing$. Orey and Taylor's discovery of Eq. (1.3) relied on special properties of Brownian motion. In particular, they used the strong Markov property in an essential way. This approach has been refined in [3, 4, 11], in order to extend (1.3) in several different directions.

Our goal is to provide an alternative proof of Eq. (1.3) which is robust enough to apply to non-Markovian situations. We will do so by (i) viewing $\mathrm{F}_1(\lambda)$ as a random set and considering its hitting probabilities; and (ii) establishing (within these proofs) links between Eqs. (1.2) and (1.3).

To keep from generalities, we restrict our attention to fractional Brownian motion. With this in mind, let us fix some $\alpha \in]0, 2[$ and define $X \triangleq \left(X(t); t \geqslant 0\right)$ to be a one-dimensional Gaussian process with stationary increments, mean zero and incremental standard deviation given by,

$$\left\| X(t) - X(s) \right\|_2 = |t - s|^{\alpha/2}.$$

See (1.8) for our notation on $L^p(\mathbb{P})$ norms.

The process X is called **fractional Brownian motion** with index α — hereforth written as fBM(α). We point out that when $\alpha = 1$, X is Brownian motion.

Let $\overline{\dim}_M(E)$ denote the upper Minkowski dimension of a Borel set $E \subset \mathbb{R}^1$; see references [17, 24]. Our first result, which is a fractal analogue of Eq. (1.2), is the following limit theorem:

Theorem 1.1. *Suppose X is fBM(α) and $E \subset [0, 1]$ is closed. With probability one,*

$$\limsup_{h \to 0^+} \sup_{t \in E} \frac{|X(t+h) - X(t)|}{h^{\alpha/2}\sqrt{\ln(1/h)}} \leqslant \sqrt{2\, \overline{\dim}_M(E)}. \tag{1.4}$$

On the other hand, with probability one,

$$\sup_{t \in E} \limsup_{h \to 0^+} \frac{|X(t+h) - X(t)|}{h^{\alpha/2}\sqrt{\ln(1/h)}} \geqslant \sqrt{2 \dim(E)}. \tag{1.5}$$

Loosely speaking, when $\alpha = 1$, Theorem 1.1 is a converse to (1.3). For all $\lambda \geqslant 0$, define $\mathcal{F}_\alpha(\lambda)$ to be the collection of all closed sets $E \subset [0,1]$ such that

$$\limsup_{h\to 0+} \sup_{t\in E} \frac{|X(t+h) - X(t)|}{h^{\alpha/2}\sqrt{\ln(1/h)}} \geqslant \lambda\sqrt{2}.$$

One can think of the elements of $\mathcal{F}_\alpha(\lambda)$ as λ–**fast sets**. Theorem 1.1 can be recast in the following way.

Corollary 1.2. *Suppose X is fBM(α) and $E \subset [0,1]$ is closed. If $\overline{\dim}_M(E) < \lambda^2$, then $E \notin \mathcal{F}_\alpha(\lambda)$ almost surely. On the other hand, if $\dim(E) \geqslant \lambda^2$, then $E \in \mathcal{F}_\alpha(\lambda)$.*

Remark 1.2.1. An immediate consequence of Theorem 1.1 is the following extension of (1.2):

$$\limsup_{h\to 0+} \sup_{0\leqslant t\leqslant 1} \frac{|X(t+h) - X(t)|}{h^{\alpha/2}\sqrt{\ln(1/h)}} = \sqrt{2}, \qquad \text{a.s.}$$

When $\alpha \in]0,1]$, this is a consequence of [15, Theorem 7]. When $\alpha \in]1,2[$, the existence of such a modulus of continuity is mentioned in [16, Section 5].

A natural question is: can one replace E by a random set? The first random set that comes to our mind is the zero set. When $\alpha = 1$, the process is Brownian motion. Its Markovian structure will be used to demonstrate the following.

Theorem 1.3. *Suppose W is Brownian motion. Let $\mathcal{Z} \triangleq \{s \in [0,1] : W(s) = 0\}$. Then, with probability one,*

$$\limsup_{h\to 0+} \sup_{t\in\mathcal{Z}} \frac{|W(t+h)|}{\sqrt{h\ln(1/h)}} = \sup_{t\in\mathcal{Z}} \limsup_{h\to 0+} \frac{|W(t+h)|}{\sqrt{h\ln(1/h)}} = 1.$$

Thus, the escape of Brownian motion from zero is slower than Lévy's modulus (1.2).

Next, we come to dimension theorems; see (1.3) for an example. Define the λ–**fast points** for fBM(α) as follows:

$$\mathrm{F}_\alpha(\lambda) \triangleq \Big\{t \in [0,1] : \limsup_{h\to 0+} \frac{|X(t+h) - X(t)|}{h^{\alpha/2}\sqrt{\ln(1/h)}} \geqslant \lambda\sqrt{2}\Big\}. \tag{1.6}$$

In particular, $\mathrm{F}_1(\lambda)$ denotes the λ–fast points of Brownian motion. In [9], R. Kaufman has shown that for any closed $E \subset [0,1]$ and every $\lambda > 0$, with probability one,

$$\dim(E) - \lambda^2 \leqslant \dim\big(E \cap \mathrm{F}_1(\lambda)\big). \tag{1.7}$$

Moreover, there exists a certain fixed closed set $E \subset [0,1]$ for which the above is an equality. Our next aim is to show that the inequality in (1.7) is an equality for many sets $E \subset [0,1]$, and that this holds for all $\mathrm{fBM}(\alpha)$'s. More precisely, we offer the following:

Theorem 1.4. *Suppose X is $\mathrm{fBM}(\alpha)$, $E \subset [0,1]$ is closed and $\lambda > 0$. Then, with probability one,*

$$\dim(E) - \lambda^2 \leqslant \dim\left(E \cap \mathrm{F}_\alpha(\lambda)\right) \leqslant \dim_{\mathrm{P}}(E) - \lambda^2,$$

where $\dim_{\mathrm{P}}$ denotes packing dimension.

See [17] and [24] for definitions and properties of $\dim_{\mathrm{P}}$.

In particular, (1.3) holds for the fast points of any fractional Brownian motion. Moreover, when $E = \mathcal{Z}$ and $\alpha = 1$, we have the following dimension analogue of Theorem 1.3. Note that $\mathrm{F}_1(\lambda)$ is the set of fast points of W as defined earlier. In other words, it is defined by (1.6) with X replaced by W.

Theorem 1.5. *Suppose W is Brownian motion and $\mathcal{Z} \triangleq \{s \in [0,1]: W(s) = 0\}$. Then, for all $\lambda > 0$, with probability one,*

$$\dim\left(\mathcal{Z} \cap \mathrm{F}_1(\lambda)\right) = \frac{1}{2} - \lambda^2.$$

A natural question which we have not been able to answer is the following:

Problem 1.5.1. *Does Theorem 1.5 have a general analogue for all $\mathrm{fBM}(\alpha)$'s?*

The proofs of Theorems 1.1, 1.3, 1.4 and 1.5 rely on parabolic capacity techniques and entropy arguments. The entropy methods follow the arguments of [18] closely. On the other hand, the parabolic capacity arguments rely on relationships between the Hausdorff dimension of random sets and stochastic co–dimension (see §2). The latter is a formalization of a particular application of [23, Theorem 4], which can be found in various forms within the proofs of [1, 7, 14, 19]. We suspect our formulation has other applications. In §3, we demonstrate (1.4) while (1.5) and the first inequality (i.e., the lower bound) of Theorem 1.4 are derived in §4. The proof of the upper bound of Theorem 1.3 appears in §5; the upper bounds of Theorems 1.4 and 1.5 can be found in §6 and §7, respectively; and the lower bounds for Theorems 1.3 and 1.5 are proved simultaneously in §8.

We conclude the Introduction by mentioning some notation which will be utilized throughout this article. Define the function ψ as

$$\psi(h) \triangleq \sqrt{2\ln(1/h)}, \qquad h \in]0,1[.$$

By $\overline{\Phi}$ we mean the tail of a standard normal distribution, i.e.

$$\overline{\Phi}(x) \triangleq \frac{1}{\sqrt{2\pi}} \int_x^\infty e^{-u^2/2}\, du, \qquad x \in]-\infty, \infty[\,.$$

Furthermore, $(\Omega, \mathcal{F}, \mathbb{P})$ denotes our underlying probability space. For any real variable Z on $(\Omega, \mathcal{F}, \mathbb{P})$ and for every $p > 0$ we write the $L^p(\mathbb{P})$–norm of Z by,

$$\|Z\|_p \triangleq \left(\int_\Omega |Z(\omega)|^p \, \mathbb{P}(d\omega) \right)^{1/p}. \tag{1.8}$$

Throughout, X denotes fBM(α) for any $\alpha \in]0,2[$. However, when we wish to discuss Brownian motion specifically (i.e., when $\alpha = 1$), we write W instead. In accordance with the notation of Theorem 1.3, $\mathcal{Z}$ will always denote the zero set of W restricted to $[0,1]$. Finally, the collection of all atomless probability measures on a set E is denoted by $\mathcal{P}_+(E)$.

Remark. Since the first circulation of this paper, many of the 'gaps' in the inequalities of this paper have been bridged. For instance, in Theorem 1.1, both constants of (1.4) and (1.5) can be computed. This and related material can be found in [10].

Acknowledgements. Much of this work was done while the first author was visiting Université Paris VI. We thank L.S.T.A. and Laboratoire de Probabilités for their generous hospitality. Our warmest thanks are extended to Steve Evans, Mike Marcus, Yuval Peres and Yimin Xiao. They have provided us with countless suggestions, references, corrections and their invaluable counsel. In particular, it was Yuval Peres who showed us the rôle played by packing dimensions as well as allowing us to use his argument (cf. [10]) in the proof of Theorem 2.5.

§2. Preliminaries on Dimension

In this section, we discuss a useful approach to estimating Hausdorff dimensions of random sets via intersection probabilities.

Let $\mathbf{S}_0^1$ denote the collection of all Borel measurable subsets of $[0,1]$. We say that $E : \Omega \mapsto \mathbf{S}_0^1$ is a **random set**, if $\mathbf{1}_E(\omega) : \Omega \times \mathbf{S}_0^1 \ni (\omega, E) \mapsto \{0,1\}$ is a random variable in the product measure space. An important class of random sets are the closed stochastic images $E \triangleq \overline{S[0,1]} \triangleq \overline{\{S(t);\ t \in [0,1]\}}$, where S is a stochastic process with càdlàg sample paths.

Let us begin with some preliminaries on Hausdorff dimension; see [8, 16, 24] for definitions and further details. Given $s \geqslant 0$ and a Borel set $E \subset [0,1]$, let $\Lambda^s(E)$ denote the s–dimensional Hausdorff measure of E. Recall that the **Hausdorff dimension** — $\dim(E)$ — of E is defined by: $\dim(E) \triangleq \inf\{s > 0 : \Lambda^s(E) < \infty\}$. When it is finite, $\Lambda^s(E)$ extends nicely to a Carathéodory outer measure on analytic subsets of $[0,1]$. By a slight abuse of notation, we continue to denote this outer measure by Λ^s.

Suppose E is a random set. Since we can economically cover E with intervals with rational endpoints, $\dim(E)$ is a random variable. We will use this without further mention.

Typically, computing upper bounds for $\dim(E)$ is not very difficult: find an economical cover (I_j) of E, whose diameter is h or less and compute $\sum_j |I_j|^s$. Obtaining good lower bounds for $\dim(E)$ is the harder of the two bounds. The standard method for doing this is to utilize the connections between Hausdorff dimension and potential theory. For any $\mu \in \mathcal{P}_+(E)$ and all $\beta > 0$, define,

$$A_\beta(\mu) \triangleq \sup_{0<h\leqslant 1/2} \sup_{t\in[h,1-h]} \frac{\mu[t-h,t+h]}{h^\beta}. \tag{2.1}$$

(It is possible that $A_\beta(\mu) = \infty$.) We need the following connection between Hausdorff dimension and potential theory; while it is only half of Frostman's lemma of classical potential theory, we refer to it as 'Frostman's lemma' for brevity.

Lemma 2.1. **(Frostman's Lemma**; [8, p. 130]) *Suppose $E \in \mathbf{S}_0^1$ satisfies $\beta < \dim(E)$. Then there exists $\mu \in \mathcal{P}_+(E)$ for which $A_\beta(\mu) < \infty$.*

Thus, a method for obtaining lower bounds on $\dim(E)$ is this: find a probability measure μ which lives on E and show that $A_\beta(\mu) < \infty$. If this can be done for some $\beta > 0$, then $\dim(E) \geqslant \beta$. In general, this is all which can be said. However, if the set E in question is a random set in the sense of the first paragraph of this section, there is an abstract version of [23, Theorem 4] which can be used to bound $\dim(E)$ from below; see also [1]. We shall develop this next. Define the **upper stochastic co-dimension** ($\overline{\text{co-dim}}$) of a random set E by

$$\overline{\text{co-dim}}(E) \triangleq \inf\left\{\beta \in [0,1] : \ \forall\, G \in \mathbf{S}_0^1 \text{ with } \dim(G) > \beta,\ \mathbb{P}(E \cap G \neq \varnothing) = 1\right\}. \tag{2.2}$$

In order to make our definition sensible and complete, we need to define $\inf \varnothing \triangleq 1$.

Remark 2.1.1. In applications, we often need the following fact: if $G \in \mathbf{S}_0^1$ satisfies $\dim(G) > \overline{\text{co-dim}}(E)$, then $\mathbb{P}(E \cap G \neq \varnothing) = 1$.

In this section we present two results about stochastic co-dimension, the first of which is the following.

Theorem 2.2. *Suppose E is a random set. Then, for all $G \in \mathbf{S}_0^1$,*

$$\dim(E \cap G) \geqslant \dim(G) - \overline{\text{co-dim}}(E), \qquad \text{a.s.}$$

As mentioned earlier, Theorem 2.2 is an abstract form of a part of [23, Theorem 4]. This kind of result has been implicitly used in several works. For example, see [7, 14, 19, 20]. To prove it, let us introduce an independent symmetric stable Lévy process $S_\gamma \triangleq (S_\gamma(t); t \geqslant 0)$ of index $\gamma \in]0,1[$. The following two facts are due to J. Hawkes; cf. [6].

Lemma 2.3. *Suppose* $G \in \mathbf{S}_0^1$ *satisfies* $\dim(G) < 1-\gamma$. *Then, with probability one,* $\overline{S_\gamma[0,1]} \cap G = \varnothing$.

Lemma 2.4. *Suppose* $G \in \mathbf{S}_0^1$ *satisfies* $\dim(G) > 1-\gamma$. *Then,* $\mathbb{P}(\overline{S_\gamma[0,1]} \cap G \neq \varnothing) > 0$. *Furthermore, on* $(\overline{S_\gamma[0,1]} \cap G \neq \varnothing)$,

$$\dim\big(\overline{S_\gamma[0,1]} \cap G\big) = \dim(G) + \gamma - 1, \qquad \text{a.s.} \tag{2.3}$$

Historically, the above results are stated with $\overline{S_\gamma[0,1]}$ replaced by $S_\gamma[0,1]$. By symmetry, semi-polar sets are polar for S_γ. Therefore, the same facts hold for $\overline{S_\gamma[0,1]}$.

We can now proceed with Theorem 2.2.

Proof of Theorem 2.2. Without loss of generality, we can assume that the compact set G satisfies $\dim(G) > \overline{\text{co-dim}}(E)$. With this reduction in mind, let us choose a number $\gamma \in]0,1[$ satisfying,

$$\gamma > 1 - \dim(G) + \overline{\text{co-dim}}(E). \tag{2.4}$$

Choose the process S_γ as in the earlier part of this section. Since $\gamma > 1-\dim(G)$, it follows from Lemma 2.4 that $\kappa \triangleq \mathbb{P}(\overline{S_\gamma[0,1]} \cap G \neq \varnothing) > 0$. By (2.3),

$$\begin{aligned}\kappa &= \mathbb{P}\big(\overline{S_\gamma[0,1]} \cap G \neq \varnothing\ ,\ \dim\big(\overline{S_\gamma[0,1]} \cap G\big) = \dim(G) + \gamma - 1\big)\\ &\leqslant \mathbb{P}\big(\overline{S_\gamma[0,1]} \cap G \neq \varnothing\ ,\ \dim\big(\overline{S_\gamma[0,1]} \cap G\big) > \overline{\text{co-dim}}(E)\big),\end{aligned}$$

where we have used (2.4) in the last inequality. In view of Remark 2.1.1, κ is bounded above by $\mathbb{P}(\overline{S_\gamma[0,1]} \cap G \cap E \neq \varnothing)$. Applying Lemma 2.3 gives

$$\begin{aligned}\kappa &\leqslant \mathbb{P}\big(\overline{S_\gamma[0,1]} \cap G \cap E \neq \varnothing\ ,\ \dim(G \cap E) \geqslant 1-\gamma\big)\\ &\leqslant \mathbb{P}\big(\overline{S_\gamma[0,1]} \cap G \neq \varnothing\ ,\ \dim(G \cap E) \geqslant 1-\gamma\big)\\ &= \kappa\, \mathbb{P}\big(\dim(G \cap E) \geqslant 1-\gamma\big).\end{aligned}$$

The last line utilizes the independence of S_γ and E. Since $\kappa > 0$, it follows that for all γ satisfying (2.4), $\dim(G \cap E) \geqslant 1-\gamma$, almost surely. Let $\gamma \downarrow 1-\dim(G) + \overline{\text{co-dim}}(E)$ along a rational sequence to obtain the result. ◇

Next, we present the second result of this Section. It is an immediate consequence of the estimates of [10, Section 3] and Theorem 2.2 above.

Theorem 2.5. *Suppose* $(E_n;\ n \geqslant 1)$ *is a countable collection of open random sets. If* $\sup_{n \geqslant 1} \overline{\text{co-dim}}(E_n) < 1$, *then*

$$\overline{\text{co-dim}}\Big(\bigcap_{n=1}^{\infty} E_n\Big) = \sup_{n \geqslant 1} \overline{\text{co-dim}}(E_n).$$

In particular, $\mathbb{P}\big(\cap_{n \geqslant 1} E_n \neq \varnothing\big) = 1$.

Informally speaking, this is a dual to the fact that for all $F_n \in \mathbf{S}_0^1$ $(n = 1, 2, \ldots)$, $\dim(\cup_{n=1}^{\infty} F_n) = \sup_{n \geqslant 0} \dim(F_n)$.

§3. Theorem 1.1: Upper Bound

Define the set of "near–fast points" as follows: for all $\lambda, h > 0$,

$$\mathrm{F}_\alpha(\lambda, h) \triangleq \big\{t \in [0,1] : \sup_{t \leqslant s \leqslant t+h} |X(s) - X(t)| \geqslant \lambda h^{\alpha/2} \psi(h)\big\}. \tag{3.1}$$

Next, for any $R, \eta > 1$, all integers $j \geqslant 1$ and every integer $0 \leqslant m < R^{\eta j} + 1$, define

$$I_{m,j}^{\eta} \triangleq \big[mR^{-\eta j}, (m+1)R^{-\eta j}\big]. \tag{3.2}$$

Finally, define for the above parameters,

$$P_{m,j}^{\lambda,\alpha} \triangleq \mathbb{P}\big(I_{m,j}^{\eta} \cap \mathrm{F}_\alpha(\lambda, R^{-j}) \neq \varnothing\big). \tag{3.3}$$

The main technical estimate which we shall need in this section is the following:

Lemma 3.1. *Let* X *be* fBM(α), *where* $\alpha \in \,]0, 2[$. *For all* $\lambda > 0$, $\varepsilon \in \,]0,1[$, $\eta > 1$ *and all* $R > 1$, *there exists* $J_1 = J_1(\varepsilon, \alpha, \eta, \lambda, R) \in [2, \infty[$ *such that for all* $j \geqslant J_1$ *and all* $m \geqslant 0$,

$$P_{m,j}^{\lambda,\alpha} \leqslant R^{-\lambda^2(1-\varepsilon)j}.$$

Remark 3.1.1. Part of the assertion is that J_1 does not depend on the choice of m.

Proof. By stationarity and scaling,

$$P_{m,j}^{\lambda,\alpha} = \mathbb{P}\big(\sup_{0 \leqslant t \leqslant R^{-(1-\eta)j}} \sup_{0 \leqslant s \leqslant 1} |X(s+t) - X(s)| \geqslant \lambda \psi(R^{-j})\big).$$

We obtain the lemma by applying standard estimates and [12, Lemma 3.1] to the Gaussian process $\big(X(s+t) - X(t); s, t \geqslant 0\big)$. ♢

The proof of the upper bound is close to its counterpart [18]; cf. the first part of Theorem 2 therein.

Proof of Theorem 1.1: upper bound. Recall (3.1) and (3.2). Consider a fixed closed set $E \subset [0,1]$. Fix $R, \eta > 1$ and $\lambda > 0$. Define for all integers $m \geqslant 0$,

$$\mathfrak{J}_k \triangleq \sum_{j \geqslant k} \sum_{m \geqslant 0} \mathbf{1}_{\{I^\eta_{m,j} \cap \mathrm{F}_\alpha(\lambda, R^{-j}) \neq \varnothing\}} \mathbf{1}_{\{I^\eta_{m,j} \cap E \neq \varnothing\}}. \tag{3.4}$$

By (3.3) and Lemma 3.1, for all $\varepsilon \in]0,1[$, there exists $J_1 = J_1(\varepsilon, \alpha, \eta, \lambda, R) \in [2, \infty[$ such that for all $k \geqslant J_1$,

$$\begin{aligned}\|\mathfrak{J}_k\|_1 &= \sum_{j \geqslant k} \sum_{m \geqslant 0} P^{\lambda,\alpha}_{m,j} \mathbf{1}_{\{I^\eta_{m,j} \cap E \neq \varnothing\}} \leqslant \sum_{j \geqslant k} \sum_{m \geqslant 0} R^{-\lambda^2(1-\varepsilon)j} \mathbf{1}_{\{I^\eta_{m,j} \cap E \neq \varnothing\}} \\ &\leqslant \sum_{j \geqslant k} R^{-\lambda^2(1-\varepsilon)j} M\big(R^{-\eta j}; E\big),\end{aligned}$$

where $M(\varepsilon; E)$ denotes the ε–**capacity** of E. That is, it is the maximal number of points in E which are at least ε apart; see [5]. On the other hand, by definition, for all $\delta \in]0,1[$, there exists $J_2 = J_2(\delta, R, \eta) \in [2, \infty[$, such that for all $j \geqslant J_2$, $M\big(R^{-\eta j}; E\big) \leqslant R^{\eta j (1+\delta)\, \overline{\dim}_M(E)}$. Hence, for all $k \geqslant J_1 \vee J_2$,

$$\|\mathfrak{J}_k\|_1 \leqslant \sum_{j \geqslant k} R^{-j\left(\lambda^2(1-\varepsilon) - \eta(1+\delta)\, \overline{\dim}_M(E)\right)}.$$

It may help to recall that $J_1 \vee J_2$ depends only on the parameters $\big(\lambda, \alpha, R, \eta, \delta, \varepsilon\big)$. Let us pick these parameters so that $\lambda^2(1-\varepsilon) > \eta(1+\delta)\, \overline{\dim}_M(E)$. It is easy to see that for this choice of parameters, $\sum_k \|\mathfrak{J}_k\|_1 < \infty$. By the Borel–Cantelli lemma, with probability one, there exists a finite random variable k_0 such that for all $k \geqslant k_0$, $\mathfrak{J}_k = 0$. In other words, with probability one, for all $j \geqslant k_0$, $\mathrm{F}_\alpha(\lambda, R^{-j}) \cap E = \varnothing$. Rewriting the above, we see that with probability one, for all $j \geqslant k_0$, $\sup_{t \in E} \sup_{t \leqslant s \leqslant t + R^{-j}} |X(s) - X(t)| \leqslant \lambda R^{-j\alpha/2} \psi(R^{-j})$. Take any $h \leqslant R^{-k_0}$. There exists a $j \geqslant k_0$, so that $R^{-j-1} \leqslant h \leqslant R^{-j}$. It follows that for all $h \leqslant R^{-k_0}$,

$$\begin{aligned}\sup_{t \in E} |X(t+h) - X(t)| &\leqslant \sup_{t \in E} \sup_{t \leqslant s \leqslant t + R^{-j}} |X(s) - X(t)| \\ &\leqslant \lambda R^{-j\alpha/2} \psi(R^{-j}) \leqslant \lambda R^{\alpha/2} h^{\alpha/2} \psi(h),\end{aligned}$$

since ψ is monotone decreasing. This implies that whenever $\lambda^2(1-\varepsilon) > \eta(1+\delta)\, \overline{\dim}_M(E)$,

$$\limsup_{h \to 0^+} \sup_{t \in E} \frac{|X(t+h) - X(t)|}{h^{\alpha/2} \psi(h)} \leqslant \lambda R^{\alpha/2}, \qquad \text{a.s.}$$

Along rational sequences, let $\varepsilon, \delta \to 0^+$, $\eta, R \to 1^+$ and $\lambda^2 \downarrow \overline{\dim}_M(E)$ — in this order — to see that with probability one,

$$\limsup_{h \to 0^+} \sup_{t \in E} \frac{|X(t+h) - X(t)|}{h^{\alpha/2} \psi(h)} \leqslant \sqrt{\overline{\dim}_M(E)}\,.$$

This proves the desired upper bound. ◊

§4. Theorems 1.1 and 1.4: Lower Bounds

For each closed set $E \subset [0,1]$ and for every $\mu \in \mathcal{P}_+(E)$ and $h, \lambda > 0$, define,

$$I_\mu(h;\lambda) \triangleq \int_0^1 \mu(dt)\, \mathbf{1}_{\{X(t+h)-X(t)>\lambda h^{\alpha/2}\psi(h)\}}. \tag{4.1}$$

The key technical estimate of this section is the following:

Lemma 4.1. *Suppose $E \subset [0,1]$ is compact, $\mu \in \mathcal{P}_+(E)$. For any $\varepsilon \in]0,1[$ and $\lambda > 0$, there exists a small $h_0 \in]0,1[$ such that for all $h \in]0,h_0[$,*

$$\frac{\|I_\mu(h;\lambda)\|_2^2}{\|I_\mu(h;\lambda)\|_1^2} \leqslant 1+\varepsilon+4 \sup_{h^{1-\varepsilon}\leqslant t\leqslant 1-h^{1-\varepsilon}} \frac{\mu\big([t-h^{1-\varepsilon}, t+h^{1+\varepsilon}]\big)}{\overline{\Phi}\big(\lambda\psi(h)\big)}.$$

Proof. Since X has stationary increments,

$$\|I_\mu(h;\lambda)\|_1 = \overline{\Phi}\big(\lambda\psi(h)\big). \tag{4.2}$$

We proceed with the estimate for the second moment. Define,

$$\begin{aligned} a &\triangleq \lambda\psi(h), \\ U &\triangleq \frac{X(s+h)-X(s)}{h^{\alpha/2}}, \\ V &\triangleq \frac{X(t+h)-X(t)}{h^{\alpha/2}}, \\ \rho &\triangleq \|UV\|_1. \end{aligned}$$

Then, ignoring the dependence on (h,s,t),

$$\|I_\mu(h;\lambda)\|_2^2 = Q_1 + Q_2 + Q_3, \tag{4.3}$$

where,

$$\begin{aligned} Q_1 &\triangleq \int_0^1 \mu(dt) \int_0^1 \mu(ds)\, \mathbf{1}_{\{\rho\leqslant(\ln(1/h))^{-2}\}} \mathbb{P}(U\geqslant a, V\geqslant a), \\ Q_2 &\triangleq 2\int_0^1 \mu(dt) \int_0^t \mu(ds) \mathbf{1}_{\{\rho>(\ln(1/h))^{-2}\}} \mathbf{1}_{\{t-s>2h\}} \mathbb{P}(U\geqslant a, V\geqslant a), \\ Q_3 &\triangleq 2\int_0^1 \mu(dt) \int_0^t \mu(ds) \mathbf{1}_{\{\rho>(\ln(1/h))^{-2}\}} \mathbf{1}_{\{t-s\leqslant 2h\}} \mathbb{P}(U\geqslant a, V\geqslant a). \end{aligned}$$

We estimate each term separately. The critical term is Q_1. Write $x^+ = \max(x, 0)$ for any x. If $\rho < 1/4$,

$$\begin{aligned}\mathbb{P}(U \geqslant a, V \geqslant a) &= \frac{1}{2\pi\sqrt{1-\rho^2}} \int_a^\infty \int_a^\infty \exp\Big(-\frac{x^2+y^2-2\rho xy}{2(1-\rho^2)}\Big)\,dx\,dy \\ &\leqslant \frac{1}{2\pi\sqrt{1-\rho^2}} \int_a^\infty \int_a^\infty \exp\Big(-\frac{(1-4\rho^+)(x^2+y^2)}{2(1-\rho^2)}\Big)\,dx\,dy \\ &= \frac{\sqrt{1-\rho^2}}{1-4\rho^+}\Big(\overline{\Phi}\Big(a\sqrt{\frac{1-4\rho^+}{1-\rho^2}}\Big)\Big)^2.\end{aligned}$$

According to Mill's ratio for Gaussian tails ([22, p. 850]), for any $x > 1$,

$$\frac{1-x^{-2}}{\sqrt{2\pi}\,x}\exp\Big(-\frac{x^2}{2}\Big) \leqslant \overline{\Phi}(x) \leqslant \frac{1}{\sqrt{2\pi}\,x}\exp\Big(-\frac{x^2}{2}\Big). \tag{4.4}$$

Therefore, using the fact that μ is an atomless probability measure, we have,

$$\begin{aligned}Q_1 &\leqslant \sup_{r:\,-1<r\leqslant(\ln(1/h))^{-2}} \frac{\sqrt{1-r^2}}{1-4r^+}\Big(\overline{\Phi}\Big(\lambda\psi(h)\sqrt{\frac{1-4r^+}{1-r^2}}\Big)\Big)^2 \\ &\leqslant \big[\overline{\Phi}(\lambda\psi(h))\big]^2 \sup_{r:\,-1<r\leqslant(\ln(1/h))^{-2}} \Big(1-\frac{1}{(\lambda\psi(h))^2}\Big)^{-2}\frac{(1-r^2)^{3/2}}{(1-4r^+)^2}\times \\ &\qquad \times\exp\Big(\frac{(\lambda\psi(h))^2(4r^+-r^2)}{1-r^2}\Big).\end{aligned}$$

Since $(\psi(h))^2(\ln(1/h))^{-2} = o(1)$ (as h goes to 0), this leads to:

$$Q_1 \leqslant B(h;\lambda)\big[\overline{\Phi}(\lambda\psi(h))\big]^2, \tag{4.5}$$

where $B(h;\lambda)$ is such that, for any $\lambda > 0$,

$$\lim_{h\to 0^+} B(h;\lambda) = 1. \tag{4.6}$$

To estimate Q_3, use the trivial inequality $\mathbb{P}(U \geqslant a, V \geqslant a) \leqslant \mathbb{P}(U \geqslant a)$, to see that

$$\begin{aligned}Q_3 &\leqslant 2 \sup_{2h\leqslant t\leqslant 1} \mu([t-2h,t])\,\overline{\Phi}(\lambda\psi(h)) \\ &\leqslant 2 \sup_{2h\leqslant t\leqslant 1-2h} \mu([t-2h,t+2h])\,\overline{\Phi}(\lambda\psi(h)).\end{aligned} \tag{4.7}$$

Finally, we need to approximate Q_2. Directly computing, note that when $s < t - 2h$,

$$\begin{aligned}\rho &= \frac{|t-s-h|^\alpha + |t-s+h|^\alpha - 2|t-s|^\alpha}{2h^\alpha} \\ &= \frac{1}{2}\Big(\frac{t-s}{h}\Big)^\alpha\Big[\Big(1-\frac{h}{t-s}\Big)^\alpha + \Big(1+\frac{h}{t-s}\Big)^\alpha - 2\Big].\end{aligned}$$

By Taylor expansion of $(1 \pm x)^\alpha$, we see that for all $|x| \leqslant \frac{1}{2}$,

$$(1-x)^\alpha + (1+x)^\alpha - 2 = \frac{\alpha(\alpha-1)}{2}\Big[(1-\xi_1)^{\alpha-2} + (1-\xi_2)^{\alpha-2}\Big]x^2,$$

where $|\xi_i| \leqslant \frac{1}{2}$ for $i = 1, 2$. In particular, for all $|x| \leqslant \frac{1}{2}$,

$$\Big|(1-x)^\alpha + (1+x)^\alpha - 2\Big| \leqslant 2^{3-\alpha}x^2.$$

In other words, when $s < t - 2h$, $\rho \leqslant \{2h/(t-s)\}^{2-\alpha}$. On the other hand, if we also know that $\rho > \big(\ln(1/h)\big)^{-2}$, it follows that for all h small, $|t-s| \leqslant 2h\big(\ln(1/h)\big)^{2/(2-\alpha)} \leqslant h^{1-\varepsilon}$. Since $\mathbb{P}(U \geqslant a, V \geqslant a) \leqslant \overline{\Phi}(\lambda\psi(h))$, we obtain the following: for all $\varepsilon > 0$, there exists $h_1 \in]0,1[$ such that for all $h \in]0, h_1[$,

$$Q_2 \leqslant 2 \sup_{h^{1-\varepsilon} \leqslant t \leqslant 1-h^{1-\varepsilon}} \mu\big([t-h^{1-\varepsilon}, t+h^{1-\varepsilon}]\big)\overline{\Phi}(\lambda\psi(h)).$$

Together with (4.7), we obtain: for all $\varepsilon > 0$, there exists $h_2 \in]0,1[$ such that for all $h \in]0, h_2[$,

$$Q_2 + Q_3 \leqslant 4 \sup_{h^{1-\varepsilon} \leqslant t \leqslant 1-h^{1-\varepsilon}} \mu\big([t-h^{1-\varepsilon}, t+h^{1-\varepsilon}]\big)\,\overline{\Phi}(\lambda\psi(h)).$$

Combining this with (4.2)–(4.3) and (4.5)–(4.7), we obtain the result. ◇

Now we can prove the lower bounds in Theorems 1.1 and 1.4.

Proof of Theorems 1.1 and 1.4: lower bounds. By Frostman's lemma (Lemma 2.1), for any $\beta < \dim(E)$, there exists a $\mu \in \mathcal{P}_+(E)$, such that for all $h \in]0,1[$ small,

$$\sup_{h \leqslant t \leqslant 1-h} \mu\big([t-h, t+h]\big) \leqslant h^\beta.$$

For such a μ, use Lemma 4.1 to see that for all $\varepsilon \in]0,1[$ and $\lambda > 0$, there exists $h_3 \in]0,1[$ such that whenever $h \in]0, h_3[$,

$$\frac{\|I_\mu(h;\lambda)\|_2^2}{\|I_\mu(h;\lambda)\|_1^2} \leqslant 1 + \varepsilon + \frac{4h^{(1-\varepsilon)\beta}}{\overline{\Phi}(\lambda\psi(h))}.$$

According to Mill's ratio for Gaussian tails (see (4.4)), for any $\lambda > 0$, there exists a small $h_4 > 0$, such that for all $h \in]0, h_4[$, $\overline{\Phi}(\lambda\psi(h)) \geqslant h^{\lambda^2}/4\lambda\sqrt{\ln(1/h)}$. Hence, for all $h \in]0, h_3 \wedge h_4[$,

$$\frac{\|I_\mu(h;\lambda)\|_2^2}{\|I_\mu(h;\lambda)\|_1^2} \leqslant 1 + \varepsilon + 16\lambda h^{(1-\varepsilon)\beta-\lambda^2}\sqrt{\ln(1/h)}.$$

By choosing λ such that $(1-\varepsilon)\beta > \lambda^2$ we can deduce that for all $h \in]0, h_3 \wedge h_4[$, $\|I_\mu(h;\lambda)\|_2^2 \, \|I_\mu(h;\lambda)\|_1^{-2} \leqslant 1 + 2\varepsilon$. Applying the Paley–Zygmund inequality ([8, p. 8]), we see that

$$\mathbb{P}\big(I_\mu(h;\lambda) > 0\big) \geqslant \frac{1}{1+2\varepsilon}. \tag{4.8}$$

We are ready to complete the proof. Define,

$$\mathcal{A}_\alpha(\lambda, h) \triangleq \Big\{ t \in [0,1] : \sup_{0 \leqslant r \leqslant h} \frac{|X(t+r) - X(t)|}{r^{\alpha/2}\psi(r)} > \lambda \Big\}. \tag{4.9}$$

This is the collection of all h–**approximate** λ–**fast points**. Observe that for each $h > 0$, $\mathcal{A}_\alpha(\lambda, h)$ is an open subset of $[0,1]$. If $I_\mu(h;\lambda) > 0$ ($\mu \in \mathcal{P}_+(E)$), then $E \cap \mathcal{A}_\alpha(\lambda, h) \neq \varnothing$. By (4.8), we see that as long as $(1-\varepsilon)\beta > \lambda^2$, then for all $h \in]0, h_3 \wedge h_4[$, $\mathbb{P}\big(E \cap \mathcal{A}_\alpha(\lambda, h) \neq \varnothing\big) \geqslant (1+2\varepsilon)^{-1}$. Note that if $h \leqslant h'$, then $\mathcal{A}_\alpha(\lambda, h) \subset \mathcal{A}_\alpha(\lambda, h')$. Hence, for all $\lambda^2 \in [0, \dim(E)[$, $\mathbb{P}\big(\mathcal{A}_\alpha(\lambda, h) \cap E \neq \varnothing, \ \forall h > 0\big) = 1$. By Theorem 2.5, for all $\lambda^2 \in [0, \dim(E)[$, $\mathbb{P}\big(\cap_{h>0} \mathcal{A}_\alpha(\lambda, h) \cap E \neq \varnothing\big) = 1$. Since $\bigcap_{h>0} \mathcal{A}_\alpha(\lambda, h) = \mathrm{F}_\alpha(\lambda)$, we have shown that

$$\overline{\text{co-dim}}\big(\mathrm{F}_\alpha(\lambda)\big) \leqslant \lambda^2.$$

Unravelling the notation, this implies Eq. (1.5) (i.e., the lower bound in Theorem 1.1). It also implies the lower bound in Theorem 1.4. ◊

§5. Theorem 1.3: Upper Bound

For $\eta, R > 1$, and $j \geqslant 1$, define,

$$\mathcal{G}(j) \triangleq \{ 0 \leqslant m \leqslant R^{\eta j} : \ |W(mR^{-\eta j})| \leqslant 2\sqrt{\eta j R^{-\eta j} \ln R} \}. \tag{5.1}$$

The notation is motivated by the following description: we think of $I^\eta_{m,j}$ (see (3.2)) as "good" if $m \in \mathcal{G}(j)$. Otherwise, $I^\eta_{m,j}$ is deemed "bad". We also recall Eq. (3.1) with $\alpha = 1$ (thus replacing X by W in (3.1)). In analogy with the definition of $\mathcal{J}_k$ (see (3.4)), we define,

$$\mathcal{J}'_k \triangleq \sum_{j \geqslant k} \sum_{m \in \mathcal{G}(j)} \mathbf{1}_{\{I^\eta_{m,j} \cap \mathrm{F}_1(\lambda, R^{-j}) \neq \varnothing\}}. \tag{5.2}$$

By the independence of the increments of W, $\{I^\eta_{m,j} \cap \mathrm{F}_1(\lambda, R^{-j}) \neq \varnothing\}$ is independent of $\{m \in \mathcal{G}(j)\}$. Recalling (3.3), we see that

$$\|\mathcal{J}'_k\|_1 = \sum_{j \geqslant k} \sum_{m \geqslant 0} \mathbb{P}\big(m \in \mathcal{G}(j)\big) P^{\lambda,1}_{m,j}.$$

Since $|W(1)|$ has a probability density which is uniformly bounded above by 1, for all $0 \leqslant m \leqslant R^{\eta j}$,

$$\mathbb{P}\big(m \in \mathcal{G}(j)\big) \leqslant 4\Big(\sqrt{\frac{\eta j \ln R}{m}} \wedge 1\Big). \tag{5.3}$$

Next, fix $\varepsilon \in]0,1[$. By Lemma 3.1, there exists $J_3 \triangleq J_1(\varepsilon, 1, \eta, \lambda, R) \in [2, \infty[$, such that for all $j \geqslant J_3$ and all $m \geqslant 0$, $P_{m,j}^{\lambda,1} \leqslant R^{-\lambda^2(1-\varepsilon)j}$. Using (5.3), we see that for all $k \geqslant J_3$,

$$\|\mathcal{J}'_k\|_1 \leqslant 4 \sum_{j \geqslant k} \sum_{m=0}^{[R^{\eta j}]} \Big(\sqrt{\frac{\eta j \ln R}{m}} \wedge 1\Big) R^{-\lambda^2(1-\varepsilon)j}.$$

If we choose $\lambda^2(1-\varepsilon) > \eta/2$, then a few lines of calculations reveal that $\sum_k \|\mathcal{J}'_k\|_1 < \infty$. By the Borel–Cantelli lemma, there exists a finite random variable k_1, such that with probability one, for all $k \geqslant k_1$, $\mathcal{J}'_k = 0$. In particular, with probability one, for all $j \geqslant k_1$,

$$\mathbf{1}_{\{m \in \mathcal{G}(j)\}} \mathbf{1}_{\{I^{\eta}_{m,j} \cap \mathrm{F}_1(\lambda, R^{-j}) \neq \varnothing\}} = 0. \tag{5.4}$$

By Lévy's modulus of continuity for W (cf. (1.2)), there exists a finite random variable $k_2(\eta, R)$, such that with probability one, for all $j \geqslant k_2(\eta, R)$,

$$\mathbf{1}_{\{m \in \mathcal{G}(j)\}} \geqslant \mathbf{1}_{\{I^{\eta}_{m,j} \cap \mathcal{Z} \neq \varnothing\}}. \tag{5.5}$$

Eq. (5.4) shows that with probability one, for all $j \geqslant k_3 \triangleq k_1 \vee k_2(\eta, R)$, $\mathrm{F}_1(\lambda, R^{-j}) \cap \mathcal{Z} = \varnothing$. That is, almost surely, for all $j \geqslant k_3$,

$$\sup_{t \in \mathcal{Z}} \sup_{t \leqslant s \leqslant t + R^{-j}} |W(t+s)| \leqslant \lambda R^{-j/2} \psi(R^{-j}).$$

If $h \leqslant R^{-k_3}$, there exists $j \geqslant k_3$, such that $R^{-j-1} \leqslant h \leqslant R^{-j}$. By monotonicity,

$$\sup_{t \in \mathcal{Z}} \sup_{t \leqslant s \leqslant t+h} |W(t+s)| \leqslant \lambda R^{-j/2} \psi(R^{-j}) \leqslant \lambda R^{1/2} h^{1/2} \psi(h).$$

In particular, we have shown that as long as $\lambda^2(1-\varepsilon) > \eta/2$, then almost surely,

$$\limsup_{h \to 0^+} \sup_{t \in \mathcal{Z}} \frac{|W(t+h)|}{h^{1/2}\psi(h)} \leqslant \lambda R^{1/2}.$$

Along rational sequences (and in this order), let $\eta, R \to 1^+$, $\varepsilon \to 0^+$ and $\lambda^2 \downarrow \frac{1}{2}$ to see that with probability one,

$$\limsup_{h \to 0^+} \sup_{t \in \mathcal{Z}} \frac{|W(t+h)|}{h^{1/2}\psi(h)} \leqslant \frac{1}{\sqrt{2}}.$$

This proves the upper bound in Theorem 1.3. ◊

§6. Theorem 1.4: Upper Bound

Recall Eqs. (1.6) and (3.1). We begin with a "regularization scheme" which is used later in our good covering for dimension calculations.

Lemma 6.1. *For all integers $k \geqslant 1$ and all reals $\theta \in]0,1[$ and $R > 1$,*

$$\mathrm{F}_\alpha(\lambda) \subset \bigcup_{j \geqslant k} \mathrm{F}_\alpha(\theta\lambda R^{-\alpha/2}, R^{-j+1}).$$

Proof. From first principles, it follows that for all $\beta < \lambda$, $\mathrm{F}_\alpha(\lambda) \subset \bigcup_{h>0}\bigcup_{0<\delta<h} \mathrm{F}_\alpha(\beta,\delta)$. Let $\delta < h < 1$, say $R^{-j} \leqslant \delta \leqslant R^{-j+1}$. For all $t \in F_\alpha(\beta,\delta)$,

$$\sup_{t \leqslant s \leqslant t+R^{-j+1}} |X(s) - X(t)| \geqslant \beta R^{-j\alpha/2}\psi(R^{-j+1}) = \beta R^{-\alpha/2} R^{-(j-1)\alpha/2}\psi(R^{-j+1}).$$

We have used the monotonicity properties of ψ. This implies that $t \in F_\alpha(\beta R^{-\alpha/2}, R^{-(j-1)})$. The result follows. ◊

We are prepared to demonstrate the upper bound in Theorem 1.4.

Proof of Theorem 1.4: upper bound. Fix $R, \eta > 1$, $\theta \in]0,1[$ and recall $I^\eta_{m,j}$ from (3.2). From Lemma 6.1, it is apparent that for any integer $k \geqslant 1$, we have the following covering of $\mathrm{F}_\alpha(\lambda)$:

$$\mathrm{F}_\alpha(\lambda) \subset \bigcup_{j=k}^{\infty}\bigcup_{m=1}^{\infty} I^\eta_{m,j} \cap \mathrm{F}_\alpha(\theta\lambda R^{-\alpha/2}, R^{-j+1}). \tag{6.1}$$

By (3.3) and Lemma 3.1, for all $\varepsilon > 0$, there exists $J_4 = J_4(\varepsilon,\alpha,\eta,\lambda,R,\theta) \in [2,\infty[$, such that for all $j \geqslant J_4$,

$$\mathbb{P}\big(I^\eta_{m,j} \cap \mathrm{F}_\alpha(\theta\lambda R^{-\alpha/2}, R^{-j+1}) \neq \varnothing\big) \leqslant R^{-(1-\varepsilon)\theta^2\lambda^2 R^{-2\alpha} j}. \tag{6.2}$$

For any $s \geqslant 0$ and every integer $k \geqslant 1$, define,

$$\mathfrak{J}_k(s) \triangleq \sum_{j \geqslant k}\sum_{m \leqslant R^{\eta j}+1} \left|I^\eta_{m,j}\right|^s \mathbf{1}_{\{I^\eta_{m,j}\cap E \neq \varnothing\}} \mathbf{1}_{\{I^\eta_{m,j}\cap \mathrm{F}_\alpha(\theta\lambda R^{-\alpha/2}, R^{-j+1}) \neq \varnothing\}}.$$

By (6.2), for all $k \geqslant J_4$,

$$\left\|\mathfrak{J}_k(s)\right\|_1 \leqslant \sum_{j \geqslant k} M\big((1+R^{\eta j})^{-1}; E\big) R^{-\eta s j} R^{-(1-\varepsilon)\theta^2\lambda^2 R^{-2\alpha} j}.$$

(Recall from §3 that $M(\varepsilon; E)$ is the ε-capacity of E.) Note that if we enlarge J_4 further, then for all $\beta > \overline{\dim}_M(E)$ and all $j \geqslant J_4$, we can also ensure that $M\big((1+R^{\eta j})^{-1}; E\big) \leqslant R^{\eta\beta j}$. Suppose $\beta > \overline{\dim}_M(E)$ and

$$\eta s > \eta\beta - (1-\varepsilon)\theta^2\lambda^2 R^{-2\alpha}. \tag{6.3}$$

It follows that $\sum_k \big\|\mathfrak{J}_k(s)\big\|_1 < \infty$. By the Borel–Cantelli lemma, $\lim_{k\to\infty} \mathfrak{J}_k(s) = 0$, a.s. From (6.1) and the definition of Hausdorff dimension, it follows that for any s satisfying (6.3),

$$\Lambda^s\big(E \cap \mathrm{F}_\alpha(\lambda)\big) \leqslant \lim_{k\to\infty} \mathfrak{J}_k(s) = 0, \qquad \text{a.s.}$$

Therefore, almost surely,

$$\dim\big(E \cap \mathrm{F}_\alpha(\lambda)\big) \leqslant s.$$

Letting $\beta \downarrow \overline{\dim}_M(E)$, $\varepsilon \downarrow 0$, $\eta, R \downarrow 1$, $\theta \uparrow 1$ in this order and along rational sequences, we obtain

$$\dim\big(E \cap \mathrm{F}_\alpha(\lambda)\big) \leqslant \overline{\dim}_M(E) - \lambda^2, \qquad \text{a.s.}$$

By [17, pp. 57 and 81], for every $G \in \mathbf{S}_0^1$,

$$\dim(G) = \inf_{G=\cup_{i=1}^\infty G_i} \sup_i \dim(G_i),$$

$$\dim_{\mathrm{P}}(G) = \inf_{G=\cup_{i=1}^\infty G_i} \sup_i \overline{\dim}_M(G_i),$$

where the G_i's are assumed to be bounded. Thus,

$$\dim\big(E \cap \mathrm{F}_\alpha(\lambda)\big) \leqslant \dim_{\mathrm{P}}(E) - \lambda^2, \qquad \text{a.s.,}$$

which is the desired upper bound. ◊

§7. Proof of Theorem 1.5: Upper Bound

Recall the notation of §3 and §6, and $\mathcal{G}(j)$ from (5.1). By Lemma 6.1 and (5.5), for any $\eta, R > 1$ and $\theta \in]0,1[$, there exists a finite random variable K such that almost surely for all $k \geqslant K$,

$$\mathcal{Z} \cap \mathrm{F}_1(\lambda) \subset \bigcup_{j\geqslant k} \bigcup_{m\in\mathcal{G}(j)} I_{m,j}^\eta \cap \mathrm{F}_1(\theta\lambda R^{-1/2}, R^{-j+1}). \tag{7.1}$$

Next, we show that the above is a fairly economical covering. Since W has independent increments, for any $s > 0$,

$$\begin{aligned}
&\sum_{j\geqslant k} \sum_{0\leqslant m<R^{\eta j}+1} \big|I_{m,j}^\eta\big|^s \,\mathbb{P}\big(I_{m,j}^\eta \cap \mathrm{F}_1(\theta\lambda R^{-1/2}, R^{-j+1}) \neq \varnothing\ ,\ m \in \mathcal{G}(j)\big)\\
&= \sum_{j\geqslant k} \sum_{0\leqslant m<R^{\eta j}+1} \big|I_{m,j}^\eta\big|^s \,\mathbb{P}\big(I_{m,j}^\eta \cap \mathrm{F}_1(\theta\lambda R^{-1/2}, R^{-j+1}) \neq \varnothing\big)\, \mathbb{P}\big(m \in \mathcal{G}(j)\big).
\end{aligned}$$

By (5.3),

$$\sum_{j\geqslant k}\sum_{0\leqslant m<R^{\eta j}+1}|I^{\eta}_{m,j}|^{s}\,\mathbb{P}\big(I^{\eta}_{m,j}\cap \mathrm{F}_1(\theta\lambda R^{-1/2},R^{-j+1})\neq\varnothing\ ,\ m\in\mathcal{G}(j)\big)$$

$$\leqslant 4\sum_{j\geqslant k}\sum_{m=0}^{[R^{\eta j}]+1}|I^{\eta}_{m,j}|^{s}\,\mathbb{P}\big(I^{\eta}_{m,j}\cap \mathrm{F}_1(\theta\lambda R^{-1/2},R^{-j+1})\neq\varnothing\big)\Big(\sqrt{\frac{\eta j\ln R}{m}}\wedge 1\Big).$$

Using Lemma 3.1, we see that for all $\varepsilon\in]0,1[$, there exists $J_5=J_5(\varepsilon,\theta,\eta,\lambda,R)\in[2,\infty[$, such that for all $k\geqslant J_5$,

$$\sum_{j\geqslant k}\sum_{0\leqslant m<R^{\eta j}+1}|I^{\eta}_{m,j}|^{s}\,\mathbb{P}\big(I^{\eta}_{m,j}\cap \mathrm{F}_1(\theta\lambda R^{-1/2},R^{-j+1})\neq\varnothing\ ,\ m\in\mathcal{G}(j)\big)$$

$$\leqslant 4\sum_{j\geqslant k}R^{-\eta s j}R^{-(1-\varepsilon)\theta^2\lambda^2R^{-1}j}\sum_{m=0}^{[R^{\eta j}]+1}\Big(\sqrt{\frac{\eta j\ln R}{m}}\wedge 1\Big).$$

In particular, if

$$\eta s>\frac{\eta}{2}-(1-\varepsilon)\theta^2\lambda^2R^{-1},\tag{7.2}$$

then,

$$\lim_{k\to\infty}\sum_{j\geqslant k}\sum_{m\in\mathcal{G}(j)}|I^{\eta}_{m,j}|^{s}\,\mathbf{1}_{\{I^{\eta}_{m,j}\cap F_1(\theta\lambda R^{-1/2},R^{-j+1})\neq\varnothing\}}=0,\qquad\text{a.s.}$$

Thanks to (7.1), we can deduce that for any s satisfying (7.2), $\Lambda^s\big(\mathcal{Z}\cap\mathrm{F}_1(\lambda)\big)=0$, almost surely. In particular, almost surely, $\dim\big(\mathcal{Z}\cap\mathrm{F}_1(\lambda)\big)\leqslant s$. Let $\varepsilon\downarrow 0$, $\theta\uparrow 1$ and $R,\eta\downarrow 1$ in this order to see that with probability one, $\dim\big(\mathcal{Z}\cap\mathrm{F}_1(\lambda)\big)\leqslant\frac{1}{2}-\lambda^2$. This is the desired upper bound. ◊

§8. Proofs of Theorems 1.3 and 1.5: Lower Bounds

The main result of this section is the following which may be of independent interest.

Theorem 8.1. *Fix a compact set $E\subset[0,1]$. Then,*

$$\dim(E)>\lambda^2+\frac{1}{2}\implies\mathbb{P}\big(\mathcal{Z}\cap\mathrm{F}_1(\lambda)\cap E\neq\varnothing\big)=1.$$

As the lower bounds in Theorems 1.3 and 1.5 follow immediately from the above, the rest of this section is devoted to proving Theorem 8.1.

Suppose $E\subset[0,1]$ is compact, $\mu\in\mathcal{P}_+(E)$ and $h,\lambda>0$. Define,

$$J_\mu(h;\lambda) \triangleq \int_0^1 \mu(ds)\, \mathbf{1}_{\{|W(s)|<h\}} \mathbf{1}_{\{W(s+h)-W(s)>\lambda h^{1/2}\psi(h)\}}.$$

For all $\mu \in \mathcal{P}_+(E)$ and any $h, \beta > 0$, define the following:

$$S_h(\mu) \triangleq \sup_{0\leqslant s\leqslant h} \int_s^h \frac{\mu(dt)}{\sqrt{t-s}},$$

$$\widetilde{S}_h(\mu) \triangleq \sup_{0\leqslant s\leqslant 1} \int_s^{(s+h)\wedge 1} \frac{\mu(dt)}{\sqrt{t-s}}.$$

The proof of Theorem 8.1 is divided into several steps. In analytical terms, our first result is an estimate, uniform in $s \in [0,h]$, for the $1/2$–potential of a measure μ restricted to an interval $[s,h]$. Indeed, recalling $A_\beta(\mu)$ from (2.1), we have the following:

Lemma 8.2. *Suppose $\mu \in \mathcal{P}_+(E)$. Then for all $\beta > 1/2$ and all $h > 0$,*

$$S_h(\mu) \leqslant \frac{2e^\beta}{2\beta-1} A_\beta(\mu)\, h^{\beta-\frac{1}{2}}, \tag{8.1}$$

$$\widetilde{S}_h(\mu) \leqslant \frac{2e^\beta}{2\beta-1} A_\beta(\mu)\, h^{\beta-\frac{1}{2}}. \tag{8.2}$$

Proof. Without loss of generality, we can assume that $A_\beta(\mu) < \infty$ for some $\beta > 1/2$ (otherwise, there is nothing to prove). We proceed with an approximate integration by parts: for all $0 \leqslant s \leqslant h$,

$$\begin{aligned}\int_s^h \mu(dt)(t-s)^{-1/2} &= \sum_{j=0}^\infty \int_{s+(h-s)e^{-j-1}}^{s+(h-s)e^{-j}} \mu(dt)(t-s)^{-1/2}\\ &\leqslant (h-s)^{-1/2} \sum_{j=0}^\infty e^{(j+1)/2} \mu[s+(h-s)e^{-j-1}, s+(h-s)e^{-j}]\\ &\leqslant e^{1/2}(h-s)^{\beta-1/2} A_\beta(\mu) \sum_{j=0}^\infty \exp\big(-j(\beta-\frac{1}{2})\big),\end{aligned}$$

which yields (8.1). The estimate (8.2) can be checked in the same way. ◊

Our next two lemmas are moment estimates for J_μ.

Lemma 8.3. *Suppose $\mu \in \mathcal{P}_+(E)$ is fixed. For every $\varepsilon > 0$, there exists an $h_\varepsilon > 0$, such that for all $h \in]0, h_\varepsilon[$, and all $\lambda > 0$,*

$$\|J_\mu(h;\lambda)\|_1 \geqslant \sqrt{\frac{2}{\pi e}}(1-\varepsilon) h\, \overline{\Phi}(\lambda\psi(h)).$$

Proof. By the independence of the increments of W,

$$\big\|J_\mu(h;\lambda)\big\|_1 = \overline{\Phi}\big(\lambda\psi(h)\big)\int_0^1 \mu(ds)\,\mathbb{P}\big(|W(s)|\leqslant h\big).$$

A direct calculation reveals that if $r\in\,]0,1[\,$, $\mathbb{P}\big(|W(1)|\leqslant r\big)\geqslant\sqrt{2/(\pi e)}\,r$. Therefore, by Brownian scaling, for every $h\in\,]0,1[\,$,

$$\begin{aligned}\big\|J_\mu(h;\lambda)\big\|_1 &\geqslant \sqrt{\frac{2}{\pi e}}\,h\,\overline{\Phi}\big(\lambda\psi(h)\big)\int_{h^2}^1\frac{\mu(ds)}{\sqrt{s}}\\ &\geqslant\sqrt{\frac{2}{\pi e}}\,\mu[h^2,1]\,h\,\overline{\Phi}\big(\lambda\psi(h)\big).\end{aligned}$$

The lemma follows upon taking $h>0$ small enough. ◇

Lemma 8.4. *Fix $\mu\in\mathcal{P}_+(E)$. Suppose $A_\beta(\mu)<\infty$ for some $\beta>1/2$. Then, for all $h,\lambda>0$,*

$$\big\|J_\mu(h;\lambda)\big\|_2^2\leqslant\frac{8e^{2\beta}}{(2\beta-1)^2}A_\beta^2(\mu)\Big\{2\,h^{\beta+3/2}\overline{\Phi}\big(\lambda\psi(h)\big)+h^2\,\overline{\Phi}^2\big(\lambda\psi(h)\big)\Big\}.$$

Proof. To save space, for all $h,t\geqslant 0$, define,

$$\Delta_hW(t)\triangleq W(t+h)-W(t).$$

By the independence of the increments of W,

$$\begin{aligned}\big\|J_\mu(h;\lambda)\big\|_2^2 &= 2\Big\|\int_0^1\mu(dt)\mathbf{1}_{\{|W(t)|\leqslant h\}}\mathbf{1}_{\{\Delta_hW(t)\geqslant\lambda h^{1/2}\psi(h)\}}\times\\ &\qquad\times\int_0^t\mu(ds)\mathbf{1}_{\{|W(s)|\leqslant h\}}\mathbf{1}_{\{\Delta_hW(s)\geqslant\lambda h^{1/2}\psi(h)\}}\Big\|_1\\ &=2\overline{\Phi}\big(\lambda\psi(h)\big)\Big\|\int_0^1\mu(dt)\mathbf{1}_{\{|W(t)|\leqslant h\}}\int_0^t\mu(ds)\mathbf{1}_{\{|W(s)|\leqslant h\}}\mathbf{1}_{\{\Delta_hW(s)\geqslant\lambda h^{1/2}\psi(h)\}}\Big\|_1\\ &\leqslant 2\overline{\Phi}\big(\lambda\psi(h)\big)\,[T_1+T_2],\end{aligned}\tag{8.3}$$

where,

$$\begin{aligned}T_1&\triangleq\Big\|\int_h^1\mu(dt)\mathbf{1}_{\{|W(t)|\leqslant h\}}\int_0^{(t-h)^+}\mu(ds)\mathbf{1}_{\{|W(s)|\leqslant h\}}\mathbf{1}_{\{\Delta_hW(s)\geqslant\lambda h^{1/2}\psi(h)\}}\Big\|_1,\\ T_2&\triangleq\Big\|\int_0^1\mu(dt)\mathbf{1}_{\{|W(t)|\leqslant h\}}\int_{(t-h)^+}^t\mu(ds)\mathbf{1}_{\{|W(s)|\leqslant h\}}\Big\|_1.\end{aligned}$$

We will estimate T_1 and T_2 in turn.

We will need the following consequence of Gaussian laws: for all $s, h > 0$,

$$\mathbb{P}\big(|W(s)| \leqslant h\big) \leqslant s^{-1/2} h. \tag{8.4}$$

First, we estimate T_1. Note that,

$$T_1 = \int_h^1 \mu(dt) \int_0^{t-h} \mu(ds) \mathbb{P}\big(|W(t)| \leqslant h, |W(s)| \leqslant h, \Delta_h W(s) \geqslant \lambda h^{1/2} \psi(h)\big). \tag{8.5}$$

Suppose $t \in [h, 1]$ and $s \in [0, t-h]$. Then $s \leqslant s + h \leqslant t$, and we have,

$$\begin{aligned} \mathbb{P}\big(|W(t)| \leqslant h \mid W(r); r \leqslant s+h\big) & \\ &= \mathbb{P}\big(|W(t) - W(s+h) + W(s+h)| \leqslant h \mid W(r); r \leqslant s+h\big) \\ &\leqslant \sup_{\zeta \in \mathbb{R}} \mathbb{P}\big(|W(t-s-h) + \zeta| \leqslant h\big). \end{aligned}$$

On the other hand, $W(t-s-h)$ has a Gaussian distribution. By unimodality of the latter,

$$\mathbb{P}\big(|W(t)| \leqslant h \mid W(r); r \leqslant s+h\big) \leqslant \mathbb{P}\big(|W(t-s-h)| \leqslant h\big).$$

(This actually is a particular case of T.W. Anderson's inequality for general Gaussian shifted balls). Using (8.5) and the principle of conditioning,

$$\begin{aligned} T_1 &\leqslant \int_h^1 \mu(dt) \int_0^{t-h} \mu(ds)\ \mathbb{P}\big(|W(t-s-h)| \leqslant h\big) \times \\ &\qquad \times \mathbb{P}\big(|W(s)| \leqslant h, \Delta_h W(s) \geqslant \lambda h^{1/2} \psi(h)\big) \\ &\leqslant \overline{\Phi}\big(\lambda \psi(h)\big) \int_h^1 \mu(dt) \int_0^{t-h} \mu(ds) \mathbb{P}\big(|W(t-s-h)| \leqslant h\big) \mathbb{P}\big(|W(s)| \leqslant h\big). \end{aligned}$$

By (8.4),

$$T_1 \leqslant h^2\, \overline{\Phi}\big(\lambda \psi(h)\big) \int_h^1 \mu(dt) \int_0^{t-h} \mu(ds) \frac{1}{\sqrt{s(t-s-h)}}.$$

Changing the order of integration, we arrive at the following estimate:

$$\begin{aligned} T_1 &\leqslant h^2\, \overline{\Phi}\big(\lambda \psi(h)\big) \int_0^{1-h} \frac{\mu(ds)}{\sqrt{s}} \int_{s+h}^1 \frac{\mu(dt)}{\sqrt{t-s-h}} \\ &\leqslant h^2\, \overline{\Phi}\big(\lambda \psi(h)\big) S_1^2(\mu) \\ &\leqslant \frac{4e^{2\beta}}{(2\beta-1)^2} A_\beta^2(\mu) h^2\, \overline{\Phi}\big(\lambda \psi(h)\big), \end{aligned} \tag{8.6}$$

by Lemma 8.2. Next, we estimate T_2. By another unimodality argument, for all $t \geqslant s$ and all $h > 0$,

$$\mathbb{P}\big(|W(s)| \leqslant h, |W(t)| \leqslant h\big) \leqslant \mathbb{P}\big(|W(s)| \leqslant h\big)\ \mathbb{P}\big(|W(t-s)| \leqslant h\big).$$

Applying (8.4), $\mathbb{P}(|W(s)| \leqslant h, |W(t)| \leqslant h) \leqslant h^2/\sqrt{s(t-s)}$. Therefore,

$$T_2 \leqslant h^2 \int_0^1 \mu(dt) \int_{(t-h)^+}^t \mu(ds) \frac{1}{\sqrt{s(t-s)}} \\ = h^2 (T_{2,1} + T_{2,2}), \tag{8.7}$$

where,

$$T_{2,1} \triangleq \int_0^h \mu(dt) \int_0^t \mu(ds) \frac{1}{\sqrt{s(t-s)}}$$
$$T_{2,2} \triangleq \int_h^1 \mu(dt) \int_{t-h}^t \mu(ds) \frac{1}{\sqrt{s(t-s)}}.$$

To estimate $T_{2,1}$, reverse the order of integration:

$$T_{2,1} = \int_0^h \frac{\mu(ds)}{\sqrt{s}} \int_s^h \frac{\mu(dt)}{\sqrt{t-s}} \leqslant S_h^2(\mu) \\ \leqslant \frac{4e^{2\beta}}{(2\beta-1)^2} A_\beta^2(\mu) h^{2\beta-1}, \tag{8.8}$$

by Lemma 8.2. Similarly,

$$T_{2,2} = \int_0^1 \frac{\mu(ds)}{\sqrt{s}} \int_s^{(s+h)\wedge 1} \frac{\mu(dt)}{\sqrt{t-s}} \\ \leqslant S_1(\mu) \widetilde{S}_h(\mu) \\ \leqslant \frac{4e^{2\beta}}{(2\beta-1)^2} A_\beta^2(\mu) h^{\beta-1/2}.$$

Since $h < 1$ and $\beta > 1/2$, using the above, (8.8) and (8.7), we arrive at the following:

$$T_2 \leqslant \frac{8e^{2\beta}}{(2\beta-1)^2} A_\beta^2(\mu) h^{\beta+3/2}.$$

Use this, together with (8.6) and (8.3) in this order to get the result. ◇

We are ready for the main result of this section:

Proof of Theorem 8.1. For $\lambda, h > 0$, recall (4.9) and define,

$$\mathcal{Z}(h) \triangleq \{s \in [0,1] : |W(s)| < h\}.$$

Path continuity of W alone implies that $\mathcal{Z}(h) \cap \mathcal{A}_1(\lambda, h)$ is an open random set. We estimate the probability that it intersects E. By Frostman's lemma, for all $\beta < \dim(E)$, there exists $\mu \in \mathcal{P}_+(E)$ such that $A_\beta(\mu) < \infty$. Let us fix a μ corresponding to an arbitrary but fixed choice of β satisfying:

$$\lambda^2 + \frac{1}{2} < \beta < \dim(E). \tag{8.9}$$

Applying Lemmas 8.3 and 8.4 to this choice of μ, we see that for all $\varepsilon > 0$, there exists $h_\varepsilon > 0$, such that for all $h \in]0, h_\varepsilon[$,

$$\frac{\|J_\mu(h;\lambda)\|_1^2}{\|J_\mu(h;\lambda)\|_2^2} \geqslant \gamma_{\varepsilon,\beta}\left[\frac{2\,h^{\beta-1/2}}{\overline{\Phi}(\lambda\psi(h))}+1\right]^{-1},$$

where,

$$\gamma_{\varepsilon,\beta} \triangleq \frac{(2\beta-1)^2(1-\varepsilon)^2}{4\pi e^{1+2\beta}A_\beta^2(\mu)}.$$

According to Mill's ratio for Gaussian tails (see (4.4)), we can pick h_λ so small that for each and every $h \in]0, h_\lambda[$, $\overline{\Phi}(\lambda\psi(h)) \geqslant h^{\lambda^2}/4\lambda\sqrt{\ln(1/h)}$. Therefore, for all $h \in]0, h_\varepsilon \wedge h_\lambda[$,

$$\frac{\|J_\mu(h;\lambda)\|_1^2}{\|J_\mu(h;\lambda)\|_2^2} \geqslant \gamma_{\varepsilon,\beta}\left[8\lambda\, h^{\beta-\lambda^2-1/2}\sqrt{\ln(1/h)}\,+1\right]^{-1}.$$

By (8.9), $\liminf_{h\to 0^+}\ \|J_\mu(h;\lambda)\|_1^2\ \|J_\mu(h;\lambda)\|_2^{-2} \geqslant \gamma_{\varepsilon,\beta} > 0$. By the Paley–Zygmund inequality ([8, p. 8]),

$$\liminf_{h\to 0^+}\ \mathbb{P}(J_\mu(h;\lambda) > 0) \geqslant \gamma_{\varepsilon,\beta} > 0.$$

Note that the event $(J_\mu(h;\lambda) > 0)$ implies that $\mathcal{Z}(h)\cap\mathcal{A}_1(\lambda,h)$ intersects E. Hence,

$$\liminf_{h\to 0^+}\ \mathbb{P}(\mathcal{Z}(h)\cap\mathcal{A}_1(\lambda,h)\cap E \neq \varnothing) \geqslant \gamma_{\varepsilon,\beta} > 0.$$

However, as $h \downarrow 0$, the (random) open set $\mathcal{Z}(h)\cap\mathcal{A}_1(\lambda,h)$ decreases to $\mathcal{Z}\cap \mathrm{F}_1(\lambda)$. Adapting Theorem 2.5 to the positive probability case, we can conclude that $\mathbb{P}(\mathcal{Z}\cap\mathrm{F}_1(\lambda)\cap E\neq\varnothing) \geqslant \gamma_{\varepsilon,\beta} > 0$. Note that the only requirement on E was that $\dim(E) > \lambda^2 + 1/2$. Since Hausdorff dimension is scale invariant, we see that for any real number $s \in]0,1[$,

$$\mathbb{P}(\mathcal{Z}\cap\mathrm{F}_1(\lambda)\cap s^{-1}E \neq \varnothing) \geqslant \gamma_{\varepsilon,\beta},$$

where $s^{-1}E \triangleq \{r/s : r \in E\}$. We finish the proof by showing that this probability is actually 1. Fix $s \in]0,1[$ and observe from Brownian scaling that $\mathcal{Z}\cap\mathrm{F}_1(\lambda)$ has the same distribution as $\mathcal{Z}\cap\mathrm{F}_1(\lambda)\cap[0,s]$ in the sense that for all $G \in \mathbf{S}_0^1$,

$$\mathbb{P}(\mathcal{Z}\cap\mathrm{F}_1(\lambda)\cap G\neq\varnothing) = \mathbb{P}(\mathcal{Z}\cap\mathrm{F}_1(\lambda)\cap[0,s]\cap sG\neq\varnothing).$$

In particular, for all $s \in]0,1[$, $\mathbb{P}(\mathcal{Z}\cap\mathrm{F}_1(\lambda)\cap[0,s]\cap E\neq\varnothing) \geqslant \gamma_{\varepsilon,\beta} > 0$, Note that $\mathcal{Z}\cap\mathrm{F}_1(\lambda)\cap E\cap[0,s]$ is increasing in s. Thus,

$$\mathbb{P}\Big(\bigcap_{s\in]0,1[}\big\{\mathcal{Z}\cap\mathrm{F}_1(\lambda)\cap E\cap[0,s]\neq\varnothing\big\}\Big) \geqslant \gamma_{\varepsilon,\beta}.$$

Let $C \triangleq \cap_{s\in]0,1[}(\mathcal{Z}\cap\mathrm{F}_1(\lambda)\cap E\cap[0,s]\neq\varnothing)$. Observe that C is measurable with respect to the germ field of W at 0 and we have just argued that $\mathbb{P}(C) \geqslant \gamma_{\varepsilon,\beta} > 0$. By Blumenthal's 0–1 law, $\mathbb{P}(C) = 1$. Since $(\mathcal{Z}\cap\mathrm{F}_1(\lambda)\cap E\neq\varnothing) \supset C$, the result follows for F_1. ◇

References

[1] M.T. Barlow and E. Perkins (1984). Levels at which every Brownian excursion is exceptional. *Sém. Prob.* XVIII, Lecture Notes in Math. **1059**, 1–28, Springer–Verlag, New York.

[2] M. Csörgő and P. Révész (1981). *Strong Approximations in Probability and Statistics*, Academic Press, New York.

[3] P. Deheuvels and M.A. Lifshits (1997). On the Hausdorff dimension of the set generated by exceptional oscillations of a Wiener process. *Studia Sci. Math. Hung.*, **33**, 75–110.

[4] P. Deheuvels and D.M. Mason (1997). Random fractal functional laws of the iterated logarithm. (preprint)

[5] R.M. Dudley (1984). A Course on Empirical Processes. *École d'Été de St. Flour 1982.* Lecture Notes in Mathematics **1097**. Springer, Berlin.

[6] J. Hawkes (1971). On the Hausdorff dimension of the range of a stable process with a Borel set. *Z. Wahr. verw. Geb.*, **19**, 90–102.

[7] J. Hawkes (1981). Trees generated by a simple branching process. *J. London Math. Soc.*, **24**, 373–384.

[8] J.-P. Kahane (1985). *Some Random Series of Functions*, second edition. Cambridge University Press, Cambridge.

[9] R. Kaufman (1974). Large increments of Brownian Motion. *Nagoya Math. J.*, **56**, 139–145.

[10] D. Khoshnevisan, Y. Peres and Y. Xiao (1998). Limsup random fractals. In preparation.

[11] N. Kôno (1977). The exact Hausdorff measure of irregularity points for a Brownian path. *Z. Wahr. verw. Geb.*, **40**, 257–282.

[12] M. Ledoux and M. Talagrand (1991). *Probability in Banach Space, Isoperimetry and Processes*, Springer–Verlag, Heidelberg-New York.

[13] P. Lévy (1937). *Théorie de l'Addition des Variables Aléatoires.* Gauthier–Villars, Paris.

[14] R. Lyons (1980). Random walks and percolation on trees. *Ann. Prob.*, **18**, 931–958.

[15] M.B. Marcus (1968). Hölder conditions for Gaussian processes with stationary increments. *Trans. Amer. Math. Soc.*, **134**, 29–52.

[16] M.B. Marcus and J. Rosen (1992). Moduli of continuity of local times of strongly symmetric Markov processes via Gaussian processes. *J. Theoretical Prob.*, **5**, 791–825.

[17] P. MATILLA (1995). *Geometry of Sets and Measures in Euclidean Spaces, Fractals and Rectifiability*, Cambridge University Press, Cambridge.

[18] S. OREY AND S.J. TAYLOR (1974). How often on a Brownian path does the law of the iterated logarithm fail? *Proc. London Math. Soc.*, **28**, 174–192.

[19] Y. PERES (1996). Remarks on intersection–equivalence and capacity–equivalence. *Ann. Inst. Henri Poincaré: Physique Théorique*, **64**, 339–347.

[20] E. PERKINS AND S.J. TAYLOR (1988), Measuring close approaches on a Brownian path, *Ann. Prob.*, **16**, 1458–1480.

[21] D. REVUZ AND M. YOR (1994). *Continuous Martingales and Brownian Motion*, second edition. Springer, Berlin.

[22] G.R. SHORACK AND J.A. WELLNER (1986). *Empirical Processes with Applications to Statistics.* Wiley, New York.

[23] S.J. TAYLOR (1966). Multiple points for the sample paths of the symmetric stable process, *Z. Wahr. ver. Geb.*, **5**, 247–64.

[24] S.J. TAYLOR (1986). The measure theory of random fractals. *Math. Proc. Camb. Phil. Soc.*, **100**, 383–406.

Davar Khoshnevisan
Department of Mathematics
University of Utah
Salt Lake City, UT. 84112
U.S.A.
E-mail: davar@math.utah.edu

Zhan Shi
Laboratoire de Probabilités
Université Paris VI
4, Place Jussieu
75252 Paris Cedex 05, France
E-mail: shi@ccr.jussieu.fr

Some invariance properties (of the laws) of Ocone's martingales

L. Vostrikova[1] et *M. Yor*[2]

(1) *Université d'Angers - Faculté des Sciences - Département de Mathématiques 2, Boulevard Lavoisier - 49045 ANGERS CEDEX 01*

(2) *Laboratoire de Probabilités et Modèles Aléatoires - Université Pierre et Marie Curie - Tour 56 - 3ème Etage - 4, place Jussieu - F - 75252 PARIS CEDEX 05*

In this note, some properties of continuous martingales shall be investigated, starting from the following important remarks :

from Lévy's characterization of Brownian motion as the continuous martingale (M_t) with increasing process $\langle M\rangle_t = t$, one readily deduces the following invariance properties :

i) if (ε_t) is a predictable process which only takes the values $+1$ and -1, then :

$$M^{\varepsilon} \overset{(law)}{=} M, \text{ where : } M^{\varepsilon}_t \overset{def}{=} \int_0^t \varepsilon_s dM_s \;;$$

ii) for every bounded predictable process (φ_t),

$$D^{\varphi}_t = \exp\left(\int_0^t \varphi_s dM_s - \frac{1}{2}\int_0^t \varphi_s^2 d\langle M\rangle_s\right)$$

is a martingale, and if we denote by $Q \equiv Q^{\varphi}$ the probability such that :

$$Q_{|\mathcal{F}_t} = D^{\varphi}_t \cdot P_{|\mathcal{F}_t}$$

then :

$$\tilde{M}^{\varphi} \equiv M - \int_0^{\cdot} \varphi_s \, d\langle M\rangle_s \quad \text{satisfies : } \{\tilde{M}^{\varphi}, Q\} \overset{(law)}{=} \{M,P\}$$

iii) if $S^M_t = \sup_{s\le t} M_s$, then : $S^M - M \overset{(law)}{=} |M|$

To avoid any confusion, let us emphasize again that these identities in law are true for M a $(\mathcal{F}_t)$ Brownian motion. (Indeed, for a general continuous local martingale M, D^{φ} may only be a local martingale...).

In fact, in the sequel, where M is not in general a Brownian motion, it will be convenient for our discussion to consider some adequate variants of ii), precisely :

ii)$_{det}$ Same as ii), but φ is now a deterministic, Borel, bounded, process ;

ii)$_{\langle M\rangle}$ Same as ii), but φ is now a bounded predictable process, which depends only on $\langle M\rangle$.

The rest of this paper consists in discussing which continuous martingales M, other than $(\mathcal{F}_t)$ Brownian motions, satisfy i), or some of the above variants of ii) or iii).

For instance, it is not difficult to prove that, more generally, if (M_t) is a Gaussian martingale, which is, as is well-known, equivalent to :

$$(\langle M\rangle_t, t \geq 0) \quad \text{is a deterministic process,}$$

then all three properties are still valid.

Pushing those arguments a little further, it is not difficult again to show that these properties are still valid for (M_t) an Ocone martingale that is : a martingale whose Dubins-Schwarz representation : $M_t = \beta_{\langle M\rangle_t}$ features independent β (: Brownian motion) and $\langle M\rangle$.

The reason for our terminology is that Ocone [2] showed that this independence property is equivalent to the above property i). Moreover, a discussion of the interest of Ocone martingales in relation with Lévy's transformation (in other terms, property iii)) is made in [1].

Concerning property ii), we shall now show the

Theorem 1 : *The following properties are equivalent :*

a) (M_t) *is an Ocone martingale.*

b) Property $ii)_{<M>}$ *holds ; c) Property* $ii)_{det}$ *holds.*

Proof : · c) $\Longrightarrow$ a) We assume that $ii)_{det}$ holds, and we consider a deterministic integrand φ, always assumed to be Borel, bounded.

We then use that, for positive functionals F, one has, simply from the definition of Q :

$$E_Q[F(<M>_s, s \le t)] = E_P[F(<M>_s, s \le t)D_t^{\varphi}].$$

Now, as a consequence of $ii)_{det}$, one has :

$$E_Q[F(<M>_s, s \le t)] = E_P[F(<M>_s, s \le t)],$$

so that :

$$E_P[F(<M>_s, s \le t)] = E_P[F(<M>_s, s \le t)D_t^{\varphi}].$$

Obviously, this is equivalent to : $E_P[D_t^{\varphi} \,|\, <M>_s, s \le t] = 1,$

hence also to :

$$(1) \qquad E_P\left[\exp\left(\int_0^t \varphi_s dM_s\right) \,\middle|\, <M>_s, s \le t\right] = \exp\left(\frac{1}{2}\int_0^t \varphi_s^2 \, d<M>_s\right).$$

The right-hand side of *(1)* is equal to :

$$E_P\left(\exp\left(\int_0^t \varphi_s d(\gamma_{<M>_s})\right) \,\middle|\, <M>_s, s \le t\right),$$

where $(\gamma_u, u \ge 0)$ is a Brownian motion independent of $(<M>_s, s \ge 0)$.

Hence, the identity *(1)* yields :

(2) $$(M_t, \langle M\rangle_t \; ; \; t \ge 0) \overset{(law)}{=} (\gamma_{\langle M\rangle_t}, \langle M\rangle_t \; ; \; t \ge 0).$$

Recall that : $M_t = \beta_{\langle M\rangle_t}$; hence, time-changing both sides of *(2)*

with the inverse of $(\langle M\rangle_t, t \ge 0)$, we obtain :

$$\{(\beta_u, u \ge 0) \; ; \; (\langle M\rangle_t, t \ge 0)\} \overset{(law)}{=} \{(\gamma_u, u \ge 0) \; ; \; (\langle M\rangle_t, t \ge 0)\}$$

which shows precisely that β and $\langle M\rangle$ are independent.

· a) $\Longrightarrow$ b) : We start from $(M_t, t \ge 0)$ an Ocone martingale, and we consider an integrand $\varphi(s)$ of the form : $\Phi(\langle M\rangle_u, u \le s)$, which is predictable and bounded.

Then, we have, denoting simply $\tilde{M}$ for $\tilde{M}^{\varphi}$:

$$E_Q(F(\tilde{M}_u, u \le t)]$$

$$= E_P\left[F(\tilde{M}_u, u \le t) \exp\left(\int_0^t \varphi(s) dM_s - \frac{1}{2}\int_0^t \varphi^2(s) d\langle M\rangle_s\right)\right].$$

We then recall :

$$\tilde{M}_u = \beta_{\langle M\rangle_u} - \int_0^{\langle M\rangle_u} \varphi(\tau_v) dv \; , \; u \le t$$

and we also perform the time-change in the exponential.

Next, within the latter expectation, we condition with respect to the σ-field generated by $(\langle M\rangle_u, u \ge 0)$; then, as a consequence of the well-known property ii) for Brownian motion, we obtain that :
this conditional expectation is equal to

$$E_P[F(\beta_{\langle M\rangle_u}, u \le t) \mid \langle M\rangle]$$

and we denote :

$$G_t = \gamma + \int_0^t \varphi(\langle M\rangle_u) dM_u .$$

On the other hand, we associate to F the $\{\mathcal{N}_t\}$ martingale :

$$F_t \equiv E[F|\mathcal{N}_t] = E[F|\mathcal{M}_t] ,$$

using the first part of the theorem.

We now apply Itô's formula :

$$F_t G_t = F_o \gamma + \int_0^t F_{s-} dG_s + \int_0^t G_s dF_s + [F,G]_t ,$$

but, since G is continuous, and (F_t) and (G_t) are orthogonal, we have :

$$[F,G]_t = \langle F^c, G\rangle_t \equiv 0.$$

Thus, finally, $\Phi \equiv FG = F_o \gamma + \int_0^\infty F_{s-} \, \varphi(\langle M\rangle_s) dM_s + \int_0^\infty G_s dF_s$,

which proves the second point. □

We now give some examples of Ocone, and non-Ocone martingales.

Theorem 3 : *Let* $(B_t, t \geq 0)$ *be a* $(\mathcal{F}_t)$ *Brownian motion, and (for simplicity) let* $(\mu_t, t \geq 0)$ *be a* $(\mathcal{F}_t)$ *adapted, continuous process such that* $\mu_s \neq 0$, $ds\, dP$ *a.s., and* $\int_0^\infty \mu_s^2 ds = \infty$ *a.s.*

Then, $\left\{M_t = \int_0^t \mu_s \, dB_s, \ t \geq 0\right\}$ *is an Ocone martingale iff the Brownian motion*

so that finally we have obtained :

$$E_Q[F(\tilde{M}_u, u \le t)] = E_P[F(M_u, u \le t)].$$

· b) $\Longrightarrow$ c) : This is obvious. □

Comment 1 : Note that property ii) only involves martingale densities (D_t^{φ}) which are stochastic integrals with respect to dM_t. In [3] it is remarked in Exercise (1.41), Chap. VIII, that the only martingales (M_t) such that the reinforcement of ii) holds with any possible martingale density are the martingales with deterministic bracket, i.e : the Gaussian martingales, denoted by $\mathcal{G}$ below.

Comment 2 : α) It is quite doubtful that the general property ii) is satisfied for an Ocone martingale ; we postpone investigating this equation in depth.

β) To avoid lengthening the statement of Theorem 1, we did not add there the following equivalent property d), which is nonetheless worth mentioning :

d) for every deterministic bounded process φ, $\{\langle M\rangle, Q^{\varphi}\} \overset{(law)}{=} \{\langle M\rangle, P\}$.

The proof of the equivalence : a) $\Longleftrightarrow$ d) uses the same arguments as : a) $\Longleftrightarrow$ c).

Comment 3 : Although an Ocone martingale shares properties i), $\text{ii})_{\langle M\rangle}$ and iii) with Brownian motion, it does not share a priori the important martingale representation property, that is, precisely : every martingale (N_t), with respect to the natural filtration of (M_t), is not necessarily a stochastic integral with respect to M.

Indeed, there is the following

Proposition : *An Ocone martingale* $(M_t, t \ge 0)$ *enjoys the martingale representation property (with respect to its natural filtration) iff* $(\langle M\rangle_t, t \ge 0)$ *is a deterministic process.*

Proof : Since, from the definition of an Ocone martingale, $\langle M\rangle$ is independent of β, the DDS Brownian motion associated to M, we can write :

$$(*) \qquad P_M = \int P(\langle M\rangle \in da)\, W^a ,$$

where P_M , resp : W^a , denotes the law of M, resp : the law of the continuous martingale with (deterministic) increasing process $a(\cdot)$.

We now recall (see, e.g. [3], Chap. V) that M enjoys the (martingale) representation property iff P_M is extremal among the set of laws of (continuous) martingales.

Now, from (*) , it follows that P_M is extremal iff $P(\langle M\rangle \in da)$ reduces to a Dirac measure ; in other terms, there exists a deterministic increasing function $a(\cdot)$ such that : $P(\langle M\rangle = a(\cdot)) = 1$. □

At this point, it seems interesting to draw the following diagram, which indicates 4 remarkable classes of continuous (local) martingales :

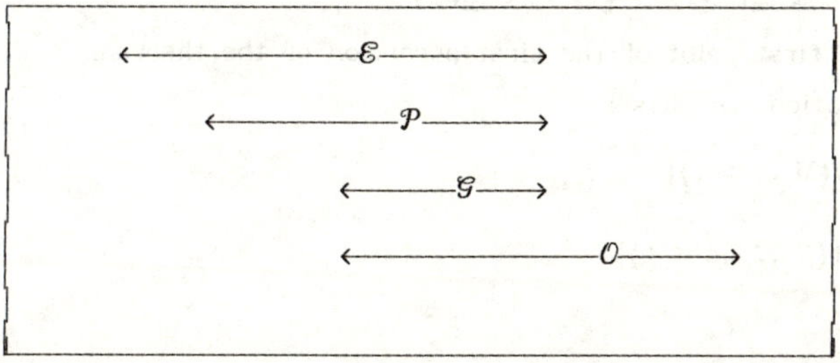

A classification of continuous local martingales.

The four letters stand for : $\mathcal{E}$: extremal, $\mathcal{P}$: pure, $\mathcal{G}$: gaussian, $\mathcal{O}$: Ocone. And the diagram indicates that : $\mathcal{G} \subset \mathcal{P} \subset \mathcal{E}$, where $\subset$ denotes strict inclusion.

On the other hand, $\mathcal{O} \cap \mathcal{E} = \mathcal{O} \cap \mathcal{P} = \mathcal{G}$.

We now wish to complete the above Proposition by describing all martingales with respect to the natural filtration $\{\mathcal{M}_t\}$ of (M_t). It will be useful to introduce $\{\mathcal{N}_t\}$ the natural filtration of $\{\langle M\rangle_t, t \geq 0\}$.

We now prove the

Theorem 2 : *Let* (M_t) *be an Ocone martingale :*

1) Every $\{\mathcal{N}_t\}$ *martingale* (N_t) *is an* $\{\mathcal{M}_t\}$ *martingale, and it is orthogonal to* (M_t) *, that is :* $(N_t M_t, t \geq 0)$ *is a* $\{\mathcal{M}_t\}$ *local martingale ;*

2) The space of square integrable $\{\mathcal{M}_t\}$ *martingales is the direct sum of the stable space generated by* $\{\mathcal{N}_t\}$ *martingales and of the stable space generated by* $\{M_t\}$.

Proof : 1) Consider (N_t) a uniformly integrable $\{\mathcal{N}_t\}$ martingale.

We shall show : $E[N_\infty | \mathcal{M}_t] = N_t$ $(= E[N_\infty | \mathcal{N}_t])$

which proves the first point of the first assertion of the theorem.

With obvious notation, one has :

$$E[N_\infty f(M_s, s \leq t)]$$

$$= E[N_\infty f(\beta_{\langle M\rangle_s}, s \leq t)]$$

$$= \int W(d\omega)\, E[N_\infty f(\omega(\langle M\rangle_s), s \leq t)]$$

$$= \int W(d\omega)\, E[N_t f(\omega(\langle M\rangle_s), s \leq t)]$$

$$= E[N_t f(M_s, s \leq t)].$$

Similarly, we now show that $(N_t M_t, t \geq 0)$ is a $\{\mathcal{M}_t\}$ martingale. Let $s < t$.

Then, we have :

$$E[N_t M_t\ f(M_u, u \le s)]$$

$$= \int W(d\beta)\ E[N_t\ \beta_{\langle M\rangle_t}\ f(\beta_{\langle M\rangle_u}, u \le s)].$$

We then use the martingale property for β, and the independence of $\langle M\rangle$ and β ; this yields :

$$E[N_t\ M_t\ f(M_u, u \le s)] = \int W(d\beta)\ E[N_t\ \beta_{\langle M\rangle_s}\ f(\beta_{\langle M\rangle_u}, u \le s)].$$

Next, we use the martingale property for (N_t), with respect to $\{\mathcal{N}_t\}$; we obtain :

$$E[N_t\ M_t\ f(M_u, u \le s)] = \int W(d\beta)\ E[N_s\ \beta_{\langle M\rangle_s}\ f(\beta_{\langle M\rangle_u}, u \le s)]$$

$$= E[N_s\ M_s\ f(M_u, u \le s)].$$

2) To show the second point, it suffices to consider variables $\Phi \in L^2(\mathcal{M}_\infty, P)$, of the form : $\Phi = FG$, where $F \in L^2(\mathcal{N}_\infty, P)$ and $G \in L^2(\mathcal{B}_\infty, P)$, where $\mathcal{B}_\infty = \sigma\{\beta_s, s \ge 0\}$.

As is well known, G may be written in the form :

$$G = \gamma + \int_0^\infty \varphi(s) d\beta_s ,$$

for some $\gamma \in \mathbb{R}$, and some $\{\mathcal{B}_t\}$ predictable process φ such that :

$$E\left[\int_0^\infty \varphi^2(s) ds\right] < \infty.$$

Making the time change $s = \langle M\rangle$, we obtain :

$$G = \gamma + \int_0^\infty \varphi(\langle M\rangle_u) dM_u ,$$

$\left\{\theta_t \overset{\mathrm{def}}{=} \int_0^t \mathrm{sgn}(\mu_s)dB_s, t \geq 0\right\}$ *is independent from the* σ*-field*

$$\mathcal{N}_\infty = \sigma\{|\mu_s|, s \geq 0\}.$$

Proof : As previously discussed, $\{M_t, t \geq 0\}$ is an Ocone martingale iff

$$(3) \qquad E\left[\exp\left(i\int_0^\infty \varphi(s)dM_s\right)\Big|\mathcal{N}_\infty\right] = \exp\left(-\frac{1}{2}\int_0^\infty \varphi^2(s)\ d\langle M\rangle_s\right)$$

for any $\{\mathcal{N}_s\}$ predictable process φ, such that : $\int_0^\infty \varphi^2(s)d\langle M\rangle_s < \infty$.

The identity *(3)* is obviously satisfied if θ is independent from $\mathcal{N}_\infty$.

Conversely, assuming that *(3)* holds, we now take : $\varphi(s) = f(s)\dfrac{1}{|\mu_s|}$ with f a generic, simple, deterministic function, with compact support. Thus, we deduce from *(3)* :

$$E\left[\exp\left(i\int_0^\infty f(s)d\theta_s\right)\Big|\mathcal{N}_\infty\right] = \exp\left(-\frac{1}{2}\int_0^\infty f^2(s)ds\right),$$

which is obviously equivalent to the independence of θ and $\mathcal{N}_\infty$. □

Here is another (fairly general) variant of Theorem 3.

Theorem 3' : *Assume that* $\{\mathcal{N}_t\}$ *is the natural filtration of a* $\{\mathcal{F}_t\}$ *martingale* (N_t) *which is pure, i.e :* $N_t = \gamma_{\langle N\rangle_t}, t \geq 0$, *with* $(\langle N\rangle_t, t \geq 0)$ *measurable with respect to the* σ*-field* $\sigma\{\gamma_u, u \geq 0\}$ *of the Brownian motion* $(\gamma_u, u \geq 0)$.

Then, $(M_t, t \geq 0)$ *is an Ocone martingale as soon as* N *and* M *are orthogonal.*

Proof : It follows immediately from our hypothesis and Knight's theorem on continuous orthogonal martingales (see, e.g, [3]) that β and γ, the respective DDS Brownian motions of M and N are independent. Now, since $\mathcal{N}_\infty$ is, again under our hypothesis equal to the σ-field generated by γ, $\mathcal{N}_\infty$ and β are independent, which finishes the proof. □

To conclude this work, we present a number of simple examples of Ocone, resp : non-Ocone, martingales.

Of course, to avoid trivialities, when looking for Ocone martingales, we exclude the Gaussian examples (one might call the non-Gaussian Ocone martingales "strictly Ocone" martingales).

a) Perhaps, the most simple example of an Ocone martingale is

$$M_t^{(1)} = \int_0^t C_s dB_s \quad , \ t \geq 0,$$

where B and C are two independent Brownian motions.

The stochastic area of the planar Brownian motion (B_t, C_t), defined as :

$$\mathcal{A}_t = \frac{1}{2}\int_0^t (C_s dB_s - B_s dC_s)$$

is another example of an Ocone martingale.

This follows readily from Theorem 3, since :

$$\langle \mathcal{A} \rangle_t = \frac{1}{4}\int_0^t ds\ R_s^2 \quad , \quad R_s^2 \equiv B_s^2 + C_s^2 \ ,$$

and we can write :

$$\mathcal{A}_t = \frac{1}{2}\int_0^t R_s d\gamma_s \ ,$$

where $\gamma_t \overset{\text{def}}{=} \int_0^t \frac{C_s dB_s - B_s dC_s}{R_s}$, $t \geq 0$, is a real-valued Brownian motion independent from $(R_t, t \geq 0)$; see, e.g., Yor [9].

b) Here are now some examples of non-Ocone martingales, among which :

$$M_t^{(2)} = \int_0^t B_s dB_s \ , \ t \geq 0$$

and

$$\pi_t \overset{\text{def}}{=} B_t C_t = \int_0^t (C_s dB_s + B_s dC_s).$$

Indeed, $(M_t^{(2)}, t \geq 0)$ is a pure martingale, which is easily seen by time-changing it with the inverse of

$$\langle M^{(2)} \rangle_t = \int_0^t ds\ B_s^2$$

(see, e.g., Stroock-Yor [6]). Since $\langle M^{(2)} \rangle$ is not deterministic, it follows from the above classification that $M^{(2)}$ is not an Ocone martingale. Also, it is easily seen that the property d) in Comment 2 above is not satisfied for $\varphi(s) \equiv \lambda$ ($\in \mathbb{R}$). In fact, under $Q^{\varphi} \equiv Q^{\lambda}$, the process (B_t) is an Ornstein-Uhlenbeck process with parameter λ, hence :

$$(\langle M^{(2)} \rangle, Q^{\lambda}) \neq (\langle M^{(2)} \rangle, P).$$

The argument we shall use for $\{\pi_t\}$ is somewhat different : first of all, one finds :

$$\langle \pi \rangle_t = \int_0^t ds\ R_s^2 \ ,$$

but the process $\{R_u, u \geq 0\}$ is certainly not independent from the Dubins-Schwarz Brownian motion attached to $\{\pi_t\}$ (hence, $\{\pi_t\}$ is not an Ocone martingale).

We see this non-independence property as follows : since one has, obviously : $|2B_tC_t| \le R_t^2$, then, conditionally on $(R_u, u \ge 0)$, the variable (B_tC_t) cannot be Gaussian, since it is bounded.

Thus, $\{\pi_t\}$ is not an Ocone martingale ; we also remark that it is of the form $\int_0^t (A\,\mathbb{B}_s, d\mathbb{B}_s)$, where $\mathbb{B}_s = \begin{pmatrix} B_s \\ C_s \end{pmatrix}$, and $A = \begin{pmatrix} 0 & 1 \\ 1 & 0 \end{pmatrix}$. Hence, since A is symmetric and has two non-zero, distinct eigenvalues (+1 and -1), it follows from [7] that the natural filtration of $\{\pi_t\}$ is that of a 2-dimensional Brownian motion.

We end up with an example of an Ocone martingale $M_t = \int_0^t \varphi_s \, dB_s$ within the filtration of a 1-dimensional Brownian motion $(B_t, t \ge 0)$, and we also assume that $\varphi_s \ne 0$, ds dP a.s. (otherwise, there are some quite easy examples).

We write : $M_t = \int_0^t \tilde{\varphi}_s \, d\tilde{B}_s$, with $\tilde{B}_s = \int_0^s \mathrm{sgn}(B_u) dB_u$ (hence : $\varphi_s = \tilde{\varphi}_s \, \mathrm{sgn}(B_s)$)

With the help of Theorem 3, we choose $(\tilde{\varphi}_s, s \ge 0)$ to be a strictly positive process, independent from the Brownian motion $\{\tilde{B}_t\}$, whose natural filtration is identical to that of $\{|B_t|, t \ge 0\}$. This implies that $\{M_t, t \ge 0\}$ is an Ocone martingale ; to be more explicit, we may take, as an example of process $\{\tilde{\varphi}_s\}$:

$$\tilde{\varphi}_s = a_s 1_{(s<t_o)} + (1 + 2\,\mathrm{sgn}(B_{t_o}))1_{(t_o \le s)} ,$$

for some $t_o > 0$, and $\{a_s\}$ a strictly positive deterministic function.

We also note that, as a consequence of Theorem 2, no strictly Ocone martingale (M_t) , such that $(d\langle M\rangle_t)$ is equivalent to Lebesgue measure dt on $\mathbb{R}_+$ can generate the filtration of a 1-dimensional Brownian motion.

Final comment : As a consequence of Theorem 2, the filtration of an Ocone martingale, when it is not Gaussian, corresponds, in Mathematical finance, to an incomplete market. However, the above representation theorem should be useful to discuss further constructions of probability measures now familiar in such cases in Mathematical finance, e.g : the variance-optimal martingale measure for continuous processes [5].

References

[1] **L. Dubins, M. Emery, M. Yor** : On the Lévy transformation of Brownian motions and continuous martingales.
Sém. Probas. XXVII, Lect. Notes in Maths. n° 1577. Springer (1993), p. 122-132.

[2] **D. Ocone** : A symmetry characterization of conditionally independent increment martingales.
Proceedings of the San Felice Workshop on Stochastic Analysis. D. Nualart et M. Sanz, eds. Birkhaüser (1993), p. 147-167.

[3] **D. Revuz, M. Yor** : Continuous martingales and Brownian motion.
Springer, Third edition : 1999.

[4] **J. Azéma, C. Rainer, M. Yor** : Une propriété des martingales pures.
Séminaire de Probabilités XXX, Lect. Notes in Maths. n° 1626, Springer (1996), p. 243-254.

[5] **F. Delbaen, W. Schachermayer** : The variance-optimal martingale measure for continuous processes.
Bernoulli 2 (1), 1996, p. 81-105.

[6] **D.W. Stroock, M. Yor** : Some remarkable martingales.
In : Sém. Probas XV, Lect. Notes in Maths. 850, Springer (1981).

[7] M. Yor : Les filtrations de certaines martingales du mouvement brownien dans $\mathbb{R}^n$, *p. 427-440.*
In : Sém. Probas XIII, Lect. Notes in Maths. 721, Springer (1979).

[8] M. Yor : Sur les martingales continues extrémales.
Stochastics, vol. 2, n° 3 (1979), p. 191-196.

[9] M. Yor : Remarques sur une formule de P. Lévy.
Sém. Proba. XIV, Lect. Notes in Maths n° 784, Springer (1980), p. 343-346.

Printing: Weihert-Druck GmbH, Darmstadt
Binding: Buchbinderei Schäffer, Grünstadt

Vol. 1630: D. Neuenschwander, Probabilities on the Heisenberg Group: Limit Theorems and Brownian Motion. VIII, 139 pages. 1996.

Vol. 1631: K. Nishioka, Mahler Functions and Transcendence. VIII, 185 pages. 1996.

Vol. 1632: A. Kushkuley, Z. Balanov, Geometric Methods in Degree Theory for Equivariant Maps. VII, 136 pages. 1996.

Vol. 1633: H. Aikawa, M. Essén, Potential Theory – Selected Topics. IX, 200 pages. 1996.

Vol. 1634: J. Xu, Flat Covers of Modules. IX, 161 pages. 1996.

Vol. 1635: E. Hebey, Sobolev Spaces on Riemannian Manifolds. X, 116 pages. 1996.

Vol. 1636: M. A. Marshall, Spaces of Orderings and Abstract Real Spectra. VI, 190 pages. 1996.

Vol. 1637: B. Hunt, The Geometry of some special Arithmetic Quotients. XIII, 332 pages. 1996.

Vol. 1638: P. Vanhaecke, Integrable Systems in the realm of Algebraic Geometry. VIII, 218 pages. 1996.

Vol. 1639: K. Dekimpe, Almost-Bieberbach Groups: Affine and Polynomial Structures. X, 259 pages. 1996.

Vol. 1640: G. Boillat, C. M. Dafermos, P. D. Lax, T. P. Liu, Recent Mathematical Methods in Nonlinear Wave Propagation. Montecatini Terme, 1994. Editor: T. Ruggeri. VII, 142 pages. 1996.

Vol. 1641: P. Abramenko, Twin Buildings and Applications to S-Arithmetic Groups. IX, 123 pages. 1996.

Vol. 1642: M. Puschnigg, Asymptotic Cyclic Cohomology. XXII, 138 pages. 1996.

Vol. 1643: J. Richter-Gebert, Realization Spaces of Polytopes. XI, 187 pages. 1996.

Vol. 1644: A. Adler, S. Ramanan, Moduli of Abelian Varieties. VI, 196 pages. 1996.

Vol. 1645: H. W. Broer, G. B. Huitema, M. B. Sevryuk, Quasi-Periodic Motions in Families of Dynamical Systems. XI, 195 pages. 1996.

Vol. 1646: J.-P. Demailly, T. Peternell, G. Tian, A. N. Tyurin, Transcendental Methods in Algebraic Geometry. Cetraro, 1994. Editors: F. Catanese, C. Ciliberto. VII, 257 pages. 1996.

Vol. 1647: D. Dias, P. Le Barz, Configuration Spaces over Hilbert Schemes and Applications. VII. 143 pages. 1996.

Vol. 1648: R. Dobrushin, P. Groeneboom, M. Ledoux, Lectures on Probability Theory and Statistics. Editor: P. Bernard. VIII, 300 pages. 1996.

Vol. 1649: S. Kumar, G. Laumon, U. Stuhler, Vector Bundles on Curves – New Directions. Cetraro, 1995. Editor: M. S. Narasimhan. VII, 193 pages. 1997.

Vol. 1650: J. Wildeshaus, Realizations of Polylogarithms. XI, 343 pages. 1997.

Vol. 1651: M. Drmota, R. F. Tichy, Sequences, Discrepancies and Applications. XIII, 503 pages. 1997.

Vol. 1652: S. Todorcevic, Topics in Topology. VIII, 153 pages. 1997.

Vol. 1653: R. Benedetti, C. Petronio, Branched Standard Spines of 3-manifolds. VIII, 132 pages. 1997.

Vol. 1654: R. W. Ghrist, P. J. Holmes, M. C. Sullivan, Knots and Links in Three-Dimensional Flows. X, 208 pages. 1997.

Vol. 1655: J. Azéma, M. Emery, M. Yor (Eds.), Séminaire de Probabilités XXXI. VIII, 329 pages. 1997.

Vol. 1656: B. Biais, T. Björk, J. Cvitanic, N. El Karoui, E. Jouini, J. C. Rochet, Financial Mathematics. Bressanone, 1996. Editor: W. J. Runggaldier. VII, 316 pages. 1997.

Vol. 1657: H. Reimann, The semi-simple zeta function of quaternionic Shimura varieties. IX, 143 pages. 1997.

Vol. 1658: A. Pumarino, J. A. Rodrıguez, Coexistence and Persistence of Strange Attractors. VIII, 195 pages. 1997.

Vol. 1659: V, Kozlov, V. Maz'ya, Theory of a Higher-Order Sturm-Liouville Equation. XI, 140 pages. 1997.

Vol. 1660: M. Bardi, M. G. Crandall, L. C. Evans, H. M. Soner, P. E. Souganidis, Viscosity Solutions and Applications. Montecatini Terme, 1995. Editors: I. Capuzzo Dolcetta, P. L. Lions. IX, 259 pages. 1997.

Vol. 1661: A. Tralle, J. Oprea, Symplectic Manifolds with no Kähler Structure. VIII, 207 pages. 1997.

Vol. 1662: J. W. Rutter, Spaces of Homotopy Self-Equivalences – A Survey. IX, 170 pages. 1997.

Vol. 1663: Y. E. Karpeshina; Perturbation Theory for the Schrödinger Operator with a Periodic Potential. VII, 352 pages. 1997.

Vol. 1664: M. Väth, Ideal Spaces. V, 146 pages. 1997.

Vol. 1665: E. Giné, G. R. Grimmett, L. Saloff-Coste, Lectures on Probability Theory and Statistics 1996. Editor: P. Bernard. X, 424 pages, 1997.

Vol. 1666: M. van der Put, M. F. Singer, Galois Theory of Difference Equations. VII, 179 pages. 1997.

Vol. 1667: J. M. F. Castillo, M. González, Three-space Problems in Banach Space Theory. XII, 267 pages. 1997.

Vol. 1668: D. B. Dix, Large-Time Behavior of Solutions of Linear Dispersive Equations. XIV, 203 pages. 1997.

Vol. 1669: U. Kaiser, Link Theory in Manifolds. XIV, 167 pages. 1997.

Vol. 1670: J. W. Neuberger, Sobolev Gradients and Differential Equations. VIII, 150 pages. 1997.

Vol. 1671: S. Bouc, Green Functors and *G*-sets. VII, 342 pages. 1997.

Vol. 1672: S. Mandal, Projective Modules and Complete Intersections. VIII, 114 pages. 1997.

Vol. 1673: F. D. Grosshans, Algebraic Homogeneous Spaces and Invariant Theory. VI, 148 pages. 1997.

Vol. 1674: G. Klaas, C. R. Leedham-Green, W. Plesken, Linear Pro-*p*-Groups of Finite Width. VIII, 115 pages. 1997.

Vol. 1675: J. E. Yukich, Probability Theory of Classical Euclidean Optimization Problems. X, 152 pages. 1998.

Vol. 1676: P. Cembranos, J. Mendoza, Banach Spaces of Vector-Valued Functions. VIII, 118 pages. 1997.

Vol. 1677: N. Proskurin, Cubic Metaplectic Forms and Theta Functions. VIII, 196 pages. 1998.

Vol. 1678: O. Krupková, The Geometry of Ordinary Variational Equations. X, 251 pages. 1997.

Vol. 1679: K.-G. Grosse-Erdmann, The Blocking Technique. Weighted Mean Operators and Hardy's Inequality. IX, 114 pages. 1998.

Vol. 1680: K.-Z. Li, F. Oort, Moduli of Supersingular Abelian Varieties. V, 116 pages. 1998.

Vol. 1681: G. J. Wirsching, The Dynamical System Generated by the 3n+1 Function. VII, 158 pages. 1998.

Vol. 1682: H.-D. Alber, Materials with Memory. X, 166 pages. 1998.

Vol. 1683: A. Pomp, The Boundary-Domain Integral Method for Elliptic Systems. XVI, 163 pages. 1998.

Vol. 1684: C. A. Berenstein, P. F. Ebenfelt, S. G. Gindikin, S. Helgason, A. E. Tumanov, Integral Geometry, Radon Transforms and Complex Analysis. Firenze, 1996. Editors: E. Casadio Tarabusi, M. A. Picardello, G. Zampieri. VII, 160 pages. 1998.

Vol. 1685: S. König, A. Zimmermann, Derived Equivalences for Group Rings. X, 146 pages. 1998.

Vol. 1686: J. Azéma, M. Émery, M. Ledoux, M. Yor (Eds.), Séminaire de Probabilités XXXII. VI, 440 pages. 1998.

Vol. 1687: F. Bornemann, Homogenization in Time of Singularly Perturbed Mechanical Systems. XII, 156 pages. 1998.

Vol. 1688: S. Assing, W. Schmidt, Continuous Strong Markov Processes in Dimension One. XII, 137 page. 1998.

Vol. 1689: W. Fulton, P. Pragacz, Schubert Varieties and Degeneracy Loci. XI, 148 pages. 1998.

Vol. 1690: M. T. Barlow, D. Nualart, Lectures on Probability Theory and Statistics. Editor: P. Bernard. VIII, 237 pages. 1998.

Vol. 1691: R. Bezrukavnikov, M. Finkelberg, V. Schechtman, Factorizable Sheaves and Quantum Groups. X, 282 pages. 1998.

Vol. 1692: T. M. W. Eyre, Quantum Stochastic Calculus and Representations of Lie Superalgebras. IX, 138 pages. 1998.

Vol. 1694: A. Braides, Approximation of Free-Discontinuity Problems. XI, 149 pages. 1998.

Vol. 1695: D. J. Hartfiel, Markov Set-Chains. VIII, 131 pages. 1998.

Vol. 1696: E. Bouscaren (Ed.): Model Theory and Algebraic Geometry. XV, 211 pages. 1998.

Vol. 1697: B. Cockburn, C. Johnson, C.-W. Shu, E. Tadmor, Advanced Numerical Approximation of Nonlinear Hyperbolic Equations. Cetraro, Italy, 1997. Editor: A. Quarteroni. VII, 390 pages. 1998.

Vol. 1698: M. Bhattacharjee, D. Macpherson, R. G. Möller, P. Neumann, Notes on Infinite Permutation Groups. XI, 202 pages. 1998.

Vol. 1699: A. Inoue, Tomita-Takesaki Theory in Algebras of Unbounded Operators. VIII, 241 pages. 1998.

Vol. 1700: W. A. Woyczyński, Burgers-KPZ Turbulence, XI, 318 pages. 1998.

Vol. 1701: Ti-Jun Xiao, J. Liang, The Cauchy Problem of Higher Order Abstract Differential Equations, XII, 302 pages. 1998.

Vol. 1702: J. Ma, J. Yong, Forward-Backward Stochastic Differential Equations and Their Applications. XIII, 270 pages. 1999.

Vol. 1703: R. M. Dudley, R. Norvaiša, Differentiability of Six Operators on Nonsmooth Functions and p-Variation. VIII, 272 pages. 1999.

Vol. 1704: H. Tamanoi, Elliptic Genera and Vertex Operator Super-Algebras. VI, 390 pages. 1999.

Vol. 1705: I. Nikolaev, E. Zhuzhoma, Flows in 2-dimensional Manifolds. XIX, 294 pages. 1999.

Vol. 1706: S. Yu. Pilyugin, Shadowing in Dynamical Systems. XVII, 271 pages. 1999.

Vol. 1707: R. Pytlak, Numerical Methods for Optical Control Problems with State Constraints. XV, 215 pages. 1999.

Vol. 1708: K. Zuo, Representations of Fundamental Groups of Algebraic Varieties. VII, 139 pages. 1999.

Vol. 1709: J. Azéma, M. Émery, M. Ledoux, M. Yor (Eds), Séminaire de Probabilités XXXIII. VIII, 418 pages. 1999.

Vol. 1710: M. Koecher, The Minnesota Notes on Jordan Algebras and Their Applications. IX, 173 pages. 1999.

Vol. 1711: W. Ricker, Operator Algebras Generated by Commuting Projections: A Vector Measure Approach. XVII, 159 pages. 1999.

Vol. 1712: N. Schwartz, J. J. Madden, Semi-algebraic Function Rings and Reflectors of Partially Ordered Rings. XI, 279 pages. 1999.

Vol. 1713: F. Bethuel, G. Huisken, S. Müller, K. Steffen, Calculus of Variations and Geometric Evolution Problems. Cetraro, 1996. Editors: S. Hildebrandt, M. Struwe. VII, 293 pages. 1999.

Vol. 1714: O. Diekmann, R. Durrett, K. P. Hadeler, P. K. Maini, H. L. Smith, Mathematics Inspired by Biology. Martina Franca, 1997. Editors: V. Capasso, O. Diekmann. VII, 268 pages. 1999.

Vol. 1715: N. V. Krylov, M. Röckner, J. Zabczyk, Stochastic PDE's and Kolmogorov Equations in Infinite Dimensions. Cetraro, 1998. Editor: G. Da Prato. VIII, 239 pages. 1999.

Vol. 1716: J. Coates, R. Greenberg, K. A. Ribet, K. Rubin, Arithmetic Theory of Elliptic Curves. Cetraro, 1997. Editor: C. Viola. VIII, 260 pages. 1999.

Vol. 1717: J. Bertoin, F. Martinelli, Y. Peres, Lectures on Probability Theory and Statistics. Saint-Flour, 1997. Editor: P. Bernard. IX, 291 pages. 1999.

Vol. 1718: A. Eberle, Uniqueness and Non-Uniqueness of Semigroups Generated by Singular Diffusion Operators. VIII, 262 pages. 1999.

Vol. 1719: K. R. Meyer, Periodic Solutions of the N-Body Problem. IX, 144 pages. 1999.

Vol. 1720: D. Elworthy, Y. Le Jan, X-M. Li, On the Geometry of Diffusion Operators and Stochastic Flows. IV, 118 pages. 1999.

Vol. 1721: A. Iarrobino, V. Kanev, Power Sums, Gorenstein Algebras, and Determinantal Loci. XXVII, 345 pages. 1999.

Vol. 1722: R. McCutcheon, Elemental Methods in Ergodic Ramsey Theory. VI, 160 pages. 1999.

Vol. 1723: J. P. Croisille, C. Lebeau, Diffraction by an Immersed Elastic Wedge. VI, 134 pages. 1999.

Vol. 1724: V. N. Kolokoltsov, Semiclassical Analysis for Diffusions and Stochastic Processes. VIII, 347 pages. 2000.

Vol. 1725: D. A. Wolf-Gladrow, Lattice-Gas Cellular Automata and Lattice Boltzmann Models. IX, 308 pages. 2000.

Vol. 1726: V. Marić, Regular Variation and Differential Equations. X, 127 pages. 2000.

Vol. 1727: P. Kravanja, M. Van Barel, Computing the Zeros of Analytic Functions. VII, 111 pages. 2000.

Vol. 1728: K. Gatermann, Computer Algebra Methods for Equivariant Dynamical Systems. XV, 153 pages. 2000.

Vol. 1729: J. Azéma, M. Émery, M. Ledoux, M. Yor, Séminaire de Probabilités XXXIV. VI, 431 pages. 2000.